CHEMICAL ANALYSIS

(*continued on back*)

X-Ray Absorption

CHEMICAL ANALYSIS

A SERIES OF MONOGRAPHS ON
ANALYTICAL CHEMISTRY AND ITS APPLICATIONS

VOLUME 92

A WILEY-INTERSCIENCE PUBLICATION

JOHN WILEY & SONS

New York / Chichester / Brisbane / Toronto / Singapore

X-Ray Absorption

PRINCIPLES, APPLICATIONS, TECHNIQUES OF EXAFS, SEXAFS AND XANES

Edited by

D. C. KONINGSBERGER AND R. PRINS

Eindhoven University of Technology
Eindhoven, The Netherlands

A WILEY-INTERSCIENCE PUBLICATION

JOHN WILEY & SONS

New York / Chichester / Brisbane / Toronto / Singapore

Library of Congress Cataloging in Publication Data:
X-ray absorption.

   (Chemical analysis; v. 92)
   "A Wiley-Interscience publication."
   1. Extended X-ray absorption fine structure.
2. X-ray absorption near edge structure. 3. Surface
extended X-ray absorption fine structure.
I. Koningsberger, D. C. II. Prins, Roelof.
III. Series. [DNLM: 1. Radiation, Ionizing.
2. Radiometry. 3. Spectrum Analysis—methods.
WN 100 X1]

QD96.X2X195   1988   543′.08586   86-28991
ISBN 0-471-87547-3

CONTRIBUTORS

A. Bianconi, Dipartimento di Fisica, Universita di Roma "La Sapienza," Rome, Italy

B. A. Bunker, Department of Physics, University of Notre Dame, Notre Dame, Indiana

Stephen P. Cramer, Schlumberger-Doll Research, Ridgefield, Connecticut

E. D. Crozier, Department of Physics, Simon Fraser University, Burnaby, British Columbia, Canada

P. J. Durham, Daresbury Laboratory, Daresbury Warrington, United Kingdom

Steven M. Heald, Brookhaven National Laboratory, Upton, New York

R. Ingalls, Department of Physics, University of Washington, Seattle, Washington

D. C. Koningsberger, Laboratory for Inorganic Chemistry and Catalysis, Department of Chemical Technology, Eindhoven University of Technology, Eindhoven, The Netherlands

R. Prins, Laboratory for Inorganic Chemistry and Catalysis, Department of Chemical Engineering, Eindhoven University of Technology, Eindhoven, The Netherlands

J. J. Rehr, Department of Physics, University of Washington, Seattle, Washington

D. E. Sayers, Department of Physics, North Carolina State University, Raleigh, North Carolina

Edward A. Stern, Department of Physics, University of Washington, Seattle, Washington

Joachim Stöhr, IBM Alamaden Research Center, San Jose, California

PREFACE

Interest in x-ray absorption spectroscopy and particularly in EXAFS and XANES has increased enormously since Stern, Lytle, and Sayers published their classic series of papers in 1971–1974. [See Chapter 1, refs. (2), (25), (37), (39), (46).] The use of synchrotron radiation for performing x-ray absorption measurements that started at Stanford Synchrotron Radiation Laboratory in 1974 was of significant importance for the development of the x-ray absorption technique. As a result, many articles were published in the past decade, which describe either the theory of EXAFS, the experimental setup and data analysis, or the application of EXAFS and XANES in physics, chemistry, biology, and material sciences. International conferences were held in Daresbury (1980), Frascati (1982), and Stanford (1984). Meanwhile, a substantial number of review articles have been published.

Most of these review articles contain many points of interest. They present an overview of the theory used to describe EXAFS, as well as examples of applications, and as such the review articles are of value to scientists who want to be informed about the possibilities that EXAFS and XANES can offer. The conference proceedings are of course of more value to the experienced user of EXAFS, who likes to be informed at an early stage about what colleagues have recently accomplished.

Although conference proceedings and review articles are of great value, they can do in general no more for the unexperienced user than to make him or her enthusiastic about the possibilities of a particular technique described. Before a potential user can embark on measurements, however, he or she has to acquire much more detailed knowledge about experimentation and data analysis. It is not easy and is sometimes difficult to extract this kind of information from the EXAFS articles published in the literature. This need of information is even more serious because x-ray absorption measurements are normally carried out at an electron storage ring, in most cases not situated near one's own laboratory. This makes it more necessary to be well prepared in order to make an optimal use of the scarcely allocated beamtime. Only a few experienced users have laboratory EXAFS apparatus at their disposal. But for the unexperienced user it would be extremely helpful to acquire the necessary amount of knowledge before carrying out experiments on synchrotron facilities.

It has been shown that laboratory EXAFS spectrometers can handle experiments on samples with moderate concentration of the material to be investigated. Using the techniques and technologies now available, laboratory EXAFS spectrometers can be optimized to make it possible to carry out a large class of experiments on these facilities.

With these considerations in mind we came to the conclusion that there is a need for a book on EXAFS and especially for a book that addresses the setup of x-ray absorption experiments and the details of analyzing EXAFS spectra. Today most experienced users have taught themselves and each other. With the steadily increasing number of users this is no longer the most sensible way to spread information. Therefore, we decided to publish a book on EXAFS with the main emphasis on experimentation and details of analyzing spectra. In addition, this book discusses the theory of EXAFS and the applications in chemistry, biochemistry, catalysis, disordered systems, surface EXAFS (SEXAFS), and XANES.

A book about EXAFS would not be complete without the addition of SEXAFS and XANES. Although a joint discussion of EXAFS and SEXAFS is certainly possible, in practice there are many differences in experimentation. Most of these differences occur because SEXAFS experiments have to be carried out in ultrahigh vacuum, which is not the case for EXAFS. For this reason, experimentation and applications of SEXAFS are presented in one chapter. XANES (x-ray absorption near-edge structure) deals with the structure of a few electron volts to several tens of electron volts above the edge. This structure needs to be described by multiple-scattering theory, a band calculation, or a cluster approximation. We have decided to include a chapter on multiple-scattering theory in the theoretical part of this book. Other descriptions of XANES and explanations of other structures in the x-ray absorption edge are presented in Chapter 11.

The strength of EXAFS is that it can provide structural information on systems that cannot be studied with other conventional techniques. It is therefore not surprising that the EXAFS technique has found wide applications in the field of disordered systems like catalysts and also in the field of metal-containing enzymes. Chapters on these applications of EXAFS, as well as a more general chapter dealing with the special problems concerning the analysis of EXAFS measurements on disordered systems, are included here.

We hope that this book will be of help to the scientist beginning work with EXAFS and XANES. We especially hope that the beginning graduate student will find answers to the many questions that arise during performance of experiments, analyses, and writing publications.

We thank all coauthors for their contributions to this book and especially for the patience of the authors who finished their manuscripts at an early stage and consequently had to wait some time before seeing the results of their efforts.

D. C. KONINGSBERGER
R. PRINS

Eindhoven University of Technology
Eindhoven, The Netherlands
August 1987

CONTENTS

PART A
Theory

PART B
Instrumental and Data Analysis

PART C
Applications

X-Ray Absorption

PART

A

THEORY

CHAPTER

I

THEORY OF EXAFS

EDWARD A. STERN

Department of Physics
University of Washington
Seattle, Washington

1.1. INTRODUCTION

1.1.1. EXAFS

The extended x-ray absorption fine structure (EXAFS) is the fine structure in the x-ray absorption coefficient starting somewhat past an absorption edge (1)

3

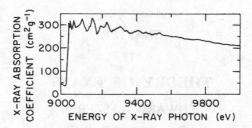

Figure 1.1. The x-ray absorption coefficient for the K-edge of copper metal.

and extending typically 1000 eV further, as shown in Fig. 1.1, for the K-edge of copper metal. Fine structure also exists nearer the absorption edge, but because the interaction of the ejected photoelectron with the potential of the surrounding atoms is still strong in this region, the simplifying single-scattering assumption leading to EXAFS cannot be made. This is discussed later in this chapter and in Chapter 2. The x-ray absorption near-edge structure (XANES) is treated in detail in Chapter 2.

The surge in interest during the last decade in EXAFS, which had been an obscure phenomenon for the previous 45 years, was caused by the realization that EXAFS can be used to obtain information about the arrangement of atoms in the locality of the absorbing atom (2). Several years after this appreciation, the technique became experimentally accessible to the general scientific community with the advent of EXAFS beamlines at synchrotron radiation sources (3). Since then EXAFS has found wide applicability in many diverse areas as a tool to determine the atomic arrangement in many classes of materials whose structures defied analysis by standard techniques such as diffraction or diffuse scattering (4–10).

The properties of EXAFS that make it so useful are

1. Long-range order is not required, so that noncrystalline and crystalline solids can be treated on the same basis.

2. The local atomic arrangement can be determined about each type of atom separately. To do the same with diffuse elastic scattering requires isotopic substitution (11) or anomalous scattering (12), which is tedious and decreases the accuracy. EXAFS is usually able to determine the atomic arrangements with greater resolution than these other techniques (11–13).

3. Structural information is obtained from EXAFS by a simple and direct analysis.

4. The measurement is relatively easy and rapid.

The breakdown of the simplifying approximations of EXAFS in the XANES region means that the EXAFS analysis discussed in Chapter 6 cannot naively be extended to the XANES region. The different analysis required in the XANES region is discussed in Chapter 11. This unfortunately limits the ability of EXAFS to give structural information when the root mean square (rms) disorder is greater than about 0.3 Å (see Chap. 9 also). In addition, because EXAFS is inherently a short-range order probe it does not give any information on long-range order and thus does not replace diffraction techniques but only complements them.

1.1.2. Historical Development

EXAFS has been known for over 50 years, but only recently has its power for structure determination been appreciated. The first experimental detection of fine structure past absorption edges was by Fricke (14) and Hertz (15). The first structure detected was the near-edge structure (16–18), which could be explained by the theory of Kossel (19). However, because the experimental measurements extended the detected fine structure to hundreds of electron volts past the edge (20, 21) (EXAFS), a new explanation was required. The temperature dependence in EXAFS was first experimentally noted by Hanawalt (22). Kronig (23) first attempted an explanation of EXAFS in condensed matter using newly developed quantum mechanics. His explanation used the energy gaps at the Brillouin zone boundaries and thus depended explicitly on the long-range order in the solid. Following Azaroff (24) we shall call this theory a long-range order (LRO) theory and the other class of theories short-range order (SRO). LRO theory is fundamentally in error, but it took more than 40 years for the error to be discovered (25).

Kronig (26) also germinated the idea of SRO theory, which he employed to explain EXAFS in molecules. Some elements of the modern theory, which are explained in the next section, were missing in his original theory, but the basic physical idea was correct. Kronig apparently never realized that the same basic physics explains EXAFS in both solids and molecules. Kronig's SRO theory explains EXAFS by the modulations of the final state wave function of the photoelectron caused by backscattering from the surrounding atoms. Petersen (27) developed the Kronig ideas further for molecules by adding the phase shift in the photoelectron wave function caused by both the potentials of the excited atom and the backscattering atoms. The next advance was made by Kostarev (28), who realized that the Petersen SRO theory was also applicable to matter in the condensed state. The lifetime of the excited photoelectron and core hole state was first calculated by Sawada et al. (29) through a mean free path. The remaining missing element was supplied by Shmidt (30), who pointed out that the interference of the backscattered waves from atoms at a given average distance (a coordination shell of atoms) will not be all in phase because of the disorder in their distances owing to thermal vibrations or structure variations. He introduced a Debye–Waller type factor to account for the thermal disorder based on the Debye theory of lattice vibrations.

The historical development outlined previously is not an exhaustive description of all the contributions to EXAFS theory. The review by Azaroff and Pease (31) covers the field until 1970 (even though it was published in 1974) and gives a more complete exposition. It is our purpose here to take advantage of hindsight, which was not available at the time of the Azaroff and Pease review, and emphasize just those contributions that we now know are correct. At the time of these developments, the issue was quite confused. Although the various elements of the modern theory were around, no one put them completely together. In fact, there was even confusion about whether the SRO or the LRO theory was the correct one. In their review to 1970, nine years after the various elements of the theory had been proposed, Azaroff and Pease (31) stated: "It is premature to draw any conclusions regarding the most appropriate calculational approach to employ for EXAFS." A clever experimental study of EXAFS that attempted to distinguish between the SRO and LRO theories stated in its summary (32): "Neither of the (EXAFS) theories discussed here (SRO and LRO) can account for even the 'gross' characteristics of the absorption spectra. . . ."

In spite of this confusion, there were some applications made of EXAFS as an experimental tool. Chemical bond information using near-edge structure was obtained by Mitchell and Beaman (33) and by Van Nordstrand (34). Nearest-neighbor distance determinations were made by Lytle (35) and by Levy (36). However, these efforts did not attract general attention because of the confusion surrounding the subject.

A major reason for the confusion was the lack of detailed agreement between any theory and the experiments. The experimental measurements were not always reliable themselves, but this was not a critical factor since there were many reliable measurements available that did not agree with the theories. The SRO theory, though correct in principle, suffered because the atomic parameters that enter it were not calculated accurately.

The situation changed when Sayers et al. (37) pointed out, based on a theoretical expression (38) of the EXAFS that has since become the accepted modern form, that a Fourier transform of EXAFS with respect to the photoelectron wave number should peak at distances corresponding to nearest-neighbor coordination shells of atoms. The introduction of the Fourier transform changed EXAFS from a confusing scientific curiosity to a quantitative tool for structure determination. Instead of comparing EXAFS measurements to theoretical calculations based on atomic parameters whose values were difficult to calculate, it was now possible to use EXAFS to extract structure information directly and to determine experimentally all the required atomic parameters. The correctness of the SRO theory was now obvious since the transform revealed only the first few nearest-neighbor shells of atoms.

The accessibility of EXAFS measurements was greatly enhanced by the availability of synchrotron radiation sources of x-rays several years after the potential of EXAFS was first shown. Because synchrotron sources typically have x-ray intensities three or more orders of magnitude greater in the continuum energies than do the standard x-ray tube sources, the time for measuring a spectrum for concentrated samples dropped from the order of a week (39) to the order of minutes (3). At the same time these sources expanded possibilities by making feasible the measurement of dilute samples that could not even be contemplated before.

1.2. THE EXAFS EQUATION

1.2.1. Basic Physics

The basic physics of EXAFS is easy to understand. EXAFS does not occur for isolated atoms but only appears when atoms are in a condensed state. Thus, one has to understand how the absorption is modified in the condensed state to produce EXAFS.

The absorption edge corresponds to an x-ray photon having enough energy to just free a bound electron in the atom. When the electrons are in the most tightly bound $n = 1$ shell the edge is called the K-edge. For the next most tightly bound shell of atoms, the $n = 2$ shell, the corresponding edges are called the L-edges. At present these edges are the only ones used to observe EXAFS, though in principle $n = 3$ or higher shells could be used.

X-ray absorption in the photon range up to 40 keV, the range of most importance for EXAFS, is dominated by photoelectron absorption where the photon is completely absorbed, transferring its energy to excite a photoelectron and leaving behind a core hole in the atom. The excited atom with its core hole has some probability of having additional excitations, but we neglect this multi-electron excitation for now. Its effects are discussed in further detail later.

Assuming that all the absorbed photon's energy goes into exciting a single core electron, the kinetic energy of the excited photoelectron is given by the difference between the photon energy and the electron's binding energy in the atom. When the photoelectron has about 15 eV or greater kinetic energy (i.e., for photon energies of 15 eV or more above the edge), this energy is large compared with its interaction energy with the surrounding atoms (~ 3 eV). In that case the interaction with the surrounding medium can be treated as a perturbation about an isolated atom. Obviously, only the final state of the photoelectron is perturbed by the surroundings. A deep core level, the initial state, is unaffected to a very good approximation.

The final-state photoelectron is modified to first order by a single scattering from each surrounding atom. Quantum mechanically the photoelectron must be treated as a wave whose wavelength λ is given by the de Broglie relation

$$\lambda = \frac{h}{p} \tag{1}$$

where p is the momentum of the photoelectron and $\hbar$ is Planck's constant. In the EXAFS regime p can be determined by the free electron relation

$$\frac{p^2}{2m} = h\nu - E_0 \tag{2}$$

where the x-ray photon of frequency ν has an energy $h\nu$ and E_0 is the binding energy of the photoelectron.

For an isolated atom the photoelectron can be represented as an outgoing wave as shown in Fig. 1.2 by the solid lines. The surrounding atoms will scatter the outgoing waves as indicated by the dashed lines. The final state is the superposition of the outgoing and scattered waves.

The absorption of the x-rays is given quantum mechanically by a matrix element between the initial and final states. In our case the initial state is the

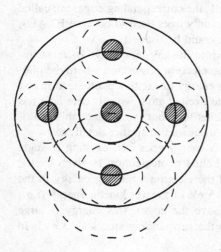

Figure 1.2. Schematic of the radial portion of the photoelectron wave. The solid line indicates the outgoing portion, and the dashed lines indicate portions of the scattered wave from the surrounding atoms.

electron in the atomic core and the final state is this electron excited to the escaping photoelectron. The matrix element is nonzero only in the region where the core state is nonzero—that is, near the center of the absorbing atom. Thus, it is only necessary to determine how the photoelectron is modified by the surrounding atoms at the center of the absorbing atom to determine the modification leading to EXAFS.

The backscattered waves in Fig. 1.2 will add or subtract from the outgoing wave at the center depending on their relative phase. The total amplitude of the electron wave function will be enhanced or reduced, respectively, thus modifying the probability of absorption of the x-ray correspondingly. As the energy of the photoelectron varies, its wavelength varies as indicated in Eqs. (1) and (2), changing the relative phase. Thus the variation of the fine structure in EXAFS as shown in Fig. 1.1 is a direct consequence of the wave nature of the photoelectron. The peaks correspond to the backscattered wave being in phase with the outgoing part while the valleys appear when the two are out of phase.

How the phase varies with the wavelength of the photoelectron depends on the distance between the center atom and backscattering atom. The variation of the backscattering strength as a function of energy of the photoelectron depends on the type of atom doing the backscattering. Thus EXAFS contains information on the atomic surroundings of the center atom. With a more quantitative description of EXAFS it is possible to obtain this information.

1.2.2. Heuristic Derivation

We can directly translate the qualitative picture into an expression for EXAFS. As discussed previously, the absorption is proportional to the amplitude of the photoelectron at the origin and the oscillatory part of this, EXAFS, is produced by the interference between the outgoing and the backscattered wave dependent on their relative phases. The spherical outgoing wave of wave number $k \equiv 2\pi/\lambda$, where λ is the wavelength of Eq. (1), is proportional to $r^{-1} \exp(ikr)$. The backscattered wave $\exp(ik|\mathbf{r} - \mathbf{r}_i|) |\mathbf{r} - \mathbf{r}_i|^{-1}$ is proportional to the product of the amplitude of the outgoing wave at the position $\mathbf{r}_i$ of the backscatterer, and a backward scattering amplitude $T_i(2k)$ characteristic of the backscatter (40), that is,

$$T_i(2k) \frac{e^{ikr_i}}{r_i} \frac{e^{ik|\mathbf{r} - \mathbf{r}_i|}}{|\mathbf{r} - \mathbf{r}_i|} \tag{3}$$

The complex form of the spherical waves is employed where it is understood that the real part of the wave is to be taken at the end. The backscattered wave in Eq. (3) emanates from $\mathbf{r}_i$ instead of from the origin of the outgoing wave.

At the origin ($r = 0$) the backscattered spherical wave in Eq. (3) has an amplitude proportional to

$$T_i(2k)\, \frac{e^{i2kr_i}}{r_i^2}$$

The factor $2kr_i$ is the phase shift introduced by a wave of wave number $k = 2\pi/\lambda$ in traveling the distance $2r_i$ from the origin and back from the backscatterer. This expression would be adequate if the electron were moving in a constant potential. However, the electron moves out of and into the varying potential of the center atom and also senses the varying potential of the backscattering atom. These potentials add additional phase shifts that we call $[\delta_i(k) - (\pi/2)]$ (40).

The complex expression for the backscattered wave becomes

$$\frac{T_i(2k)e^{i[2kr_i + \delta_i(k) - \pi/2]}}{r_i^2} \qquad (4)$$

The backscattered wave modifies the absorption as it interferes with the outgoing wave, and this modification is, by definition, the EXAFS. The real part of Eq. (4) is thus proportional to EXAFS, which we denote by $\chi_i(k)$,

$$\chi_i(k) = K\, \frac{T_i(2k)}{r_i^2} \sin\left[2kr_i + \delta_i(k)\right] \qquad (5)$$

Here K is a constant of proportionality. Finally, we incorporate K into $T(2k)$ by defining

$$\frac{m}{2\pi\hbar^2 k^2}\, t_i(2k) = K\, T_i(2k) \qquad (6)$$

The expression for the EXAFS then becomes

$$\chi_i(k) = \frac{m}{2\pi\hbar^2}\, t_i(2k)\, \frac{\sin\left[2kr_i + \delta_i(k)\right]}{(kr_i)^2} \qquad (7)$$

In the single-scattering approximation, the effect of many scatterers can be obtained by simply adding the effects of each scatterer, and the total EXAFS becomes

$$\chi(k) = \sum_i \chi_i(k) = \sum_i \frac{m}{2\pi\hbar^2}\, \frac{t_i(2k)}{(kr_i)^2} \sin\left[2kr_i + \delta_i(k)\right] \qquad (8)$$

Up to this point our derivation has left out one important physical effect, namely, the lifetime of the excited photoelectron state. There are two contributions to this lifetime. The hole in the atom, left behind after the excitation of the photoelectron, is filled in a time of the order of 10^{-15} s for the K-shell of copper. The hole lifetime varies from atom to atom and from shell to shell in a manner that has been tabulated in ref. (41). The photoelectron itself has a finite lifetime because of scattering from the surrounding electrons and atoms. These two lifetimes contribute to determine the finite lifetime of the excited state consisting of the photoelectron together with the core hole from which it came. The lifetime is important because in order for the backscattered wave to interfere with the outgoing wave the two must be coherent; that is, the phase difference between the two must be well defined, as in Eqs. (4) and (5). The processes that determine the lifetime destroy the coherence of the final state and thus the interference that produces the EXAFS. The lifetime effect can phenomenologically be taken into account by a mean free path term $\exp(-2r_i/\lambda)$, which represents the probability that the photoelectron travels to the backscattering atom and returns without scattering or the hole being filled. The term λ is the mean free path and is, in general, a function of k. Adding the lifetime contribution to $\chi(k)$, Eq. (8) becomes

$$\chi(k) = \frac{m}{2\pi\hbar^2 k^2} \sum_i \frac{t_i(2k)\exp(-2r_i/\lambda)}{r_i^2} \sin\left[2kr_i + \delta_i(k)\right] \qquad (9)$$

The sum in Eq. (9) is over the surrounding atoms at a distance r_i from the x-ray absorbing atom. If, instead of summing over all atoms, we divide the sum first over atoms that have approximately the same average distance R_i from the center, that is, over coordination shells, and then sum over coordinate shells, we obtain the more common form of the EXAFS expression. In any coordination shell the atoms will not all be exactly at the same distance because of thermal vibrations or structural disorder. Thus the contributions of these atoms to the interference will not all be exactly in phase. If this disorder is small and has a Gaussian distribution about the average distance R_j, that is, it has a probability of deviating from the average by

$$(2\pi\sigma_i^2)^{-1/2} \exp\left(-\frac{(r_i - R_j)^2}{2\sigma_i^2}\right)$$

then the dephasing it produces adds a factor $N_i \exp(-2k^2\sigma_i^2)$ to the EXAFS expression instead of N_i, where N_i is the number of atoms of the type i in the shell and σ_i is the rms deviation from the average distance R_i. The EXAFS in Eq. (9) then becomes

$$\chi(k) = \frac{m}{2\pi\hbar^2 k^2} \sum_i N_i \frac{t_i(2k)}{R_i^2} \exp\left(-2k^2\sigma_i^2\right)$$

$$\cdot \exp\left(\frac{-2R_i}{\lambda}\right) \sin\left[2kR_i + \delta_i(k)\right] \qquad (10)$$

Shorter wavelengths or larger k sense a larger dephasing effect and the EXAFS is correspondingly smeared out. The sum remaining in Eq. (10) is over coordination shells at average distance R_i. We assume that all the same atoms are in each coordination shell so that $t_i(2k)$ is the same. If there are different atoms approximately the same average distance apart, then each type of atom would constitute a separate coordination shell and the sum in Eq. (10) is over each of these coordination shells, even though they may have the same R_i.

This expression (10), as we shall note from the more rigorous derivation given later, is valid for all polycrystalline samples, cubic crystals, and amorphous and liquid materials.

1.2.3. Simple Derivation

Several elegant derivations of the EXAFS equation have been presented that show how to correct Eq. (10) beyond its simplifying approximations (42, 43). However, these corrections make the EXAFS equation quite cumbersome. Fortunately, the corrections are usually not very large, and even when they are, they can be calibrated away by using appropriate standards, as will be discussed later. For that reason Eq. (10) is the form in which EXAFS is usually analyzed. We shall present a simple derivation of this EXAFS equation, which has built into it all the simplifying assumptions. An advantage of this approach, besides its mathematical simplicity, is that its derivation follows closely the physical description of EXAFS.

In this derivation a center atom absorbs the photon and the effect of neighboring atoms is considered to only first order in their scattering because the kinetic energy of the photoelectron is large compared with its interaction energy with the sample. To first order in the scattering the effect of neighboring atoms is a simple sum of their isolated contributions. It is assumed that the K-shell of the center atom is excited. If the photon has its electric field polarization E in the z direction it perturbs the Hamiltonian of the center atom in the semiclassical approximation by a term $eE_0 z \cos \omega t$, where E_0 is the amplitude of the electric field and ω is the angular frequency of the wave (44). The dipole approximation is assumed where the electric field E of the electromagnetic wave is assumed to be constant over the dimensions of the K shell so it can be approximated by a potential $-Ez$. This assumption is valid when $(qa)^2 \ll 1$, where $q = \omega/c$

for the photon and a is the dimension of the core state. For the K-edge of copper $(qa)^2 \simeq 0.01$, justifying the approximation. In the time-dependent term

$$\cos \omega t = \tfrac{1}{2} \left[\exp (i\omega t) + \exp (-i\omega t) \right]$$

only the $\tfrac{1}{2} \exp (-i\omega t)$ term causes transitions that absorb energy. The other term would induce emissions if any atoms had a K-shell hole, which is not the case here. The term that thus causes absorption in Fermi's Golden Rule is $(E_0 z / 2) \exp (-i\omega t)$, and the transition rate is (44)

$$W = \frac{2\pi}{\hbar} \left(\frac{eE_0}{2} \right)^2 \left| \langle i | z | f \rangle \right|^2 \rho(E_f) \tag{11}$$

Here $\rho(E_f)$ is the density of final states, the initial K-shell core state is $| i \rangle$ and the final state is $| f \rangle$, and their respective energies are E_i and $E_f = E_i + \hbar\omega$, where $\hbar\omega$ is the photon energy.

The power being absorbed from the EM wave per unit volume is $du/dt = \hbar\omega W N_a$, where N_a is the number of atoms per unit volume, so that

$$\frac{du}{dt} = -4\pi^2 \omega e^2 N_a u \left| \langle i | z | f \rangle \right|^2 \rho(E_f) \tag{12}$$

The energy density u of the EM wave is $E_0^2 / 8\pi$.

Since the EM wave travels at the velocity of light c, $dx = c\, dt$; and since the attenuation coefficient μ is defined by

$$\frac{du}{dx} = -\mu u \tag{13}$$

Eq. (12) yields from Eq. (13)

$$\mu = \frac{4\pi^2 \omega e^2}{c} N_a \left| \langle i | z | f \rangle \right|^2 \rho(E_f) \tag{14}$$

As argued previously, the only term in Eq. (14) that varies in a nonmonotonic fashion as the photon energy $\hbar\omega$ varies is $| f \rangle$. Because $z = r \cos \theta$ and $\langle i |$ is an s wave that has no angular dependence, $| f \rangle$ must have a $\cos \theta$ angular dependence to give a nonzero matrix element. This corresponds to $| f \rangle$ having $\ell = 1$ and $m = 0$ near the origin.

The potentials in the sample are treated in the so-called muffin-tin approximation, where within a sphere centered about each atom the core potential is spherically symmetric and outside the spheres the potential is averaged and set to a constant value. In this model the outgoing part of the wave function $|f\rangle$ can be written outside the center atom sphere as the real part of $h_1^+ (kr) \cos \theta$, where

$$h_1^+ (kr) \cos \theta = \left[(kr)^{-1} + i(kr)^{-2} \right] e^{ikr} \cos \theta \qquad (15)$$

is the outgoing Hankel function for $\ell = 1$. Usually a phase shift factor $e^{i\delta_1}$ is added to $h_1^+ (kr)$ to account for the center atom potential, but we can always multiply the wave function of the outgoing state by an arbitrary phase factor $e^{i\phi} = e^{-i\delta_1}$ so as to cancel this phase shift without changing any physics.

The outgoing state will scatter from the neighboring atom centered at r_j, and to first order the scattering amplitude will be

$$\psi_{sc} = h_1^+ (kr_j) \frac{e^{ik|r - r_j|}}{k \, |r - r_j|} \cos \theta_j f(\alpha) \qquad (16)$$

The general scattered amplitude is a spherical wave emanating from r_j with an amplitude proportional to the value of the outgoing wave at r_j and a general dependence on the angle α about the direction r_j (as defined in Fig. 1.3) is explicitly expressed in Eq. (16).

The form for ψ_{sc} given in Eq. (16) is valid only in the constant muffin-tin region surrounding the atom at r_j. As this scattered wave propagates into the spherical core about the center atom it will be modified in a significant manner, and we desire to account for this interaction exactly. This can be done using standard scattering theory. The wave ψ_{sc} can be decomposed into spherical harmonic partial waves about the center atom. Only the $\ell = 1$, $m = 0$ harmonic is of interest because only that partial wave will interfere with the outgoing wave that has the same symmetry.

Partial wave scattering theory states that after scattering the incoming partial wave produces an outgoing partial wave of the same symmetry and magnitude but with a phase shift factor of $e^{i2\delta_1}$ in the constant muffin-tin region outside the center atom (45). Denoting the $\ell = 1$, $m = 0$ partial wave of ψ_{sc} by $\psi_{sc}^{(1,0)}$, then just outside the muffin-tin sphere about the center atom

$$\psi_{sc}^{(1,0)} = A(1, 0) \, h_1^- (kr) \cos \theta \qquad (17)$$

where

$$h_1^- (kr) = \left[(kr)^{-1} + i \, (kr)^{-2} \right] e^{-ikr}$$

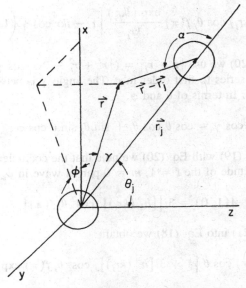

Figure 1.3. Schematic defining the angles between the coordinate axes, the bond to the neighboring atom at r_j, the general point at r, and the scattered direction $r - r_j$.

is the incoming Hankel function for $\ell = 1$ and $A(1, 0)$ is the amplitude of that function present in ψ_{sc}. The outgoing partial wave produced by the scattering of $\psi_{sc}^{(1,0)}$ by the center potential is then $A(1, 0) \, h^+(kr) e^{i2\delta_1} \cos \theta$. The final state $|f\rangle$ is the sum of this outgoing partial wave and the original outgoing part of $|f\rangle$ given in by Eq. (15). Thus

$$|f\rangle = h_1^+(kr) \cos \theta \left[1 + A(1, 0) \exp(i2\delta_1) \right] \tag{18}$$

It remains to determine $A(1, 0)$. We employ an elementary derivation to do so. We note that as $r \to 0$ the real part of $h_1^+(kr)$ of Eq. (15) becomes

$$\text{Re } h_1^+ = \frac{\cos kr}{kr} - \frac{\sin kr}{(kr)^2} \to -\frac{kr}{3} \tag{19}$$

as $r \to 0$. Only the $\ell = 1$ partial wave is linear in r as $r \to 0$.

We can thus pick out the $\ell = 1$ partial wave in ψ_{sc} by finding the coefficient of the term linear in r as $r \to 0$. As $r \to 0$, ψ_{sc} in Eq. (16) becomes

$$\psi_{sc} \to h_1^+(kr_j) \cos \theta_j f(\pi) \frac{\exp(ikr_j)}{kr_j} \left[1 - ikr \cos \gamma \left(1 + \frac{i}{kr_j} \right) \right] \quad (20)$$

In obtaining Eq. (20) we use $|\mathbf{r} - \mathbf{r}_j| = (|r^2 + r_j^2 - 2rr_j \cos \gamma)^{1/2}$ and expand ψ_{sc} by a Taylor's series to first order in r. The angle γ is between $\mathbf{r}$ and $\mathbf{r}_j$ as shown in Fig. 1.3. In terms of θ and ϕ,

$$\cos \gamma = \cos \theta \cos \theta_j + \sin \theta_j \sin \theta \cos \phi$$

By comparing Eq. (19) with Eq. (20) we note that the coefficient of $[-(kr \cos \theta)/3]$ is the amplitude of the $\ell = 1$, $m = 0$ partial wave in ψ_{sc} and thus

$$A(1, 0) = 3i\left[\left[h_1^+(kr_j) \right]^2 \cos^2 \theta_j f(\pi) \right] \quad (21)$$

Substituting Eq. (21) into Eq. (18) we obtain

$$|f\rangle = h_1^+(kr) \cos \theta \left[1 + 3i\left[h^+(kr_j) \right]^2 \cos^2 \theta_j f(\pi) \exp(2i\delta_1) \right] \quad (22)$$

Substituting Eq. (22) into Eq. (14) we obtain to first order in the scattering that

$$\mu = \mu_0 \left[1 - 3 \, \text{Im} \left\{ \left[h_1^+(kr_j) \right]^2 \cos^2 \theta_j f(\pi) \exp(2i\delta_1) \right\} \right] \quad (23)$$

where μ_0 is the absorption coefficient of an isolated atom and the effect of the jth atom is calculated to first order in its scattering. In obtaining Eq. (23) the relationship was employed that in an average over space $(\text{Re } h_1^+)^2 = \frac{1}{2} \text{Re}(h_1^+)^2$. To obtain the effect of all the surrounding atoms requires summing over j, and μ becomes

$$\mu = \mu_0 \left[1 - 3 \, \text{Im} \sum_j \left[h_1^+(kr_j) \right]^2 \cos^2 \theta_j f(\pi) \exp(2i\delta_1) \right] \quad (24)$$

and

$$\begin{aligned} \chi(k) &= \frac{\mu - \mu_0}{\mu_0} \\ &= 3 \, \text{Im} \sum_j \left\{ \left[h_1^+(kr_j) \right]^2 \cos^2 \theta_j f(\pi) \exp(2i\delta_1 - i\pi) \right\} \end{aligned} \quad (25)$$

where the minus sign is incorporated as $e^{-i\pi}$.

The additional approximations made in obtaining the standard EXAFS expression are the following:

1. $kr_j \gg 1$, so that $h_1^+(kr_j) \to \exp(ikr_j)/kr_j$.
2. The small atom approximation—that is, $D/r_j \ll 1$, where D is the atomic diameter—so that $f(\pi)$ can be calculated (45) by

$$f_0(\pi) = \sum_\ell (2\ell + 1) \sin \delta_l \exp(i\delta_l)(-1)^\ell$$

where the curvature of the spherical wave is neglected and the incoming and scattered waves can be approximated by plane waves.

With these approximations

$$\chi(k) = 3 \sum_j \frac{m}{2\pi\hbar^2} \frac{t_i(2k)}{(kr_j)^2} \cos^2 \theta_j \sin\left[2kr_j + \delta_j(k)\right] \qquad (26)$$

where

$$t(2k)\, e^{i\beta} = \frac{2\pi\hbar^2}{m} f_0(\pi) \quad \text{and} \quad \delta_j(k) = 2\delta_1 + \beta - \pi$$

and $t(2k)$ and β are real.

In the plane wave approximation $f_0(\pi)$ is independent of r_j. When this approximation cannot be made, $f(\pi)$ is a function of r_j.

1.3. ADDITIONS TO THE SIMPLE DERIVATION

1.3.1. Lifetime

The lifetime of the state $|f\rangle$ has been neglected in this derivation. It is added phenomenologically in terms of a mean free path λ, which adds the term $\exp(-2r_j/\lambda)$ to $\chi(k)$, giving

$$\chi(k) = \operatorname{Im} \sum_j \frac{3m}{2\pi\hbar^2} t_j(2k) \frac{\cos^2 \theta_j \exp\{i[2kr_j + \delta_j(k)]\}}{k^2 r_j^2} \exp\left(\frac{-2r_j}{\lambda}\right) \qquad (27)$$

1.3.2. Disorder

Consider a material with a large number of center and neighboring atoms. The χ will be a superposition of the χ from each environment, leading to the need to average χ over the distribution of atoms in the material. From all these distributions the ones corresponding to a given A-type center atom and B-type neighboring atom will first be considered in our averaging. The r_j will vary because of either structural variations or thermal motion, and the measured χ of Eq. (27) can be rewritten

$$\chi(k) = \text{Im } 3 \sum_j A_j(k)\, G(r_j) \exp\left[i(2kr_j + \delta_j)\right] \cos^2 \theta_j \qquad (28)$$

where

$$A_j(k) = \frac{m}{2\pi\hbar^2}\frac{t_j(2k)}{k^2} \qquad G(r_j) = \frac{\exp(-2r_j/\lambda)}{r_j^2}$$

The A_j and G_j are slowly varying functions of both photoelectron wave number k and distance of the jth atom at r_j. Rapid variation occurs only in the exponential function. If we consider only the B-type atoms in the ith coordination shell, the sum averaged over all central atoms of a given A type in the sample becomes

$$x_i(k) = \text{Im } \sum_B{}' A_B(k)\, G(R_B) \int_B^N P_B^i(r_B) \exp\left\{i\left[2kr_B + \delta_B(k)\right]\right\} dr_B \qquad (29)$$

where the prime indicates addition over the various types of atom in only the ith shell,

$$P_B^i(r_B) = \frac{G(r_B)\, p_B^i(r_B)}{G(R_B)}$$

and $p_B^i(r_B)\, dr_B$ is the probability of finding a B type atom of the ith shell in the range to $r_B + dr_B$ and R_B is a fixed value of r_B, for example, the average value of the ith coordination shell. Letting $r_B = R_B + \Delta r_B$ and performing the integration over Δr_B, we can write Eq. (29) in the form

$$\chi_i(k) = \sum_B N_B\, Q_B^i(k)\, A_B(k)\, G(R_B) \sin\left[2kR_B + \delta_B(k) + \phi_B^i(k)\right] \qquad (30)$$

where

$$Q_B^i(k) \exp\left[i\phi_B^i(k)\right] = \int_{-\infty}^{\infty} P_B^i(r_B) \exp(i2k\Delta r_B)\, d\Delta r_B \qquad (31)$$

both $Q(k)$ and $\phi(k)$ are real, and the sum in Eq. (30) is now over all different atom types in the ith shell. We assume that there are N_B of the B-type atoms in the shell. If all atoms in the shell are the same, then there remains no sum over B and $N_B \to N$, the total number of atoms in the shell. Note that $\phi(k) = 0$ if $P_B^i(R_B + \Delta r_B)$ is symmetric about R_B. General expressions for $Q(k)$ and $\phi(k)$ can be obtained from cumulant theory, which shows that (40)

$$\int P_B^i(r_B) \exp(i2k\Delta r_B) \, d\Delta r_B = \exp\left(\sum_{n=1}^{\infty} \frac{(2ik)^n}{n!} C_n\right) \qquad (32)$$

where C_n is the nth-order cumulant avverage. In terms of ordinary averages, the leading cumulants are

$$C_1 = \langle \Delta r_B \rangle \equiv \bar{r}_B - R_B \qquad C_3 = \langle (r_B - \bar{r}_B)^3 \rangle \qquad (33)$$
$$C_2 = \langle (r_B - \bar{r}_B)^2 \rangle \equiv \sigma_B^2 \qquad C_4 = \langle (r_B - \bar{r}_B)^4 \rangle - 3\sigma_B^4$$

The average over $P_B^i(r_B)$ is denoted by $\langle \ \rangle$. By comparing Eq. (32) with Eq. (31) we note that

$$Q_B^i(k) = \exp \sum_{n=1}^{\infty} \frac{(-1)^n (2k)^{2n}}{(2n)!} C_{2n} \qquad (34)$$
$$2kR_i + \phi_B^i(k) = \sum_{n=0}^{\infty} \frac{(-1)^n (2k)^{2n+1}}{(2n+1)!} C_{2n+1}$$

Another general approach to the disorder problem has been used by Hayes and Boyce (10). They assume a particular parameterized form for $P_B^i(r_B)$ and vary the parameters to obtain the best computer fit to the Fourier transform of Eq. (29) in r space. If an accurate guess to the form of $P_B^i(r_B)$ is used, this method correctly includes all the effects of disorder discussed in the following section.

1.3.2.1. Thermal Vibrations and Gaussian Disorder

In the case where the disorder is induced by thermal vibrations or any other cause that induces a Gaussian disorder of the form

$$P_B^i(r_B) = \frac{\exp\left[-(r_B - R_B)^2 / 2\sigma_B^2\right]}{\sqrt{2\pi\sigma_B^2}}$$

the form for the disorder is a Debye–Waller type term (46):

$$Q_B^i(k) = \exp(-2k^2\sigma_B^2) \qquad \phi_B^i(k) = 0 \qquad (35)$$

where σ_B^2 is the mean square deviation of the B-type atoms about the average value R_B of the ith shell of atoms.

An expression for σ as a function of temperature was first given by Schmidt (30) based on the Debye approximation for lattice vibrations. The Debye approximation for σ was rederived and somewhat generalized to include anisotropy by Beni and Platzman (47). Going beyond the Debye approximation the exact expression for σ in the harmonic approximation was numerically calculated (48) for several force constant models of copper, iron, and platinum. Rehr (43) has simplified the expressions in the Debye model and has developed techniques for a more accurate calculation of σ in general cases including amorphous solids. It is important to note that the σ that enters into EXAFS is not the same as that which enters the Debye–Waller term in diffraction. In diffraction the mean square deviation, u^2, that enters is one about the lattice site for each atom, while for EXAFS the deviation is a relative one between the absorbing and backscattering atoms. Long-wavelength phonons, for example, will give a much smaller contribution in EXAFS than in diffraction.

Empirically it is found (49), and the theoretical work of Rehr and collaborators (43) has corroborated this, that the temperature dependence of σ can be accurately described by a simple Einstein model of lattice vibrations. In fact, the Einstein model for the nearest-neighbor shell of atoms is even more accurate for nonprimitive lattices where the optical modes contribute. For example, the exact calculation for crystalline germanium is much closer to an Einstein model than the Debye model. To illustrate this, Fig. 1.4 shows the phonon modes of germanium [as obtained by Rehr (43)] that contribute to the diffraction u^2 and the weighting that contributes to the σ^2 of the first and second shells of germanium. In the Einstein approximation the weighting is a δ function at a single frequency, and it can be seen that the σ^2 weighting for the first shell is a strikingly close approximation to that distribution. Rabe and collaborators (50) have checked the Debye model against various materials and have found agreement in some cases and disagreement in others. The disagreement occurs in nonprimitive lattices in agreement with the case of germanium just given.

The simple Einstein model predicts a rms u^2 about the average position of the ith atom of the form:

$$u_i^2 = \frac{\hbar}{M_i \omega_i} \left(\frac{1}{\exp\left(\hbar \omega_i / k_B T\right) - 1} + \frac{1}{2} \right) \tag{36}$$

where M_i is the mass of the atom and ω_i is its frequency of vibration.

In the simple Einstein model the motion between adjacent atoms is uncor-

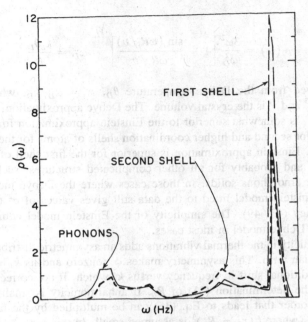

Figure 1.4. Effective density of modes as a function of radial frequency in germanium. Shown are the density of phonon modes (appropriate for u^2), as well as the weighting of modes that contribute to σ^2 between the center atom and the first and second neighbors in germanium.

related, and σ_{ij}^2 between the ith and jth atoms is given by $u_i^2 + u_j^2$, so that we obtain

$$\sigma_{ij}^2 = \frac{\hbar^2}{M_r \omega} \left(\frac{1}{\exp\left(\hbar\omega/k_B T\right) - 1} + \frac{1}{2} \right) \tag{37}$$

where M_r, the reduced mass, is given by $M_r = M_i M_j/(M_i + M_j)$. The frequency ω is obtained by fitting Eq. (37) to the experimental measurements of the temperature dependence.

The Debye approximation for the σ^2 of EXAFS has a simple form given by (48)

$$\sigma^2 = \frac{\hbar}{2M_r} \int \rho_j(\omega) \coth\left(\frac{\hbar\omega}{2k_B T}\right) \frac{d\omega}{\omega} \tag{38}$$

where

$$\rho_j(\omega) = \frac{3\omega^2}{\omega_D^2} \left[1 - \frac{\sin(\omega R_j/v)}{\omega R_j/v} \right] \qquad \omega_D = \frac{k_B \theta_D}{\hbar}$$

and is derived from the Debye temperature θ_D, $v = \omega_D/k_D$, where $k_D = (6\pi^2 N/V)^{1/3}$ and V is the crystal volume. The Debye approximation, according to Rehr (43), is somewhat superior to the Einstein approximation for primitive lattices and for second and higher coordination shells of atoms for the nonprimitive lattices. Einstein approximation is superior for the first shell of nonprimitive lattices and probably for all other complicated structures, such as large molecules or amorphous solids. In those cases where the Debye model is superior, an Einstein model fitted to the data still gives values of σ^2 to 10% or better accuracy (43, 49). The simplicity of the Einstein model commends its use over the Debye model in most cases.

Anharmonicity in the thermal vibrations adds an asymmetric distribution term to the Gaussian term. This asymmetry makes ϕ nonzero and its k dependence causes an additional shift in frequency versus k, which, if not corrected, gives an error in the determination (51) of R_B. If anharmonicity is small, then the Gaussian disorder that leads to Eq. (35) can be multiplied by the factor $1 + b(r_B - R_B)^3$, where $b(r_B - R_B)^3$ is assumed small. In this case,

$$\begin{aligned} Q_B^i(k) &\simeq \exp(-2k^2\sigma_B^2) \\ \phi_B^i(k) &\simeq 2b\sigma_B^3 [3k\sigma_B - 4(k\sigma_B)^3] \end{aligned} \qquad (39)$$

We note that the anharmonicity to lowest order affects only the phase and not the amplitude of the $\chi(k)$. The term in ϕ linear in k corresponds to a shift in distance of $R_{anh} = 3b\sigma_B^4$, which is the correct expansion in $\bar{r}_B$ due to the anharmonic factor. However, the anharmonic term also contributes the k^3 dependence. Because of interference with band structure or density of states contributions the EXAFS regime begins about 15 eV past the edge, which means that the minimum usable EXAFS data start at $k \simeq 2$ Å^{-1}. If σ is large enough, then in the EXAFS regime of $k > 2$ Å^{-1} the k^3 term can be as large as the linear term and can obscure it. The average value of the derivative of the phase shift is consequently decreased, moving the peak position in the Fourier transform to lower r. This would lead to a too small value for R_i if the anharmonic effect is neglected, as was noted in the case of zinc metal (51).

1.3.2.2. Structural Disorder

As occurs in many structures, for example, those with a complicated basis or with symmetry less than cubic, the average position of each atom in a given coordination shell is not equal. This brings up the question of the definition of a coordination shell: What criterion is used to decide when the average position of a given atom is far enough away from that of other atoms to warrant classifying that atom in another coordination shell? The definition is somewhat arbitrary, but it is convenient to base this criterion on the Fourier transform in r space. If atoms are well enough separated so that their transforms in r space show clearly separated peaks, we shall consider the atoms to be in different coordination shells, otherwise they will be considered to be in the same coordination shell. In practice, this criterion means that atoms more than about 0.6-Å apart are considered to be in separate shells.

Martens et al. (52) first calculated explicitly the modification to $\chi(k)$ produced by structural disorder. More general formulations have been given since. Consider first the case of atoms of the same type, N_1 of which are at r_1 and N_2 at r_2. Initially, we neglect vibrational disorder so that

$$
P_B^i(r_B) = \frac{R_B^2 \exp\left[-2(r_1 - R_B)/\lambda\right]}{r_1^2} \delta(r_B - r_1)
$$
$$
+ \frac{R_B^2 \exp\left[-2(r_2 - R_B)/\lambda\right]}{r_2^2} \delta(r_B - r_2)
$$

(40)

By our assumption, the atoms at r_1 and r_2 are the same type so $A_j(k) = A(k)$ and $\delta_j(k) = \delta(k)$ are the same, and Eq. (30) becomes (52–54)

$$
\chi_i(k) = A(k)\left[B_1^2 + B_2^2 + 2B_1 B_2 \cos 2k(r_1 - r_2)\right]^{1/2}
$$
$$
\times \sin\left[2kR_B + \delta(k) + \phi(k)\right]
$$

(41)

where

$$
\tan \phi(k) = \frac{B_1 \sin 2k(r_1 - R_B) + B_2 \sin 2k(r_2 - R_B)}{B_1 \cos 2k(r_1 - R_B) + B_2 \cos 2k(r_2 - R_B)}
$$

$$
B_i = \frac{N_i \exp(-2r_i/\lambda)}{r_i^2} \qquad i = 1, 2
$$

If vibrational disorder is added so that

$$
P_B^i(r_B) = \frac{R_B^2 \exp\left[-2(r_1 - R_B)/\lambda\right] \exp\left[-(r_B - r_1)^2/2\sigma_1^2\right]}{r_1^2 \left(2\pi\sigma_1^2\right)^{1/2}}
$$
$$
+ R_B^2 \frac{\exp\left[-2(r_2 - R_B)/\lambda\right] \exp\left[-(r_B - r_2)^2/2\sigma_2^2\right]}{r_2^2 \left(2\pi\sigma_2^2\right)^{1/2}}
$$

(42)

then Eq. (41) is modified (53) by replacing B_i with B_i', where

$$B_i' = \frac{N_i \exp(-2r_i/\lambda)}{r_i^2} \exp(-2k^2\sigma_i^2) \qquad i = 1, 2 \qquad (43)$$

All the information on the disorder is contained in the $\phi(k)$ term and the expression in the square root term multiplying $A(k)$. Minima in the square root term occur when $\cos 2k(r_1 - r_2) = -1$ or when $k = (2n + 1)\,\pi[2(r_1 - r_2)]^{-1}$.

The first minimum occurs at $k = \pi/2(r_1 - r_2)$ and subsequent minima occur at fixed intervals of $\pi/(r_1 - r_2)$. By comparing the ith shell with a standard whose absorbing and backscattering atoms are the same as the unknown, it is possible to isolate just those two terms containing the disorder information, as is discussed in further detail in Chapter 6.

The most general situation is when the disorder in a given shell is distributed among more than one type of atom. If the atoms centered at r_B in Eq. (40) have an $A_i(k)$ and $\delta_i(k)$ associated with them ($i = 1, 2$), then

$$
\begin{aligned}
\chi_i(k) = \Big\{ &[A_i(k)\,B_1']^2 + [A_2(k)\,B_2']^2 + 2A_1 A_2 B_1' B_2' \\
&\times \cos\left[2k(r_1 - r_2) + \delta_1(k) - \delta_2(k)\right] \Big\}^{1/2} \\
&\times \sin\left| 2kR_i + \tfrac{1}{2}[\delta_1(k) + \delta_2(k)] + \phi(k) \right|
\end{aligned} \qquad (44)
$$

where

$$
\begin{aligned}
\tan\phi = \Big\{ &A_1(k)B_1' \sin\left[2k(r_1 - R_i) + \tfrac{1}{2}(\delta_1 - \delta_2)\right] \\
&+ A_2(k)B_2' \sin\left[2k(r_2 - R_i) + \tfrac{1}{2}(\delta_2 - \delta_1)\right] \Big\} \\
&\times \Big\{ A_1(k)B_1' \cos\left[2k(r_1 - R_i) + \tfrac{1}{2}(\delta_1 - \delta_2)\right] \\
&+ A_2(k)B_2' \cos\left[2k(r_2 - R_i) + \tfrac{1}{2}(\delta_2 - \delta_1)\right] \Big\}^{-1}
\end{aligned}
$$

In this case the minima in the square root term occur at $2k(r_1 - r_2) + \delta_1(k) - \delta_2(k) = (2n + 1)\pi$, and they are no longer equally spaced in k, nor is the first maximum at half of the interval between minima.

1.3.3. Angular Dependence

If one has a randomly ordered sample such as a polycrystalline sample with no preferred orientation or an amorphous solid, the expression for $\chi(k)$ in Eq. (27) must be averaged over θ_j, the angle between a particular bond and the x-ray polarization direction. A random average over θ_j in three dimensions results in

$$\langle 3 \cos^2 \theta_j \rangle = 1 \tag{45}$$

For an unoriented sample with a Gaussian disorder [combining Eqs. (28) and (30) with Eqs. (35) and (45)] this leads to

$$\chi(k) = \frac{m}{2\pi\hbar^2} \sum_i \sum_B \frac{N_B t_B(2k)}{k^2 R_B^2} \sin \left[2kR_B \right.$$

$$\left. + \delta_B(k) \right] \exp \left(\frac{-2R_B}{\lambda} \right) \exp \left(-2k^2 \sigma_B^2 \right) \tag{46}$$

Comparing Eq. (46) with the heuristic expression in Eq. (10) we note that these are essentially the same except that Eq. (46) makes the distinction between different B-types of atoms in the ith shell. Thus, the heuristic expression (10) is appropriate for an unoriented sample with Gaussian disorder. The generally used EXAFS formula for an unoriented sample with Gaussian disorder with only one type of atom present in the jth coordination shell can now be derived from Eq. (46):

$$\chi(k) = \sum_j \frac{N_j F_j(k)}{kR_j^2} \exp \left(-2k^2 \sigma_j^2 \right) \exp \left(\frac{-2R_j}{\lambda} \right) \sin \left[2kR_j + \delta_j(k) \right] \tag{47a}$$

with

$$F_j(k) = \frac{m}{2\pi\hbar^2 k} t_j(2k) \tag{47b}$$

If the sample has some preferred orientation such as a single crystal or diatomic molecules adsorbed on a surface, then it is possible for EXAFS to have some dependence on the angle between the preferred sample direction and the x-ray polarization. This angular dependence could be calculated by performing the appropriate averaging over $\cos^2 \theta_j$ in Eq. (27). However, it is possible to use a more general argument for the angular dependence of EXAFS that will be valid for all x-ray absorption edges and not be limited to the K-edge as is the case in Eq. (27).

The more general argument is based on a macroscopic argument relating x-ray absorption to Maxwell's equation (4a, 55). A general electromagnetic plane wave, of wave number q, and radial frequency ω, propagating in the x direction can be represented as

$$E = E_0 \exp{(iqx)} \tag{48}$$

where E_0 is the amplitude of the wave. Maxwell's equations state that $q^2 = n^2\omega^2/c^2$, where the refractive index n is related to the relative permittivity of the medium ε_r (usually a complex number) at the frequency ω by $\varepsilon_r = n^2$, and c is the speed of light in a vacuum. In the x-ray regime we can assume that the magnetic permeability of the medium is that of a vacuum and $\varepsilon_r \simeq 1 - \delta + i\varepsilon_2$, where δ and ε_2 are real and much less than 1. Substituting this value of $\varepsilon_r = n^2$ into the expression for q^2 one finds

$$q = \frac{\omega}{c}\left(1 - \frac{\delta}{2} + \frac{i\varepsilon_2}{2}\right)$$

Substituting this expression for q into Eq. (48) and taking the absolute square of both sides gives

$$I = I_0 \exp{\left(\frac{-\omega\varepsilon_2 x}{c}\right)} \tag{49}$$

Comparing Eqs. (49) and (13) leads to the relation for the x-ray absorption coefficient

$$\mu = \frac{\omega\varepsilon_2}{c} \tag{50}$$

Since ε_2 is a tensor of rank two, μ also is a tensor of rank two in the x-ray range. The orientation dependence of ε_2 (that is, the variation of ε_2 with the angle between the polarization vector of the electromagnetic wave and the sample axes) is well known. If a rotation axis has threefold or higher symmetry, ε_2, and thus μ is independent of the angle between the polarization vector of the electromagnetic wave and the sample axes and remains constant during rotation. In the case of hexagonal symmetry, $\mu = \mu_{\parallel} + \mu_{\perp} \sin^2{\theta}$, where $\mu_{\parallel}$ and $\mu_{\perp}$ are the absorption coefficients of the sample when the polarization is parallel and perpendicular to the c axis, respectively, and θ is the angle the polarization makes with the c axis.

1.3.4. L-Edge

Expression (27) for EXAFS is valid for the K-edge where the initial state has s symmetry and angular momentum $\hbar l = 0$. In that case the dipole selection rule of $\Delta l = 1$ permits only a single final state of $l = 1$ with p symmetry.

For the case of the L-edge there are three initial states, the $2s$, $2p_{3/2}$, and $2p_{1/2}$ states. The $2s$ state that produces the L_1-edge makes transitions to only a p state similar to the case of the K-edge, and the expression in Eq. (27) applies there with the initial state $|i\rangle$ being a $2s$ instead of a $1s$ state. Transitions from the $2p_{3/2}$ and $2p_{1/2}$ states, which produce the L_3- and L_2-edges, respectively, are to two types of final states with s and d symmetry. The expression for $\chi(k)$ in this case, for unoriented samples or samples with cubic and higher symmetry, is (55, 56)

$$\chi(k) =$$

$$\sum \frac{A_j(k)\, G(r_j)\left\{ \left|\langle 2|z|1\rangle\right|^2 \sin\left[2kr_j + \delta'_{2j}(k)\right] + \tfrac{1}{2}\left|\langle 0|z|1\rangle\right|^2 \sin\left[2kr_j + \delta'_{0j}(k)\right]\right\}}{\left|\langle 2|z|1\rangle\right|^2 + \tfrac{1}{2}\left|\langle 0|z|1\rangle\right|^2}$$

$$(51)$$

where $\langle l|z|1\rangle$ is the dipole matrix elements of the radial portions of the wave functions between the final state of angular momentum $\hbar l$ and the initial $2p$ state of an isolated atom, and

$$\delta'_{2j}(k) = 2\delta_2(k) + \beta_j(k)$$

$$\delta'_{0j}(k) = 2\delta_0(k) + \beta_j(k) \qquad\qquad (52)$$

Note that the definition of δ_j in Eq. (26) for the K- and L_1-edges differ from the δ'_{2j} and δ'_{0j} by a $-\pi$ term.

Here δ_l ($l = 0, 2$) is the phase shift of the partial wave of angular momentum $\hbar l$ introduced by the central atom potential, and β_j is the phase shift introduced by backscattering from the ith atom.

Calculations (56) and a measurement (55) have shown that

$$\frac{|\langle 0|z|1\rangle|^2}{2|\langle 2|z|1\rangle|^2} \simeq 0.02 \tag{53}$$

and to a reasonable approximation for the L-edge EXAFS in unoriented samples one can neglect the contributions to the final s states and the expression reduces to Eq. (46) of the K-edge with the phase shift δ'_{2j} of Eq. (52) replacing δ_j.

All the modifications of disorder discussed in Section 1.3.2 apply equally well to the L-edge. Thus in unoriented samples the EXAFS about the L-edge can be treated, in practice, just as the K-edge.

However, for samples with a preferred orientation a new effect can manifest itself at the L_2- and L_3-edges (57). In the calculation of the square of the matrix element a cross term arises between the matrix element from a p to an s state and from a p to a d ($l = 2$) state. This leads to an angular dependence of the EXAFS of the form (55)

$$\chi(k) = \sum_j A_j(k)\, G(r_j) \Big\{ (1 - 3\cos^2\theta_j) M_{02} \sin\left[2kr_j + \delta'_{02j}(k)\right]$$

$$+ \tfrac{1}{2}(1 + 3\cos^2\theta_j) \sin\left[2kr_j + \delta'_{2j}(k)\right] \Big\} \tag{54}$$

where

$$M_{02} = \left[(\langle 2|z|1\rangle\,\langle 0|z|1\rangle + cc))\langle 2|z|1\rangle\right]^{-2}$$

$$\delta'_{02j}(k) = \delta_2(k) + \delta_0(k) + \beta_j(k) \tag{55}$$

where $\delta'_{2j}(k)$ is defined by Eq. (52), and we have used relation (53) to neglect the terms in the square of the matrix element containing $|\langle 0|z|1\rangle|^2$, and cc means complex conjugate of the preceding term. The quantitatively new term in Eq. (54) is a cross term of matrix elements to final d and s states. Note that the relation (53) does not imply that the cross term is negligible. On the contrary $|M_{02}| \simeq 0.4$. Because the angular average of the cross term is zero for samples whose symmetry produces no angular variation in $\chi(k)$, the cross term becomes discernible only when anisotropy is present. Equation (54) satisfies the macroscopic requirements of the angular variation of a second rank tensor; as it must.

The presence of the cross term proportional to M_{02} causes difficulties in interpreting the angular dependence of the EXAFS at $L_{2,3}$-edges. Besides the obvious mistake of neglecting the M_{02} term as was done in ref. (58), the different k dependence of δ'_{02j} and δ'_{2j} causes interference between the sine functions of the two terms of Eq. (54), which further complicates the interpretation (59).

1.4. CALCULATION OF BACKSCATTERING AMPLITUDE $F(k)$ AND PHASE SHIFT $\delta_l(k)$

Calculation of $\delta_l(k)$ and thus, by the equation between Eqs. (25) and (26), $f_0(\pi)$, have to be performed numerically. Knowledge of $f_0(\pi)$ determines $t(2k)$ as per Eq. (26). The most extensive calculations are those of Teo and Lee (56) based on the formulism of Lee and Beni (60). The major difficulty of the calculation is to determine the best effective potential sensed by the photoelectron. This problem is different from the one usually encountered because of the high energy of the photoelectron. In the usual band theory calculation all the electrons have energies of the order of the binding energy, and the potential appropriate to this case has had an extensive development. The exchange and correlation contributions to the potential are the ones that are difficult to include and in the usual low-energy theory (61) they are taken to be proportional to the local electron density to the $\frac{1}{3}$ power. For EXAFS the high energy of the photoelectron requires an additional dependence on the energy of the photoelectron to be added to the usual formulation. In addition, the energy loss suffered by the photoelectron, which has been added phenomenologically through a mean free path, is a new element in the calculation not present in the band theory calculations.

The Lee and Beni formulation (60) uses a simple energy-dependent approximation to the exchange and correlation potentials. More sophisticated approximations to these potentials have been proposed by Rehr and Chou (62). A calculation of EXAFS in bromine has been made using the more sophisticated approximations (62), which obtain an improvement over the Lee and Beni cal-

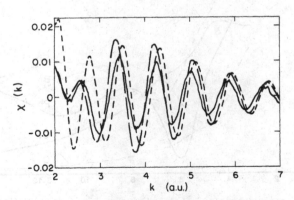

Figure 1.5. Comparison between calculated and measured EXAFS for bromine vapor. The short dashed line is theory in the small-atom approximation, and the long dashed is theory accounting for the spherical nature of the photoelectron wave. Energy-dependent exchange is used in the theoretical scattering potential. The solid line is the experimental measurements.

culations and give a better match to the low k dependence of the experimental results, as illustrated in Fig. 1.5. However, note that the theoretical values are generally larger in magnitude than the experiment. This discrepancy is a many-body effect, as discussed in Section 1.5.3.

In Fig. 1.6 $F(k)$, as calculated by Teo and Lee (56) for several representative atoms, is plotted. In Figs. 1.7 and 1.8 the calculations of Teo and Leo (56) for δ_l and β of Eq. (26) are shown for some representative atoms. Measurements have shown that the calculations of Teo and Lee have to be corrected for many-body effects (49, 63) and that additional corrections are necessary at low-k values (49). Many-body effects are discussed in Section 1.5.3.

However, the Teo and Lee calculations do show the general features of the dependence of $F(k)$ and $\delta(k)$ with atomic number Z in the EXAFS range. For the low-Z atoms $F(k)$ is large at low k, decreasing rapidly with increasing k. As the Z increases a maximum appears at intermediate k value. This maximum then occurs at larger k values as Z increases. The highest Z atoms show two maxima with appreciable values of $F(k)$ extending to large values of k. Char-

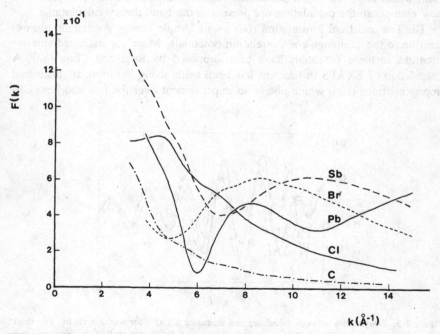

Figure 1.6. Backscattering amplitude $F(k)$ as a function of k [from ref. (56)] for carbon, chlorine, lead bromine, and antimony atoms. Note the characteristic variation as the atomic number of the atoms increases.

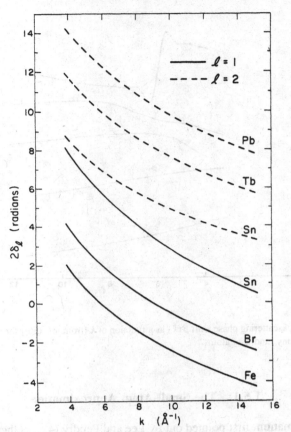

Figure 1.7. Twice the center atom phase shift as a function of k for some typical atoms as calculated by ref. (56). Note that in Eq. (26) $\delta_j = 2\delta_1 + \beta - \pi$, while $\delta_{2j} = 2\delta_2 + \beta$ in Eq. (52); that is, the phase shifts at the K- and L_1-edges differ from those at the $L_{2,3}$-edges by π in addition to the center atom phase shifts.

acteristic behavior of $\delta(k)$ as a function of Z also occurs. This Z dependence of $F(k)$ and $\delta(k)$ can be used to distinguish the type of atoms surrounding the center atom about whose edge EXAFS is measured.

1.5. CORRECTIONS TO THE SIMPLE PICTURE

The standard formula for EXAFS given in Eq. (26) makes several simplifying assumptions. These assumptions will be reviewed in this section and the errors they introduce will be assessed.

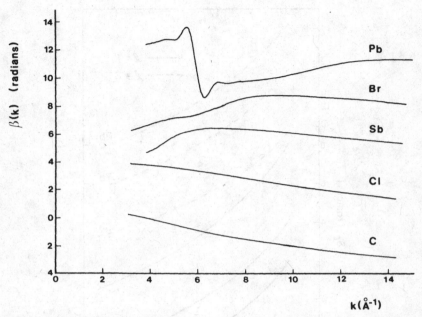

Figure 1.8. Backscattering phase shift $\beta(k)$ as a function of k [from ref. (56)] for carbon, chlorine, bromine, antimony, and lead atoms.

1.5.1. The Small-Atom Approximation

This approximation, first pointed out by Lee and Pendry (42), is the most serious of the approximations made in the independent particle model. Serious errors in the amplitude and phase are introduced (62) at low values of k as illustrated in Fig. 1.5 for bromine (Br_2). The seriousness of the errors introduced by the breakdown of the small-atom approximation was first emphasized by Pettifer (63a). In this approximation the spherical wave incident on a backscattering atom is approximated by a plane wave. The approximation is valid if the effective size of the atom in backscattering is small compared with the distance between the atom and the center atom. At low k this is not the case, because the effective atomic size is about the same as the interatomic distance. As the k of the photoelectron increases, the effective size of the atom decreases as the photoelectron penetrates deeper into the atom before scattering. The electron will scatter significantly only when the spatial variation of the atomic potential has a significant fractional change in the distance of $1/k$. At high k the diminishing effective size of the backscattering atom makes the small-atom approximation satisfactory. A formalism to calculate EXAFS in the single-scattering approximation but without the small-atom approximation was presented by Müller and Schaich (64).

The effect of the breakdown of the small-atom approximation on the phase of EXAFS can be approximately compensated for by a shift of E_0, as discussed later. Here E_0 is the binding energy of the photoelectron and is related to k by the relation in Eq. (2). In fact, as discussed later in regard to transferability, varying E_0 can cover many "blemishes" in the standard EXAFS expression, Eq. (26). However, E_0 shifts cannot correct for the amplitude error introduced by the small-atom approximation, as is evident in Fig. 1.5.

Eliminating the small-atom approximation makes EXAFS less convenient because $t(2k)$ and $\delta(k)$ then depend on R_j in addition to k, and their tabulation would become prohibitive because of the additional parameter of R_j. Another approach is to expand the correction to the small-atom approximation in terms of an asymptotic expansion in $1/R_j$. The first correction term in this expansion is

$$t'(2k) = t(2k)\left\{1 - M[\beta(2k)]\right\}$$

$$\beta'(2k) = \beta(2k) + \tan^{-1}\left\{t^{-1}(2k)M[t(2k)]\right\} \tag{56}$$

where M is the operator

$$\frac{1}{R_j}\left| 2\frac{\partial}{\partial k} - \frac{1}{k}\frac{\partial^2}{\partial\theta^2}\right|_{\theta=\pi}$$

and the phase factor is

$$\delta_j = 2\delta_2(k) + \beta'(2k)$$

From Fig. 1.6 it is noted that as k goes from large values toward zero a large value of $\partial F/\partial k$ for bromine atoms occurs, first at around $k = 6$ Å^{-1}. This coincides with the region where the phase and amplitude of bromine EXAFS in Fig. 1.5 starts deviating appreciably from the plane wave approximation. To calculate the full correction to $t(2k)$ and δ_j to order $1/R_j$ requires knowledge also of

$$\left.\frac{\partial^2 t}{\partial\theta^2}\right|_{\theta=\pi}, \quad \left.\frac{\partial\beta}{\partial k}\right|_{\theta=\pi}, \quad \text{and} \quad \left.\frac{\partial^2\beta}{\partial\theta^2}\right|_{\theta=\pi}$$

1.5.2. Multiple Scattering

The derivation of the EXAFS expression (26) was based on the single-scattering approximation. The outgoing wave function scattered backward only once from the surrounding atoms before being combined with the unscattered wave. The single-scattering approximation is valid as long as the scattering is small. The addition of further scatterings such as scattering from one neighbor to the next before combining with the unscattered wave was first considered with a physically realistic atomic potential by Lee and Pendry (42). Since then other techniques for including the multiple-scattering effects have been developed (65) and are discussed in detail in Chapter 2.

We can estimate the size of the multiple-scattering contributions by using the method of the heuristic derivation leading to Eq. (8). Consider the three atom systems illustrated in Fig. 1.9 with atom A being excited by absorbing the x-ray photon. Two processes are compared. In Fig. 1.9*a*, the single scattering case, the photoelectron is backscattered from atom B and interferes with the outgoing state at atom A. In Fig. 1.9*b*, a double scattering, the photoelectron scatters from atom B and then atom C before returning to atom A and interfering with the outgoing state. The single-scattering contribution of Fig. 1.9*a* produces a backscattered wave, given in Eq. (4). The double scattering contribution of Fig. 1.9*b* produces a backscattered wave of the form

$$\frac{F(\theta_B) \, F(\theta_C) \, \exp i\left[k(2r_i + r_{BC}) + \delta^{(2)}(k) - \pi/2\right]}{kr_i^2 r_{BC}} \cos \gamma \qquad (57)$$

where $\delta^{(2)}(k) = 2\delta_1 + \beta_B + \beta_C$, $F(\theta_B)$ is the amplitude of the scattering from atom B through an angle θ_B, and β_B is the phase introduced by that scattering. Here γ is the angle between the first and last scattering paths (as illustrated in Fig. 1.9*b*), and the cos γ factor measures the overlap of the *p* state of the final

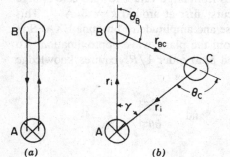

Figure 1.9. Illustrating (*a*) a single-scattering path for the photoelectron and (*b*) a double-scattering path.

incoming electron with the p state excited by the x-ray (67). Two things should be noted about the double scattering of Eq. (57) compared with the single scattering of Eq. (4). A Fourier transform of Eq. (57) will peak at the larger distance, $R \simeq r_i + (r_{BC}/2)$, neglecting the shift caused by the k dependence of the phase $\delta^{(2)}(k)$. The magnitude of the double-scattering amplitude is about F/r_{BC} times that of the single scattering. If both θ_B and θ_C are greater than about 40°, $[F(\theta)/r_{BC}] \simeq [F(\pi)/r_{BC}]$, which is small for typical distances of $r_{BC} \geqslant 2$ Å, as can be ascertained from Fig. 1.10, where $F(\theta)$ is plotted for some selected values of k for oxygen atoms. However, note that for $\theta \simeq 0°$, $F(0)$ peaks and becomes much larger. In that case $F(0)/r_{BC}$ typically becomes about one and the double scattering becomes important. This forward-scattering configuration, called the shadowing, or focusing effect, is illustrated in Fig. 1.11a and 1.11b and is the only case in the EXAFS regime where double scattering is important relative to single scattering in the same shell. In fact, the triple-scattering case illustrated in Fig. 1.11c with two forward scatterings has to be added to the double scatterings of Fig. 1.11a and 1.11b, and, because of the large value of $F(0)$, is also important.

Teo (66) was the first to point out that the forward multiple scattering could be used to determine atom configurations where three atoms are aligned or almost aligned. These ideas were further developed by Boland et al. (67). More recently, Lengeler (68) pointed out that the location of interstitial H can be obtained in crystals by exploiting the focusing effect. Since the $F(2k)$ of H is very small, H normally does not produce detectable EXAFS. However, if H is the intervening B atom between the excited A atom and a heavy C atom of Fig. 1.11, the focusing effect is large enough for H to introduce a detectable en-

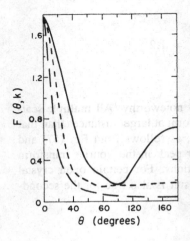

Figure 1.10. Backscattering amplitude of oxygen atoms, $F(\theta, k) = (m/2\pi\hbar^2)\, t(\theta, 2k)/k$, as a function of θ for several values of k, calculated in ref. (66). The solid curve is for $k = 3.78$, the short dashed curve is for $k = 7.56$, and the long dashed curve is for $k = 15.12$ Å^{-1}. Note that $t(180°, 2k) \equiv t(2k)$.

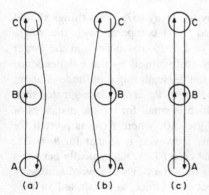

Figure 1.11. Scattering processes important in the forward direction: (*a*) and (*b*) double scattering; (*c*) triple scattering. The additional scattering occurs in the intervening B atom on the outward portion of the path in (*a*) and in the incoming portion in (*b*). In (*c*) the additional scattering in the intervening B atom occurs in both the outgoing and the incoming portions.

hancement of the EXAFS from atom C. In this manner H can be located along the line between atoms A and C.

Although it has been shown that the nonshadowing type of multiple scattering is small compared with the single-scattering contribution in the same shell, this multiple scattering may be significant compared with single scattering from a shell farther out. For example, the multiple scattering of Fig. 1.9*b* between the first-shell atoms in the face centered cubic (*fcc*) lattice may be significant compared with the single scattering from the second shell of atoms. There are 12 first-shell atoms and 6 second-shell atoms. The double scattering between the first-shell atoms would produce a Fourier transform peak at about the same distance as the single-scattering peak of the second shell. There are a total of 48 such double-scattering paths and only 6 single-scattering peaks. Since the double-scattering contribution relative to the single scattering is given by

$$\frac{F(120°)}{R_1^3} \frac{48}{6} R_2^2 \cos 60° \simeq 1.2$$

the two are comparable.

One final comment on multiple scattering is noteworthy. All multiple-scattering contributions will peak in a Fourier transform at larger distances than that of the single scattering from the first neighbor, as follows from Eq. (57) and subsequent discussions. Thus the first-neighbor peak in the Fourier transform has rigorously no multiple-scattering contributions. For certain open crystal structures such as the diamond structure this result is also true for the second-neighbor distance.

1.5.3. Many-Body Effects

The physics employed up to now to describe the EXAFS process is in the framework of the independent particle model, where each electron is acted on by an external potential whose value is dependent on only the averaging over the motion of the other electrons. In reality, electrons interact with one another through the Coulomb potential, and this interaction does depend on the instantaneous positions of the other electrons and cannot be entirely accounted for by the averaging inherent in the independent particle model. Solid state physics is blessed by the circumstance that this many-body effect is negligible in most cases. Is EXAFS as favored? The answer appears to be no. There is evidence that there are significant many-body effects that must be included for a full understanding of EXAFS.

1.5.3.1. Mean Free Path

The independent model, Eq. (27), already has some many-body effects incorporated into it. The mean free path λ has contributions from electron–electron scattering. The independent particle calculations (56, 60) include some electron–electron scattering in the backscattering atoms. However, this scattering does not contribute to λ since its effect is independent of the distance of the scattering atom and, instead, is incorporated into the $F(2k)$ dependence.

The λ effect is equivalent to a finite lifetime of the excited state. It measures the average distance an excited electron can travel before losing coherence with its initial state, that is, before it scatters into another state. This average time to lose coherence, that is, the lifetime of the state, is simply given by

$$\tau = \frac{\lambda}{v} \tag{58}$$

where $v = \hbar k/m$ is the electron velocity. The lifetime can be divided into contributions τ_h from the excited hole and τ_f from the excited photoelectron where

$$\frac{1}{\tau} = \frac{1}{\tau_h} + \frac{1}{\tau_f} \tag{59}$$

τ_h is independent of the chemical state of the atom to a very good approximation, since the inner electrons, which dominate the processes involved in τ_h, are so well shielded from the surrounding atoms. Values of τ_h as a function of Z have been tabulated in varous places (41). τ_f of the final photoelectron state senses the atomic surroundings and is dependent on the electronic structure of the material in which the atom is embedded. The behavior of τ_f has been investigated in photoemission and Auger emissions studies from the viewpoint of the mean free path. Based on these studies, a "universal" curve for $\lambda = v\tau_f$ has been proposed (69). The measured mean free path fluctuates about the "universal" curve, suggesting that it differs somewhat from material to material.

In a recent study (49, 70) on the diamond structure semiconductor germanium and the zinc blende structures of GaAs, ZnSe, and CuBr, λ was found to have a value at the minimum of 4 Å for germanium, increasing to about 7 Å for CuBr. This large variation of λ from material to material shows that caution must be used in estimating the value of λ. For reasons described in Section 1.5.3.2, the λ concept is not useful for the first shell. If it does have significance it will only be between the first and outer shells.

Recent studies (71) give an indication that the scattering effects caused by electron–electron interactions cannot always be described by an isotropic λ. The value of λ may depend on the direction from the center atom. For example, paths that go between atoms may require a different λ than do paths along bonds, especially for covalently bonded solids in which the valence charge distribution is directional.

The results on germanium and zinc blendes have the reasonable behavior of λ increasing with increasing energy gap between valence and conduction states. The approximately universal behavior of λ suggests that λ may not be too sensitive to the electronic structures of the material. This, together with the correlation of λ with energy gap in the zinc blendes, suggests that variations in λ from material to material may be correlated with as simple a characteristic as energy gap. Further studies, both theoretical and experimental, are required to clarify both the magnitude and the causes of variations in λ.

1.5.3.2. Passive Electrons

There are other many-body effects not accounted for at all in Eq. (27). The most obvious is that the atomic electrons in the excited atom see a different potential after a core electron has been knocked out. For the outer atomic electrons this new potential is closely approximated by that of the atom with its Z increased by one. This different potential is one consequence of the Coulomb repulsion between electrons, since the shielding of the positive nucleus by the core electron is removed as the atom is excited by removing that electron, increasing the attraction of the nucleus for the other electrons, "pulling in" the wave functions of the original atom, and relaxing them to a lower energy.

The electrons whose wave functions are relaxed are called "passive" electrons because they are not directly excited by the x-ray, in contrast to the core electron, the so-called "active" electron. However, these passive electrons still can be excited through their Coulomb interaction with the active electron and affect the absorption of the x-ray. The matrix element of Eq. (13) should, in the exact treatment, include for the initial state the full N-particle wave function of all the electrons in the unexcited atom, and for the final state of the wave function of the excited atom and the photoelectron. In the independent particle model, the wave function of the atom is simply a product of wave functions for each electron. If relaxation is neglected, the passive electron states would not change as the atom is excited, and their contribution to the matrix element consists of simply the product of their overlaps before and after excitation, that is,

$$\prod_i \left| \langle p_i | p_i \rangle \right|^2 = 1$$

where $|p_i\rangle$ is the wave function of the ith passive electron. Thus in this case the passive electrons give a factor of one, and the expression in Eq. (13) need only contain the active electron states. However, if the ith passive electron state relaxes to $|p_i'\rangle$ in the excited atom, and the excited photoelectron has a large enough kinetic energy so that the changes induced in it by the core hole are negligible, then the sudden approximation is valid and the overlap contribution of the passive electrons becomes (72–74)

$$S_0^2 = \prod_i \left| \langle p_i | p_i' \rangle \right|^2 < 1 \tag{60}$$

[The expression for the overlap between the initial and final passive electron requires, in actuality, a more complicated calculation of the determinant of the overlaps of the two sets of states because they are not orthogonal to one another, but the general idea of Eq. (60) is correct.]

This product S_0^2 is less than 1 because $|p_i'\rangle$ is normalized, and its overlap with $|p_i\rangle$ can only be as great as 1 if it is equal to $|p_i\rangle$. Thus the many-body relaxation effect reduces the value of μ below that given in Eq. (25) by the factor S_0^2. Estimates of this factor range between 0.7 and 0.8 in typical cases (72, 75).

There is a sum rule for the dipole matrix elements in Eq. (11) stating that the total absorption integrated over all energies must be the same regardless of many-body effects (76). Thus the reduction factor cannot be the only many-body effect, and it is not. The reduction in the single-electron excitation due to the reduced overlap is compensated for by multielectron excitations. Although the x-ray photon interacts directly with only the active electron, this active electron can excite other electrons through their Coulomb interaction.

The important aspect of the multielectron excitations is that they tend not to produce EXAFS. Consider a two-electron excitation into the continuum. Conservation of energy in this case requires that the sum of the energies of the two electrons be constant, but the energy of each electron can vary from zero up to the maximum possible energy. The EXAFS from this process will have contributions from electrons with varying energies and values of k, which tend to smear out the oscillations. Thus, one expects the main contribution to be the single-electron excitation whose contribution should be subject to the reduction factor. A quantitative calculation of the multielectron contribution verifies this qualitative discussion (73).

If this reduction in EXAFS were a constant independent of the chemical environment, it could be normalized out by measuring a standard and would not be a problem. Fortunately, there is both theoretical (73) and experimental (49, 63) evidence that the variation of S_0^2 with chemical environment is not important. This means that the atomic values of S_0^2 can be used to a reasonable approximation. Atomic values of S_0^2 have been previously calculated because of their importance in photoemission and electron capture decay of nuclei (77). Some values are tabulated in Table 1.1.

The first experimental evidence of the presence of S_0^2 in EXAFS was mea-

Table 1.1. Atomic S_0^2 as a
Function of Z

Z	Element	S_0^2
2	He	0.73
10	Ne	0.78
18	Ar	0.75
21	Sc	0.62
26	Fe	0.69
31	Ga	0.70
36	Kr	0.78
46	Pd	0.78
54	Xe	0.79
61	Pm	0.73
70	Yb	0.76
79	Au	0.80
83	Bi	0.77
92	U	0.73

Source: Carlson et al. (77).

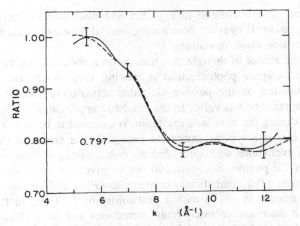

Figure 1.12. Ratio of the experimentally measured to single particle model calculation of the backscattering from bromine in bromine vapor (solid curve) and bromine liquid (dashed curves). The horizontal line is the high-k many-body overlap correction S_0^2 to the single particle model calculation.

surements on bromine vapor and liquid (63). The ratio of the measured $t(2k)$ to the theoretical independent particle model calculated values (56) is shown in Fig. 1.12. As can be seen, the experimental values fall below the theoretical ones for the higher k. It is only for such k values that the sudden approximation is valid and the effect of S_0^2 appears. The calculated value of S_0^2 for atomic bromine is indicated by the horizontal line, showing surprisingly good agreement with experiment. Later measurements on the zinc blende solids (49, 70) mentioned previously gave further evidence for an S_0^2 contribution to the EXAFS.

The consequence of the S_0^2 effect is that $t(2k)$ in the EXAFS equations should be replaced by $t''(2k) = S_0^2(k) \, t(2k)$. Whereas the independent particle model indicates that $t(2k)$ depends on the backscattering atom alone, the addition of the many-electron overlap factor, $S_0^2(k)$, adds a dependence on the center atom. We write $S_0^2(k)$ as a function of k even though the sudden approximation concludes that S_0^2 is independent of k. The k dependence is added because of breakdown of the sudden approximation at low k, as shown in Fig. 1.12 for bromine. Theoretically (73), an additional slow dependence of the effectiveness of S_0^2 in decreasing the EXAFS is expected at high k. Normally, this high k dependence will be too small to be observable.

This discussion assumed that the photoelectron had enough kinetic energy so that the sudden approximation is valid. It is clear from Fig. 1.12 that at lower k values deviations occur from the sudden approximation prediction of a constant S_0^2 reduction. This deviation has several causes. The breakdown of the small-atom approximation illustrated in Fig. 1.5 and discussed in Section 1.5.1 is certainly one cause. However, there are theoretical reasons to expect multielectron effects to also cause deviations.

There is one school of thought that reasons on a semiclassical basis that near threshold the escaping photoelectron is moving very slowly and its induced change in potential on the passive electrons acts slowly enough so that the adiabatic approximation is valid. In the adiabatic approximation, the change in the potential during the core hole excitation is assumed to be slow compared to ω_e^{-1}, where $\hbar\omega_e$ is a typical excitation energy in the atom. In this case, no multielectron excitations will occur, and the overlap integral between initial and final states of the passive electrons will not be given by Eq. (60), but will be unity. Each state $|P_i\rangle$ adiabatically converts to $|P_i'\rangle$, and an electron initially in $|P_i\rangle$ will remain in $|P_i'\rangle$, with a probability of 1. Thus, in the adiabatic approximation there are no many-body corrections and no multielectron excitations, and the independent particle model is accurate!

This picture is cited (72) as an explanation for the measured dependence (78) of the probability of multielectron excitation in neon atoms as a function of x-ray energy. At low energies near threshold the probability of excitation of the passive electrons is down because of the adiabatic change in potential. Only at large photoelectron energy does the probability of multielectron excitation rise to its constant sudden approximation limit. However, there are some conceptual difficulties in understanding the argument of an adiabatic variation of potential sensed by the passive electrons near threshold. One is the localization of the photoelectron to describe its transition as a trajectory escaping from the atom. This classical idea has difficulty from the uncertainty principle of quantum mechanics. To localize the photoelectron to the required extent introduces an uncertainty in its momentum of the same order as its momentum. A second difficulty is that even in the classical trajectory picture the photoelectron does not move slowly enough to produce an adiabatic change in potential. For the potential change to be adiabatic the photoelectron must move slowly relative to the motion of an electron in its orbit as the photoelectron passes through the electron orbit on its path to escape the atom. The photoelectron at threshold has a kinetic energy just equal to the potential energy of the atomic potential. At a passive electron orbit its kinetic energy is equal to half the atomic potential energy by the virial theorem. Thus the photoelectron, even at threshold, is moving faster than a bound electron as it passes through its orbit and the conditions for the adiabatic approximation are never attained. In spite of these difficulties, the adiabatic to sudden transition is still being proposed to explain the energy dependence of the multielectron excitation probability (79).

An energy dependence of the multielectron excitation probability occurs quantum mechanically without assuming an adiabatic to sudden transition (80).

The effect is closely related to the explanation of the edge singularity in soft x-ray emission and absorption processes (74, 81) in the presence of a Fermi electron gas. The quantum mechanical effect near the edge can be understood in a simple physical picture in terms of the Pauli exclusion principle. First consider a photoelectron excited near threshold. As discussed previously in relation to Eq. (60) the core hole potential created when the photoelectron is excited changes initial states $|p_i\rangle$ to relaxed ones $|p_i'\rangle$. The excited photoelectron is also relaxed into an unbounded state $|p_k'\rangle$ that is no longer orthogonal to the $|p_i\rangle$ states since they are eigenstates of a Hamiltonian with a different one-electron potential, one without and the other with a core hole potential. Of course $|p_k'\rangle$ is orthogonal to the $|p_i'\rangle$ since they are solutions of the same one-electron potential. The physical consequence of the nonorthogonality of $|p_k'\rangle$ to the $|p_i\rangle$ is that $|p_k'\rangle$ is composed of some parts of the $|p_i\rangle$. If $|p_k'\rangle$ were to become occupied by the photoelectron the Pauli exclusion principle would be violated, because then the $|p_i\rangle$ states would have more than one electron in them: One electron from the passive electron initially present in the neutral atom and the fraction of an electron due to the overlap of $|p_k'\rangle$ with $|p_i\rangle$. To avoid this the one-electron matrix element has to be corrected (74) so that the actual one-electron state to which the photoelectron makes its transition is orthogonal to all the initially occupied passive electron states $|p_i\rangle$. This correction is greatest at threshold and goes to zero at large kinetic energy of the photoelectron. The scale of the kinetic energy where the correction becomes small is the Rydberg energy, the change in the potential energy induced by the core hole potential on the photoelectron, which can be approximated by a change in the effective positive charge of the atom from 0 to $|e|$. Thus, corrections to the constant S_0^2 reduction of the EXAFS predicted by the sudden approximation are expected for about the first 30 eV past threshold due to interactions with the core hole potential.

The rest of the variation seen in Fig. 1.12 must be due to the breakdown of the small-atom approximation, errors in the one-electron calculation of $t(2k)$, and residual many-body interactions between the photoelectron and the passive electrons. The approximations of the one-particle potential used in calculating $t(2k)$ are expected to become more accurate at higher photoelectron energies. The low k behavior of the EXAFS in bromine requires more quantitative theoretical calculations that include the many-body effect and do not employ the small-atom approximation to verify whether this picture is adequate to explain the experimental results in Fig. 1.12.

Strong experimental evidence has recently been obtained in support of a quantum mechanical mechanism to the multielectron excitations. Multielectron excitations have been measured at the L_1-, L_2-, and L_3-edges of xenon vapor (82). At the L_2- and L_3-edges the absorption is dominated by $p \rightarrow d$ transitions.

In the semiclassical picture of the adiabatic-to-sudden transition the character of the initial and final states of the photoelectron is immaterial. Only the question of the energy of the photoelectron is pertinent. However, in the quantum mechanical explanation the symmetry of the final photoelectron is pertinent since it must be orthogonalized to all of the initially and finally occupied states. If it is, for example, in a final d state, then it is already orthogonalized to p, s, and so on, states due to the angular term, and it only has to be orthogonalized to other d states. As the final state changes, it has to be orthogonalized to other states of the same angular momentum, producing a mechanism for a differing correction for different final state angular momenta. The experiment (82) shows different behavior of the same multielectron excitation with different photoelectron final angular momenta. The experiment also shows a more rapid onset of the multielectron transition than predicted by the adiabatic to sudden transition. Thus the experimental evidence is against the classical adiabatic to sudden turn-on in atomic transition.

The suggestion has been made that a mean free path mechanism can contribute to the decrease in the EXAFS produced by the center atom, in addition to or in place of the overlap factor, S_0^2. Fortunately, the experiments show that this complication does not occur. Multielectron excitations have been measured in rare gas atoms as a function of photon excitation energy (78). These show a characteristic feature of building up to a constant value at high energies, which implies that the mean free path mechanism is negligible. The fact that the multielectron excitations build up from the threshold monotonically to their high-energy constant value is the behavior expected from the sudden approximation and not from a mean free path. A mean free path would have a decreasing probability of excitation with increasing photoelectron energy in the high energy regime. This argument is originally due to Carlson (72).

This discussion has an important implication with respect to the meaning of the mean free path of the previous section. The overlap mechanism in the center atom leads to an energy loss of the photoelectron due to multielectron excitations. The calculations by Lee and Beni (60) and Teo and Lee (56) include an energy loss mechanism in the backscattering amplitude that is physically equivalent to the mean free path within the backscattering atomic potential. Thus energy loss mechanisms in the center atom and the backscattering atoms are already included in the $S_0^2(k)$ and the $t(2k)$. The mean free path, λ, then must include all additional energy loss factors. The mean free path contribution to energy loss is distinguished from the other mechanisms by its dependence on distance. For the first shell, the S_0^2 and the backscattering energy loss include most of the energy losses. Additional losses can occur because of the modification of the atomic levels as the atoms are combined into the condensed state. The broadening of the energy levels lowers the excitation energy for electron–

hole pairs, opening up additional excitation possibilities not included in the atomic calculations. It is clear, therefore, that the mean free path that accounts for just the additional losses of the first shell will correspond to only a fraction of the interatomic distance. In the second shell the overlap and backscattering energy losses are the same as in the first shell, but now the photoelectron has to travel an extra distance of $2(R_2 - R_1)$, and the mean free path factor should decrease the amplitude by this full distance $\{\exp[-2(R_2 - R_1)/\lambda]\}$. In this case the contribution of λ is clearly separated from the other factors. Thus, the mean free path concept is only unambiguously defined when comparing the amplitudes between different shells. In the first shell the additional energy loss mechanism of the mean free path can be accounted for by using a smaller effective distance than the first-neighbor distance that we denote by $(R_1 - \Delta)$. For molecules or molecular solids $\Delta \simeq R_1$, that is, practically all of the energy losses are accounted for by the atomic calculations and S_0^2, since there is no significant modification of the energy spectrum from the atomic values. Even in some other types of solids it is found empirically (49) that $\Delta \simeq R_1$.

These considerations can be incorporated into the $\chi(k)$ expression, for example, Eq. (27), to correct the independent particle model by replacing $t(2k)$ $\exp(-2r_j/\lambda)$ by $S_0^2(k)t(2k)\exp[-2(r_j - \Delta)/\lambda]$. This changes Eq. (27) to

$$\chi(k) = \sum_j \frac{N_i}{kR_j^2} S_0^2(k) F_j(k) \exp(-2k^2 \sigma_j^2)$$
$$\cdot \exp\left(\frac{-2(R_j - \Delta)}{\lambda}\right) \sin[2kR_j + \delta_j(k)]$$

(61)

Expression (61) represents the EXAFS formula for an unoriented sample with Gaussian disorder and incorporating corrections for many-body effects.

1.6. TRANSFERABILITY OF PHASE SHIFT AND BACKSCATTERING AMPLITUDE

A crucial question concerning the use of EXAFS for quantitative structure determination is how accurately the nonstructural parameter such as $\delta_j(k)$ and $t(2k)$ in Eq. (27) can be determined and thus their contributions eliminated. If they can be exactly accounted for, then the remaining information in the EXAFS signal of Eq. (27) depends only on the structure, and by proper analysis, as discussed in Chapter 6, the structure can be determined within an accuracy limited only by noise in the signal.

An ideal situation would occur if the nonstructural parameters were transferable, that is, if they were independent of the environment in the condensed state. They could therefore be determined once and for all, tabulated, and employed in determining the structure of all unknowns. Unfortunately, as our theoretical discussion indicates, this is not the case. The breakdown of the small-atom approximation adds an R_j dependence to $\delta_j(k)$ and $t(2k)$; see Eq. (56). The addition of many-electron effects (Section 1.5.3.) adds a dependence of $t(2k)$ on the center atom in addition to the backscattering atom. Multiple-scattering effects that depend on the details of the atomic arrangements affect further-out shells. Variations of the mean free path, λ, are experimentally (49, 70, 71) found to occur, which affect the transferability of $t(2k)$ from the farther-out shells. Even in the independent-particle small-atom approximation model of Eq. (27), both the phase $\delta_j(k)$ and the backscattering amplitude $t(2k)$ depend on the chemical states of the center and backscattering atoms, as shown by the calculations of Teo and Lee (56). An experimental study (83) that incorrectly claimed to show the transferability of phase used a different definition of transferability from the one in general use. As discussed in Chapter 6, the usual way to determine distance is from the k dependence of the phase of the sine function of Eq. (27), which is affected by the k dependence of $\delta_j(k)$. Thus the pertinent character of $\delta_j(k)$ that must be proven transferable is the k dependence of $\delta_j(k)$. The transferability criterion used in the experimental study of Eq. (83) was the value of the absolute phase at $k \simeq 9$ Å^{-1}. Using the generally employed criterion of the k dependence, the data of ref. (83) show a nontransferability of phase for Ge$-$C pairs, which introduces an error in the distance of 0.08 Å.

Fortunately, most of the nontransferability of $\delta_j(k)$ can be compensated for by varying the value of E_0 of Eq. (2) as first suggested by Lee and Beni (60). This point is more fully discussed in Chapter 6. A careful experimental study (84) of the causes of phase nontransferability and how well they can be compensated for by changes in E_0 shows that the largest effect is from the breakdown of the small-atom approximation. The breakdown of the small-atom approximation can be approximately compensated for by an E_0 shift of -13 eV for the first-neighbor shell in solid germanium, while ionicity in the bonding requires an E_0 shift of about 5 eV/electron charge transfer. The compensation of nontransferability by shifting E_0 is accurate to about 0.02 Å in first-shell distance determination.

The reason why shifting E_0 can approximately compensate for the nontransferability of phase is that all of the causes of nontransferability affect low k values more than high k values and a shift in E_0 does similarly. Taking both sides of Eq. (2) and differentiating, one obtains

$$\Delta k \simeq -\frac{m}{\hbar^2} \frac{\Delta E}{k} \tag{62}$$

The ensuing shift Δk introduces a change in the total phase of the sine function in Eq. (27) which, if incorporated into $\delta_j(k)$, would add a term

$$\Delta\delta_j(k) \simeq 2\Delta k r_j = -\frac{2mr_j}{\hbar^2}\frac{\Delta E_0}{k} \tag{63}$$

This phase shift has the similar feature of affecting low k values of δ_j more than high k values, and thus an appropriately chosen ΔE_0 can approximately compensate for nontransferability.

However, shifting E_0 does not compensate for nontransferability in $t(2k)$. The effects of nontransferability on $t(2k)$ are more complicated than those on $\delta_j(k)$ and cannot be compensated simply by shifting k, as can be seen, for example, in Fig. 1.5. In addition, the EXAFS results are not affected as much by similar percentage changes in $\delta(k)$ as they are in $t(2k)$. The $\delta(k)$ produces a typical distance shift of 0.5 Å, and a 10% change in $\delta(k)$ would produce only a 0.05-Å error in distance, typically about 2% of interatomic spacings. However, a 10% error in $t(2k)$ produces a 10% change in the final result. Thus the lack of transferability of $t(2k)$ has greater impact on the accuracy of structure determination by EXAFS than that of $\delta(k)$.

The nontransferability of, particularly, $t(2k)$ can be greatly ameliorated by use of goods standards. An example of a very good standard is crystalline germanium (c-Ge) for analyzing amorphous germanium (a-Ge). The first shell of a-Ge is very similar to that of c-Ge. The only discernible difference is a slight increase of disorder. In that case the distance and the coordination number can be determined (85) to an accuracy of 0.002 Å and 1%, respectively, about 10 times more accurate than with a more usual standard. An ideal standard would have the same center and backscattering atoms in the same chemical state, at about the same distance apart, and with the same coordination number. The closer the actual standard is to the ideal, the smaller the errors introduced by nontransferability of $t(2k)$ and $\delta(k)$.

A necessary condition for transferability is dominance of single-scattering processes so that the backscattering is given by $t(2k)$ independently of the surroundings. Multiple scattering would give a dependence of the backscattering on the environment surrounding the backscattering atom. As mentioned in Section 1.5.3, the nearest-neighbor peak in the Fourier transform of the EXAFS rigorously consists of only a single-scattering process, and this is also true for the second peak in the diamond structure and may be true in some other open lattice structures. During the development of the multiple-scattering theory, it was realized that multiple scattering, instead of being just a nuisance, can be used to obtain further information about the structure of solids such as bond angles (66, 67), sensing the location of hydrogen atoms (68), and obtaining a characteristic signature for the imidazole complex (86).

The challenge for the future in the theory of EXAFS is to understand more quantitatively the factors that contribute to the nontransferability of EXAFS. This will have the obvious result of improving the accuracy of structure determination by EXAFS while relaxing the requirement of finding standards close to the ideal. At the same time this deeper understanding may suggest new information to be obtained by EXAFS analogous to what occurred in multiple scattering.

REFERENCES

1. It is usually stated that EXAFS starts about 30 eV past an absorption edge. However, a recent experimental study by G. Bunker and E. A. Stern, *Phys. Rev. Lett.*, **52**, 1990 (1984), has shown that the single-scattering mechanism and focusing forward scattering of EXAFS, described in Section 1.5.2, usually remain dominant down to the edge region, where the absorption has its rapid rise. In this book (Chapter 2) the conventional view that XANES extends to somewhat higher energy beyond the edge is maintained. The calculated spectra of XANES reduce to that of EXAFS in the region where the two phenomena overlap, so no error is introduced.

2. D. E. Sayers, E. A. Stern, and F. Lytle, *Phys. Rev. Lett.*, **27**, 1204 (1971).

3. P. Eisenberger, B. Kincaid, S. Hunter, D. Sayers, E. A. Stern, and F. Lytle, in *Proceedings of the IV International Conference on Vacuum Ultraviolet Radiation Physics*, E. E. Koch, R. Haensel and C. Kunz (Eds.), Pergamon, Oxford, 1974, p. 806.

4. E. A. Stern, *Contemp. Phys.*, **19**, 289 (1978).

4a. E. A. Stern and S. M. Heald, in *Handbook on Synchrotron Radiation*, Vol. 1, E. E. Koch (Ed.), North-Holland, New York, 1983, p. 957.

5. P. M. Eisenberger and B. M. Kincaid, *Science*, **200**, 1441 (1978).

6. D. R. Sandstrom and F. W. Lytle, *Ann. Rev. Phys. Chem.*, **30**, 215 (1979).

7. S. P. Cramer and K. O. Hodgson, in *Progress in Inorganic Chemistry*, Vol. 25, S. J. Lippard (Ed.), Wiley, New York, 1979, p. 1.

8. L. Powers, *Biochem. Biophys. Acta*, **683**, 1 (1982).

9. P. A. Lee, P. H. Citrin, P. Eisenberger, and B. M. Kincaid, *Rev. Mod. Phys.*, **53**, 769 (1981).

10. T. M. Hayes and J. B. Boyce, in *Solid State Physics*, Vol. 37, H. Ehrenreich, F. Seitz, and D. Turnbull (Eds.), Academic, New York, 1982, p. 173.

11. J. E. Enderby, D. M. North, and P. A. Egelstaff, *Philos. Mag.*, **14**, 961 (1966).

12. P. Fuoss, P. Eisenberger, W. K. Warburton, and A. Bienenstock, *Phys. Rev. Lett.*, **46**, 1537 (1981).

13. E. A. Stern, C. E. Bouldin, B. von Roedern, and J. Azoulay, *Phys. Rev. B*, **27**, 6557 (1983).

14. H. Fricke, *Phys. Rev.*, **16**, 202 (1920).

15. G. Hertz, *Z. Phys.*, **3**, 19 (1920).

16. D. Coster, *Z. Phys.*, **25**, 83 (1924).

17. A. E. Lindh, *Z. Phys.*, **6**, 303 (1921); **31**, 210 (1925).

18. G. A. Lindsay, *C. R. Acad. Sci. (Paris)*, **175**, 150 (1922).

19. W. Kossel, *Z. Phys.*, **1**, 119 (1920); **2**, 470 (1920).

20. B. B. Ray, *Z. Phys.*, **55**, 119 (1929).

21. B. Kievet and G. A. Lindsay, *Phys. Rev.*, **36**, 648 (1930).

22. J. D. Hanawalt, *Z. Phys.*, **70**, 20 (1931); *Phys. Rev.*, **37**, 715 (1931).

23. R. de L. Kronig, *Z. Phys.*, **70**, 317 (1931).

24. L. V. Azaroff, *Rev. Mod. Phys.*, **35**, 1012 (1963).

25. E. A. Stern, *Phys. Rev. B*, **10**, 3027 (1974).

26. R. de L. Kronig, *Z. Phys.*, **75**, 468 (1932).

27. H. Peterson, *Z. Phys.*, **76**, 768 (1932); **80**, 258 (1933); **98**, 569 (1936).

28. A. I. Kostarev, *Zh. Eksp. Teor. Fiz.*, **11**, 60 (1941); **19**, 413 (1949).

29. M. Sawada, *Rep. Sci. Workshop Osaka Univ.*, **7**, 1 (1959); T. Shiraiwa, T. Ishimura, and M. Sawada, *J. Phys. Soc. Jpn.*, **12**, 788 (1957).

30. V. V. Shmidt, *Bull. Acad. Sci. USSR, Phys. Ser.*, **25**, 998 (1961); **27**, 392 (1963).

31. L. V. Azaroff and D. M. Pease, in *X-Ray Spectroscopy*, L. V. Azaroff (Ed.), McGraw-Hill, New York, 1974, Chap. 6.

32. J. Perel and R. D. Deslattes, *Phys. Rev. B*, **2**, 1317 (1970).

33. G. Mitchell and W. W. Beaman, *J. Chem. Phys.*, **20**, 1298 (1952).

34. R. A. Van Nordstrand, *Adv. Catal.*, **12**, 149 (1960).

35. F. W. Lytle, in *Physics of Non-Crystalline Solids*, J. A. Prins (Ed.), North-Holland, Amsterdam, 1965, p. 12; *Adv. X-Ray Anal.*, **9**, 398 (1966).

36. R. M. Levy, *J. Chem. Phys.*, **43**, 1846 (1965).

37. D. E. Sayers, E. A. Stern, and F. W. Lytle, *Phys. Rev. Lett.*, **27**, 1204 (1971).

38. D. E. Sayers, F. W. Lytle, and E. A. Stern, *Adv. X-Ray Anal.*, **13**, 248 (1970).

39. F. W. Lytle, D. E. Sayers, and E. A. Stern, *Phys. Rev. B*, **11**, 4825 (1975).

40. In general, both the backscattering amplitude $t_i(2k)$ and the additional phase shift $\delta_i(k)$ are functions of k, and we show this dependence. The actual functional forms versus k of these two quantities have to be calculated or determined empirically, as we discuss later.

41. Keski-Rahkomen and M. O. Krause, *At. Data Nucl. Data Tables*, **14**, 140 (1974).

42. P. A. Lee and J. B. Pendry, *Phys. Rev. B*, **27**, 95 (1975).

43. J. J. Rehr, unpublished.

44. J. T. Sakurai, *Advanced Quantum Mechanics*, Addison-Wesley, Reading, MA, 1967, Chap. 2.

45. L. I. Schiff, *Quantum Mechanics*, McGraw-Hill, New York, 1968, Chap. 5.

46. E. A. Stern, D. E. Sayers, and F. W. Lytle, *Phys. Rev. B*, **11**, 4836 (1975).

47. G. Beni and P. M. Platzman, *Phys. Rev. B*, **14**, 1514 (1976).

48. E. Sevillano, H. Meuth, and J. J. Rehr, *Phys. Rev. B*, **20**, 4908 (1979).

49. E. A. Stern, B. Bunker, and S. M. Heald, *Phys. Rev. B*, **21**, 5521 (1980).

50. P. Rabe, G. Tolkiehn, and A. Werner, *J. Phys. C*, **12**, L545 (1979).

51. P. Eisenberger and G. S. Brown, *Solid State Commun.*, **29**, 481 (1979).

52. G. Martens, P. Rabe, N. Schwentner, and A. Werner, *Phys. Rev. Lett.*, **39**, 1411 (1977).

53. S. M. Heald and E. A. Stern, *J. Synth. Met.*, **1**, 249 (1980).

54. P. Rabe, *J. Appl. Phys.*, **17-2**, 22 (1978).

55. S. M. Heald and E. A. Stern, *Phys. Rev. B*, **16**, 5549 (1977).

56. B. K. Teo and P. A. Lee, *J. Am. Chem. Soc.*, **101**, 2815 (1979).

57. P. A. Lee, *Phys. Rev. B*, **13**, 5261 (1976).

58. P. H. Citrin, P. Eisenberger, and R. C. Hewitt, *Phys. Rev. Lett.*, **45**, 1948 (1980); **47**, 1567(E) (1981).

59. J. Stohr and R. Jaeger, *Phys. Rev. B*, **27**, 5146 (1983).

60. P. Lee and G. Beni, *Phys. Rev. B*, **15**, 2862 (1977).

61. J. C. Slater, *Phys. Rev.*, **81**, 385 (1951); *The Self-Consistent Field for Molecules and Solids, Quantum Theory of Molecules and Solids*, Vol. 4, McGraw-Hill, New York, 1974.

62. S.-H. Chou, Ph.D. thesis, University of Washington, Seattle, 1982; S.-H. Chou, J. J. Rehr, E. A. Stern, and E. Davidson *Phys. Rev.*, accepted for publication.

63. E. A. Stern, S. M. Heald, and B. Bunker, *Phys. Rev. Lett.*, **42**, 1372 (1979).

63a. R. F. Pettifer, in *Trends in Physics*, Adam Hilger, Bristol, England, p. 522.

64. J. E. Müller and W. L. Schaich, *Phys. Rev. B*, **27**, 6489 (1983).

65. P. J. Durham, J. B. Pendry, and C. H. Hodges, *Comput. Phys. Commun.*, **25**, 193 (1982).

66. B.-K. Teo, *J. Am. Chem. Soc.*, **103**, 3990 (1981).

67. J. J. Boland, S. E. Crane, and J. D. Baldeschwieler, *J. Chem. Phys.*, **77**, 142 (1982).

68. B. Lengeler, *Phys. Rev. Lett.*, **53**, 74 (1984).

69. C. J. Powell, *Surf. Sci.*, **44**, 29 (1974).

70. B. Bunker, Ph.D. thesis, University of Washington, Seattle, 1980.

71. Kyungha Kim and E. A. Stern, 3rd International Conference on EXAFS, Stanford, CA, July 1984.

72. T. A. Carlson, *Photoelectron and Auger Spectroscopy*, Plenum, New York, 1975, Chap. 3.

73. J. J. Rehr, E. A. Stern, R. L. Martin, and E. R. Davidson, *Phys. Rev. B*, **17**, 560 (1978).

74. E. A. Stern and J. J. Rehr, *Phys. Rev. B*, **27**, 3351 (1983).

75. R. L. Martin and E. R. Davidson, *Phys. Rev. A*, **16**, 1341 (1977).

76. H. A. Bethe and E. E. Salpeter, *Quantum Mechanics of One- and Two-Electron Atoms*, Academic, New York, 1957.

77. T. A. Carlson, C. W. Nestor, Jr., T. C. Tucker, and F. B. Malik, *Phys. Rev.*, **169**, 7 (1968).

78. T. A. Carlson and M. O. Krause, *Phys. Rev. A*, **140**, 1057 (1965).

79. T. D. Thomas, *Phys. Rev. Lett.*, **52**, 417 (1984).

80. J. J. Rehr, in J. Stohr, R. Jaeger, and J. J. Rehr, *Phys. Rev. Lett.*, **51**, 821 (1983).

81. G. D. Mahan, in *Solid State Physics*, Vol. 29, H. Ehrenreich, F. Seitz, and D. Turnbull (Eds.), Academic, New York, 1974, p. 75; M. Combescot and P. Nozieres, *J. Phys. Paris*, **32**, 9193 (1971).

82. K. Zhang, E. A. Stern, and J. J. Rehr, 3rd International Conference on EXAFS, Stanford, CA, July 1984.

83. P. H. Citrin, P. Eisenberger, and B. M. Kincaid, *Phys. Rev. Lett.*, **36**, 1346 (1976).

84. B. A. Bunker and E. A. Stern, *Phys. Rev. B*, **27**, 1017 (1983).

85. C. E. Bouldin, E. A. Stern, B. von Roedern, and J. Azoulay, *J. Non-Cryst. Solids*, **66**, 105 (1984).

86. G. Bunker, E. A. Stern, R. E. Blankenship, and W. W. Parson, *Biophys. J.*, **37**, 539 (1982).

CHAPTER

2

THEORY OF XANES

P. J. DURHAM

Daresbury Laboratory
Daresbury Warrington, United Kingdom

2.1. INTRODUCTION

This chapter deals with the theoretical analysis of x-ray absorption near-edge structure, or XANES. The near-edge structure in an absorption spectrum covers the range between the threshold and the point at which the extended x-ray absorption fine structure, EXAFS, begins. This is admittedly a rather loosely defined spectral range, its limits being in principle different for each system, but one normally thinks of the XANES extending to an energy of the order of 50 eV above the edge. Bianconi (1) suggests that the energy dividing the XANES from the EXAFS is roughly that at which the wavelength of the excited electron is equal to the distance between the absorbing atom and its nearest neighbors. Figure 2.1 shows some experimental data in which the different character of the two spectral regimes can clearly be appreciated.

The first questions that arise, then, are (a) how does the XANES regime differ from the EXAFS?, and (b) what information does it contain? To answer these questions we must focus on the nature of the excited electron states populated in the absorption process. Of all the kinds of electron states in condensed systems, the most easily understood are those of very low and very high energy. Core states are, of course, essentially the same as in free atoms—interesting questions relate, for example, to the shifts of core-level binding energies in

53

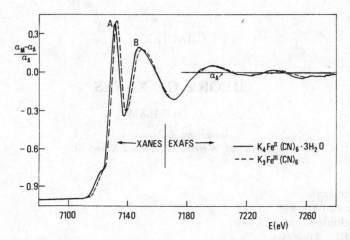

Figure 2.1. K-edge x-ray absorption spectra of iron in $K_3Fe(CN)_6$ and $K_4Fe(CN)_6$. The relative absorption with respect to the high-energy continuum background, α_A, is plotted.

different chemical environments (2), but the nature of the states is not in question. On the other hand, at the very high energies in the continuum of the electrons participating in EXAFS, the effect of the atoms in the system becomes quite small; the electron states approximate to plane waves that are only weakly scattered by the atoms. Thus practically all the interest in the theory of the electronic structure of condensed matter centers on the low-lying extended states. These are the states that bond the system together and determine its electronic properties. It is to handle these states that the whole apparatus of electronic structure theory (3), in particular the band theory of regular crystals, has been assembled. The states populated by the excited electrons in XANES (call these the XANES states) belong to this last category. They encompass all the unoccupied states from the Fermi level up to the EXAFS limit. Thus in a transition metal the XANES states include the unoccupied part of the narrow d bands, just above the Fermi level, as well as the less tightly bound s and p bands. For an isolated molecule or complex ion the XANES states include unoccupied bound states as well as the low-lying continuum states. Clearly, a theoretical analysis of XANES involves solving the Schrödinger equation for a range of energies at the lower end of which the interaction of the electrons with the atoms is very strong, becoming weaker as the high energy EXAFS limit is approached. In the language of scattering theory, one can say that while at high energies the scattering of the excited electrons is so weak that the only significant contributions to the final state wave function in the vicinity of the absorbing atom come from paths in which the electron is scattered once only (single scattering), as the photoelectron energy is lowered into the XANES region multiple scattering becomes more and more important. Because the EXAFS regime is dominated by single scattering (apart from occasional shadowing effects) the information it contains is purely geometrical and rather easily extracted; the pair correlation function can be obtained essentially by Fourier transforming the EXAFS oscillations (4). In the XANES regime, however, multiple scattering of the excited electron confers sensitivity to the details of the spatial arrangement of atoms neighboring the absorbing one; not only their radial distances but also their orientations relative to one another, bond angles, and so on. Any multiple-scattering process is sensitive to multiatom correlations. In the extreme case of covalent systems, multiple-scattering contributions build up to give bonding and antibonding states whose charge density is strongly concentrated and directed into certain regions of space. Even in the continuum quite strongly localized and directional states can arise, perhaps the most famous being the so-called shape resonances found in diatomic molecules (5). It is also in the near-edge region where quasiatomic effects such as "white lines" occur, for example, in the $L_{2,3}$-edges of transition metal and rare earth systems. These so-called white lines have a large intensity and are due to $2p$ to nd transitions (see Chapter 11, Section 11.5.7). Again, changes in the charge distribution around a given atom in different chemical environments can alter core-level binding energies and thus produce absorption edge shifts that show up in the XANES. Thus a variety of physical effects come into play in the near-edge region. A full theoretical description of XANES should account for all these effects on the same footing and thereby reveal which are the important factors in any given system. In this chapter recent progress toward such a theory will be outlined. The richness of phenomena occurring in XANES means that many fundamental theoretical questions remain unanswered at present, but a good deal of progress has been made in the last few years, and we are reaching the point at which a quantitative analysis of near-edge structures in quite complex systems is feasible. Present and possible future applications of XANES are quite various; here most attention will be focused on the extent to which the spectra are sensitive to the local atomic geometry. This looks like a fruitful field of application, capable of yielding information not available from EXAFS or from standard diffraction techniques. A practical point worthy of note is that XANES structures are usually much stronger than the EXAFS oscillations (e.g., see Fig. 2.1), and this may make XANES particularly useful in dilute systems. The current experimental situation is reviewed by Bianconi in Chapter 11, and so at this point it will simply be mentioned that XANES has been studied in metals (6), alloys (7), glasses (8), molecules (9), complex ions (10, 11), biological systems (12), surfaces (1), valence fluctuation compounds (13), and so on. The field is very wide.

Current theories of XANES in condensed systems employ an independent particle or one-electron approximation, and the nature of this approximation is worth discussing at this stage. The x-ray absorption transition rate depends on a matrix element between the initial and final states of the many-electron system. To construct these states one would have to solve the many-electron Schrödinger equation, and this is, of course, not a practical proposition. The one-electron approximation represents the many-particle wave functions as products of single-particle wave functions, which are eigenfunctions of an effective one-electron Hamiltonian. Such one-electron approaches, in particular the Hartree and Hartree–Fock approximations, have for years been central to electronic structure theory, and the self-consistent way they operate will be familiar to the reader. Somewhat more recently, density functional theory (14) has provided another variety of effective one-electron theory that gives the correct total energy and charge density of the ground state rigorously. The so-called local density approximation (LDA) to density functional theory (15) is particularly simple to use, since exchange and correlation effects are included in the one-electron Hamiltonian by means of a local potential function (unlike, e.g., the exchange potential in Hartree–Fock theory). In band structure calculations the LDA has given impressive and illuminating results for the energetics of a variety of crystals (16) and is now very widely used. Variants of it are employed in most modern x-ray absorption calculations. Unfortunately, there is no rigorous density functional theory for excited states (17). Indeed, in using one-electron states to discuss x-ray absorption, or any other excitation process, one is implicitly hoping (a) that a quasiparticle description (18) of the excited state is appropriate and (b) that the one-electron states are in some sense good approximations to the quasiparticle states. In density functional theory the one-electron eigenvalues are certainly not quasiparticle energies; in fact, they play a role auxiliary to the central objective of obtaining the correct total energy and charge density of the ground state. Thus, in principle, one should not determine excitation energies from LDA one-electron eigenvalues, and while the errors in this procedure seem not to be too great for extended states, for localized states they can be substantial (19). In the case of x-ray absorption this means that the binding energy of the core state may not be given accurately by the LDA core eigenvalue. However, the primary interest is not in the absolute threshold energy, but in the relative energies of spectral features above the edge, and for this LDA should be much better. Using LDA one-electron eigenstates in this way to calculate x-ray absorption spectra is a kind of static approximation, in that the electrons not directly involved in the transition do not respond to the external field associated with the photon. Recently, a time-dependent extension of the LDA has been proposed, and, while this does not seem to have the formal rigor of ground-state density functional theory, it has been extremely successful in describing atomic photoabsorption (20). To my knowledge such an approach has not yet been realistically applied to condensed systems, but this could represent an important development in the theory of x-ray absorption.

If, in XANES calculations, we adopt an independent particle approximation of the LDA type, the major remaining issue is the more technical one of how best to solve the effective one-electron Schrödinger equation. For ordered crystals the whole battery of band theoretical methods is available and has for some time been used successfully to calculate x-ray absorption spectra in the near-edge region, especially for metals (21). This approach of course, is limited to systems with translational symmetry, for which a k-space description is possible. On the other hand, there has recently been an increase of interest in the use of real space techniques, which represent an extended system by a finite cluster of atoms surrounding that at which the absorption occurs. This cluster method, made possible by the localized nature of the transition, has the great advantage of being very flexible regarding the spatial arrangement of atoms in the cluster. No translational symmetry, indeed no symmetry at all, is required. Such geometrical flexibility is essential if one wants to extract structural information from XANES in complex systems. In such an application one has to be able to change the positions of the atoms in the calculation to estimate the corresponding effect on the XANES. (Actually, even in ordered crystals the presence of the core hole breaks translational symmetry, so that when the effect of the hole on the spectrum is large, cluster technique may be more attractive than band structure calculations.) For finite systems, for example, molecules and complex ions, quantum chemical techniques of varying sophistication, depending on the size of the system, have been used, especially semiempirical molecular orbital methods (22). In the remainder of this chapter a variant of the cluster approach to XANES theory will be described, which uses multiple-scattering theory to solve the effective one-electron Schrödinger equation. It is an extension of the usual single-scattering EXAFS theory to the lower-lying near-edge regime and provides the geometrical flexibility called for previously. Its relationship to band theory calculations for ordered extended systems and to one-electron molecular orbital theory for finite systems will be demonstrated, details of the calculations will be discussed, and some examples of applications will be given. But first some general results will be established.

2.2. GENERAL THEORY

In this section an expression for the x-ray transition rate will be derived, which is equivalent to the usual Golden Rule formulation [cf. Chapter 1, Section 1.2.3, Eq. (11)], but which describes the excited electron states by means of the one-particle Green's function.

Consider the evolution of the system under the Hamiltonian

$$H = H_0 + H'$$

where H_0 is the Hamiltonian for the electrons plus the (uncoupled) photons, and H' is the electron–photon interaction

$$H' = \frac{e}{c} \int d\mathbf{r}\, \mathbf{j}(\mathbf{r}) \cdot \mathbf{A}(\mathbf{r}) + \frac{e^2}{2mc^2} \int d\mathbf{r}\, \rho(\mathbf{r})\, A^2(\mathbf{r}) \tag{1}$$

Here $\rho(\mathbf{r})$ and $\mathbf{j}(\mathbf{r})$ are the electron number and current densities, respectively, and $\mathbf{A}(\mathbf{r})$ is the vector potential. In x-ray absorption only the first term of H', that linear in $\mathbf{A}(\mathbf{r})$, comes into operation; the second term only enters into higher-order processes such as x-ray scattering (23).

At time $= 0$ the system consists of the electrons in their ground state plus a single photon in state $(\mathbf{q}, \lambda)$, with wave vector $\mathbf{q}$, energy ω, and polarization $\mathbf{e}^\lambda(\mathbf{q})$. For simplicity the photon state will be given as $|\mathbf{q}\rangle$, with the polarization implicit. Furthermore, since we are deploying the independent particle approximation for the electrons, we need only specify the state of the electron involved in the transition. In the initial state this electron is in a core state $|c\rangle$ with eigenvalue ε_c. Thus at time $= 0$ we have

$$|\psi(t = 0)\rangle = |c, \mathbf{q}\rangle$$

At some later time $= t$, the system has evolved into state

$$|\psi(t)\rangle = U(t)|\psi(0)\rangle = U(t)|c, \mathbf{q}\rangle$$

where the evolution operator $U(t)$ is related to the noninteracting Hamiltonian H_0 by

$$U(t) = e^{-iH_0 t} \tag{2}$$

At this instant t, an electron–photon interaction occurs. At some much later time t_0, then, the system is in state $|\psi(t_0)\rangle$ given, to first order in the interaction, by

$$|\psi(t_0)\rangle = -i \int_0^{t_0} dt\, U(t_0 - t)\, H' U(t)|c, \mathbf{q}\rangle$$

Thus the probability $P_n(t_0)$ that at time t_0 the photon has been absorbed and the excited electron is in state $|n\rangle$ is just

$$P_n(t_0) = \left| \langle n | \psi(t_0) \rangle \right|^2 \tag{3}$$

Since the final states $|n\rangle$ are not themselves observed, they must be summed over:

$$P(t_0) = \sum_n P_n(t_0)$$

and the total transition rate W is given by

$$W = \frac{d}{dt_0} \left\{ \lim_{t_0 \to \infty} P(t_0) \right\}$$

Now we introduce the Green's function or resolvent, given as an operator function of the complex variable z by

$$G(z) = (z - H_0)^{-1}$$

This is related to the evolution operator by (24)

$$U(t) = -\frac{1}{2\pi i} \int_{-\infty}^{\infty} dx\, e^{-ixt} (G^+(x) - G^-(x)) \tag{4}$$

with $G^{\pm}(x) = G(x \pm i\eta)$, η being a positive infinitesimal. Substituting into Eq. (3), we obtain

$$W(\omega) = -2 \left\langle c \left| h^* \operatorname{Im} G^+(\varepsilon_c + \omega)h \right| c \right\rangle \tag{5}$$

where

$$h = \frac{e}{c} \int d\mathbf{r}\, \mathbf{j}(\mathbf{r}) \cdot \mathbf{A}_{\mathbf{q}}^{\lambda} e^{i\mathbf{q} \cdot \mathbf{r}}$$

is an operator on the electron states only, $\mathbf{A}_{\mathbf{q}}^{\lambda} e^{i\mathbf{q} \cdot \mathbf{r}}$ being the expectation value of the vector potential in photon state $|\mathbf{q}\rangle$.

Equation (5) is our fundamental expression for the transition rate. Its equivalence to the Golden Rule is clear from the derivation but can be made explicit by inserting the eigenfunction expansion of the Green's function:

$$G^+(\varepsilon) = \sum_n \frac{|n\rangle\langle n|}{\varepsilon - \varepsilon_n + i\eta} \tag{6}$$

hence,

$$W(\omega) = 2\pi \sum_n \left|\langle n|h|c\rangle\right|^2 \delta(\varepsilon_c + \omega - \varepsilon_n)$$

Frequently, the wave functions $\langle r|n\rangle$ and eigenvalues ε_n are obtainable from an electronic structure calculation (e.g., the Bloch functions in a band structure calculation), and in such cases this standard form of the Golden Rule is easy to apply. Indeed, if $G^+(\varepsilon)$ could only be obtained via Eq. (6) we would, of course, have gained nothing. The point is that there exist very convenient methods of calculating $G^+(\varepsilon)$ energy by energy, that is, without reference to any states at energies other than ε. Multiple-scattering theory is one such method, and one which, moreover, can be used even in systems whose wave functions are difficult to obtain.

The Green's function in Eq. (5) is also a useful way to make rough and ready contact with the many-body theory of x-ray absorption. The final state really contains not only an electron in state $|n\rangle$ but a hole in a core state $|c\rangle$. Both the excited electron state and the core hole state decay at finite rates; the core hole tends to be filled quite rapidly, especially by Auger transitions (25), while the high-energy electron can excite electron–hole pairs, plasmons, and so on, in inelastic collisions. Thus the time evolution in state $|n\rangle$ should contain a damping term. Now, if $|n\rangle$ were an exact eigenstate of H_0

$$\langle n|U(t)|n\rangle = e^{-i\varepsilon_n t}$$

and the suggestion is that this be modified to

$$\langle n|U(t)|n\rangle = e^{-i(\varepsilon_n - i\Gamma_n)t} \tag{7}$$

where Γ_n is, to a first approximation, the sum of the decay rates of the excited electron and of the core hole. This can be achieved if the Green's function adopts the modified form

$$G^\pm(x) = \left[x - H_0 - \Sigma^\pm(x)\right]^{-1} \tag{8}$$

where $\Sigma(x)$ is the (complex) self-energy

$$\Sigma^{\pm}(x) = \Delta(x) \mp i\Gamma(x)$$

(Δ and Γ being real). If $\Delta(x)$ and $\Gamma(x)$ are not too large and are slowly varying functions of x, then, from Eq. (4)

$$\langle n|U(t)|n\rangle \approx \exp\left(-i(\varepsilon_n + \Delta(\varepsilon_n) - i\Gamma(\varepsilon_n))t\right)$$

which has the same form as Eq. (7). The real part, $\Delta(x)$, of the self-energy produces a shift of the eigenvalue ε_n, and the imaginary part, $\Gamma(x)$, represents the decay rate of the excited state. A good deal of many-body theory has been devoted to these self-energies (26), but the results are seldom sufficiently quantitative to be of much use in the present context. The Γ parameter can at least be estimated from other experiments (e.g., from the widths of core-level XPS lines), and so the most common practice has been to use empirical values for Γ and to omit Δ altogether (27), that is, to use

$$G^+(x) = \frac{1}{x - H_0 + i\Gamma(x)} \qquad (9)$$

Equation (5) can then easily be rewritten

$$W(\omega) = \frac{1}{\pi} \int d\omega' \frac{\Gamma(\varepsilon_c + \omega)}{(\omega - \omega')^2 + \Gamma^2(\varepsilon_c + \omega)} W_0(\omega')$$

where $W_0(\omega)$ is the original one-particle spectrum. Thus, naturally, the effect of the decay of the final state is to broaden out $W_0(\omega)$ by folding it with a Lorentzian of width $\Gamma(\varepsilon_c + \omega)$. Often this folding is carried out numerically (21), but it can also be convenient to use Eq. (9) directly, that is, to calculate the Green's function at the complex energy $\varepsilon_c + \omega + i\Gamma(\varepsilon_c + \omega)$ (28).

It should be noted that the foregoing discussion is not at all a satisfactory or consistent way of including many-electron effects in this or any other electron spectroscopy. In general, a systematic treatment of such effects requires more sophisticated many-body techniques (29). But, as regards the practical matter of including the most visible many-body effects in a realistic one-electron calculation, the phenomenology just outlined suffices at present.

We finish this section by putting the main result, Eq. (5), into a spatial representation:

$$W(\omega) = -2 \int d\mathbf{r} \int d\mathbf{r}' \, \phi_c^*(\mathbf{r}) \, h^*(\mathbf{r}) \, \mathrm{Im} \, G^+(\mathbf{r}, \mathbf{r}'; \varepsilon_c + \omega) \, h(\mathbf{r}')$$
$$\cdot \phi_c(\mathbf{r}') \, \theta(\varepsilon_c + \omega - \mu) \tag{10}$$

where $\phi_c(\mathbf{r})$ is the core state wave function. We have added a θ function to ensure that the final-state energy lies above the Fermi level μ. In a formally more complete theory (23) this would emerge automatically; here it has to be added by hand.

2.3. MULTIPLE-SCATTERING THEORY

2.3.1. Principles

In this section the basic multiple-scattering formulas needed to evaluate Eq. (10) are given, and used to obtain an expression for the transition rate that is convenient for numerical calculations. We use atomic units ($\hbar = m = e = 1$, $c = 137.036$).

First, we assume that the one-electron potential can be approximated by the nonoverlapping muffin-tin form commonly used in solid state physics

$$V(\mathbf{r}) = \sum_i v_i(\mathbf{r} - \mathbf{R}_i) \tag{11}$$

where v_i is the potential of the ith atom as position $\mathbf{R}_i$. The Green's function is related to the potential by the operator equation

$$G = G_0 + G_0 V G = G_0 + G_0 T G_0$$

G_0 being the free electron propagator and T the total T-matrix of the system. We can express G and T in terms of the scattering amplitudes t^i of each atom if we introduce the scattering path operator (30) or τ-matrix

$$T = \sum_{i,j} \tau^{ij}$$

$$\tau_{LL'}^{ij} = t_l^i \delta_{ij} \delta_{LL'} + \sum_{\substack{k \neq i \\ L''}} t_l^i g_{LL''}^{ik} \tau_{L''L'}^{kj} \tag{12}$$

Equation(12) gives the on-shell matrix elements of the scattering path operator; L stands for the pair (l, m) of angular momentum quantum numbers, t_l^i is the l-wave t-matrix of atom i given in terms of the phase shift δ_l^i by

$$t_l^i(\varepsilon) = \frac{i}{2\kappa} [e^{2i\delta_l^i(\varepsilon)} - 1] \tag{13}$$

and $g_{LL'}^{ij}$ is a real space structure constant

$$g_{LL'}^{ik}(E) = -4\pi i\kappa(1 - \delta_{ik})$$

$$\cdot \sum_{L''} i^{l-l'-l''}(-1)^{m'+m''} h_l^{(1)}(\kappa R_{ik}) Y_{L''}^*(\hat{\mathbf{R}}_{ik}) \tag{14}$$

$$\cdot \int Y_L Y_{L''} Y_{L'}^* d\Omega$$

$[\kappa = \sqrt{2\varepsilon}, h_l^{(1)}(\kappa r)$ is a spherical Hankel function of the first kind, $\mathbf{R}_{i(k)}$ is the position of atom $i(k)$, and $\mathbf{R}_{ik} = \mathbf{R}_i - \mathbf{R}_k]$.

Now it is clear from Eq. (10) that, owing to the highly localized nature of the core state wave function ϕ_c, we need to evaluate $G(\mathbf{r}, \mathbf{r}'; \varepsilon)$ only when $\mathbf{r}$ and $\mathbf{r}'$ lie within the same muffin tin. Indeed, let us now take the origin of coordinates to be at the nucleus of the excited atom, to which we give the site index $i = 0$. It is easy to show (31) that when $r < R_{MT}$ and $r' < R_{MT}$ (R_{MT} is the muffin-tin radius)

$$G(\mathbf{r}, \mathbf{r}'; \varepsilon^+) = \sum_{L,L''} Z_l^0(r, \varepsilon) Y_L^*(\hat{\mathbf{r}}) [i\kappa\tau_{LL'}^{00}(\varepsilon)] Z_{l'}^0(r', \varepsilon) Y_l(\hat{\mathbf{r}}')$$

$$- \sum_L Z_l^0(r_<, \varepsilon) J_l^0(r_>, \varepsilon) Y_L^*(\hat{\mathbf{r}}) Y_L(\hat{\mathbf{r}}') \tag{15}$$

Here $Z_l^0(r, \varepsilon)$ is a regular solution of the radial Schrödinger equation containing only the muffin-tin potential $v_0(\mathbf{r})$ of the atom at the origin and matches smoothly on to

$$j_l(\kappa r) (t^0)_l^{-1} - i\kappa h_l^{(1)}(\kappa r)$$

at R_{MT} [$j_l(\kappa r)$ is a spherical Bessel function]. Similarly, $J_l^0(r, \varepsilon)$ is an irregular solution of the same Schrödinger equation that matches smoothly on to $j_l(\kappa r)$ at R_{MT}. Note that t_l^0, Z_l^0, and J_l^0 are atomic quantities in the sense that they are determined by $v^0(\mathbf{r})$ alone, and that Z_l^0 and J_l^0 are real.

Using Eq. (1) and (9) we find that, putting $E = \varepsilon_c + \omega$

$$W_c(\omega) = -2 \sum_{LL'} m_L(E) \, \text{Im} \, \tau^{00}_{LL'}(E) \, m_{L'}^*(E) \, \theta(E - \mu) \qquad (16)$$

where the matrix element is just

$$m_L(E) = \int d\mathbf{r} \, \phi_c(\mathbf{r}) \, h(\mathbf{r}) \, Z_l^0(r, E) \, Y_L^*(\hat{\mathbf{r}}) \qquad (17)$$

Clearly this matrix element is also an atomic quantity, and all the effects of the surrounding atoms are contained in the τ-matrix, $\tau^{00}_{LL'}$. Indeed $m_L(E)$ is a smooth and slowly varying function of energy, without, for example, any peaks around l-wave resonances (23).

2.3.2. Physical Meaning of the τ-Matrix

What is the meaning of this τ-matrix? Multiple-scattering theory represents the Green's function as a sum of contributions from every possible scattering path. As may easily be seen by expanding out Eq. (12), τ^{ij} gives the sum of all scattering paths that begin at atom i and terminate at atom j, possibly visiting sites i and j at intermediate steps. Thus τ^{00} gives the sum of all scattering paths that begin and end on the site at the origin. This quantity appears naturally in expressions for all observables which, like x-ray absorption, represent strictly local aspects of the electronic structure. We shall therefore give a brief dicussion of $\tau^{00}_{LL'}$ itself, before returning to the details of x-ray absorption.

For an ordered system (i.e., one with translational periodicity and in which $t^i = t$ for all i) one can solve Eq. (12) by lattice Fourier transform (the $\mathbf{k}$ integral runs over the first Brillouin zone, BZ, whose volume is Ω_{BZ})

$$\tau^{00}_{LL'}(\varepsilon) = \frac{1}{\Omega_{BZ}} \int_{BZ} d\mathbf{k} \, \left[t^{-1}(\varepsilon) - G(\mathbf{k}, \varepsilon) \right]^{-1}_{LL'} \qquad (18)$$

with

$$g^{ij}_{LL'}(\varepsilon) = \frac{1}{\Omega_{BZ}} \int_{BZ} d\mathbf{k} \, e^{ik \cdot (\mathbf{R}_i - \mathbf{R}_j)} \, G_{LL'}(\mathbf{k}, \varepsilon) \qquad (19)$$

Now the eigenenergies of a system, usually obtained by diagonalizing the Hamiltonian, can also be found by locating the poles of the Green's function or τ-matrix. We can see from Eq. (18) that the condition for such a pole is just

$$\det\left[t^{-1}(\varepsilon) - G(\mathbf{k}, \varepsilon)\right] = 0 \qquad (20)$$

which is the equation for the band energies ε_k in KKR theory (32). In this way we again make contact, for ordered systems, with the band structure approach.

Another useful way of writing Eq. (16) follows if we define a local density of states for the atom at the origin by

$$n^0(\varepsilon) = -\frac{2}{\pi} \int_{\text{cell}} d\mathbf{r} \, \text{Im} \, G(\mathbf{r}, \mathbf{r}; \varepsilon^+)$$

where the r integral runs over the cell occupied by the atom at the origin. Again we may use Eq. (15) to define angular momentum components of $n^0(\varepsilon)$

$$n^0(\varepsilon) = \sum_L n_L^0(\varepsilon)$$

$$n_L^0(\varepsilon) = -\frac{2}{\pi} \, \text{Im} \, \tau_{LL}^{00}(\varepsilon) \int_{\text{cell}} dr \, 4\pi r^2 \left[Z_l^0(r, \varepsilon)\right]^2$$

From this expression it can easily be seen that if we define a new atomic matrix element by

$$M_L(\varepsilon) = \int d\mathbf{r} \, \phi_c(\mathbf{r}) \, h(\mathbf{r}) \, X_l^0(r, \varepsilon) \, Y_L^*(\hat{\mathbf{r}})$$

where X_l^0 is a radial function proportional to Z_l^0 but normalized to unity within the atomic cell, we may write

$$W_c(\omega) = \pi \sum_L \left|M_L(E)\right|^2 n_L^0(E) \qquad (21)$$

(We have assumed here that $\tau_{LL'}^{00}$ is diagonal in L; this is true for cubic systems up to $l = 2$, if real spherical harmonics are used.) Since the dipole interaction has been used in the matrix elements $m_L(E)$ [and $M_l(E)$], the usual $l = l_{\text{core}} \pm 1$ selection rule applies. Moreover, the matrix elements $M_L(E)$ usually vary smoothly and slowly with energy E. Thus we can see from Eq. (21) that, for example, in K-shell excitation the ω dependence of the transition rate $W_c(\omega)$ mainly reflects the local density of unoccupied p-type states at the excited atom.

At high photon energies quadrupole transitions may occur with appreciable intensity. Wakoh and Kubo (33) have estimated that for K-shell excitation in $3d$ transition metals the magnitude of the quadrupole term in the electron–photon interaction is about 9% of the dipole contribution. We note, however, that in such systems quadrupole transitions couple the $1s$ core state with the high density of unfilled d states, and may therefore be more important than this estimate suggests.

One can also use Eq. (16) to identify a purely atomic contribution to the x-ray absorption transition rate. In fact we may write this atomic term, using Eq. (17), as

$$W_c^a(\omega) = -2 \sum_L \left| m_L(E) \right|^2 \operatorname{Im} t_l^0(E) \, \theta(E - \mu)$$

$$= 2\kappa \sum_L \left| \int d\mathbf{r} \, \phi_c(\mathbf{r}) \, h(\mathbf{r}) \, R_l(r, E) \, Y_L^*(\hat{\mathbf{r}}) \right|^2 \theta(E - \mu) \quad (22)$$

in which $R_l(r, E)$ is the usual regular solution of the radial Schrödinger equation, that is, that which matches smoothly on to

$$j_l(\kappa r) \cos \delta_l(\varepsilon) - n_l(\kappa r) \sin \delta_l(E)$$

at R_{MT} [$n_l(\kappa r)$ is a spherical Neumann function]. We can thus write the total transition rate as

$$W_c(\omega) = W_c^a(\omega) - 2 \sum_{L,L'} m_L(E)$$

$$\cdot \operatorname{Im} \left\{ \tau_{LL'}^{00} - t_l^0 \delta_{LL'} \right\} m_{L'}^*(E) \, \theta(E - \mu) \quad (23)$$

It is instructive at this stage to examine the weak scattering limit of Eq. (23). Recall that the multiple-scattering terms are described by Eq. (12). The weak scattering condition is $t_l^i(E) \, g_{LL'}^{ij}(E) \ll 1$, which leads to

$$\tau_{LL}^{00} - t_l^0 \simeq \sum_{i \neq 0} \sum_{L'} t_l^0 g_{LL'}^{0j} \cdot t_{l'}^i g_{L'L}^{i0} t_l^0 \quad (24)$$

(For simplicity we consider only the L-diagonal term of $\tau_{LL'}^{00}$; this is sufficient for K-shell excitation, for which $l = 1$.) Equations (13) and (22)–(24) now allow us to write, in the weak scattering limit,

$$W_c(\omega) = W_c^a(\omega) \left[1 + \chi(\omega) \right] \quad (25)$$

where

$$\chi(\omega) = \text{Im} \left\{ e^{2i\delta_l} \kappa \sum_{i \neq 0} \sum_{L'} g_{LL'}^{0i} t_{l'}^i g_{L'L}^{i0} \right\} \tag{26}$$

For the case of K-shell excitation ($l = 1$) it is easy to show that, if $S^i(\kappa)$ is the backscattering amplitude of atom i,

$$\chi(\omega) \simeq \frac{1}{\kappa} \sum_{i \neq 0} \frac{1}{R_i^2} S^i(\kappa) \sin(2\kappa R_i + 2\delta_1) \tag{27}$$

[One uses the asymptotic form $h_l^{(1)}(\kappa R) \simeq -e^{i\kappa r}/\kappa r$, valid when $\kappa r > 1$, together with $R = |\mathbf{R}_i + \mathbf{r}| \simeq R_i + (\mathbf{r} \cdot \mathbf{R}_i)/R_i$ if $r \ll R_i$.] Apart from a Debye–Waller factor and a damping term coming from the finite mean free path of the photoelectron, Eq. (27) is just that used by Stern, Sayers, and Lytle (34) to describe the oscillatory part of the absorption rate at high electron energies; in short, EXAFS. Note also that Eq. (26) itself, without the simplifications needed to obtain Eq. (27), leads straightforwardly to the "curved-wave" formulation of EXAFS (35). Thus, as the energy of the photoelectron increases, and the weak scattering condition is met, the XANES spectrum merges smoothly into the EXAFS regime (36).

2.3.3. Calculation of the τ-Matrix

Let us now return to the full expression, Eq. (16), for the transition rate, and consider the calculation of the τ-matrix, $\tau_{LL'}^{00}$. The fundamental multiple-scattering formula, Eq. (12), can be rewritten (dropping the angular momentum subscripts, for clarity)

$$\sum_k \left[(t^i)^{-1} \delta_{ik} - g^{ik} \right] \tau^{kj} = \delta_{ij} \tag{28}$$

Thus, to calculate τ^{ij} one has to invert a matrix that contains the inverse t-matrices, $(t^i)^{-1}$, of each atom on the diagonal, and propagators $-g^{ij}$ between the atoms as off-diagonal terms ($g^{ii} = 0$ by definition for all i). At this stage it is useful to distinguish the central (excited) atom at site zero from the surrounding atoms, to which we give the collective label c. The τ-matrix τ^{00} is then given as the (0, 0) element of the inverse of this matrix

$$\begin{bmatrix} t^{0-1} & -g^{0c} \\ -g^{c0} & T_c^{-1} \end{bmatrix} \tag{29}$$

Here T_c is the full T-matrix of the surrounding atoms (i.e., excluding the central atom), and g^{0c} is a vector

$$g^{0c} = [\, g^{01} \quad g^{02} \quad \cdots \quad g^{0i} \quad \cdots \,] \tag{30}$$

This is a simple application of the technique known as partitioning Eq. (37), and the inversion gives

$$\tau^{00} = [t^{0-1} - R^{OI}]^{-1} \tag{31}$$

where

$$R^{OI} = g^{0c} T_c g^{c0}$$

$$= \sum_{i,j \neq 0} g^{0i} T_c^{ij} g^{i0} \tag{32}$$

The quantity $R^{OI}_{LL'}$ gives the amplitude with which an outgoing spherical wave of angular momentum L about the central atom is scattered into an incoming wave of angular momentum L' about the central atom; we call it the "out–in" reflection matrix of the surrounding atoms (hence the superscript OI). This reflection matrix summarizes in an economical way the influence of the surrounding atoms on the τ-matrix, τ^{00}.

To calculate R^{OI} it is natural to divide the surrounding atoms into shells. For an extended system the number of such shells is infinite, but in practice the major contributions to R^{OI} come from a rather small number of shells of small radius. This is essentially a question of distance from the central atom. The amplitudes at the central atom of waves scattered from distant atoms are in any case small, and this effect is compounded by the various loss processes that limit the mean free path of the excited electron. So, let us first consider the reflection matrix R^{OI} of a pair of shells, A and B, A being the inner shell. R^{OI} is given by Eq. (26) in which T_c has the following structure

$$T_c = \begin{bmatrix} T_A^{-1} & -g^{AB} \\ -g^{BA} & T_B^{-1} \end{bmatrix}^{-1} \tag{33}$$

Here T_A and T_B are full T-matrices of each shell, and g^{AB}, g^{BA} are propagator matrices linking the two shells. The inversion in Eq. (33) can easily be performed and combined with Eq. (32) to give R^{OI}. However, the result can be cast into a more transparent form if we introduce the following scattering matrices for each shell:

$$T_A^{OO} = \sum_{i,j} g^{0i} T_A^{ij} \gamma^{j0}$$

$$T_A^{OI} = \sum_{i,j} g^{0i} T_A^{ij} g^{j0}$$

$$T_A^{IO} = \sum_{i,j} \gamma^{0i} T_A^{ij} \gamma^{j0}$$

$$T_A^{II} = \sum_{i,j} \gamma^{0i} T_A^{ij} g^{j0} \qquad (34)$$

together with similar quantities for shell B. The modified propagators γ^{ij} are defined just as in Eq. (14) for g^{ij}, except that the Hankel function $h_{l''}^{(1)}(\kappa R_{ij})$ is replaced by a Bessel function $j_{l''}(\kappa R_{ij})$. These propagators g^{ij} and γ^{ij} always appear in multiple-scattering theory in consequence of the following useful relations (38)

$$Y_{L'}(\hat{\mathbf{r}}_j) \, h_{l'}^{(1)}(\kappa r_j) = \sum_L Y_L(\hat{\mathbf{r}}_i) \, j_l(\kappa r_i) \, g_{LL'}^{ij} \qquad r_i < R_{ij}$$

$$Y_{L'}(\hat{\mathbf{r}}_j) \, j_{l'}(\kappa r_j) = \sum_L Y_L(\hat{\mathbf{r}}_i) \, j_l(\kappa r_i) \, \gamma_{LL'}^{ij}$$

$$Y_{L'}(\hat{\mathbf{r}}_j) \, h_{l'}^{(1)}(\kappa r_j) = \sum_L Y_L(\hat{\mathbf{r}}_i) \, h_l^{(1)}(\kappa r_i) \, \gamma_{LL'}^{ij} \qquad r_i > R_{ij} \qquad (35)$$

in which $\mathbf{r}_i = \mathbf{r} - \mathbf{R}_i$, $\mathbf{r}_j = \mathbf{r} - \mathbf{R}_j$, $\mathbf{R}_{ij} = \mathbf{R}_i - \mathbf{R}_j$.

The significance of the quantities introduced in Eq. (34) is as follows. T_A^{II} and T_A^{OO} are the transmission matrices of shell A for incoming and outgoing waves, respectively, resolved about the central atom, while T_A^{OI} and T_A^{IO} are the "out-in" and "in-out" reflection matrices, respectively (note the similarity of T_A^{OI} to R^{OI} in Eq. (32). In short, these quantities represent the T-matrix of shell A resolved in angular momentum about the central atom. This point is schematically depicted in Fig. 2.2.

It is now straightforward, using Eq. (32)–(35), to show that the "out-in" reflection matrix of the pair of shells is given by

$$R^{OI} = T_A^{OI} + (1 + T_A^{OO})\left[(T_B^{OI})^{-1} - T_A^{IO}\right]^{-1} (1 + T_A^{II}) \qquad (36)$$

This equation has a simple interpretation. The first term, T_A^{OI}, represents direct backscattering by the inner shell, A. The second term has three factors; (a) (1

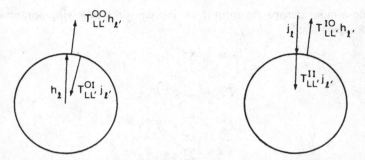

Figure 2.2. Schematic representation of the shell transition and reflection matrices T^{OO}, T^{OI}, T^{IO}, and T^{II}.

$+ T_A^{OO})$ represents outward transmission through shell A, (b) $[(T_B^{OI}) - 1 - T_A^{IO}]^{-1}$ represents repeated scattering between shells A and B, (c) $(1 + T_A^{II})$ represents inward transmission through shell A. A useful feature of Eq. (36) is that it contains only one quantity referring to shell B, namely, the "out-in" reflection matrix, T_B^{OI}. This means that Eq. (36) can be applied to N shells if we interpret T_A as the T-matrix of the innermost shell and T_B^{OI} as the reflection matrix of the outer $(N - 1)$ shells. Thus, our method operates by considering a finite system in which the central atom is surrounded by N shells, using Eq. (36) to find the "out-in" reflection matrix of the outermost pair of shells, N and $N - 1$, and thereafter applying Eq. (36) successively to each of the remaining $N - 2$ shells until the reflection matrix R^{OI} of the entire surrounding cluster has been calculated. Equation (31) then gives the τ-matrix, τ^{OO}, and Eq. (16) the transition rate. As a point of technique, computer codes developed at Daresbury exploit any point group symmetry the cluster may possess (although this is not essential), and a full discussion of this and the other numerical advantages of this scheme has been given elsewhere (28). As mentioned in the last section, the lifetime broadening can be included by performing the calculations at a complex energy whose imaginary part is equal to one half of the total core hole and photoelectron broadening. We can also take account of instrumental resolution in the same way.

Finally, we note that the transition rate for x-ray emission is also given by Eq. (16), but with $\theta(E - \mu)$ replaced by $\theta(\mu - E)$ (39); it is the states below the Fermi level that take part in the emission process. There is also the interesting possibility of using techniques such as this to calculate anomalous x-ray scattering amplitudes. The imaginary part of this scattering amplitude is just the absorption cross section, and the real part is its Hilbert transform (23).

I want to emphasize, at the end of this rather formal section on multiple scattering, that what has been described is entirely equivalent to solving the one-electron Schrödinger equation for the wave functions of states at XANES energies. The same Schrödinger equation could be solved in other ways, for example, by methods based on linear combinations of atomic orbitals (LCAO), by empirical molecular orbital methods, and so on. Different methods have different accuracy, but their physical content is the same; the mathematical connection is embodied in Eq. (6). In the present context of XANES calculations the multiple-scattering method is accurate, economical, flexible, and has its own physical appeal.

2.4. NUMERICAL CALCULATIONS—INGREDIENTS

In the last section multiple-scattering theory was used to describe the motion of electrons in a potential having the form of a finite cluster of nonoverlapping muffin tins, and thereby to calculate the x-ray absorption rate. In fact, the solution of this muffin-tin model can be carried out essentially to arbitrary numerical accuracy, and so the question of the validity and accuracy of the whole theory really turns on the adequacy of the underlying potential. The reasons for using a one-particle potential at all have been outlined in Section 2.1 and so now a comment on the form and construction of the potentials that are actually used is necessary.

In the band theory of ordered crystals, particularly in its earlier days, a great deal of attention was paid to the "best" method of constructing potentials, that is, to finding the procedure that gave the best agreement with experiment (on Fermi surfaces, e.g.). One of the simplest and most generally reliable methods goes by the name of the Mattheiss prescription (40), and consists of the following steps.

1. Charge densities for neutral atoms are obtained from a self-consistent field (SCF) atomic calculation, often of the Hartree–Fock–Slater kind.

2. These atomic charge densities are then placed on each site of the chosen geometry, and the superposed charge density is spherically averaged about the atom whose potential is required.

3. Taking this spherical, superposed charge density $\rho(r)$, the Poisson equation is solved for the Coulomb part of the potential, and the exchange-correlation potential is found using a local approximation of the $X\alpha$ type $[\sim \rho(r)^{1/3}]$ with the α parameter equal to 1.

In the context of band theory this Mattheiss prescription is generally regarded as an approximation to the LDA (15), mentioned in Section 2.1, which requires the self-consistent solution of the following set of equations:

$$\left[-\tfrac{1}{2} \nabla^2 + V_{\text{eff}}(\mathbf{r}) \right] \psi_n(\mathbf{r}) = \varepsilon_n \psi_n(\mathbf{r})$$

$$V_{\text{eff}}(\mathbf{r}) = V_{\text{nuc}}(\mathbf{r}) + \int dr' \, \frac{\rho(r')}{|\mathbf{r} - \mathbf{r}'|} + V_{xc}(\mathbf{r})$$

$$V_{xc}(\mathbf{r}) \sim \rho(\mathbf{r})^{1/3}$$

$$\rho(\mathbf{r}) = \sum_{n,\,\text{occupied}} \left| \psi_n(\mathbf{r}) \right|^2$$

[Here the exchange-correlation potential $V_{xc}(\mathbf{r})$ is written as varying as $\rho(\mathbf{r})^{1/3}$. In fact, more sophisticated functional forms are now used (16), but the dominant variation is still approximately $\rho(\mathbf{r})^{1/3}$. V_{nuc} is the nuclear potential.] Thus the Mattheiss prescription short circuits the repeated calculation of the charge density by guessing that the final result will not be too far away from the superposition of atomic charge densities. Therefore, if one is interested in the fine details of the distribution of electronic charge around an atom in any given environment a full self-consistent calculation is advisable. But, if it is sufficient to be able to model the gross changes in charge density and potential as the environment of the atom changes, then the Mattheiss prescription may well be good enough. Both approaches produce similar band structures when applied to close-packed metallic crystals (40). Regarding the XANES of such systems, self-consistency is unlikely to be important in view of the rather strong broadening to which the spectra are subject, and both kinds of potential have successfully been used (6, 41). In ionic materials, on the other hand, charge transfer effects described in the self-consistent LDA calculations (but not in the Mattheiss prescription) may become important. Good results have been achieved for the XANES of small inorganic clusters using the self-consistent $X\alpha$ scattered wave method (9, 11). There are, however, two reasons for wishing to avoid self-consistent calculations as a route to XANES. The first is the fundamental problem, mentioned in Section 2.1, that the local density eigenvalue spectrum is not equivalent to the quasiparticle spectrum revealed in XANES. In principle some many-body self-energy correction (generally unknown, at least in its quantitative details) should be applied. The second difficulty is the purely operational one that self-consistent calculations are very time consuming (and also require considerable expertise), especially for the large, low-symmetry systems often of interest in applications to XANES. It is in the light of this problem that the attraction of a simple, quick, and fairly realistic procedure such as the Mattheiss prescription becomes clear. It shortcomings have to be recognized, however, and it may be that in unfavorable cases it may have to be jettisoned in favor of a self-consistent calculation.

We have been assuming that the output from either the Mattheiss prescription or a self-consistent calculation is a set of one-electron potentials of muffin-tin form. From these, phase shifts can be generated by standard methods (42), and then the t-matrices, Eq. (13), which are the components of the multiple-scattering equations in Section 2.3, can be found. The nonoverlapping muffin-tin form of the potentials is very important for the application of multiple-scattering theory, (a) because the multiple-scattering expansions are formally not valid if the potentials overlap (38), and (b) because the spherical form of the potentials within each muffin-tin well causes the t-matrices to be diagonal in angular momentum. The muffin-tin approximation has been thoroughly tested for close-packed metals (43), and found to be rather accurate—certainly accurate enough for XANES applications. The more open the structure, however, the worse the muffin-tin approximation is bound to become, because the interstitial region between the muffin-tin spheres is larger. It is in this intersphere region that the muffin-tin approximation makes its worst errors, especially when the exact interstitial potential undergoes large variations about its mean value. Many interesting XANES applications will be to complex organic or biological systems, which are very far from being close-packed metals, and for these the muffin-tin approximation has to come under suspicion. On the other hand, the energies of XANES electrons are similar to those common in LEED experiments, and it is known that the muffin-tin model leads to accurate analyses of LEED data even for complex, open structures at crystal surfaces (42). It seems that, at the rather high XANES or LEED energies, scattering of the electrons by the ion cores becomes relatively more important than that by the peripheral regions of the atoms. The suggestion is, therefore, that in XANES the multiple scattering of the photoexcited electron is most strongly dependent on the geometrical arrangement of atoms in its vicinity, rather than the fine details of their potentials, and this is borne out by most of the applications so far made (some of which are described in the next section). It is only to be expected, however, that the muffin-tin approximation will break down in the most unfavorable (i.e., open, low-symmetry) cases. In such instances one could resort to an approach sometimes employed in band calculations, namely, to attempt to describe the potential in the interstitial region by packing it with "empty spheres" each of which carries it own muffin-tin potential. An alternative would be to generalize the multiple-scattering theory to treat nonoverlapping, nonspherical potentials; then the space around each atom could be divided into nonoverlapping, nonspherical regions, and the interstitial region eliminated altogether. Such a generalized multiple-scattering theory can be written down (44), but numerical applications have so far been few.

The examples of XANES calculations in the next section all use muffin-tin potentials constructed by means of the Mattheiss prescription. The results are usually sensible, and sometimes surprisingly good. However, it is likely that future improvements of current XANES theory will include a more accurate treatment of the potential.

A final factor entering the construction of potentials concerns the role of the core hole. The so-called "final-state rule" (45) suggests that for calculations of x-ray emission spectra the potential for the emitting atom should be taken as that describing the unexcited ground state, while for absorption calculations the potential should contain a contribution from a self-consistently screened core hole. This view emerges from model calculations of the dynamic response of the conduction electrons to the suddenly switched on (or off) core hole potential. These model calculations may be applicable directly to simple metals, but in more complex systems the final-state rule probably needs to be systematically tested. Band theory calculations of absorption spectra for transition metals seem to agree reasonably well with experiment without including the core hole potential at all (21, 33), while for systems with rather localized unoccupied levels in the ground state (e.g., the f levels in rare earths) the core hole potential can induce new features in the spectra (13, 46). The cluster method outlined in Section 2.3 is, of course, ideally suited for investigating such effects, since the presence of the core hole breaks translational symmetry even in an ordered crystal. Band structure techniques would thus be difficult to apply in this case. The full dynamical response problem mentioned previously is just that considered, within a simplified model of a metal, by Nozieres and De Dominicis (47), who found at the threshold the celebrated x-ray edge singularities. In practice such threshold effects are difficult to observe, except for the lighter metals (48), and no attempt has been made to include them into our XANES calculations.

As mentioned in Section 2.2, broadening effects due to the finite decay rates of both core hole and photoelectron can be included by performing the calculations at a complex energy whose imaginary part, Γ, is approximately equal to the sum of the inverse lifetimes of hole and electron. There exist tabulations of core hole widths (25), but the lifetimes of the photoelectrons are somewhat less well characterized. Some measurements of the mean free paths of excited electrons have been made (49), and have been used to determine broadening parameters in XANES calculations (50). In general, such data are not available. Local approximations to the self-energies of excited electrons have successfully been used in photoemission calculations (51), and may well prove useful in XANES applications. Often in the XANES regime the photoelectron contributes significantly less to Γ than the core hole, and so its importance is lessened. At this stage the photoelectron inverse lifetime is essentially a parameter to be fitted to experiment; useful values tend to lie in the range 1–4 eV.

2.5. NUMERICAL CALCULATIONS—EXAMPLES

This section describes the results of a few calculations on systems of varying complexity. These examples have been chosen to exhibit certain features that will be commented on, but certainly do not amount to a comprehensive list of current applications (see ref. 5–12).

The first system is a very simple one, namely, pure iron in the ordered body-centered cubic (bcc) structure (23). Figure 2.3 shows calculations of the near-

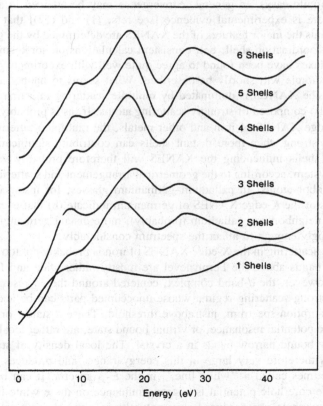

Figure 2.3. Multiple-scattering cluster calculations of the *K*-edge absorption spectra of bcc iron, including various numbers of shells in the cluster.

edge structure at the K-edge, using potentials constructed according to the Mattheiss prescription and the multiple-scattering method outlined in Section 2.3, and illustrates clearly how the XANES converges as increasing numbers of shells are included in the cluster. This convergence depends on the broadening parameters used in the calculation. If no broadening is included, then in principle the calculation never converges in the number of shells—one simply obtains structure on a finer and finer energy scale. With finite broadening, a spatial limit is reached at which no further structure in the spectrum appears. In band theoretical calculations the complementary convergence parameter is the number of k points included in the Brillouin zone integral [see e.g. Eq. (18)].

Figure 2.3 shows that after five shells are included the K spectrum has converged, and agrees satisfactorily with both band structure calculations and experiment (33). Five shells of atoms in the bcc structure constitutes quite a large cluster (although the symmetry of the system is sufficiently high to make the calculation quite easy). In general, convergence may be attained in a smaller cluster. There is experimental evidence [see refs. (1) and (52)] that in some ionic materials the major features of the XANES are determined by the geometry of the first coordination shell. Self-consistent calculations on some single-shell ionic complexes have been found to agree quite well with experiment (11), and a similar example will shortly be described. What seems to happen is that in these cases the XANES is dominated by multiple scattering in a first shell (or pair of shells) composed of strongly scattering atoms. If, as is probably the case for the K-edge XANES of iron and other metals, the multiple scattering is not particularly strong, then more distant shells can contribute significantly. The number of shells influencing the XANES will therefore probably vary from system to system, according to the geometrical arrangement and scattering power of the neighbor atoms. In palladium–germanium glasses, for instance, model calculations of the K-edge XANES of germanium indicate (8) that even though the nearest neighbors are palladium (probably), more distant germanium atoms scatter strongly enough to affect the spectrum considerably.

Multiple scattering in the K-edge XANES of iron is probably not too strong—the energy bands above the Fermi level are mostly rather wide and free-electronlike. However, the d-band complex, centered around the $3d$ resonance energy, is a strong scattering regime whose unoccupied parts can be accessed in the $L_{2,3}$ absorption spectrum, just above threshold. These d states, originating in an atomic potential resonance, or virtual bound state, are rather localized and form tightly bound narrow bands in a crystal. The local density of states of d character is therefore very large in this energy range, and produces a strong peak, sometimes called a "white line," in the $L_{2,3}$ spectra. It can be argued (53) that the core hole potential is a strong influence on these white lines, but they are, nevertheless, reproduced quite naturally in a one-electron calculation.

The results of Müller et al. (50) have shown this to be the case in transition metals. Munoz et al. (7) have made XANES calculations (based on the KKRCPA) for copper–nickel alloys, and were able to relate the nickel $L_{2,3}$ white line to the average number of d holes on the nickel atoms as a function of alloy concentration (this is an important parameter for the magnetic properties of the system). Since the d states in transition metals are rather localized, they are frequently described by LCAO (tight-binding) methods (54). I note this to emphasize once more the connection between the multiple-scattering approach and the perhaps more familiar methods that expand the wave function in an atomic orbital basis set.

Some of the strongest multiple-scattering effects occur in very small systems, particularly in certain diatomic molecules, where they are manifested as so-called "shape resonances." These usually occur in systems where the atoms are very close together (as in multiply bonded diatomics), so that the potential between the atoms is rather deep and attractive. The calculations of Dehmer and Dill (5) on the nitrogen molecule (N_2) showed that electrons of high angular momentum about the center of the molecule could be trapped by the combination of this attractive molecular potential and the centrifugal potential $l(l + 1)/2r^2$, and this virtual bound state (55), or shape resonance, occurs at quite a high energy in the continuum. For dinitrogen, Dehmer and Dill found a resonance in the $l = 3$ channel at an energy of about 12 eV above threshold, as Fig. 2.4 indicates. Note that while this resonance is characterized (approximately) as having an angular momentum of three units about the center of the molecule, it has an $l = 1$ (i.e., p) component about the center of any one of the atoms and so can be excited in the K-shell absorption spectrum.

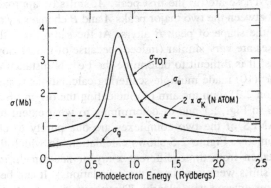

Figure 2.4. Calculated K-shell photoionization cross section for the nitrogen molecule (56), showing a shape resonance of σ_u symmetry at ~ 0.8 Ry. Two times the atomic cross section is shown for comparison.

In the language of molecular physics, one describes the (continuum) molecular wave function corresponding to the shape resonance as being concentrated, or localized, within the molecule. In the language of scattering theory, one describes this state of affairs as strong multiple scattering of the electron between the two atoms; it is the interference of the multiply scattered waves that produces the concentration of the wave function. Indeed, although Dehmer and Dill used a variant of the self-consistent field $X\alpha$ method to perform their calculations, the important features of their model (a local approximation for the exchange and correlation, muffin-tin potentials) are just those entering the multiple-scattering calculations described in Section 2.3.

The effects of shape resonances can also be seen when multiply bonded diatomic molecules or ligands form part of a larger system. Davenport (57) found such resonances in calculations on carbon monoxide chemisorbed on to the surface of nickel, and recently Stöhr et al. (58) have used the polarization dependence of the x-ray absorption cross section associated with the shape resonances to determine the orientation at which CO and NO molecules are chemisorbed on to nickel surfaces.

The absorption spectra of iron atoms in ferro- and ferricyanide complex ions also exhibit related features. The structure of these complexes is approximately octahedral, with six CN groups surrounding a central iron atom (in a formal oxidation state of II or III), the $Fe-C-N$ bond being very nearly linear. The $C-N$ distance is quite short (~ 1.1 Å), so that each cyanide ligand forms a quasidiatomic system of the kind for which shape resonances can be expected. Indeed, two very strong peaks are seen in the K-edge XANES of iron (Fig. 2.1), in marked contrast to the much weaker EXAFS oscillations. In addition, certain distinct differences between the XANES of the Fe^{II} and Fe^{III} complexes can be observed. These are (a) the first peak, A, shifts by approximately 1 eV, (b) the splitting between the two major peaks A and B changes by approximately 0.5 eV, and (c) the shape of peak B alters. At the same time, the EXAFS of the two complexes are very similar (indeed, because of the shadowing effect of the carbon atoms, it is difficult to determine the $Fe-N$ distance accurately).

Bianconi et al. (10) made multiple-scattering calculations (using the method outlined in Section 2.3) with the aim of elucidating the origin of the two strong features A and B in Fig. 2.1 and of determining to what extent the differences between the XANES of the two complexes are due purely to changes in the coordination geometry. Figure 2.5 shows calculations in which the atomic coordinates were taken from three different x-ray or neutron diffraction studies (the same phase shifts were used in all the calculations). It can be seen that the two strong features emerge very clearly. That they originate in strong multiple scattering between the carbon and nitrogen atoms is demonstrated by Fig. 2.6— without the nitrogen atoms only a very weak EXAFS-like structure occurs in

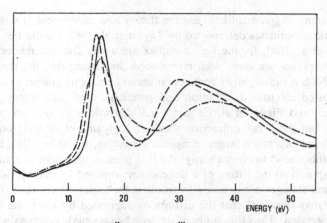

Figure 2.5. Calculated XANES of $Fe^{II}(CN)_6$ and $Fe^{III}(CN)_6$ using atomic coordinates derived from the diffraction data of Figgis et al. (60) for $Fe^{III}(CN)_6$ (full curve), Taylor et al. (59) for $Fe^{II}(CN)_6$ (dashed curve), and Kiriyama et al. (61) also for $Fe^{II}(CN)_6$.

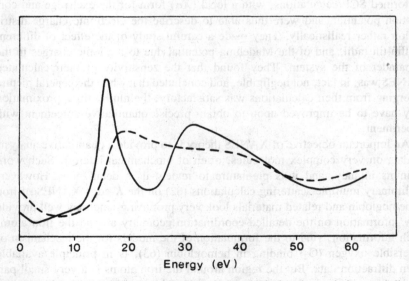

Figure 2.6. Calculated XANES for an $Fe(CN)_6$ cluster with both carbon and nitrogen shells included (full curve), and with the carbon shell only included (dashed curve).

the calculations. Agreement between the theory and experiment is rather good, if the atomic coordinates determined by Taylor et al. (59), for the Fe^{II} complex and Figgis et al. (60), for the Fe^{III} complex are used. The differences between the two complexes are quite well reproduced, indicating that the major effect on the XANES is the slightly altered coordination geometry, rather than changes in the detailed electron distribution and potential. The coordinates given by Taylor et al. and Figgis et al. contain small distortions away from octahedral symmetry. But the x-ray diffraction work of Kiriyama et al. (61) on the Fe^{II} complex indicates a much larger tetragonal distortion. In the calculated XANES (Fig. 2.5) this shows up very clearly; the first peak in particular is actually split (this corresponds to the lifting of a degeneracy imposed by octahedral symmetry). No such changes are in fact discernible in the experimental data (Fig. 2.1), and so we may conclude that the distortions suggested by Kiriyama et al. are rather exaggerated. Thus this study of the iron hexacyanide complexes provides an illustration not only of strong multiple-scattering effects in XANES, but also of the way one can obtain structural information complementary to that available from other techniques.

Similar shape resonances have also been reported in tetrahedral molecules such as the carbon, silicon, and germanium tetrahalides (9). Figure 2.7 shows the calculation of Kutzler et al. (11) on the tetrahedral CrO_4^{2-} ions. It accounts very well for the major features of the experimental spectrum. Kutzler et al. performed SCF calculations, with a local ($X\alpha$) form for the exchange and correlation potential, and were thus able to describe the electronic charge distribution rather realistically. They made a careful study of the effect of different muffin-tin radii, and of the Madelung potential due to the ionic charges in the remainder of the system. They found that the sensitivity of their calculated XANES was, in fact, not negligible, and concluded that while the general picture emerging from their calculations was satisfactory, the muffin-tin approximation may have to be improved upon to obtain precise quantitative agreement with experiment.

An important objective of XANES theory is to provide a quantitative analysis of data on very complex molecules, often of biochemical interest. Such work is in its infancy, and it is premature to review it in detail now. However, preliminary multiple-scattering calculations (62) on the K-edge XANES of iron in hemoglobin and related materials look very promising, and may well provide new information on the detailed coordination geometry around the iron atom. Such information, vital to the formulation of the models for the mechanism of reversible oxygen (O_2) binding in hemoglobin (63), is in principle available from diffraction data. But the region around the iron atoms is a very small part of the hemoglobin molecule; x-ray absorption is sensitive to just this region, while x-ray or neutron diffraction samples the entire molecule. Furthermore, diffraction requires artificial crystalline samples. It may be a real advantage of

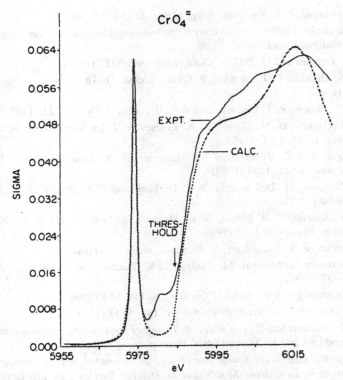

Figure 2.7. Calculated and experimental K-shell XANES of chromium in the CrO_4^{2-} ion (11).

x-ray absorption studies that experiments on *in vivo* systems can be contemplated.

Thus XANES holds out the possibility of obtaining new information, both electronic and structural, for a wide variety of systems. Current work on surfaces (64) or amorphous materials (8) have hardly been mentioned. Nor has the possibility that many-body interactions may introduce new features (e.g., shake-up and shake-down effects) into the spectra (65) been discussed. Much in the theory remains to be clarified, but the recent advances outlined in this chapter demonstrate that accurate one-electron calculations will be a vital factor in making XANES a powerful and useful technique.

REFERENCES

1. A. Bianconi, *Appl. Surf. Sci.*, **6**, 392 (1980).
2. A. R. Williams and N. D. Lang, *Phys. Rev. Lett.*, **40**, 954 (1978).

3. For example, J. S. Faulkner, *Prog. Mater. Sci.*, **27**, 3 (1982).

4. Other chapters in this volume correct this oversimplification and show how EXAFS is actually analyzed.

5. J. L. Dehmer and D. Dill, *J. Chem. Phys.*, **65**, 5327 (1976).

6. G. N. Greaves, P. J. Durham, P. Quinn, and G. Diakun, *Nature (London)*, **294**, 139 (1981).

7. M. C. Munoz, P. J. Durham, and B. L. Gyorffy, *J. Phys. F*, **12**, 1497 (1982).

8. P. H. Gaskell, D. M. Glover, A. K. Livesey, P. J. Durham, and G. N. Greaves, *J. Phys. C*, **15**, L597 (1982).

9. For example, C. R. Natoli, F. W. Kuntzler, D. K. Misemer, and S. Doniach, *Phys. Rev. A*, **22**, 1104 (1980).

10. A. Bianconi, M. Dell'Ariccia, P. J. Durham, and J. B. Pendry, *Phys. Rev. B.*, **26**, 6052 (1982).

11. F. W. Kutzler, C. R. Natoli, D. K. Misemer, S. Doniach, and K. O. Hodgson, *J. Chem. Phys.*, **73**, 3274 (1980).

12. For example, P. J. Durham, A. Bianconi, and S. S. Hasnain,.

13. A. Bianconi, S. Modesti, M. Campagna, K. Fischer, and S. Stizza, *J. Phys. C*, **14**, 4737 (1981).

14. P. Hohenberg and W. Kohn, *Phys. Rev. B*, **136**, 864 (1964).

15. W. Kohn and L. J. Sham, *Phys. Rev. A*, **140**, 1133 (1965).

16. A. R. Williams and U. von Barth, in *Theory of the Inhomogeneous Electron Gas*, S. Lundqvist and N. March (Eds.), Plenum, New York, 1982.

17. Unless the state is the lowest one of a given symmetry—see O. Gunnarsson, in *Electrons in Disordered Metals and at Metallic Surfaces*, P. Phariseau, B. L. Gyorffy, and L. Scheire (Eds.), Plenum, New York, 1979.

18. P. Noziéres, *Theory of Interacting Fermi Systems*, Benjamin, New York, 1964.

19. U. von Barth, in *The Electronic Structure of Complex Systems*, P. Phariseau and W. Temmerman (Eds.), Plenum, New York, 1984.

20. A. Zangwill and P. Soven, *Phys. Rev. A*, **21**, 1561 (1980).

21. For some recent work see J. E. Müller, O. Jepsen, O. K. Andersen, and J. W. Wilkins, *Phys. Rev. Lett.*, **40**, 720 (1978).

22. For example, W. Seka and H. P. Hanson, *J. Chem. Phys.*, **50**, 344 (1969).

23. P. J. Durham, in *The Electronic Structure of Complex Systems*, P. Phariseau and W. Temmerman (Eds.), Plenum, New York, 1984.

24. A. Messiah, *Quantum Mechanics*, North-Holland, Amsterdam, 1981, Chap. XXI.

25. O. Keski-Rahkonen and M. O. Krause, *At. Data Nuc. Data Tables*, **14**, 139 (1974).

26. For example, L. Hedin and S. Lundqvist, *Solid State Phys.*, **23**, 84 (1969).

27. See, for example, P. J. Durham, *J. Phys. F*, **12**, 1539 (1982).

28. P. J. Durham, J. B. Pendry, and C. H. Hodges, *Comput. Phys. Commun.*, **25**, 193 (1982).

29. See, for example, D. C. Langreth, in *Linear and Nonlinear Electron Transport in Solids*, J. T. Devreese and V. Van Doren (Eds.), Plenum, New York, 1976.

30. B. L. Gyorffy and M. J. Stott, in *Band Structure Spectroscopy of Metals and Alloys*, D. J. Fabian and L. M. Watson, Eds., Academic, London, 1973.

31. J. S. Faulkner and G. M. Stocks, *Phys. Rev. B*, **21**, 3222 (1980).

32. J. Korringa, *Physica*, **13**, 392 (1947).

33. S. Wakoh and Y. Kubo, *Japanese J. Appl. Phys.*, **17**, (S17-2), 193 (1978).

34. E. A. Stern, D. E. Sayers, and F. W. Lytle, *Phys. Rev. B*, **11**, 4836 (1975).

35. S. J. Gurman and J. B. Pendry, *Solid State Commun.*, **20**, 287 (1976).

36. P. A. Lee and J. B. Pendry, *Phys. Rev. B*, **11**, 2795 (1975).

37. For example, P. J. Durham, B. L. Gyorffy, and A. J. Pindor, *J. Phys. F*, **10**, 661 (1980).

38. P. Lloyd and P. V. Smith, *Adv. Phys.*, **21**, 69 (1975).

39. For example, P. J. Durham, D. Ghaleb, B. L. Gyorffy, C. F. Hague, J.-M. Mariot, G. M. Stocks, and W. Temmerman, *J. Phys. F*, **9**, 1719 (1979).

40. L. Mattheiss, *Phys. Rev. A*, **134**, 970 (1964).

41. J. E. Müller, O. Jepsen, O. K. Andersen, and J. W. Wilkins, *Phys. Rev. Lett.*, **40**, 720 (1978).

42. For example, J. B. Pendry, *Low Energy Electron Diffraction*, Academic, New York, 1974.

43. G. S. Painter, J. S. Faulkner, and G. M. Stocks, *Phys. Rev. B*, **9**, 2488 (1974).

44. J. S. Faulkner, *Phys. Rev. B*, **19**, 6186 (1979).

45. U. von Barth and G. Grossmann, *Solid State Commun.*, **32**, 645 (1979) and *Phys. Rev. B*, **25**, 5150 (1982); G. D. Mahan, *Phys. Rev. B*, **21**, 1421 (1980).

46. N. D. Lang and A. R. Williams, *Phys. Rev. B*, **16**, 2408 (1977); K. Schönhammer and O. Gunnarsson, *Z. Physik B*, **30**, 297 (1978).

47. P. Nozieres and C. T. De Dominicis, *Phys. Rev.*, **178**, 1097 (1969).

48. P. H. Citrin, G. K. Wertheim, and M. Schlüter, *Phys. Rev. B*, **20**, 3067 (1980).

49. I. Lindau and W. Spicer, *J. Electron Spectrosc.*, **3**, 409 (1974).

50. J. E. Müller, O. Jepsen, and J. W. Wilkins, *Solid State Commun.*, **42**, 365 (1982).

51. P. O. Nilsson and C. G. Larsson, *Phys. Rev. B*, **24**, 1917 (1981).

52. M. Belli, A. Scafati, A. Bianconi, S. Mobilio, L. Palladino, A. Reale, and E. Burattini, *Solid State Commun.*, **35**, 355 (1980).

53. G. Wendin, *Phys. Scr.*, **21**, 535 (1980).

54. For example, J. Callaway and C. S. Wang, *Phys. Rev. B*, **16**, 2095 (1979). These authors use Gaussians rather than atomic orbitals centered on each site, but the basis functions are still localized, and this is the main point.

55. This is just the usual kind of resonance encountered in potential scattering (e.g., see J. R. Taylor, *Scattering Theory*, Wiley, New York, 1972), completely analogous to the atomic *d* resonances of the transition materials.

56. J. L. Dehmer and D. Dill, *Phys. Rev. Lett.*, **35**, 213 (1975).

57. J. W. Davenport, *Phys. Rev. Lett.*, **36**, 945 (1976).

58. J. Stöhr, K. Baberschke, R. Jaeger, R. Treichler, and S. Brennan, *Phys. Rev. Lett.*, **47**, 381 (1981).

59. J. C. Taylor, M. H. Mueller, and R. L. Hitterman, *Acta Cryst.*, **A26**, 559 (1970).

60. B. N. Figgis, M. Gerloch, and R. Mason, *Proc. R. Soc. London Sec. A*, **309**, 91 (1969).

61. R. Kiriyama, H. Kiriyama, T. Wada, N. Nuzeki, and H. Hiribayashi, *J. Phys. Soc. Jpn.*, **19**, 540 (1964).

62. P. J. Durham, A. Bianconi, A. Congiu-Castellano, A. Giovanelli, S. S. Hasnain, L. Incoccia, S. Morante, and J. B. Pendry, *EMBO Journal*, **2**, 1441 (1983).

63. For example, J. J. Hopfield, in B. Lundqvist and S. Lundqvist (Eds.), *Collective Properties of Physical Systems (Nobel Symposium 24)*, Academic, New York, 1973, p. 238.

64. P. J. Durham, J. B. Pendry, and D. Norman, *J. Vac. Sci. Technol.*, **20**, 665 (1982).

65. G. Wendin, *Struct. Bonding Berlin*, **45**, (1981).

PART

B

INSTRUMENTAL AND DATA ANALYSIS

DESIGN OF AN EXAFS EXPERIMENT

STEVEN M. HEALD

Brookhaven National Laboratory
Upton, New York

3.1. INTRODUCTION

As the EXAFS field matures and data analysis procedures become standardized, the most important aspect in the success of an EXAFS experiment will be its conception and design. There are several stages in the process of obtaining EXAFS data. The first is an identification of a problem for which EXAFS is appropriate. This should be done carefully, keeping in mind the possibility of employing other types of scattering or diffraction measurement to obtain structural information. Once EXAFS is chosen as the preferred technique, appropriate standards should be identified. Although EXAFS theory has improved, comparison to standards remains the best analysis method. The next step is the choice of detection technique. This is done largely on the basis of statistical arguments that can also be employed to evaluate the feasibility of the proposed experiments. Finally, the data are taken with due precaution to avoid experimental pitfalls common to the various detection methods.

This chapter concerns itself with the final three steps in the process. After a brief discussion of standards it continues with a discussion of the different types of detection technique and gives some simple statistical arguments as a basis for choosing between them. It ends with a discussion of some common experimental difficulties—how they are detected, avoided, and, if necessary, corrected for.

3.2. STANDARDS

Crucial to any EXAFS experiment is the calibration of the measured phase and amplitudes. This can be done either with theory (1) or empirically from measured standards (2). Both techniques have been successful, although theoretical amplitudes are likely to be seriously in error (3, 4). The reasons for this are discussed in Chapters 1 and 6. In this section the choice of empirical standards is discussed.

The basic goal in choosing standards is to match as closely as possible the environment of the unknown. In some cases this is quite simple. For example, the crystal analogue to an amorphous material is likely to have nearly an identical local environment. For many experiments, however, the choice is not as simple. Either the unknown is not sufficiently characterized in advance or well-defined analogues do not exist. The former case might be a metalloprotein with unknown local environment while an example of the latter case is krypton adsorbed on graphite for which no Kr-C compounds exist. In these cases two strategies can be employed to improve on theory.

For an unknown environment it is usually necessary to measure a variety of standards for comparison. Since chemical and other analyses can usually provide some idea of the constituents, the number of standards needed to cover the likely range of local environments is generally not unwieldy. Comparison with the standards then provides a first-order guess to the structure. Direct examination of the data can also provide useful clues; low-Z and high-Z neighbors are easily distinguished by their differing backscattering amplitudes, and the use of approximate bond lengths to suggest a particular configuration are examples. The determination can then be refined by measuring new standards chosen to best represent the proposed environment.

For cases where more is known about the local structure in advance the problem may be due to a lack of well-characterized analogues. A useful technique in this situation is to measure standards of neighboring elements to calibrate the theory and then to use the theory to estimate the difference between the standard system and the unknown. Returning to the example of Kr-C interactions, possible standards are Br-C compounds. The theoretical calculations are then used to estimate the difference needed to correct the Br-C EXAFS to obtain a Kr-C standard. This procedure is likely to be better than relying completely on theory, especially for coordination number and Debye–Waller factor determinations.

Standards can also be used to estimate the accuracy of a final result. By comparing standards among themselves the degree of nontransferability can be estimated and folded into the estimated experimental errors to estimate the reliability of a proposed result.

3.3. EXPERIMENTAL DETECTION OF EXAFS

Once an experiment has been proposed and a list of possible standards identified, the actual design of the experiment begins. Foremost in the design process is a determination of the experimental technique. This section discusses the detection methods available along with their advantages and weaknesses. In most cases statistical estimates can be used to estimate the feasibility of a proposed experiment and to compare the possible detection methods. As EXAFS is applied to more difficult and exotic systems it becomes more important to perform such estimates in advance.

3.3.1. Detection Methods

Several techniques for measuring EXAFS have been developed, each with its advantages and disadvantages. The most obvious and widely used method is to measure the absorption of the sample by monitoring the incoming, I_0, and transmitted, I, flux. The absorption is then given by

$$\mu x = \ln \left(\frac{I_0}{I} \right) \tag{1}$$

where μ is the linear absorption coefficient and x is the sample thickness. This method is simple to apply and makes use of all the photons incident on the sample. It is often possible to apply it in conjunction with other detection methods.

There are two techniques for measuring the absorption signal. The most common uses a monochromator to select a small wavelength range $\Delta\lambda$ and monitors I and I_0 with suitable detectors. The full spectrum is acquired by sequentially stepping the monochromator through the required energies. A second technique allows the full EXAFS energy range to impinge upon the sample and uses a crystal to disperse the different energies spatially (5–7). Examples of the various geometries possible are shown in Fig. 3.1. A position sensitive

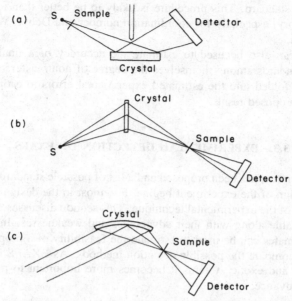

Figure 3.1. Three possible geometries for a dispersive EXAFS measurement: Arrangements (*a*) and (*b*) are most appropriate for laboratory sources for which the x-rays are not collimated while (*c*) is appropriate for the more collimated output of a synchrotron.

x-ray detector is used to detect the signal. The resulting energy resolution is determined by the spatial resolution of the detector and the dispersive power of the crystal.

In principle the statistical errors of the two methods are the same since the total number of photons collected per unit energy is the same. In the step mode the total output of the synchrotron at a particular energy is counted for a small portion of the scan time while in the energy dispersive mode an equivalently small fraction of the output is counted for the total scan time. In practice, the time overhead incurred in stepping the monochromator can be significant for concentrated samples. If the available intensity is large enough significant time savings can be realized using the energy dispersive technique, and its greatest applications are in the study of transient events or short lived samples. To date the sequential mode yields the most accurate data and is most widely used.

In some cases it is advantageous to measure the absorption by monitoring processes that are proportional to the absorption. Two of these, x-ray fluorescence and emission of Auger electrons, are illustrated in Fig. 3.2. Monitoring

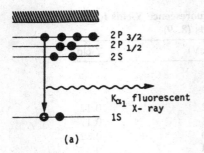

(a)

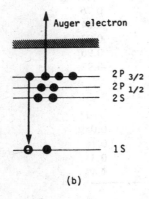

(b)

Figure 3.2. Two decay mechanisms for the core hole left after the absorption of an x-ray photon: (a) K_α fluorescence and (b) Auger electron emission.

these processes is useful when the EXAFS signal is only a small fraction of the total absorption. Isolating the EXAFS from the background in transmission then requires the subtraction of two nearly equal signals, a process requiring very accurate data. In these cases, it is desirable to exhance the signal of interest by using the detector to discriminate it from the background.

Auger emission and x-ray fluorescence are competing processes. Their relative strengths depend on the atomic number of the absorber. In light elements Auger emission is more probable, while for heavy elements fluorescence becomes more likely. For the same element fluorescence is more likely for K-shell holes than for L-shell holes. Table 3.1 gives examples of the fluorescence yields for various elements.

As seen in Fig. 3.2, the fluorescence radiation results from the filling of the core hole generated by the absorption of an x-ray photon. For the K shell this is dominated by the production of K_α radiation. The energy of this radiation is characteristic of the absorbing element and is less than the original exciting

Table 3.1.   K- and L-Shell Fluorescence Yields for Various Elements (8, 9)

Element	Energy (keV)	Yield
	K-edges	
$_{18}$Ar	3.202	0.12
$_{20}$Ca	4.038	0.17
$_{25}$Mn	6.540	0.32
$_{30}$Zn	6.659	0.49
$_{32}$Ge	11.104	0.54
$_{35}$Br	13.474	0.63
$_{42}$Mo	19.999	0.76
$_{48}$Cd	26.711	0.84
$_{64}$Gd	50.240	0.93
$_{80}$Hg	83.103	0.96
$_{92}$U	115.603	0.97
	L_3-edges	
$_{32}$Ge	1.218	0.015
$_{48}$Cd	3.537	0.059
$_{64}$Gd	7.243	0.16
$_{80}$Hg	12.284	0.33
$_{92}$U	21.756	0.49

radiation. The background consists of elastically and Compton scattered radiation, both of which are at higher energies than the fluorescent line. Thus, a suitable energy dispersive detector can discriminate the background from the signal. It will be shown that the fluorescence technique gives better data than simple transmission when the absorption from the element of interest is less than a few percent of the total absorption in the sample.

Application of the fluorescence technique is not as straightforward as for absorption since, in general, corrections have to be made to the measured signal, $\mu'(E)$. For a thick sample it can be shown

$$\mu'(E) = \frac{I_f}{I_0} \sim \frac{\mu(E)\sin\theta}{\mu_T(E)/\sin\theta + \mu_T(E_f)/\sin\phi} \qquad (2)$$

where $\mu_T(E)$ is the total absorption coefficient at energy E, E_f is the fluorescence energy, θ is the entrance angle of the incident x-rays, and ϕ is the exit angle of the fluorescent x-rays. This expression should be integrated over the angles ϕ

subtended by the detector. For dilute samples $\mu_T(E)$ is nearly constant and the correction to $\mu'(E)$ to obtain $\mu(E)$ is small and can be made using tabulated absorption coefficients. Other possible corrections are discussed in the section on experimental difficulties.

For K-edges below about 2000 eV in energy, the probability for fluorescence becomes small, and detection of electrons emitted by the sample is often advantageous. Electron detection also has the property of enhancing the signal from the near surface region making it suitable for surface studies. It is, thus, also desirable for higher-energy edges when enhancement of the surface signal is necessary (10). For comparison purposes a brief discussion of electron detection techniques is given in Section 3.4.3. Chapter 10 should be consulted for a detailed discussion of the strengths and weaknesses of these techniques.

Detection of Auger electrons (11–13) is analogous to fluorescent detection in that a single line characteristic of a particular element is measured. Again, a good discrimination of background is possible, and since the Auger electrons originate within about 30 Å of the surface, a strong surface enhancement is achieved. The Auger yield is related to μ by a relation similar to Eq. (2) with $\mu_T(E_f)$ becoming the attenuation coefficient for the Auger electrons, μ_A. However, in this case $\mu_A \gg \mu_T(E)$, which means I_{Auger}/I_0 is proportional to μ to a good approximation. There is some unexplained amplitude behavior in the original experiments of Citrin et al. (13) and other more recent experiments. Possible causes will be considered in Section 3.4.3.

Other electron detection channels can be used to obtain EXAFS information. An obvious choice is the detection of the primary photoelectron emitted as the x-ray is absorbed. However, the angular distribution of the photoelectrons varies as the photon energy is changed. This introduces an additional variation that causes a different dependence with energy than EXAFS, unless the photoelectron is collected over a full 4π solid angle. Since it is only possible to collect at most a 2π solid angle, such a photoelectron monitoring is unreliable (11, 12). A better technique is to measure the total yield of electrons emitted by the surface. In this case the complicated cascade processes involved in producing the secondary electrons provides a form of angular averaging. Such total yield (14, 15) and related partial yield (16) measurements have been shown to give fine structure analogous to EXAFS, although as discussed in Chapter 10 the amplitudes show some deviations. It remains to be demonstrated that this fine structure contains the full information of EXAFS, but it certainly contains much useful information, especially concerning the near neighbor distances. Also, uncertainties in absolute amplitudes do not rule out measuring the amplitude dependence on polarization from which important coordination information can be inferred.

Electron detection is not the only method of enhancing surface sensitivity. Since the index of reflection for x-rays is slightly less than 1, at glancing incidence angles they will undergo total external reflection. When this occurs the x-rays penetrate only 20–30 Å, and it was first pointed out by Parratt (17) that this gives the potential of using total external reflection for surface studies. A major advantage is the greater penetration ability of x-rays, which means a vacuum environment is not required. This opens up new experiments such as surfaces in gaseous environments, an important advantage for studies such as chemical reactions on surfaces. Similarly the reflection can be made to occur at a solid–solid interface if there is a significant density difference, allowing interface structures to be probed.

As with bulk EXAFS the signal can be measured either by monitoring the reflected x-rays (17–19) or the fluorescence signal (20). The signals are larger for the reflected beam, but recent work (20) indicates that fluorescence detection is more useful for surface EXAFS studies. To understand why it is necessary to consider the reflection process in greater detail the index of refraction for x-rays can be written $n = 1 - \delta - i\beta$ where the absorption coefficient is contained in β. The reflectivity, however, depends on n as given by the Fresnel equations (19, 21). Therefore, to extract μ from the measured reflectivity it is necessary to determine δ. It was found by Martens and Rabe (17) that the EXAFS oscillations also show up in δ, which makes a determination of β a difficult task. A proper experiment requires a detailed measurement of the reflectivity both as a function of energy and angle.

The fluorescence signal, on the other hand, is directly proportional to the probability for absorption. The absorption probability is determined by the strength of the evanescent wave that penetrates into the substrate (22). Below the critical angle the evanescent wave penetration depth is typically 30 Å, but its amplitude depends on the incident angle, θ. This can be determined by matching the evanescent wave field to the standing wave field set up by the incoming and reflected beams. It is found that for an ideal surface (100% reflectivity) the evanescent wave amplitude varys from 0 at $\theta = 0$ to twice the incident amplitude for $\theta = \theta_c$, where θ_c is the critical angle. The absorption probability depends on the intensity of the evanescent wave, and is, therefore, enhanced by a factor of four near θ_c over the case where no reflection is occurring.

These points are illustrated in Fig. 3.3, which shows data taken for approximately monolayer gold films evaporated onto glass and silver substrates. The reflectivities are plotted in Fig. 3.3a. Since the gold films are thin the reflectivity is determined by the substrate, and the more dense silver substrate has a larger critical angle. In Fig. 3.3b the fluorescent and background signals are plotted as a function of angle. As expected the fluorescent signal increases up to the critical angle at which point it drops by about a factor of three. The enhancement in this case is not fourfold because the surface is not ideal. Above the critical

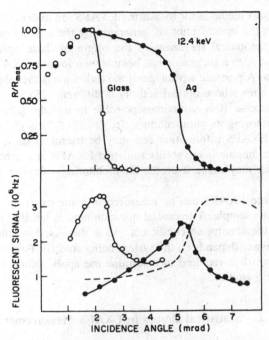

Figure 3.3. Results for glancing incidence fluorescence detection of gold monolayers on glass and silver surfaces. (*a*) The angle dependence of the reflectivity at 12.4 keV. Since the gold is thin these curves are determined by the substrate. (*b*) The angle dependence of the fluorescence signals for the silver (solid points) and glass (open points) samples. The dashed curve is the background level for the silver substrate. For these curves the ion chamber currents have been converted to equivalent counting rates.

angle the incoming beam begins to penetrate and the scattered background increases, but the signal to background ratio is still respectable. With these counting rates good EXAFS spectra can be obtained in a few seconds per point.

In this example of monolayer films, the problem of converting the reflected signal into an absorption is reduced because the reflectivity and δ are dominated by the substrate. However, this means that the change in reflectivity in scanning over the gold L_3-edge is small reducing the signal to noise. A more serious problem, however, is that small changes in the substrate reflectivity with energy will dominate the EXAFS, effectively rendering it unusable. It was concluded that because of these problems fluorescence detection is always better than monitoring the reflected signal for determining the EXAFS in the glancing incidence case (20).

Some additional methods for measuring EXAFS are under development. The first is to detect the optical photons generated by the absorption of the x-rays (23, 24). The theoretical treatment of the origin of these optical photons is complicated, but such a technique has been shown to give EXAFS information in some samples. A possible advantage of the technique is the selective detection of inequivalent sites whose optical activity is different (25).

In solution studies it is sometimes possible to use the solution itself as a detector by monitoring its photoconductivity (26, 27). This technique has been shown to give EXAFS information and may be useful in some cases. An advantage of the technique is the ability to detect EXAFS in a nontransmissive or black cell, which may make it useful for monitoring solutions in complicated environments.

Finally, surface EXAFS can be measured in some cases by monitoring the ion yield from the sample. The crucial question here is whether the ion yield is proportional to the absorption coefficient. As is discussed in Chapter 10 it appears that the answer depends on the system being studied. Further experimental and theoretical work is required to determine the applicability of ion yield detection to EXAFS studies.

3.3.2. Statistical Errors in EXAFS Measurements

Fundamental to any EXAFS measurement technique are the statistical errors due to the finite number of photons available. The statistical errors in an absorption measurement are discussed in detail by Rose and Shapiro (28) for a variety of cases. For EXAFS measurements the primary considerations are the amount of signal used to determine the incoming beam intensity I_0 and the sample thickness μx. If a fraction f of the incoming beam is used to determine the intensity of the incoming beam, then the intensity hitting the sample is $(1 - f)I_0$ and the transmitted intensity is $(1 - f)e^{-\mu x}I_0$. Thus, the measured ratio is

$$R = \frac{f}{(1 - f)} e^{-\mu x} \tag{3}$$

For EXAFS the optimum choice of f and μx maximizes the accuracy in measuring the changes in R due to small changes in μ. The important quantity is

$$S = \frac{dR}{d\mu} = xR$$

The statistical noise N in R gives

$$\frac{N}{S} = \frac{1}{x I_0^{1/2}} \left(\frac{1}{f} + \frac{e^{\mu x}}{(1-f)} \right)^{1/2} \tag{4}$$

Minimizing N/S with respect to x and f gives $f = 0.24$ and $\mu x = 2.6$. However, as will be shown in the next section, other considerations often make it desirable to use a smaller μx. In this case the optimum f becomes

$$f = \frac{1}{1 + e^{\mu x/2}} \tag{5}$$

For fluorescence measurements the situation is similar. In this case the fluorescent signal is $p I_0 (1 - f)$ where p is the probability of detecting a fluorescent photon for each photon absorbed in the sample. It depends on such factors as the fluorescent efficiency, the probability of a fluorescent photon escaping the sample, and the detector solid angle. The optimum f becomes

$$f = \frac{1}{1 + p^{-1/2}} \tag{6}$$

Since p is usually only a few percent, the optimum f for fluorescence is smaller than for transmission measurements. This is apparently a problem when it is desired to apply both techniques simultaneously. Fortunately, the minimum in N/S is rather broad and an intermediate f can be used without significantly degrading the statistics of either technique.

These statistical considerations can also be used to compare the fluorescence detection technique with absorption. Simplifying Eq. (2) the error in a fluorescence measurement can be estimated as

$$\frac{\delta \mu}{\mu} = \left(\frac{\mu_T + \mu_F}{\alpha I_0 \mu} \right)^{1/2} \tag{7}$$

where μ_T is the total absorption coefficient, μ_F is the absorption coefficient for the fluorescent photons, and α is the proportionality constant that is primarily composed of the fluorescent yield and the collection efficiency of the detector. The corresponding error for a dilute absorption measurement is

$$\frac{\delta \mu}{\mu} = \left(\frac{2}{I_0} \right)^{1/2} \frac{\left(e^{\mu_T x/2} + 1 \right)}{\mu x} \tag{8}$$

The two techniques are equal when

$$\mu x = \frac{2(e^{\mu_T x/2} + 1)^2 \alpha}{(\mu_T + \mu_F)x} \tag{9}$$

For the iron K-edge the fluorescent yield is 0.3 and a typical x-ray filter detector (see the following section) might collect approximately 15% of 4π and attenuate the fluorescence signal by a factor of 2. This gives $\alpha \approx 0.022$ and if $\mu_T x \approx \mu_F x = 2.6$ the two techniques have equal signal to noise ratios when $\mu/\mu_T = 0.07$. Thus, ignoring a small background contribution that makes it through to the detector, in this case the fluorescence technique is advantageous when $\mu x < 0.18$.

The background in a fluorescence experiment has two main contributions: scattering and photoelectron induced bremsstrahlung (29, 30). The most important is scattering, which for a single electron is described classically by the Thomson formula (31).

$$d\sigma = \frac{3\sigma_T}{8\pi} \sin^2 \psi \, d\Omega \qquad \sigma_T = \frac{8\pi}{3}\left(\frac{e^2}{mc^2}\right)^2 \tag{10}$$

where ψ is the angle between the polarization direction and the direction of observation. When quantum effects are included the total scattering is the same, but it has both a coherent and an incoherent component. The incoherent component is the Compton scattering and is shifted in wavelength by

$$d\lambda = 0.024(1 - \cos \theta) \text{ Å} \tag{11}$$

where θ is the angle between the beam direction and the scattering direction. For atoms the angular distribution of the coherent radiation is modified by interference effects, and is described by the atomic scattering factor tabulated in ref. (32). However, near $\psi = 90°$, which is the usual geometry, equation (10) is a good estimate of the total scattering with σ_T replaced by $\sigma_0 = Z\sigma_T$ where Z is the atomic number. Then the scattering distribution can be integrated to find the total cross section for scattering into the detector. For a circular detector of radius R at a distance D from the sample the result for the horizontal and vertical polarization components is

$$\sigma_h = \frac{\sigma_0}{4} (\cos^3 \varphi - 3 \cos \varphi + 2)$$

$$\tan \varphi = \frac{R}{D} \qquad (12)$$

$$\sigma_v = \frac{\sigma_0}{8} (4 - 3 \cos \varphi - \cos^3 \varphi)$$

For small φ the scattering is dominated by the vertical polarization component.

The sample bremsstrahlung has been estimated by Goulding and Jaklevic (30):

$$N(E) = 2.5 \times 10^{-6} ZI_0\sigma_t \frac{(E - E_x)}{E_x} \qquad (13)$$

where Z is the atomic number of matrix, I_0 the incident flux, σ the photoelectron cross section at excitation energy E in $cm^2 \ g^{-1}$, t the sample thickness in g cm^{-2}, and E_x the observation energy. The total background is obtained by integrating over the energies accepted by the detector. While small compared to scattering, since the bremsstrahlung appears at the fluorescence energy, it can never be entirely eliminated and represents an ultimate limit on detectability of the fluorescent signal.

To discriminate the background from the fluorescent radiation, some degree of energy resolution is needed. Early experiments (32) used solid state detectors that generally do a good job of discriminating the fluorescent signal from scattered radiation. Their chief drawback is their low counting rate, which is limited to ~40,000 Hz for the total signal. If, as often occurs, the background is many times the signal then the counting rate for fluorescent photons is quite small. Use of many solid state detectors in parallel can overcome this, but the expense and complication quickly became unreasonable.

A better technique is to remove the background before it reaches the detector. Two methods for doing this, shown schematically in Fig. 3.4, are a crystal monochromator (33) and x-ray filters (34). Wavelength selection in the crystal monochromator is achieved by setting the Bragg angle θ. This provides excellent wavelength discrimination, and with suitable designs the wavelength setting can be changed by adjusting the crystal radius (see Chapter 4, Section 4.5.2.2). However, much of the available signal is wasted, since the diffraction efficiency is in the range 10–50% for suitable crystals, and the solid angle subtended by present designs is about 1.5% of 4π (33, 35). Detectors designed to collect significantly larger solid angles will likely require very expensive crystal optics and lack tunability. They will be applied only for specialized experiments on a few absorption edges and not be available for "typical" experiments.

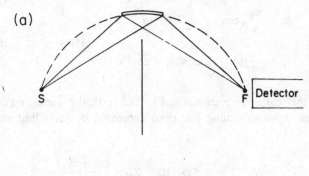

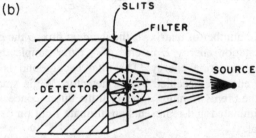

Figure 3.4. Two schemes for reducing the scattered background in a fluorescence experiment. (*a*) Bragg reflection is used for energy selection. To maximize the collection real detectors are cylindrically symmetric about the line SF. (*b*) Filter-slit combination. The fluorescence x-rays from the filter (dashed lines) are attenuated by the slits while the fluorescence from the sample is not.

X-ray filters use x-ray absorption edges to selectively attenuate the scattered radiation. As shown in Fig. 3.5 this is possible since the background radiation, which consists primarily of elastically and Compton scattered incident radiation, is of higher energy than the fluorescent signal. By choosing the appropriate filter, the absorption edge can be placed between the background and fluorescent energies preferentially absorbing the former. Some examples of suitable filters are shown in Table 3.2. The important parameter is the ratio of μ for the background to μ for the signal that is called R in the table. Most of the filters are K-edges for the next lighter element. However, for elements lighter than vanadium this is not possible, and suitable L-edges must be used. For calcium the indium L_3-edge is appropriate, and it is seen that the R factor is nearly as good as for the K-edge filters.

The main difficulty with filters is that radiation absorbed in them is reemitted as fluorescent radiation that is relatively unattenuated by the filter. However, as shown in Fig. 3.4, a suitable set of Soller-type slits can block most of this filter

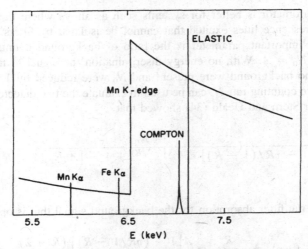

Figure 3.5. An iron fluorescence experiment using a manganese filter for background suppression. The absorption for the Compton and elastic peaks is much larger than for the iron K_α signal.

fluorescence and allow the ideal filter characteristics to be approached (34). With filters, a large collection efficiency is possible, and changing the energy setting is simply accomplished by changing filters. Although the background discrimination is not perfect, for many applications this is more than offset by the large collection efficiency.

The choice of filter or monochromator depends on the application at hand.

Table 3.2. Examples of Appropriate Filter Materials for Various Edges^a

Element	E_K (eV)	E_F (eV)	Filter	E_{Filter} (eV)	R
V	5,465	4,952	Ti	4,965	6.5
Fe	7,112	6,403	Mn	6,540	6.6
Cu	8,979	8,047	Ni	8,333	5.5
Br	13,474	11,923	Se	12,658	4.2
Mo	19,999	17,478	Zr	17,998	4.5
Ca	4,038	3,691	In-L_3	3,730	3.2
Pt L_3	11,564	9,441	Ga	10,367	4.4

^{a}The factor R is the ratio of the absorption coefficient for the elastically scattered background at an energy of 200 eV above the edge to the absorption coefficient for the fluorescence radiation. For the background due to Compton scattering, R is slightly larger. E_K and E_F are the K-edge and fluorescence energies, respectively, and E_{Filter} is the filter edge energy.

The monochromator is better for systems such as alloys where there may be several fluorescence lines excited that cannot be isolated by filters. For other systems the important parameter is the ratio of background counts to signal counts, $N_b/N_0 = A$. With no energy discrimination the signal to noise is the same as if no background were present and N_0 were reduced by $1/(1 + A)$. This effective counting rate N_e can be used to evaluate the two detector systems. For the filter Stern and Heald (34) showed that

$$N_e = \frac{N_0}{A \left[1 + \eta R/(1 - R) \right] e^{-\mu x (1 - 2/R)} + 1 - \left[A\eta R/(1 - R) \right] e^{\mu x/R}}$$

(14)

where μx is the filter absorption for the background signal that is optimum for

$$(\mu x)_1 = \frac{R}{R - 1} \ln \frac{A \left[1 + (\eta R/1 - R) \right] (R - 2)}{1 - A \left[\eta R/(1 - R) \right]}$$

(15)

η is the probability that a photon absorbed in the filter is emitted as a fluorescence photon in a direction that can reach the detector. For an ideal filter $\eta = 0$.

For the crystal monochromator, the background is greatly reduced and will be ignored in this discussion. Then N_e is just the detection efficiency for fluorescence photons:

$$N_e = N_0 \, \epsilon \, \frac{\Omega_c}{\Omega_f}$$

(16)

In this expression, N_0 is the same as for the filter, with Ω_c and Ω_f taking account of the solid angle collected by the crystal and filter, respectively. ϵ is the diffraction efficiency. For existing systems, $\Omega_c/\Omega_f = 0.1$, and taking $\epsilon \approx 0.3$ gives $N_e = 0.03 \, N_0$ for the crystal monochromator. The filter-slit system is superior when its N_e is larger. Table 3.3 compares the two systems for various cases. $\eta = 0.02$ is characteristic of existing filter systems using a line focus for the x-rays. If the x-rays are focused to a point, as is necessary for the crystal detector, then it should be possible to make $\eta \approx 0.004$. $R = 5$ and $R = 3$ are characteristic of K-edge and L-edge filters, respectively.

For most cases the filter is superior for A values up to several hundred, especially since in practice even the crystal monochromator allows some background through. To convert these A values into concentrations, an example is useful. For an iron protein sample with an iron concentration of about 3×10^{-3}

Table 3.3.   Comparison of the Crystal Monochromator and Filter-Slit Detection Systems[a]

Case	R = 5		R = 3	
	$\Omega_c = \Omega_F$	$\Omega_c = 0.1\Omega_F$	$\Omega_c = \Omega_F$	$\Omega_c = 0.1\Omega_F$
Perfect filter $\eta = 0$	9	1.28×10^5	2	274
Point focus $\eta = 0.004$	8	885	2	101
Line focus $\eta = 0.02$	6	271	2	44

[a]When A, the ratio of background to signal, equals the numbers given for each case, the two systems are equal. If A is smaller, the filter is superior. For existing systems $\Omega_c = 0.1\Omega_F$. A 30% diffraction efficiency for the crystal is assumed. η is the probability that a photon absorbed in the filter will ·roduce an unwanted fluorescent photon in the detector.

M (1 iron atom/3×10^5 molecular weight) it was found experimentally that $A = 20$. Thus, in this case the two detectors are equivalent at a concentration of about 10^{-4} M. For heavier elements, A is smaller for the same concentration because the x-rays are more penetrating. This capability means that the filter system is superior for most of the chemical and biological systems of current interest.

For biological systems, radiation damage is often a limiting factor. It can be reduced by spreading the x-rays over a larger amount of sample by using a line focus rather than a spot focus. This is not possible with the crystal system, since it requires a point focus making the radiation damage problem more severe in the crystal case.

Similar statistical consideration can be used to compare electron detection with fluorescence and absorption measurements. There are two categories of experiments. The first is the measurement of low-Z elements for which the fluorescent yields are small. In this case electron detection is definitely superior to fluorescence for both bulk and surface systems. For bulk systems, however, absorption measurements can also be made if suitably thin samples are prepared. To date experimental problems have hindered these types of absorption measurements.

For the case of heavier elements (e.g., $Z \geq 20$) absorption or fluorescence detection is obviously preferred for bulk systems and the only question concerns the measurement of surface layers. It is shown in Chapter 10 that electron yield techniques are generally superior to standard fluorescence measurements for surface layers. However, if use can be made of glancing incidence then the situation is different. As shown in Section 3.1 the signal is enhanced by at least a factor $1/\sin\theta$, and possibly more if the surface is smooth enough for standing waves to be set up. The background signal is due to scattered photons that have the same absorption probability as the incoming photons. Thus, the increase in the background in going to glancing angles is only a small factor, and may actually decrease in the total external reflection regime. The net result is an increase in the signal to noise of about two orders of magnitude.

The situation is different for electron detection. It was found that at glancing angles the signal increased, but the background increased at a faster rate (15). This is due to the escape depth of the background electrons being much shorter than the incoming photons. As the angle is decreased, more photons are absorbed within an escape depth of the surface generating electrons that contribute to the background. Also the scattering cross sections for photons are typically 50–100 times smaller than the absorption cross section. Since a substantial fraction of absorption events generate electrons that contribute to the background, the background rates for electron detection are much greater than for photon detection when glancing angles are used.

To illustrate the comparison a concrete example is useful. Consider the example shown in Fig. 3.3 of a monolayer gold on silver. The fluorescence and Auger yields are roughly the same. In glancing incidence the signal is enhanced by $1/\theta$ or about 200. However, a focused beam of 1 mm^2 would require a 200-mm long sample to use the entire beam. A more reasonable value is 20-mm long, which means the enhancement is about 20. If the sample can be polished smooth then there is an additional enhancement of about 2–4 and a reduction of the background to near zero. Also, in this energy range it is easy to construct ion changer fluorescence detectors to collect 20–30% of 4π while for Auger detectors the solid angle is significantly less. Thus, the net enhancement is from 40 to 250 depending on the exact conditions. Even if the sample is only a few squat millimeters the enhancement is significant. For the more commonly used total yield technique the signals are higher. If we assume every absorption event results in a detected electron the signal is enhanced over the Auger signal by about a factor of 10–20 from the increased solid angle and a factor of 2–3 from the increased yield. However, the background is also increased to 5–10 times the signal that gives a net increase of about 10–20. This is still less than the enhancement of the fluorescence signal for sample sizes greater than a few millimeters.

Clearly, statistical considerations are not the only factor in the design of an experiment. Equally important are the sample preparation and environment that may strongly influence the choice of a detection technique. However, the statistics discussed in this section represent the ultimate limit on the quality of data that can be obtained.

3.3.3. Properties of X-Ray Detectors

In the preceding discussion idealized detectors were assumed. Real detectors can introduce additional noise and nonlinearities. Thus, successful experimental design requires a knowledge of detector properties. This section summarizes some of the important properties of x-ray detectors. Electron detectors are discussed in Chapter 10.

The high intensities encountered at synchrotron experiments often necessitate the use of integrating detectors. The most useful integrating detector is the gas ionization detector. A good introduction to the design and construction of ionization chambers is given by Rossi and Staub (37). For EXAFS applications the most important properties are the noise and nonlinearities introduced by the chamber. Noise in an ion chamber comes chiefly from the electronics and microphonics. Microphonics are vibrations in the collecting plates owing to ambient vibrations. This changes the capacitance of the chamber and current flows in response. Microphonics are easily eliminated by employing rigid construction techniques. If thin metallized plastic films are used as x-ray transparent collecting plates, they should be mounted under tension in a stretched condition (38).

With standard electronics it is easy to obtain a current noise of about 10^{-14} A in a 1-s integration time. A typical gas mixture production of an ion pair requires about 30 eV. Therefore, 10-keV photons will produce about 300 ion pairs for each photon absorbed in the chamber. Thus, 10^{-14} A corresponds to the signal produced by about 200 photons s^{-1}. The same statistical noise is produced by a signal of 4×10^4 photons s^{-1}. When the signal is significantly larger the amplifier noise can be ignored and the dominant noise source is statistical.

At high signal levels the ion chamber response can become nonlinear. This can be different for the I and I_0 chambers and can result in imperfect cancellation of intensity fluctuations. An excellent discussion of the problem is given by Boag (39). To some extent it depends on the design of the chamber. The two most common types are shown in Fig. 3.6. Since edge effects can be ignored when the x-rays are transmitted through the plates, this type has been most extensively studied. Boag expresses the efficiency of an ion chamber by the fraction f of ion pairs actually collected by the plates, and gives as an approximate expression

$$f = \frac{1}{1 + E^2/6}, \qquad E = m \frac{d^2 \sqrt{q}}{V} \qquad (17)$$

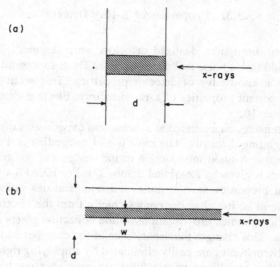

Figure 3.6. Two ion chamber geometries.

where d is the plate spacing in cm, q the charge per unit volume produced in the gas in esu cm^{-3}, V is the electric field in volt cm^{-1}, and m is characteristic of the gas mixture used. For air m is approximately 37 (40). This expression assumes that the negative charge is carried by ions and somewhat underestimates f for gases such as Ar where the negative charge is carried by free electrons. Nevertheless it is a useful guideline.

For good linearity f should be larger than .99, which implies $E < 0.25$ or $d^2\sqrt{q}/V < 0.007$. For unfocused beams the ionization intensity is of order 10^3 R s^{-1} (1 R = 1 esu cm^{-3} ionization or 2×10^9 ion pairs cm^{-3}). If $d = 1$ cm, then fields of 5000 V cm^{-1} are required, which are barely possible.

The principal source of nonlinearity is the recombination of ions before they reach the plates. This is naturally more probable for higher ion production rates. For the second type of ion chamber, in which the x-rays pass between the plates, the situation is more favorable. Recombination can only occur when the ions remain in the ionizing region of width w. Ignoring end effects, this case can be approximated by replacing d with w in the previous results. For a typical case $w = 0.1$ cm, which means fields of order 50 V cm^{-1} are required for unfocused beams. These are easily obtainable. For focused beamlines ionization rates could be as high as 10^6 R s^{-1} at the more intense facilities. These would require fields in excess of 1500 V cm^{-1} to ensure linearity.

For dispersive EXAFS techniques spatial resolution is also required. The

Table 3.4. Properties of Some Common X-Ray Detectors[a]

Detector	Maximum Counting Rate	ΔE (FWHM) at 10 keV	References
Solid state			
Si, Ge	4×10^5	200 eV	(45)–(47)
HgI$_2$	4×10^5	400 eV	(48), (49)
Gas proportional	4×10^6	1300	(50)–(52)
Scintillation			
Inorganic	4×10^5	3500	(50), (53)
(NaI)			
Organic	4×10^7	8000	(53)–(55)
Gas scintillation proportional	2×10^5	700	(56)–(58)

[a]In most cases the resolution and counting rates are related and the listed values represent typical operating conditions.

simplest detector in this case is film. It offers excellent spatial resolution, and when properly calibrated, accurate intensities. The major problem is an intrinsic fog level equivalent to 2×10^5 photons mm^2 (38). This degrades the sensitivity to low-level signals. To overcome this problem solid state detectors based on self-scanning photodiode arrays have been developed (41–44). The x-ray photons can be detected directly or converted first to light that is coupled to the diode array through a fiber optic face plate. The noise level in such devices can be as low as a few photons per pixel. Time resolved information is also easy to obtain by appropriate gating of the electronics.

When energy discrimination is required, as, for example, in a fluorescence detector, single photon counting is necessary. Table 3.4 lists the important characteristics of some common x-ray detectors. The reader is referred to the original references for details of the operation and construction of the listed detector types. Of the various types only the solid state and gas scintillation proportional detectors have resolutions sufficient for separating a fluorescence line from the background. However, as discussed in Section 3.3.2, a filter-slit system can be used to attenuate the background. The real need for energy resolution in the detector is to isolate interfering fluorescence lines from a multicomponent sample. In many cases a gas proportional counter has sufficient resolution for this task. Only for the case of neighboring elements is the full resolution of a solid state detector required.

For pulse counting detectors the primary cause of nonlinearity is dead time losses. As the counting rate increases the chance of pulses overlapping increases, and since these pulses are rejected by energy discriminating electronics the detector efficiency falls. In the simplest case of random pulses the effective counting rate is

$$N' \approx N(1 - N\tau) \tag{18}$$

where N is the true counting rate and τ the dead time of the detector system. For synchrotron radiation the source is not continuous, but takes the form of brief pulses repeating at intervals of approximately 1 μs for single bunch operation. In this case counts are rejected if two counts occur during a single radiation pulse. Various cases are considered in papers by Arndt (59) and Wescott (60). It is found that for single bunch operation there can be a significant reduction in the counting rates given in Table 3.4, while for multibunch operation the detector properties are likely to dominate.

3.4. EXPERIMENTAL DIFFICULTIES

The most difficult parameter to obtain in an EXAFS experiment is accurate amplitude information. A variety of effects can distort the amplitudes severely while leaving the phase information relatively undisturbed, and it is the amplitude information that is often crucial in developing a complete understanding of an unknown system. The amplitude contains information about coordination number, site disorder, and, when polarization studies are made, the site symmetry. As the conditions necessary for amplitude transferability become better understood (4), the experimental limitations on the accuracy of EXAFS amplitudes become more important. Fortunately, with proper experimental design inaccuracies in EXAFS amplitudes can be recognized and corrected.

3.4.1. Thickness Effects in Absorption Measurements

Most of the amplitude distorting factors in an absorption measurement come under the general heading of thickness effects (61-64). Basically a thickness effect occurs whenever some part of the incoming beam is not attenuated by the sample, through pinholes in the sample, or be at a different energy from the primary radiation. This "leakage" becomes a larger fraction of the total signal as the sample becomes thicker and, thus, the absorption signal appears to depend on sample thickness, which is the origin of the term thickness effect.

If a fraction $\alpha(E)$ of the incoming radiation is leakage, then the measured absorption coefficient is (63)

$$\mu'(E)x = \ln\left(\frac{1 + \alpha(E)}{e^{-\mu(E)x} + \alpha(E)}\right) \tag{19}$$

EXAFS is a small variation, $\delta\mu(E)$, in $\mu(E)$ about a smooth background $\mu_0(E)$ and expanding $\mu'(E)$ gives

$$
\mu'(E)x = \ln\left(\frac{1 + \alpha(E)}{e^{-\mu_0(E)x} + \alpha(E)}\right) + \frac{\delta\mu(E)x}{1 + \alpha(E)e^{\mu_0(E)x}}
$$

$$
- \frac{1}{2}\frac{\alpha(E)e^{\mu_0(E)x}}{\left[1 + \alpha(E)e^{\mu_0(E)x}\right]^2}\left[\delta\mu(E)x\right]^2 \qquad (20)
$$

If the EXAFS $\chi(k)$ is normalized with respect to the edge step $\Delta\mu_0$, then the leakage reduced the EXAFS amplitude by a factor

$$
\frac{\chi'(k, x)}{\chi(k)} = \frac{\Delta\mu_0 x}{1 + \alpha(k)e^{\mu_0(k)x}}\left[\ln\left(\frac{1 + \alpha(k_0)}{e^{-\Delta\mu_0 x} + \alpha(k_0)}\right)\right]^{-1} \qquad (21)
$$

where k_0 refers to the edge value. This amplitude reduction is shown in Fig. 3.7 for various leakage levels. Note that for all levels of leakage the effect is small for thin samples. Also, if $\Delta\mu_0 x = 2.6$, the statistically optimum value, then serious amplitude reduction can occur for small leakage. If leakage is discovered after the fact Goulon et al. (64) have given an analysis procedure for correcting the measured absorption coefficient.

Solid samples are often made from powders. For concentrated materials sample thicknesses are approximately 10 μm, which is also the grain size for typical powders. This means that powdered samples can be quite nonuniform resulting in distortion of the EXAFS signal. For spherical particles of diameter D the EXAFS is reduced by a factor (65)

$$
\frac{\chi'}{\chi} = \frac{2 - (\mu_2 D)^2 e^{-\mu_2 D}/\left[1 - (1 + \mu_2 D)e^{-\mu_2 D}\right]}{\Delta\mu_0\mu_2 \ln\left\{\mu_2^2\left[1 - (1 + \mu_1 D)e^{-\mu_1 D}\right]/\mu_1^2\left[1 - (1 + \mu_2 D)e^{-\mu_2 D}\right]\right\}}
$$

$$(22)$$

where $\Delta\mu_0 = \mu_2 - \mu_1$ is the edge step. Some typical results are shown in Fig. 3.8 for various materials. Measurements confirmed the general validity of Eq. (22), and it was concluded that the absorption edge step of a single layer of powder should be less than 0.1 to minimize the distortion. If this cannot be arranged, alternative techniques of sample preparation should be considered.

After a uniform sample is obtained a thickness effect may still persist. Most monochromators pass multiples of the fundamental wavelength. These are unaffected by the edge steps and cause behavior similar to pinholes. For a laboratory source the harmonics can be eliminated by operating the tube at voltages

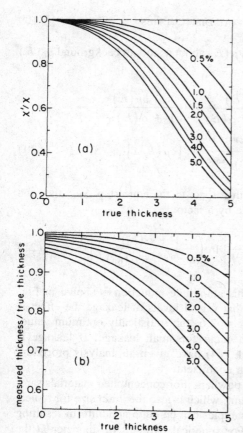

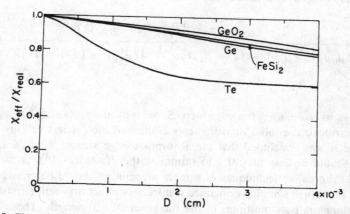

Figure 3.7. The effect of leakage on (a) χ'/χ and (b) the measured thickness. The curves are for various leakage levels expressed as a percentage of the incoming beam. [From ref. (63).]

Figure 3.8. The reduction of $\chi(k)$ due to a finite particle size for several materials. [From ref. (65).]

110

below the threshold for harmonics production. This is not possible at a synchrotron source, but two other techniques can be applied. One possibility is to design the monochromator to eliminate harmonics (66). Typically, this has not been done, but rather other considerations have resulted in the use of double crystal monochromaters. These can achieve some harmonic reduction by detuning the monochromater slightly. Since the reflection width for the harmonic is narrower, the fundamental to harmonic ratio is improved (67). At optimum detuning the fundamental intensity is reduced by about a factor of 2 while the harmonic is reduced by about $100\times$. The second technique is to employ a grazing incidence mirror. If the incidence angle is properly chosen the harmonics will be above the mirror cutoff energy and strongly suppressed. Either of these techniques can reduce the harmonic content to approximately 10^{-3}. This is satisfactory for EXAFS studies, but care still must be exercised when measuring strong near-edge white lines.

All x-ray monochromators have another property that can lead to a thickness effect. This is the presence of long tails in the monochromator transmission function (61, 68). Although weak, the tails can extend into the low absorption region below the edge and thereby constitute a significant part of the transmitted signal for thick samples. Again this is primarily a problem when measuring strong near-edge features.

Another type of leakage comes from the sample itself. When an x-ray is absorbed it can be reemitted as a fluorescent photon. If this photon is detected then the measured absorption is reduced. If the simplifying assumption is made that the fluorescent photons that reach the detector are emitted normal to the surface (i.e., the detector solid angle is small) then the leakage is given by

$$\alpha(k) = \Omega_d/4\pi \, \epsilon \, \frac{\mu(k)}{\mu(k) - \mu_f} \left(e^{-\mu_f x} - e^{-\mu(k)x} \right) \tag{23}$$

where Ω_d is the detector solid angle, ϵ is the fluorescent efficiency, and μ_f is the absorption coefficient for fluorescent photons. This result can be inserted into Eq. (21) to find the EXAFS reduction factor. For a concentrated sample a reasonable assumption is $\mu_f = \mu(k_0)/6$. Figure 3.9 shows the calculated reduction factors for various values of $\Omega_d \epsilon/4\pi$ and the assumption that $\mu(k) = \mu(k_0)$.

Again this can be a serious effect for thick samples. For example, a typical ion chamber detector might have an opening 1 cm $\times$ 5 cm. If it is 2.5 cm from the sample and $\epsilon = 0.3$ then $\Omega_d \epsilon/4\pi = 0.02$ and the reduction is 10% for $\mu x = 2.6$. The solution, of course, is to make Ω_d small, which is usually quite easy.

From this list of difficulties it might be concluded that accurate EXAFS

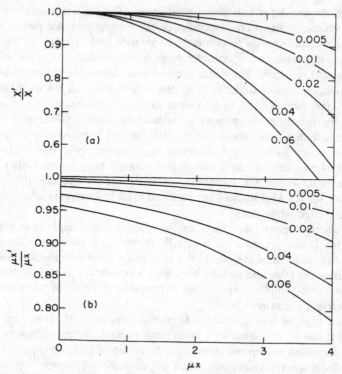

Figure 3.9. The reduction in (a) $\chi(k)$ and (b) μx from sample fluorescence for various values of $\Omega_d/4\pi$.

amplitudes are unlikely in an absorption measurement. This is true if precautions are not taken. However, all of these problems can be minimized if care is exercised in designing the experiment, and time is taken to characterize the apparatus. Unfortunately, the time pressures of carrying out experiments at synchrotron radiation facilities has sometimes resulted in unnecessary distortions of the resulting EXAFS amplitudes. This situation should improve as more synchrotron time becomes available and the experimental stations become better characterized.

All of the effects described are functions of sample thickness. Thus, thickness effects can always be ruled out by measuring the EXAFS as a function of thickness and extrapolating to zero thickness. This is tedious, however, and usually unnecessary. Empirically, it is found that if reasonable precautions are taken, the distortions usually become negligible when $\Delta \mu x < 1.5$. In this case the experimental errors in the amplitude determination should be $< 5\%$.

3.4.2. Fluorescence Measurements

For fluorescence EXAFS measurements thickness effects are unimportant both because the sample $\Delta\mu x$ is small and because leakage through the sample has little affect on the fluorescence signal. On the other hand, other corrections apply. From Eq. (2) it is seen that the measured $\mu'(E)$ must be corrected for the self-absorption in the sample to obtain the true $\mu(E)$. Also, if I_0 is measured by a partially transmitting ion chamber, the energy dependence of the chamber efficiency must be taken into acount. As the energy is increased the chamber becomes more transparent making I_0 appear to be smaller. This enhances the apparent signal. Both of these affects are relatively small (5–15%) and can be removed using tabulated absorption coefficients. The corrections are particularly important to make when the data is to be compared to standards taken by absorption measurements.

3.4.3. Electron Detection Measurements

As mentioned earlier EXAFS experiments comparing electron detection results with absorption results often find different amplitudes. These may be due to effects similar to those found in fluorescence. Auger detection is most analogous to fluorescence detection in that the detected electrons always have the same energy. Since the mean free path for the electrons is very short, the correction in Eq. (2) that is needed for fluorescence is a constant and thus not needed. The correction for the energy dependence of the I_0 is still needed, and for electrons it may be more difficult to apply. In this case the I_0 monitor is often a partially transmitting metal grid from which the electron yield is recorded. The gain of such a detector is likely to have an energy dependence requiring a correction if the data is to be compared to absorption measurements. Such a correction has apparently not been applied in most previous studies. Only if the standard and unknown samples are measured using the same technique is the correction unnecessary.

When total yield measurements are made there is another possible amplitude distortion from the energy dependence of the yield efficiency. The yield signal has several contributions: direct Auger and photoelectrons, scattered and secondary electrons induced by electrons ejected into the sample, and a small contribution from electrons induced by fluorescent x-rays. This can result in the ejection of several electrons for an absorbed photon. These are emitted simultaneously and would give a single pulse in a detector. As pointed out by Martens et al. (15) if each absorbed photon produces at least one electron that reaches the detector then the counting rate is directly proportional to the absorption. This is likely the case in most experiments.

When high signal levels are achieved the detection of the yield signal is often a current measurement that measures the total number of electrons produced. In this case the constant proportionality of the yield signal to the absorption is not assured. In particular the photoelectron energy and possibly the photoelectron angular distributions are functions of the incoming photon energy. This means the average number of emitted electrons per absorbed photon can also be energy dependent resulting in distortions of the EXAFS amplitudes. For partial yield measurements the window setting influences the relative sensitivity to the photoelectron and Auger derived signals, and the distortion can be modified.

Since these two effects have not been systematically investigated, absolute amplitudes obtained by electron detection techniques should be used with caution. A better technique is to compare relative amplitudes obtained using the same detectors. In particular, polarization-dependent studies can be most useful in determining the site location of surface species (69, 70).

3.4.4. Other Problems

A common feature in many EXAFS measurements is reproducible sharp noise features often referred to as glitches. These are generally due to variations in the intensity from the crystal monochromator. At a particular angle an additional diffraction channel can subtract intensity from the primary beam. If precautions are taken to linearize the detection system, these glitches can often be eliminated (71). The precautions necessary are essentially the same as those needed to eliminate thickness effects. It is also possible, particularly in "channel cut" monochromators to have additional reflections from the monochromator that can enter the sample (72). These can affect the experiments in a manner similar to harmonics. However, since they represent different reflection angles on the crystal, they will appear at heights different from the fundamental and can be eliminated by appropriate slits.

A related problem can occur with single crystal samples. In this case the sample itself can Bragg diffract the beam at particular angles. This gives the appearance of a sharp increase in absorption. The problem is easily eliminated by changing the sample orientation slightly.

It is often found that when data taken at different times are compared, the amplitudes are different. If thickness effects have been eliminated the most likely cause is a different energy resolution for the two experiments. This is a particular problem for higher shell data for which the structure is sharper. Corrections can be made (73) but it is often difficult to determine the energy resolution after the fact. The best practice is to measure samples for direct comparison on the same apparatus, preferably at the same time.

3.5. SUMMARY

This chapter has attempted to develop some simple results to serve as a basis for the design of an EXAFS experiment. Two basic areas were covered: the choice of detection technique and the avoidance of some common experimental problems. The choice of experimental technique is fairly simple. Statistical arguments can be used to assess the relative strengths of the various techniques and also to assess the feasibility of the proposed experiments. The most straightforward and best understood technique is the measurement of the absorption in transmission. Experiments requiring high accuracy amplitudes for success should be designed with this in mind.

The less direct techniques such as fluorescence and electron detection offer important advantages when measuring dilute or surface systems. In both cases there are various possible detection systems. For fluorescence it is simple to evaluate the relative efficiencies of the various detectors. For the electron detection methods the estimates are less certain, and experimental comparisons of the various methods may be required. Also, in the case of total and partial yield measurements, more experimental and theoretical work comparing their relation to the true absorption coefficient is desirable.

A large number of potential experimental problems were discussed. Fortunately, with proper precautions, most can be easily eliminated. The main message here is that experience helps. Experience with EXAFS in general allows one to recognize a problem when it occurs, and experience with the particular apparatus being used is helpful in avoiding problems from the beginning. With the growing availability of EXAFS sources, both in laboratory and synchrotron sources, this should become easier with time.

ACKNOWLEDGMENTS

Portions of this work benefited greatly from collaborations with Professor E. A. Stern and the author is grateful for his guidance. The author also thanks Professor J. Budnick for his comments on the manuscript. This work was supported in part by the U.S. Department of Energy, Division of Materials Sciences, Office of Basic Energy Sciences under Contracts DE-AC02-76CH00016 and DE-AS05-80-ER10742.

REFERENCES

1. B. K. Teo and P. A. Lee, *J. Am. Chem. Soc.*, **101**, 2815 (1979).
2. E. A. Stern, D. E. Sayers, and F. W. Lytle, *Phys. Rev. B*, **11**, 4836 (1975).

3. E. A. Stern, S. M. Heald, and B. Bunker, *Phys. Rev. Lett.*, **42**, 1372, (1979).

4. E. A. Stern, B. A. Bunker, and S. M. Heald, *Phys. Rev. B*, **21**, 5521 (1980).

5. P. J. Mallozzi, R. E. Schwerzel, H. M. Epstein, and B. E. Campbell, *Phys. Rev. A*, **23**, 824 (1981).

6. T. Matsushita, in E. A. Stern (Ed.), *Laboratory EXAFS Facilities—1980*, American Institute of Physics, New York, 1980, p. 109.

7. M. Hida, M. Maeda, N. Kamijo, and H. Terauchi, *Phys. Status Solidi*, **A69**, 297 (1982).

8. M. H. Chen, B. Crasemann, and H. Mark, *Phys. Rev. A*, **21**, 436 (1980).

9. M. O. Krause, *Phys. Rev. A*, **22**, 1958 (1980).

10. E. A. Stern, *J. Vac. Sci. Tech.*, **14**, 461 (1977).

11. P. A. Lee, *Phys. Rev. B*, **21**, 4507 (1980).

12. U. Landman and D. L. Adams, *Proc. Natl. Acad. Sci. USA*, **73**, 2550 (1976).

13. P. H. Citrin, P. Eisenberger, and R. C. Hewitt, *Phys. Rev. Lett.*, **41**, 309 (1978).

14. W. Gudat and C. Kunz, *Phys. Rev. Lett.*, **29**, 169 (1972).

15. G. P. Martens, P. Rabe, N. Schwentner, and A. Werner, *J. Phys. C*, **11**, 3125 (1978).

16. J. Stöhr, D. Denley, and P. Perfetti, *Phys. Rev. B*, **18**, 4132 (1978).

17. G. Martens and P. Rabe, *Phys. Status Solidi*, **58**, 415 (1980).

18. R. Barchewitz, M. Cremonese-Visicato, and G. Onori, *J. Phys. C*, **11**, 4439 (1978).

19. R. Fox and S. J. Gurman, *J. Phys. C*, **13**, L249 (1980).

20. S. M. Heald, E. Keller, and E. A. Stern, *Phys. Lett.*, **103A**, 155 (1984).

21. L. G. Parratt, *Phys. Rev.*, **95**, 359 (1954).

22. R. S. Becker, J. A. Golovchenko, and J. R. Patel, *Phys. Rev. Lett.*, **50**, 153 (1983).

23. F. W. Lytle and D. R. Sandstrom, 5th Annual SSRL User Group Meeting, Oct. 1978, SSRL Report No. 78/09.

24. A. Bianconi, D. Jackson, and K. Monahan, *Phys. Rev. B*, **17**, 2021 (1978).

25. J. Goulon, P. Tola, M. Lemonnier, and J. Dexpert-Ghys, *Chem. Phys.*, **78**, 347 (1983).

26. T. K. Sham and S. M. Heald, *J. Am. Chem. Soc.*, **105**, 5142 (1983).

27. T. K. Sham and R. A. Holroyd, *J. Chem. Phys.*, **80**, 1072 (1984).

28. M. E. Rose and M. M. Shapiro, *Phys. Rev.*, **74**, 1853 (1948).

29. B. Gordon, *Nucl. Instrum. Methods*, **204**, 223 (1982).

30. F. S. Goulding and J. M. Jaklevic, *Nucl. Instrum. Methods*, **142**, 323 (1977).

31. M. W. Woolfson, *An Introduction to X-Ray Crystallography*, Cambridge University Press, Cambridge, 1970.

32. J. M. Jaklevic, J. A. Kirby, M. P. Klein, A. S. Robertson, G. S. Brown, and P. Eisenberger, *Solid State Commun.*, **23**, 679 (1977).

33. J. B. Hastings, P. Eisenberger, B. Lengeler, and M. L. Perlman, *Phys. Rev. Lett.*, **43**, 1807 (1979).

34. E. A. Stern and S. M. Heald, *Rev. Sci. Instrum.*, **50**, 1579 (1979); E. A. Stern, B. A. Bunker, and S. M. Heald, *Phys. Rev. B*, **21**, 5521 (1980).

35. M. Marcus, L. S. Powers, A. R. Storm, B. M. Kincaid, and B. Chance, *Rev. Sci. Instrum.*, **51**, 1023 (1980).

36. J. Stöhr, *J. Vac. Sci. Technol.*, **16**, 37 (1979).

37. B. B. Rossi and H. H. Staub, *Ionization Chambers and Counters: Experimental Techniques*, McGraw-Hill, New York, 1949.

38. E. A. Stern, W. T. Elam, B. A. Bunker, K. Q. Lu, and S. M. Heald, *Nucl. Instrum. Methods*, **195**, 345 (1982).

39. J. W. Boag, *Br. J. Appl. Phys.*, **3**, 222 (1952).

40. Y. Morichi, A. Katoh, N. Takata, R. Janaka, and N. Tamura, Report No. IAEA-SM-222/44.

41. G. Rosenbaum and K. C. Holmes, in *Synchrotron Radiation Research*, H. Winick and S. Doniach (Eds.), Plenum, New York, 1980.

42. L. N. Koppel, *Rev. Sci. Instrum.*, **51**, 1669 (1980).

43. R. C. Gambel, J. D. Baldeschwieler, and C. E. Griffin, *Rev. Sci. Instrum.*, **50**, 1416 (1979).

44. C. S. Borso and S. S. Danyluk, *Rev. Sci. Instrum.*, **51**, 1669 (1980).

45. D. A. Gedke, *X-Ray Spectrosc.*, **1**, 129 (1972).

46. R. L. Heath, *Adv. X-Ray Anal.*, **15**, 1 (1972).

47. F. S. Goulding and D. A. Landis, *IEEE Trans. Nucl. Sci.*, **N5-29**, 1125 (1982).

48. G. C. Hutch, A. J. Dabrowski, M. Singh, T. E. Economu, and A. L. Turkevich, in *Proceedings of the 27th Annual Conference on Applications of X-Ray Analysis*, Denver, 1978.

49. A. Holzer, *IEEE Trans. Nucl. Sci.*, **NS-29**, 1119 (1982).

50. J. Sharpe, *Nuclear Radiation Detectors*, 2nd ed., Methuen, London, 1955.

51. G. Charpak, R. Bouclier, T. Bressoni, and J. Favier, *Nucl. Instrum. Methods*, **62**, 262 (1968).

52. I. Veress and A. Montvai, *Nucl. Instrum. Methods*, **156**, 73 (1978).

53. J. B. Birks, *The Theory and Practice of Scintillation Counting*, Pergamon, New York, 1964.

54. L. A. Erikkson, C. M. Tsai, A. H. Cho, and C. R. Hurlbutt, *Nucl. Instrum. Methods*, **122**, 373 (1974).

55. Z. H. Cho, C. M. Tsai, and L. A. Eriksson, *IEEE Trans. Nucl. Sci.*, **NS-22**, 72 (1975).

56. H. E. Palmer, *IEEE Trans. Nucl. Sci.*, **NS-22**, 100 (1975).

57. D. F. Anderson, T. T. Hamilton, W. H.-M. Ku, and R. Novick, *Nucl. Instrum. Methods*, **163**, 125 (1979).

58. W. H.-M. Ku and R. Novick, in *Low Energy X-Ray Diagnostics—1981*, D. T. Atwood and B. L. Henke (Eds.), American Institute of Physics, New York, p. 78.

59. U. W. Arndt, *J. Phys. E*, **11**, 671 (1978).

60. C. H. Westcott, *Proc. R. Soc. (London)*, **194**, 508 (1948).

61. L. G. Parratt, C. F. Hempstead, and E. L. Jossem, *Phys. Rev.*, **105**, 1228 (1957).

62. S. M. Heald and E. A. Stern, *Phys. Rev. B*, **16**, 5549 (1977).

63. E. A. Stern and K. Kim, *Phys. Rev. B*, **23**, 3781 (1981).

64. J. Goulon, C. Goulon-Ginet, R. Cortes, and J. M. Dubois, *J. Phys.*, **43m** 539 (1982).

65. E. A. Stern and K. Q. Lu, *Nucl. Instrum. Methods*, **212**, 475 (1983).

66. M. Hart and R. D. Rodriques, *J. Appl. Cryst.*, **11**, 248 (1978).

67. D. Mills, and V. Pollock, *Rev. Sci. Instrum.*, **51**, 1664 (1980).

68. D. M. Pease, L. V. Azaroff, C. K. Vaccaro, and W. A. Hines, *Phys. Rev. B*, **19**, 1576 (1979).

69. P. H. Citrin, P. Eisenberger, and J. E. Rowe, *Phys. Rev. Lett.*, **48**, 802 (1982).

70. J. Stöhr, L. Johansson, I. Lindau, and P. Pianetta, *Phys. Rev. B*, **20**, 664 (1979).

71. E. A. Stern and K-q Lu, *Nucl. Instrum. Methods*, **195**, 415 (1982).

72. V. O. Kostroun, *Nucl. Instrum. Methods*, **172**, 243 (1980).

73. B. Lengeler and P. Eisenberger, *Phys. Rev. B*, **21**, 4507 (1980).

CHAPTER

4

EXAFS WITH SYNCHROTRON RADIATION

STEVEN M. HEALD

Brookhaven National Laboratory
Upton, New York

4.1. INTRODUCTION

The development of synchrotron radiation facilities has greatly expanded the use of the EXAFS technique. Laboratory facilities just cannot compete in terms of energy resolution and intensity. Along with the advantages of synchrotron radiation, there are also some disadvantages. The large machine atmosphere often means the experimenter has little feel for what is actually happening with the apparatus. The major interaction with the apparatus is via a computer terminal directing the desired scans to be carried out. This approach is satisfactory if there are no problems, but invariably problems come up whose diagnosis and solution demands some knowledge of the machine operation. The purpose of this chapter is to introduce the various factors involved in the design of an EXAFS beamline at a synchrotron radiation facility. The goal is to give the user an understanding of what is actually involved in bringing the x-rays from the source onto the sample.

The chapter begins with a description of the development and availability of synchrotron radiation. Next the properties of the source are described. This includes such properties as the energy distribution polarization, intensity, optical properties, and time structure of the source. This discussion is on a more qualitative level since the user generally has little control over these parameters and detailed descriptions already exist in the literature.

More important to the experimenter is the operation of the optical components. Often the operation of the line requires a choice of crystals, setting of slits, or adjustment of mirrors. In this case it is desirable to have some understanding of how these choices will affect the operation of the line, and thus these topics are discussed in more detail. The chapter ends with a discussion of current and proposed beamline designs.

4.2. HISTORICAL DEVELOPMENT AND CURRENT FACILITIES

Early calculations of synchrotron radiation began as attempts to understand the radiation from electrons orbiting in atoms (1, 2). The development of quantum theory ended these calculations and it was the development of particle accelerators that rekindled interest in the problem. The solution was worked out independently in the mid-1940s by Ivanenko and Pomeranchuk (3) and by Schwinger (4, 5). At the same time accelerators were built in which detection of the radiation was possible, and experimental searches were begun. Blewett (6) was the first to detect the presence of synchrotron radiation by measuring the orbit contraction in a 100-MeV betatron. Since the vacuum chamber was opaque the radiation was not directly observed. This occurred first at the 70-MeV General Electric synchrotron (7) in 1947. The name synchrotron radiation derived from these observations and has been in general use ever since. Since then the predictions of Schwinger (5) have been accurately confirmed. More details on the early discoveries are found in a paper by Hartman (8).

About 10 years later the studies of Tomboulin and co-workers (9, 10) demonstrated the usefulness of synchrotron radiation as a vacuum UV source, and the first experiments exploiting this were carried out by Madden and Coddling (11). The use of synchrotron radiation did not take off, however, until the 1970s. This came with the development of storage rings as more stable sources of radiation, and the construction of new facilities planned and dedicated for the production of synchrotron radiation. EXAFS experiments generally require

Table 4.1. Synchrotron Radiation Sources in the X-Ray Energy Range

Machine	Location	E_c (keV)	I (mA)	E (GeV)	Type	Experimental Ports
ALADDIN	Stoughton USA	1.1	500	1.0	SR	20
LUSY	Lund Sweden	1.1	10	1.2	SY	1
SIRIUS	Tomsk USSR	1.2	20	1.3	SY	
INS-SUR I	Tokyo Japan	1.2	60	1.3	SY	1
PACHRA	Moscow USSR	1.2	100	1.3	SY	
ADONE	Frascati Italy	1.5	60	1.5	SR	2
SRS	Daresbury Great Britain	3.2	500	2.0	SR	2
DCI	Orsay France	3.4	400	1.8	SR	2
VEPP-3	Novosibirsk USSR	3.8	100	2.2	SR	
NSLS	Upton USA	4.3	500	2.5	SR	26
Photon Factory	Tsukuba Japan	4.3	500	2.5	SR	6
BONN II	Bonn W. Germany	4.5	30	2.5	SY	4
SPEAR	Stanford USA	11.2	100	4.0	SR	8
VEPP-4	Novosibirsk USSR	14.5	10	6.0	SR	
ARUS	Yerevan USSR	19.5	1.5	4.5	SY	
DORIS	Hamburg W. Germany	22.9	50	5.0	SR	10
DESY	Hamburg W. Germany	25.5	20	7.5	SY	3
CESR	Ithaca USA	35	20	8.0	SR	3

$^a E_c$ is the critical energy, I the electron current, and E the electron energy. SR and SY refers to synchrotron and storage ring, respectively. Each experimental port may contain several experiments.

121

x-ray energies of several keV, which in turn require more powerful machines than used in the early measurements. This energy region was utilized briefly at the Cambridge Electron Accelerator in 1972 (12), and on a more permanent basis at SPEAR (Stanford, USA) beginning in 1974. Since then a number of x-ray facilities have been opened up and they are listed in Table 4.1 (13–15). In the table the rather arbitrary definition is made that an x-ray machine has a critical energy (see Section 4.3) of at least 1 keV.

4.3. ORIGIN AND PROPERTIES OF SYNCHROTRON RADIATION

When a charged particle is accelerated it emits radiation. For nonrelativistic energies the radiation is emitted in a dipole pattern with the intensity varying as the sine of the angle between the observation direction and the acceleration vector. For an electron in circular motion the acceleration is perpendicular to the direction of motion and the dipole pattern has a maximum in the plane that includes the direction of motion. For relativistic velocities the dipole pattern is compressed along the direction of motion. The transformation of angles is (16)

$$\tan \theta' = \frac{\sin \theta}{\gamma(\beta + \cos \theta)} \tag{1}$$

where $\beta = v/c$ and $\gamma = E/mc^2$. The angular distribution has been narrowed down to a width of order $1/\gamma$. Thus, an observer viewing the electron along its velocity vector would see light during the time the electron travels an arc length ρ/γ where ρ is the radius. For the electron this is a time $\rho/\gamma\beta c$. During the same time light travels a distance $\rho/\gamma\beta$. This means that the duration of the pulse actually observed is

$$\frac{1}{c}\left(\frac{\rho}{\gamma\beta} - \frac{\rho}{\gamma}\right) \approx \frac{\rho}{c\gamma^3} \tag{2}$$

In order for a pulse to have this duration it must contain frequency components up to

$$\omega \approx \frac{1}{\Delta t} \approx \frac{c\gamma^3}{\rho} = \omega_0\gamma^3 \tag{3}$$

where ω_0 is the angular frequency of rotation for the electrons. A 500 MeV machine has $\gamma = 1000$, which means the radiation spectrum extends to frequencies of order $10^9 \, \omega_0$. The basic usefullness of synchrotron radiation comes from this result; for easily attainable electron energies extremely high frequency radiation can be generated.

These qualitative ideas can be put on a quantitative footing although the equations involve complicated integrals of modified Bessel functions. The work of Green (17), which is also summarized by Winick (16), is an excellent compilation of expressions needed to calculate all the properties of the synchrotron radiation. Only the more important properties are discussed here, with the emphasis on developing a feel for the behavior of the various properties rather than exact results.

Synchrotron radiation properties are often discussed in terms of a critical energy, which is usually defined as

$$E_c = \frac{3}{2} \hbar\omega_0\gamma^3 = \frac{3\hbar c\gamma^3}{\rho} \tag{4}$$

Using the critical energy (or alternatively, the corresponding critical wavelength) a universal synchrotron radiation power spectrum can be plotted as shown in Fig. 4.1. The intensities can be calculated from electron energy and current along with the horizontal acceptance angle for a given machine. As expected from the qualitative arguments the spectrum is fairly flat at low energies, but

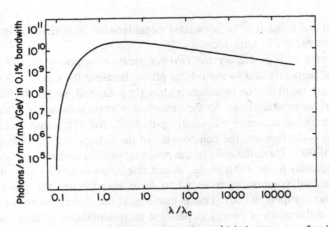

Figure 4.1. Universal synchrotron radiation spectrum from which the spectrum of a given machine can be calculated if its critical wavelength is known (15).

drops sharply above the critical energy. In fact it can be shown to a good approximation that the intensity varies as (18)

$$I(E) \approx \frac{E^{1/2}}{E_c} e^{-2E/E_c} \qquad E \gg E_c$$

$$I(E) \approx E^{1/3} \qquad E \ll E_c \qquad (5)$$

These results hold for a single electron. For a pulse of electrons collective effects can become important at low energies, giving a large enhancement in the output (19).

From the qualitative argument relating the energy spectrum to the degree of collimation, it might be expected that they are related. Indeed, this is the case as shown in Fig. 4.2. At high energies the radiation is more highly collimated with an angular spread $(1/e$ point) of (18)

$$\Delta\theta \simeq \frac{1}{\gamma} \left(\frac{E_c}{3E} \right)^{1/2} \qquad E > E_c \qquad (6)$$

Similarly, at low energies the width is larger than $1/\gamma$.

$$\Delta\theta \simeq \frac{1}{\gamma} \left(\frac{E_c}{E} \right)^{1/3} \qquad E < E_c \qquad (7)$$

The natural collimation is an important consideration in optical design as will be discussed further in later sections.

Also shown in Fig. 4.2 are the two polarization components, the **E** vector parallel and perpendicular to the orbital plane. Because the electron is confined to a single orbital plane the radiation is strongly polarized, and is, in fact, 100% polarized in the orbital plane. As the observation angle moves out of the orbital plane the radiation becomes elliptically polarized. An interesting aspect of the polarization is the fact that the components of the radiation always have a phase difference of 90°. Thus, the axes of the polarization ellipsoid are always parallel and perpendicular to the orbit plane. When integrated over all angles and energies the parallel component is found to have seven times the power of the perpendicular component (20). This is also true at the critical energy. However, just as the collimation is energy dependent the polarization is also energy dependent. Above the critical energy the radiation is more strongly polarized and it is less strongly polarized below.

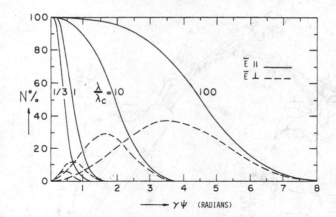

VERTICAL ANGULAR DISTRIBUTION OF PARALLEL AND PERPENDICULAR POLARIZATION COMPONENTS

Figure 4.2. Distribution of the parallel and perpendicular polarization components (courtesy of A. van Steenbergen).

4.3.1. Synchrotrons and Storage Rings

The preceding discussion applies to isolated electrons. In real sources the properties are modified somewhat. Before discussing the optical properties of the source it is instructive to consider the general properties of synchrotron radiation sources. There are two classes of sources, synchrotrons and storage rings. In synchrotrons, electrons are accelerated in a period of approximately 10 ms and then ejected, typically to be used in a high-energy experiment. Strong synchrotron radiation is only emitted at the end of the acceleration cycle, and the energy spectrum is a function of time.

In a storage ring the electrons are accelerated and then stored at energy. If the vacuum is sufficiently high (10^{-9}–10^{-10} Torr) the electrons can be maintained for many hours. The source is much steadier both in terms of the energy spectrum and the beam position. The advantages are such that all machines designed specifically as radiation sources are storage rings.

Figure 4.3 shows a schematic of a typical electron storage ring. Synchrotron radiation is produced at the bending magnets. It is coupled to the experiment through a beam port tangent to the ring. Such a port typically makes available 10–100 mrad of radiation to be divided among the user experiments via a beamline and associated optics. A beamline must contain various safety devices to protect the user from radiation hazard and the storage ring from vacuum contamination by the user.

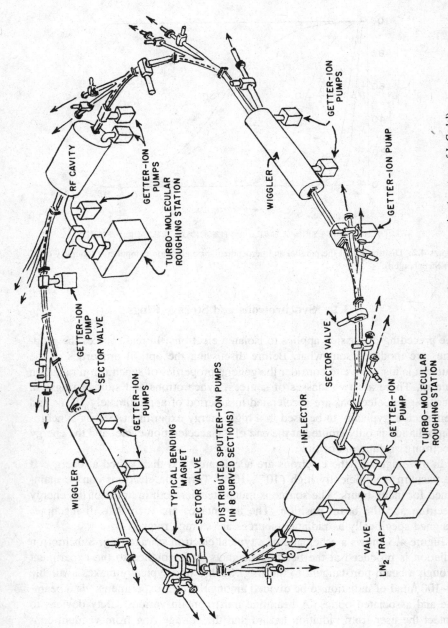

Figure 4.3. Schematic of an intermediate energy electron storage ring (courtesy of J. Godel).

GETTER-ION PUMPS

GETTER-ION PUMP

WIGGLER

GETTER-ION PUMP

SECTOR VALVE

GETTER-ION PUMP

TURBO-MOLECULAR ROUGHING STATION

INFLECTOR

VALVE

LN₂ TRAP

DISTRIBUTED SPUTTER-ION PUMPS (IN 8 CURVED SECTIONS)

SECTOR VALVE

TYPICAL BENDING MAGNET

WIGGLER

GETTER-ION PUMPS

GETTER-ION PUMP

SECTOR VALVE

TURBO-MOLECULAR ROUGHING STATION

GETTER-ION PUMPS

RF CAVITY

Some other features of the storage ring are worthy of note. Extensive pumping is available to obtain and maintain the vacuum at 10^{-9}–10^{-10} Torr in the presence of the heat and radiation load of the synchrotron radiation. A radio frequency (rf) cavity is necessary to make up the synchrotron radiation losses that in a large ring can be of the order of megawatts. Another straight section contains an inflector to couple the electron accelerator to the storage ring. In the example shown there are two other straight sections that contain wiggler magnets. These are specialized sources of radiation that will be discussed in Section 4.3.4 along with the related undulator magnets.

4.3.2. Time Structure

The time structure of the emitted synchrotron radiation is dependent on the operation and dynamics of the source. In a storage ring a cavity is used to maintain the electron energy. This means the electrons are confined in bunches whose length is determined by the rf period. Thus, the radiation is pulsed with a pulse length determined by the bunch length. For most storage rings the pulse duration is in the range of 0.1–1 ns. The repetition rate depends on the number of bunches circulating in the machine and the natural orbit period. For x-ray machines the orbit period is approximately 1 μs and the number of bunches is likely to be in the range of 1–10 (20). This time structure can have important consequences in the design of experiments. Most detectors will be unable to distinguish two events that occur during a single pulse, which modifies their saturation behavior at high count rates. On the other hand, the short pulse length means precise timing experiments can be carried out.

In synchrotrons a set of electron bunches are injected and accelerated every 0.1–1 s. The electrons do not radiate extensively at x-ray energies until the end of the acceleration cycles. Therefore, a synchrotron time structure consists of a short period of storage ringlike pulses separated by relatively long dead spaces. Because of the relative instability of the synchrotron time structure no fast timing experiments have been carried out on them.

4.3.3. Source Optical Properties

The electron beam in a real source has a finite extent and the individual electrons do not travel in parallel orbits. Rather they oscillate about the equilibrium position causing the height and width of the beam to vary. In steady state operation a particular source point in a storage ring will have constant dimensions, but due to the oscillations there are correlations between the position of the electron in the source point and the direction of its velocity vector. This is conveniently

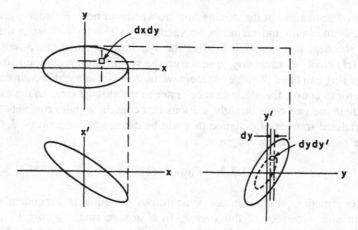

Figure 4.4. Derivation of the horizontal and vertical phase space plots from the real space source. Each element dx or dy in real space emits a range of angle x' or y', which are represented in the phase space plots (17).

displayed by plotting phase space contours, examples of which are shown in Figs. 4.4 and 4.5.

Figure 4.4 shows the derivation of the vertical (yy') and horizontal (xx') phase space contours for a single point in the electron orbit. The beam dimensions are typically Gaussian in profile with differing horizontal and vertical extent. This gives rise to the elliptical profile shown. The phase space contours derive from the physical source as shown. Each element $dx\,dy$ emits radiation over a limited range of angles. The range is a convolution of the natural radiation width and the range in orbital directions contained within the element $dx\,dy$. Since the orbital directions are correlated with the source position the phase space ellipses are tilted. An optimized source would have a minimum physical extent in x and y, and would have x' and y' dimensions dominated by the photon opening angle of an individual electron.

An actual beamline views an extended segment of the electron orbit. This results in a horizontal source as shown in Fig. 4.5. The electron phase ellipse sweeps through a range of angles as the beam sweeps through the arc of the magnet. Thus the contours become the nearly straight vertical lines as shown. The curvature is due to the bending radius and results in a slight increase in the width of the effective source if a large horizontal angle is collected. Similarly the effective source height increases when a large horizontal angle is collected. The photons from the rear of the segment have diverged by the time they reach the front.

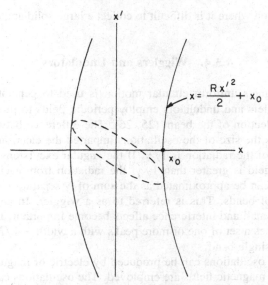

$$x = \frac{R{x'}^2}{2} + x_0$$

Figure 4.5. Horizontal phase space plot for a bending magnet. Because the beam is moving around an arc of radius R the phase space ellipse is elongated as shown (17).

As the photon beam propagates, each point on the phase space ellipse stretches along the x axis to a value $x(L) = x + x'L$. If a slit of width $2d$ is located at L it can be represented by vertical lines at $x = \pm d$. The beam can be propagated further and additional optical elements encountered. Their influence on the photon beam can be found in an analogous fashion. For example, two crystals in a dispersive diffracting arrangement behave like an angular slit that is represented by horizontal lines at $x' = \pm\alpha$ where 2α is the angular acceptance of the crystals. Details of the phase space method applied to synchrotron sources can be found in work by Hastings (21), Pianetta and Lindau (22), and Matsushita and Kaminaga (23). In addition, Matsushita and Kaminaga (24) describe an extension of the phase space representation to include a third axis representing the photon wavelength. The important point is that the correlated nature of the synchrotron source must be taken into account in calculating the performance of beamline optics.

The concept of the source brightness can be used to compare different sources. The source brightness is defined as the number of photons emitted per unit area per unit solid angle. Due to the high degree of collimation a synchrotron source has an intrinsically high brightness particularly when compared to bremsstrahlung sources that have no collimation. A high brightness is especially important in the x-ray region where it is difficult to collect a large solid angle with existing optical elements.

4.3.4. Wigglers and Undulators

In bending magnets uniform circular motion is used to generate synchrotron radiation. Wigglers and undulators employ periodic fields to produce radiation with no net deflection of the beam (25, 26). The difference between a wiggler and undulator is the size of the oscillations imparted the electrons. The natural emission angle of the radiation is $1/\gamma$. If the angular excursions introduced by the oscillating field is greater than $1/\gamma$, the radiation from each oscillation is incoherent and can be approximated as the sum of N separate sources where N is the number of bends. This is referred to as a wiggler. In an undulator the excursions are small and interference affects become important. In this case the spectrum becomes a set of one or more peaks with a width $\sim 1/N$ and intensity $\sim N^2$ that of a single bend.

The electron oscillations can be produced by electric or magnetic fields, but in practice only magnetic fields are employed. The oscillations can be planar or helical (27), but again practical considerations have limited storage ring wigglers to the planar type.

The attractiveness of wiggler and undulators comes from the additional control they give over the synchrotron radiation source. Typically magnetic fields can be much higher (up to 60 kG using superconducting magnets) than the bending magnets. This allows the spectrum to be hardened considerably allowing high-energy photons to be produced at more easily attainable electron energies. A typical wiggler might also have three complete oscillations giving a sixfold intensity increase. However, the total power generated by such a device is correspondingly higher, which may be difficult to handle with existing optics. For an undulator, on the other hand, the intensity is concentrated into narrow bands, which means large increases at specific energies can be achieved without an increase in the total energy striking the optics. This fact has led to a considerable development effort of undulators at several facilities (26, 28).

The properties of wigglers and undulators are often described by a magnetic field strength parameter K given by

$$K = \frac{\lambda_0}{2\pi\rho_0} \approx 0.93 B_0 \text{ (tesla) } \lambda_0 \text{(cm)}$$

where λ_0 is the period of the magnetic field oscillations and ρ_0 is the radius of curvature corresponding to the peak field B_0. For a wiggler $K \gg 1$ while an undulator has $K \lesssim 1$. Particular attention in recent years has been focused on structures with K slightly larger than 1. These combine the properties of both. At low energies there is the characteristic interference spectrum of an undulator that is washed out at high energies to a nearly smooth spectrum characteristic of a wiggler. The energy spectra of various undulators and wigglers proposed for the NSLS (29) are given in Fig. 4.6.

From a user standpoint the planar wiggler is very similar to an enhanced bending magnet source. The polarization is predominately linear, and the source phase space properties are also similar. Of course, the emission angle in the horizontal plane is limited due to the limited excursions of the electrons, al-

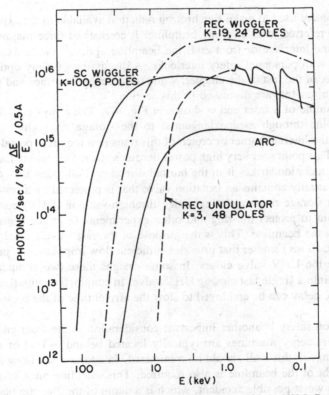

Figure 4.6. Comparison of the spectra for various sources proposed for the NSLS. Both superconducting (SC) and rare earth cobalt (REC) magnet structures are considered. In the case of the REC undulator the dashed line is an average of small oscillations that persist to high energies (30).

though mechanical constraints often limit the angles available from bending magnets to similar values.

For undulators the interference effects make the source more complicated. The polarization is a function of both the angle of observation and the energy, and may be completely polarized at certain energies. The phase space plots of the source can also become more complex. In particular, the horizontal emission angle is now of order $1/\gamma$, comparable to the vertical angle. This means that the source brightness is substantially increased. Indeed, in some cases a small beam size at the sample can be achieved without focusing optics.

4.4. INTERFACING TO SYNCHROTRON RADIATION SOURCES

The equipment used to make synchrotron radiation available to the experimenter is usually referred to as an x-ray beamline. It consists of three major sections: the machine interface or front end, the beamline optics, and the experimental apparatus with personnel safety interlocks. A discussion of x-ray optics makes up the bulk of the rest of the chapter, while the machine interface and personnel safety requirements are discussed in this section.

An example of a front end is shown in Fig. 4.7. The x-rays are admitted to the beamline through a pipe tangential to the storage ring. On high power machines the internal corner or crotch of this transition requires special attention (29–31). This point sees very high power densities since it is close to the electron beam and radiation strikes it in the normal direction. Following the crotch, the front end usually contains an isolation valve that is protected by a water cooled mask. For storage rings operating under ultrahigh vacuum (UHV) conditions it is important to protect the ring and other experiments from accidental loss of vacuum in the beamline. This is the purpose of the fast valve (32). It is a fast closing (< 10 ms) shutter that provides sufficient flow impedance to protect the ring while the UHV valve closes. In some designs these two components are combined into a single fast closing UHV valve. In conjunction with these valves an acoustic delay can be employed to slow the arrival time of the pressure wave front (33).

Radiation safety is another important consideration in the front end design. The higher-energy machines are typically located behind ~ 1 m of concrete. The openings in this wall should be minimized. In addition shielding along the line of sight of the beamline is also required. This shielding must be sufficient to stop the worst possible accident, which is a dump of the electron beam down the beamline. This requires 20–30 cm of lead and possibly some low-Z material (concrete, wax, etc.) to stop any photoinduced neutrons. For lines in which the

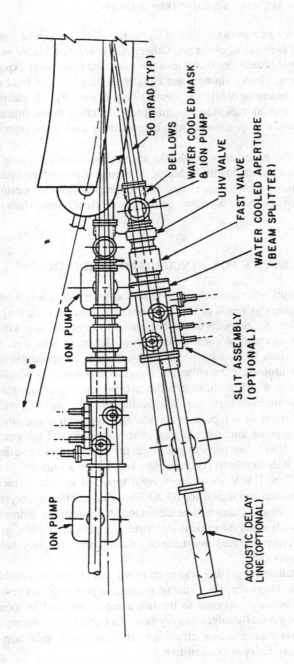

Figure 4.7. Schematic of an x-ray beamline front end.

50 mRAD (TYP)

BELLOWS

WATER COOLED MASK
& ION PUMP

UHV VALVE

FAST VALVE

WATER COOLED APERTURE
(BEAM SPLITTER)

FRONT END
X-RAY BEAM LINE

ION PUMP

SLIT ASSEMBLY
(OPTIONAL)

ION PUMP

ACOUSTIC DELAY
LINE (OPTIONAL)

optics deviate the beam away from the line of sight, the shielding can be placed immediately following the optical elements. Otherwise, it must be placed at the end of the beamline. In this case personnel must be excluded from areas exposed to line of sight radiation. Heavy shutters are usually included in the front end to allow access to the beamline while the storage ring is running. In addition, at some facilities the special radiation hazards present during beam injection may require the evacuation of personnel from certain areas during the injection process.

Radiation safety at the experiment is usually accomplished by enclosing the experiment in an interlocked enclosure or hutch. The hutch can only be opened when appropriate beamline shutters are closed and contains additional interlocks to guard against the possibility of someone being left inside a hutch when it is closed (14).

4.5. X-RAY OPTICS FOR SYNCHROTRON RADIATION

The designer of x-ray optics for a synchrotron radiation beamline has two basic goals. The first is to collect as much radiation as possible. For bending magnet sources this means collecting as much of the horizontal divergence as possible. This can be limited by mechanical constraints such as the beam tube size, but in the x-ray region is often limited by the available optical components. The second goal is to monochromatize the collected x-rays as efficiently as possible. Although the radiation is highly collimated, the acceptance angles of typical monochromator crystals are even narrower. This is illustrated in Fig. 4.8, which compares the opening angle of a typical x-ray storage ring with some silicon reflections. If the full opening angle is allowed to strike the crystal the energy resolution is degraded. If slits are used to collimate the beam intensity suffers. These facts may make it desirable to collimate the beam with a mirror.

At x-ray energies $E > 1$ keV there are several types of optical elements available. The most common and useful for EXAFS are Bragg reflecting crystals and grazing incidence mirrors. These will be discussed in depth in the following sections. In addition, such standard optical elements as zone plates (34), gratings, normal incidence mirrors, and interference coatings are currently being extended to x-ray energies (35).

As early as 1925 a reflection grating was used to diffract copper and molybdenum K radiation (36). However, the extreme grazing angles required results in a large scattered radiation component to the diffracted beam, and as monochromators, gratings are generally inferior to crystals. An exception is the region near 1 keV where crystals have a low efficiency, and improved reflection or transmission gratings may become competitive.

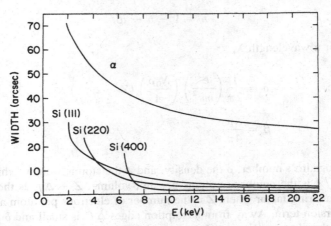

Figure 4.8. Comparison of the synchrotron radiation opening angle α for the NSLS with some typical silicon Bragg reflections.

When multiple thin layers are deposited on a smooth substrate interference can occur for x-rays just as at optical wavelengths (37, 38). The effect can be used to enhance the reflectivity of a normal incidence optic (39, 40). However, the reflectivity is optimized only for a single wavelength and again normal incidence optics are not applicable to EXAFS optics. A more promising application is the manufacture of artificial Bragg reflecting crystals with a large d spacing (38, 41). At present the obtainable energy resolution is not sufficient for EXAFS, but they have high reflectivities near 1 keV. Since they can be constructed of stable materials (e.g., tungsten and carbon), Bragg reflecting multilayers are promising as premonochromators. As such they would be employed to absorb most of the energy from the source allowing a more radiation and thermal sensitive crystal to be used as a monochromator.

4.5.1. X-Ray Mirrors

The properties of x-ray mirrors have attracted increasing attention in recent years. For the EXAFS user it is useful to understand the fundamentals in order to make optimum use of beamlines that employ mirror optics. The user may have a choice between beamlines employing different optical systems, or when using a beamline may find the performance less than expected. The next three sections describe the important properties of x-ray mirrors.

The refractive index for x-rays can be written (42, 43)

$$n = 1 - \delta - i\beta \tag{8}$$

where for a wavelength λ,

$$\delta = \frac{1}{2\pi} \left(\frac{e^2}{mc^2} \right) \left(\frac{N_0 \rho}{A} \right) (Z + \Delta f') \lambda^2$$

$$\beta = \frac{\lambda \mu}{4\pi} \tag{9}$$

N_0 is Avogadro's number, ρ the density, and A the atomic weight, which means that $N_0 \rho / A$ is the number of atoms per unit volume. $Z + \Delta f'$ is the real part of the scattering factor where Z is the number of electrons per atom and $\Delta f'$ is the dispersion term. Away from absorption edges $\Delta f'$ is small and δ is proportional to the electron density of the material. The imaginary part of the index, β, is simply proportional to μ, the linear absorption coefficient. Useful sources of $\Delta f'$ and μ are Henke et al. (44), MacGillavry et al. (45), and McMaster et al. (46).

Since the index of refraction is less than 1, total external reflection can occur if the incidence angles are small enough. Ignoring absorption ($\beta = 0$) application of Snell's law gives

$$\theta_c = \sqrt{2\delta} \approx \rho_e^{1/2} \lambda \tag{10}$$

where ρ_e is the electronic density of the material. Thus short wavelengths require more glancing angles, and dense materials will reflect to higher angles. Because of its density, platinum is a common mirror material. Some experimental reflecting curves (47) are shown in Fig. 4.9. In this case absorption is not negligible resulting in a rounding of the onset of total external reflection. The oscillations at high energies are interference effects due to the finite platinum thickness (~ 500 Å). In the region of the platinum L-edges structure also appears. At absorption edges optical dispersion theory (42) predicts a sharp decrease in reflectivity. In this data the finite energy resolution of the detector (~ 200 eV) smooths out the decrease. Also shown is a reflection curve for float glass that is characteristic of a low-Z material. The critical energy is lower and the rounding of the falloff is smaller.

An important property of any x-ray reflecting surface is the surface finish. In the optical region the surface roughness must be several times smaller than the wavelength if scatter is to be minimized. This is apparently impossible for wavelengths of atomic dimensions. However, it is pointed out by Franks (35) that the glancing angles necessary for reflection relax the finish requirements by

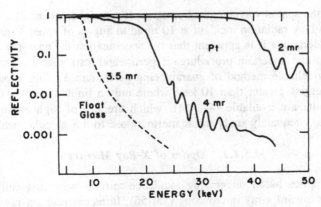

Figure 4.9. Experimental reflection curves (47) for the x-ray reflectivity of a float glass and a platinum surface. The oscillations in the platinum curves are from interference effects due to the finite thickness of the coating.

a factor $1/\sin\theta$ where θ is the incident angle. Nevertheless the surface finish for x-rays should be at least one or two orders of magnitude better than high quality optical surfaces, preferably of the order of 10 Å rms roughness.

When the surface is smooth (i.e., $\sigma\sin\theta/\lambda \ll 1$ where σ is the surface roughness) a simple scalar theory can be applied to estimate the scattering (48–50). In this theory the reflectivity is

$$R = R_0 e^{-(4\pi\sigma\sin\theta/\lambda)^2} \tag{11}$$

where R_0 is the smooth surface reflectivity. This result is found to only qualitatively represent the experimental data, but is a useful starting point. It is often used to define the surface roughness from measurements of the total integrated scatter. More complex theories are based on vector scattering formalisms (51). Bilderback and Hubbard (47) also found that a graded density theory (52, 53) gives a good fit to his platinum data.

For real surfaces the surface roughness is not necessarily random and a complete description of the surface requires a roughness power spectrum (50). Physically, what is occurring is that each spatial wavelength component is acting as a grating that diffracts the incident beam. The diffraction angle is the deviation of the scattered radiation from the specular beam and for grazing incidence is

$$\Delta\theta \simeq \frac{\lambda}{\theta d} \tag{12}$$

where d is the spatial wavelength. For example, spatial frequencies near 10 μm will scatter 1-Å radiation incident at 10 mrad to angles of about 1 mrad. From these considerations it is apparent that the specification for an x-ray optic is a complicated and uncertain procedure. Experimental tests with x-rays appear to be the only reliable method of guaranteeing performance. This is particularly true for energies greater than 10 keV where only a limited number of experimental results are available (47, 54), which are not of high enough spatial resolution to adequately study the scattering close to the specular peak.

4.5.1.1. Optics of X-Ray Mirrors

Reflection optics based on grazing incidence mirrors were originally studied with an eye toward x-ray microscopy (55, 56). In recent years, x-ray telescope applications have been the driving force behind x-ray optics studies (57, 58). Advances in fabrication technology and metrology as a result of these efforts is potentially a great boon to the synchrotron radiation community. To date, however, optical systems employed in synchrotron radiation beamlines have not been as sophisticated as in the telescope and microscope fields. This is partly a result of differing needs, and partly due to a lack of communication between the two communities.

For EXAFS beamlines the imaging quality of a focusing element need not be extremely good. The purpose is to condense the radiation onto a small sample. Thus, approximate focusing optics can often be used. When mirrors are used as collimating elements preceding crystal monochromators, on the other hand, the requirements are quite stringent. The task of collimating a significant portion of the incident energy into an angular spread of a few arc seconds requires mirrors comparable in quality to those produced for state of the art x-ray telescopes. Therefore, it is important to understand basic properties of grazing incidence optics to apply them intelligently to EXAFS beamlines.

Ideal optical elements are based on the conic sections: ellipsoids, paraboloids, and hyperboloids. Since synchrotron radiation is highly collimated it is not necessary to use a full surface of revolution, and the geometry is as shown in Fig. 4.10a. The simplest focusing case in an ellipsoidal surface with the source a distance F_1 from the mirror being focused at a distance F_2. The magnification is then $M = F_2/F_1$. It is useful to define the radii of curvature of the surface in the sagittal (R_s) and meridional (R_m) directions. For an ellipsoid on axis (point C in Fig. 4.10a) they are

$$R_s = \frac{2F_1 F_2}{F_1 + F_2} \sin \theta \qquad R_m = \frac{R_s}{\sin^2 \theta} \qquad (13)$$

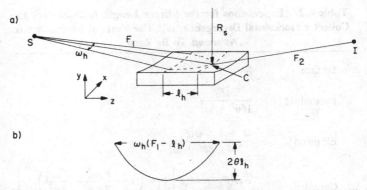

Figure 4.10. Geometry of a glancing incidence mirror. (*a*) Definition of terms. For ellipsoids or paraboloids the radius R_s varies with the position on the mirror and there is an additional curvature R_m along the length of the mirror. (*b*) Shape of the output beam for such a mirror.

For a paraboloid, which is the limit as F_2 goes to infinity, R_s becomes $2F_1 \sin \theta$.

The use of an ellipsoidal focusing element has been studied by Howell and Horowitz (59). They used a bent cylindrical approximation to focus broad band synchrotron radiation. The same mirror was later used at the Stanford Synchrotron Radiation Laboratory (SSRL) by Hastings et al. (60) where it was followed by a monochromator. However, the energy resolution of the monochromator was seriously degraded. The reason is shown in Fig. 4.10. The initial plane of radiation is output as a parabolic arc. Since the extreme rays are focused at a distance F_2 they must acquire an additional vertical divergence

$$\Delta\theta_e = \frac{2\theta\ell_h}{F_2 + \ell_h} \simeq \frac{2\theta\omega_h^2}{(8\theta^2 + \omega_h^2)} \quad (M = 1) \tag{14}$$

where θ is the incidence angle and the other parameters are defined in Fig. 4.10*a*. The mirror length ℓ_h is due only to the horizontal divergence. An actual mirror must be longer to also collect the vertical extent of the beam. Expressions for ℓ_h for some common surfaces are given in Table 4.2.

An obvious solution to this problem is to put the ellipsoid after the monochromator. This results in a good focus but does nothing to match the emittance of the storage ring with the acceptance of the monochromator. To achieve this parabolic mirrors can be used. If the mirror in Fig. 4.10 is a paraboloid the outgoing rays are parallel. A real source has a finite vertical opening angle, and a finite vertical and horizontal extent. The vertical opening angle is made par-

Table 4.2. Expressions for the Mirror Length l_h, Necessary to Collect a Horizontal Divergence, ω_h: The Vertical Divergence is Assumed To Be Zero

Surface	l_h
Paraboloid	$\dfrac{\omega_h^2 F_1}{16\theta^2 + \omega_h^2}$
Ellipsoid	$\dfrac{M+1}{2M} \dfrac{\omega_h^2 F_1}{8\theta^2 + \omega_h^2}$
Cylinder	$\dfrac{-\omega_h^2 F_1 + 4R_s\theta \left[1 - \left(1 - \dfrac{F_1 \omega_h^2}{2R_s\theta} - \dfrac{F_1^2 \omega_h^2}{4R_s^2} \right)^{1/2} \right]}{4\theta^2 + \omega_h^2}$

allel, but the finite source size results in a residual $\Delta\theta$ as shown in Fig. 4.11. The vertical source size gives directly $\Delta\theta_v = S_v/F_1$ (this contribution is also present with an ellipsoid). The contribution of the horizontal extent is less obvious and is related to the comatic aberration (discussed later) of a paraboloid. It appears only for a finite ω_h. Rays starting from point A in Fig. 4.11 parallel to extreme rays from the source center will strike the mirror displaced by a distance $\pm\Delta l$ where for $\omega_h < \theta$

$$\pm\Delta\ell \approx \ell(\omega_h \pm S_h F_1) - \ell(\omega_h) \approx \pm \frac{\omega_h S_h}{8\theta^2} \qquad (15)$$

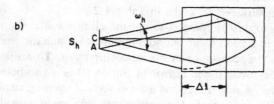

Figure 4.11. Sources of divergence for a collimating mirror. (*a*) From the vertical source size S_v. (*b*) From the horizontal source size S_h.

The vertical component of the normal has changed by an amount $\pm\Delta\ell/R_m$ resulting in

$$\Delta\theta_p \approx \frac{\omega_h S_h}{2F_1\theta} \qquad (16)$$

Since this contribution is a function of both ω_h and S_h the actual distribution of angles is found by integrating over the horizontal phase space of the source. When attempting to collimate a large ω_h the contribution from $\Delta\theta_p$ can often be as large as $\Delta\theta_v$.

Another possibility for collimation is the use of a singly curved mirror ($R_s \to \infty$) figured as a parabola. Since no horizontal collimation is done $\Delta\theta_p = 0$ and better collimation is achieved. The output is no longer an arc as in Fig. 4.10 but remains a line. This can result in a significant size reduction for the monochromator crystals. To focus the horizontal a second mirror is required following the monochromator. This mirror may be difficult to produce since it is not a surface of revolution. However, since the radiation is already well collimated in the vertical, an ellipsoidal focusing mirror gives good results.

Singly curved mirrors can be formed by bending appropriately shaped plates (61). They have other advantages. The radius of curvature can be varied *in situ* to optimize their optical properties. For example, thermal distortions can be compensated or the incident angle adjusted to adjust the critical energy of the mirror.

So far the discussion has concentrated on the collimating properties of the mirrors. Also important are the focusing properties. For grazing incidence reflectors the principal abberation is coma (56, 62). Off-axis points are imaged as circular arcs of length $2M\,d\phi$, where

$$\sin\left(\frac{\phi}{2}\right) = \frac{M+1}{M}\frac{\omega_h}{2\theta} \qquad (17)$$

and d is the distance off axis. For a reflector that is a segment of a surface of revolution (i.e., ellipsoid or paraboloid) ϕ is the azimuthal angle subtended by the reflector. For a rectangular source typical of storage rings coma results in a bow-tie image as shown in Fig. 4.12.

To minimize coma two reflections are required that is the design basis of Wolter (56) telescope optics. For synchrotron beamlines a possible configuration is a double paraboloid focusing system with the monochromator in between. In this application it is important to orient the paraboloids properly taking into account the possible image inversion by the monochromator.

Since EXAFS measurements do not require perfect imaging, approximate

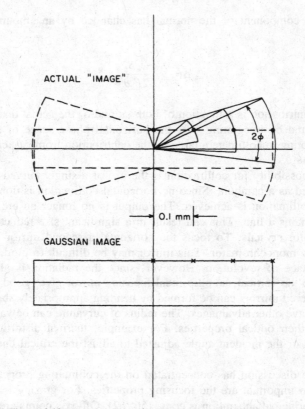

Figure 4.12. Distortion due to coma on a rectangular image (62).

optics based on bent cylinders can often be used (59, 63). For an $M = 1$ ellipsoid this approximation is most useful. If the cylinder is bent to a toroidal shape, rays originating from a source point with divergence $\pm\omega_h/2$ will strike the focal plane at

$$\Delta x = \frac{\pm 5\omega_h \ell_h^2}{8F_1} \qquad \Delta y = \frac{2\theta \ell_h^2}{F_1} \qquad (18)$$

Since ℓ_n varies as ω_h^2 these are strong functions of ω_h resulting in a sharp limiting value of ω_h above which the focus rapidly deteriorates. To keep the toroidal aberration less than a characteristic source size d the mirror length is limited to

$$\ell_h^2 < \frac{dF_1}{2\theta} \qquad (19)$$

For typical values this implies $\omega_h \approx \theta$. When $M \neq 1$ the expressions become much more complicated and are given by Heald and Hastings (63).

A bent cylinder is less successful as an approximation to a paraboloid. In this case the important property is the degree of collimation. For a point source, which is perfectly collimated by a paraboloid, the bent cylinder has

$$\Delta\theta \approx \frac{\omega_h^2}{16\theta} \qquad (20)$$

Taking as a restriction in this case $\Delta\theta < S_v/F_1$ gives a horizontal collection angle $\omega_h^2 < 16\theta\, S_v/F_1$, which is only a few milliradians in most cases.

4.5.1.2. Mirror Materials

As is evident from the preceding discussion the materials chosen for x-ray mirrors must have a number of special properties. They must be resistant to radiation, have good thermal properties, be highly polishable, and be available in large sizes. The properties of some candidate mirror materials are listed in Table 4.3.

For mirrors exposed to a significant fraction of the beam power, thermal properties are especially important. The important parameters are the thermal expansion and thermal conductivity, and a good figure of merit is the ratio of the conductivity to the expansion coefficient. This comes from the fact that an

Table 4.3. Candidate Materials for X-Ray Mirrors: E, k, and α Are the Young's Modulus, Thermal Conductivity, and Thermal Expansion Coefficient (64, 68)

Material	E (10^{10} N m^2)	Density (10^3 kg m^3)	k (W mK)	α (10^{-6} K^{-1})	k/α	Surface Finish Attained (Å rms)
Fused silica	7.32	2.2	1.37	0.56	2.45	4
Zerodur	9.1	2.53	1.64	0.05	32.8	5
Beryllium (I-70)	30.4	1.85	220	11.2	19.6	60
Al (6061-T6)	6.90	2.71	171	23.0	7.4	19
Copper (OFHC)	11.7	8.94	392	16.7	23.5	10
Molybdenum (TZM)	31.8	10.20	146	5.0	29.2	15
Electroless Nickel	14.7	7.9	5	14	0.36	5
SiC (CVD)	38	3.10	200	3.5	57	4
Si	11	2.33	100	4.0	25	4

unconstrained plate of thickness d with a temperature difference ΔT will bend to a radius (64)

$$R = \frac{d}{\alpha \, \Delta T} = \frac{kd^2}{\alpha \, \Delta q} \tag{21}$$

where k is the conductivity, α the expansion coefficient, and Δq is the heat transferred through the thickness d. R can easily become a significant source of slope errors in the mirror figure. This means that careful attention to the mirror design and supports is necessary to cope with the thermal problems. To date, no universal solution has been developed, and since synchrotron sources continue to become more powerful, thermal problems are likely to remain for some time.

One technique to minimize thermal problems is to reduce the power absorbed in the mirror. This can be done by setting the grazing angle so small that nearly all the power is reflected or by placing the mirror after the monochromator. Of course, both solutions require a thermally stable monochromator.

The best surface finishes are usually obtained on materials for which there is extensive experience. These include standard optical materials such as fused silica, zerodur, and electroless nickel (Kanigen). Electroless nickel is particularly useful since it can be plated on materials such as aluminum and also accepts a high polish (65). A class of materials that looks particularly promising are the carbide and nitride based ceramics (66). The prototype of these materials is SiC, which has recently been used in beamlines at the SRS (Daresbury) and the NSLS (Upton) (67–70).

For mirrors that see smaller heat loads, fused silica is a widely used material (60, 71). Large mirrors are readily available from a number of manufacturers. When using glass materials for mirrors, however, caution should be exercised in masking the beam so that it only strikes the reflecting surface. Radiation damage has been observed in some cases (72). In this respect mirrors made of conducting materials seem to be more rugged.

Usually the mirrors are coated with a heavy metal such as gold or platinum to enhance the high-energy reflectivity. Best results are obtained when this is done under UHV conditions. Degradation has been seen for gold coatings (73), but the causes have not been systematically investigated. In any case the coatings should be kept to a few hundred angstroms to avoid degrading the surface finish.

4.5.2. X-Ray Crystals

4.5.2.1. Perfect Crystal Diffraction

The general theory of the diffraction of x-rays by perfect crystals has been extensively studied and summarized by several authors (74–77). More recently the problem of monochromator design for synchrotron radiation has revived interest in the diffraction properties of perfect crystals (78–81). This section summarizes some of the important results.

To take advantage of the intrinsic collimation of synchrotron radiation perfect crystals are generally used. These are generally silicon or germanium because of their availability and their resistance to thermal and radiation damage. For centrosymmetric crystals in the Bragg reflection case (incoming and incidence beams on the same side of the crystal) the reflectivity is

$$R(L) = L - [L^2 - (1 + 4k^2)]^{1/2}$$

$$L = \left| [(-1 + y^2 - g^2)^2 + 4(gy - k)^2]^{1/2} \right| + y^2 + g^2 \quad (22)$$

where

$$y = \frac{(1 - b) X_{ro} + b\alpha}{2C |X_{rh}| \sqrt{|b|}} \quad (23)$$

$$g = \frac{(1 - b) X_{io}}{2C |X_{rh}| \sqrt{|b|}} \quad (24)$$

$$k = \frac{X_{ih}}{X_{rh}} \quad (25)$$

This result assumes $k \ll 1$. The electrical susceptibility is written as $X = X_r + iX_i$ with the susceptibility components given by

$$X_{ro} = -\Gamma F'_o \quad X_{io} = -\Gamma F''_o$$

$$X_{rh} = -\Gamma F'_h \quad X_{ih} = -\Gamma F''_h \quad (26)$$

with $\Gamma = r_e \lambda^2 / \pi V$. The classical electron radius is r_e, the wavelength λ, and the unit cell volume V. The structure factor is $F_h = F'_h + iF''_h$ for the $H \equiv (h, k, \ell)$ reflection. For the incident beam $F'_o = Z/V$ and $F''_o = \lambda \mu / 2\pi$ where Z is the unit cell charge and μ the absorption coefficient. The parameter C that depends on the x-ray polarization is one for σ polarization and $\cos 2\theta_B$ for π polarization. The quantity b is the asymmetry parameter and is given by $-\sin (\theta_B - \theta)/\sin (\theta_B + \theta)$, where θ_B is the Bragg angle and θ is the inclination of the crystal planes relative to the surface. θ is positive when the incident beam is more glancing than the outgoing beam.

The reflectivity depends on the parameter α. For plane parallel monochromatic radiation,

$$\alpha = 2(\theta_B - \theta_0) \sin 2\theta_B \tag{27}$$

where θ_0 is the actual incidence angle and θ_B is the corresponding Bragg angle for the incoming wavelength. Because of refraction the peak reflectivity does not occur at θ_B, but is centered about $y = 0$. This is shown in Fig. 4.13 for the (111) reflection of silicon. Because the absorption in silicon is low, it is nearly a perfect reflector in the range of $y = \pm 1$.

Synchrotron radiation is not monochromatic, and it is sometimes useful to express α on a wavelength scale keeping θ fixed. Hence we can write

$$\alpha = 4\left(\frac{\lambda - \lambda_B}{\lambda_B}\right) \sin^2 \theta_B \tag{28}$$

In this case λ_B is the wavelength that satisfies the Bragg equation for angle θ_B. This description is useful for calculating the wavelengths output by a set of crystals illuminated by a white radiation source.

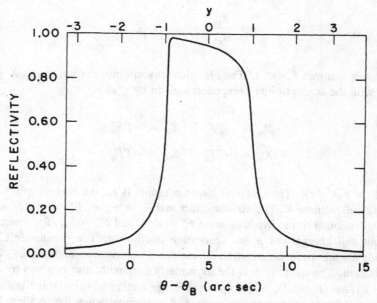

Figure 4.13. Single crystal reflectivity curve for Si(111) at 10 keV.

From these results the angular width of reflection ($y = \pm 1$) is found to be

$$\Delta\theta = \frac{2CX_{rh}}{\sqrt{|b|}\,\sin^2\theta_B} \tag{29}$$

Thus, adjustment of b can be used to provide a better match to the output of the synchrotron source (80). The output angle of the diffracted beam is

$$\theta_h - \theta_B = -b(\theta_0 - \theta_B) \tag{30}$$

Except in the symmetric case ($|b| = 1$) a perfect crystal does not act as a flat mirror. If $|b|$ is chosen to be less than 1 to increase the acceptance angle, the angular spread of the output beam is decreased from the symmetric case. If $\Delta\theta_s$ is the symmetric width then for an asymmetric reflection

$$\Delta\theta_0 = |b|^{-1/2}\Delta\theta_s, \qquad \Delta\theta_h = |b|^{1/2}\Delta\theta_s \tag{31}$$

Similarly the deviation from θ_B depends on b. Materlik and Kostroun (81) show how this affect can be used to eliminate higher-order reflections. In their monochromator an asymmetric reflection is followed by a symmetric reflection. To maximize the reflectivity of such an arrangement a slight angular offset is required. As seen in these results this offset depends on the energy and, thus, is different for the various orders. When the fundamental is maximized the higher orders are strongly suppressed.

For synchrotron radiation beamlines designed for EXAFS the rapid scanning necessary requires a fixed or nearly fixed exit beam. Multiple crystal reflections are therefore often used. The most common configuration is the nondispersive ($+$, $-$) arrangement consisting of two parallel crystals (see Fig. 4.14). In this case, the output beam is parallel to the input beam with a slight offset given by $h = 2d\cos\theta$, where d is the crystal separation. If the crystals are rotated as a unit the offset varies with energy requiring a small translation of the sample. To avoid this, a number of monochromators have been designed that translate one or both of the crystals as they are rotated to maintain a fixed output (82, 83). Since the angular tolerances are arc seconds these designs usually require some type of correction scheme to keep the crystals aligned.

Order sorting can be achieved in the ($+$, $-$) configuration by detuning the pair (84). Detuning is achieved by adjusting the pair to be slightly nonparallel as shown in Fig. 4.15. Since the higher-order reflections are narrower they are extinguished sooner leaving predominately the fundamental energy. Typically the optimum detuning results in a 50% loss of fundamental intensity.

A nondispersive monochromator has the property that all the incident rays

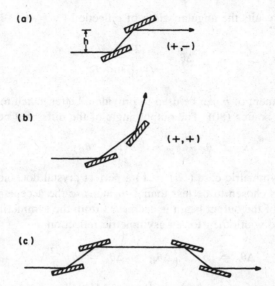

Figure 4.14. Three configurations for a multiple crystal monochromator: (*a*) nondispersive (+, −) configuration; (*b*) dispersive (+, +) configuration; and (*c*) four crystals in a dispersive nondeviating arrangement.

are reflected since every ray makes the same angle on each crystal (assuming $|b| = 1$ as is the usual case). The energy resolution is therefore determined by a convolution of the input angular range with the acceptance of the crystal. In most cases this is dominated by the input angular range, which in turn is a function of the source size and slits employed in the beamline. If the crystals are arranged in the dispersive (+, +) arrangement shown in Fig. 4.14, the energy resolution is characteristic only of the crystals. A fixed output dispersive arrangement as suggested by Beamont and Hart (78) employs four reflections as shown in Fig. 4.14. Scanning is accomplished by rotating the pairs in opposite directions. Since the reflectivity of silicon is high, four reflections result in little intensity loss for energies above about 5 keV.

4.5.2.2. Crystal Optics

Crystal diffracting elements can be used as focusing opticals elements just as mirrors. However, there is one very important constraint; to maintain the intrinsic energy resolution all of the incoming rays must make the same angle with respect to the lattice plane normal. This condition can be quite restrictive

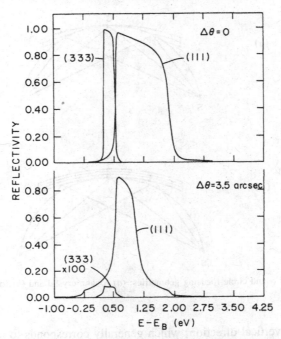

Figure 4.15. Reduction of the higher-order reflection by detuning for the Si(111) reflection. When $\Delta\theta = 3.5$ arcsec the fundamental (111) reflection at $E_B = 10$ keV has about one-half of its original intensity while the (333) reflection at 30 keV is reduced by about 10^{-3}.

on the design of crystal optics, but in many cases is offset by the fact that crystal reflections do not require the extreme grazing angles necessary for x-ray mirrors.

The most common focusing geometry using crystals is the Rowland circle configuration shown in Fig. 4.16. The crystals can be simply bent to a radius $2R$ giving the Johann (85) case, of for the Johansson case the crystal surface is also ground to a radius R to reduce aberrations (86, 87). Changing energy requires the movement of the source and focus around the Rowland circles. Such a movement is not practical for an EXAFS monochromator at a synchrotron source since the source to crystal distance must be approximately 10 m or more. Scanning the monochromator would require large motion of the crystal and sample approximately every second. This is the reason that synchrotron radiation EXAFS lines to date have used double flat crystal monochromators and relied on mirrors for focusing.

The Rowland circle configuration focuses the x-rays in the dispersive or meridian direction. For synchrotron radiation the x-rays are already highly col-

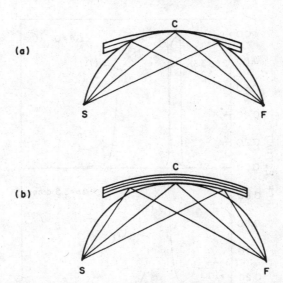

Figure 4.16. Rowland circle focusing geometries: (*a*) Johann crystal and (*b*) Johansson crystal.

limated in the vertical direction, which generally corresponds to the dispersive direction. What is required is focusing in the nondispersive or sagittal direction. The principle of sagittal focusing has been developed by Sparks et al. (88) and is shown in Fig. 4.17. One of the crystals in a double crystal monochromator is bent to focus the horizontal. The bend must be chosen appropriately if all the rays striking the flat crystal also make the same angle on the bent crystal.

Consider the geometry shown in Fig. 4.18. If the crystal is flat the extreme rays strike the crystal at a slightly different angle from the central ray. The difference is

$$\theta = \theta_0 - \frac{\omega_h^2 \tan \theta_0}{8} \tag{32}$$

Figure 4.17. Use of a bent crystal for sagittal or horizontal focusing of a synchrotron beam.

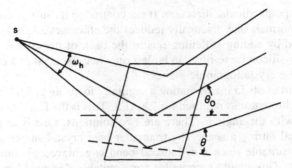

Figure 4.18. Geometry of horizontally diverging beam striking a flat crystal.

If the crystal is bent to a radius R,

$$\theta_R = \theta_0 + \frac{\omega_h^2 F_1}{4R \cos \theta_0} - \tan \theta_0 \left(\frac{\omega_h^2 F_1^2}{8R^2} + \frac{\omega_h^2}{8} \right) \tag{33}$$

which means for an extreme ray striking first a flat crystal and then a cylindrical crystal the angle difference is

$$\Delta\theta = \theta_R - \theta = \frac{\omega_h^2 F_1}{4R \cos \theta_0} - \frac{\omega_h^2 F_1^2 \tan \theta_0}{8R^2} \tag{34}$$

In order for the ray to be transmitted by the two crystals $\Delta\theta$ must be less than the crystal diffraction width. As seen in Eq. (34) $\Delta\theta = 0$ for

$$R = \frac{F_1 \sin \theta_0}{2} \tag{35}$$

Comparing with Eq. (13) this is just the focusing condition for $M = \frac{1}{3}$ or the focus is at $F_2 = F_1/3$. Thus, a cylindrically bent crystal operated in the $M = \frac{1}{3}$ case behaves in a double crystal monochromator just as a flat crystal.

For other bending radii a cylinder has a residual $\Delta\theta$. In this case it is necessary to bend the crystal into a conical shape which requires a more complicated bending apparatus. However, such a crystal has been operated at a magnification 0.9 with efficiencies of 50-90% of the corresponding flat crystal configuration (88).

The one major problem in bending a sagittally focusing crystal is anticlastic bending in the perpendicular direction. If not controlled it causes a wide variation in the crystal normals that drastically reduces the efficiency. Anticlastic bending can be reduced by adding stiffening ribs to the back of the crystal as shown in Fig. 4.19. For silicon these ribs can be left on when the crystal is cut, reducing distortion of the crystallographic planes.

One problem exists in incorporating a sagittally focusing crystal into a rapidly tunable monochromator as required for EXAFS. This is the fact that the optimum radius varies with the angle θ. There are two solutions. One is to dynamically bend the crystal during a scan. The triangular bent crystal shown in Fig. 4.19 is particularly suitable since a cylindrical bend is achieved by simply pushing on the tip (89). This should be possible for a suitably designed bender, but has yet to be demonstrated in a working monochromator.

Another solution is to keep the radius fixed but to translate the crystal to keep the focus fixed (90). The required translation for a cylindrically bent crystal is

$$F_1 = \frac{L}{2} + \frac{L}{2}\left(1 - \frac{3}{4}\frac{E}{E_0}\right)^{1/2} \tag{36}$$

where $L = F_1 + F_2$ is the total length of the beamline and E_0 is the energy for which $M = \frac{1}{3}$. At E_0 the throughput is a maximum and drops off as the energy is changed. Figure 4.20 shows the intensity variation for a crystal collecting 10 mrad of radiation. For typical scans the intensity variation is only 20–30%. Also, the outermost rays are affected first, so crystals collecting fewer milliradians would have smaller variations. The difficulty with this approach is that translations of approximately 1 m are required for EXAFS scans, and some type

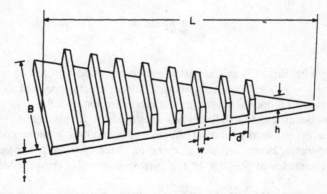

Figure 4.19. Triangular crystal designed for cylindrical bending. By clamping the base and pushing on the apex a cylindrical bend is achieved. The stiffening ribs prevent anticlastic bending.

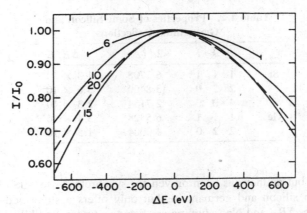

Figure 4.20. Intensity variation for a fixed radius sagitally focusing crystal scanned as described in the text. The curves are for different energies in keV. The solid lines are for Si(111) reflections and the dashed for Si(220). A horizontal collection of 10 mrad is assumed.

of monitoring system is required to compensate for imperfections in the translation, which may affect the energy reproducibility. At the time this article was written both approaches were being pursued and these questions should be answered in the near future (90).

4.5.2.3. Crystal Properties

The high thermal and radiation loads at synchrotron radiation sources call for rugged monochromator crystals. Certain common monochromator crystals such as quartz have only a limited lifetime in the beam. For this reason and the fact that perfect crystals are readily available, EXAFS beamlines have almost exclusively employed silicon and germanium. The properties of some common reflections are given in Table 4.4. The energy resolutions are calculated from Eq. (29), and are achieved for crystals with dislocation densities less than 10^4 cm^{-2} (89).

The Ge(111) reflection will operate down to about 2 keV. At lower energies the choice of crystals is less clearcut. Table 4.5 shows some candidate crystals for use at lower energies. None of these crystals is entirely satisfactory. KAP (92), beryl (93), and α-SiO$_2$ (94) have been used in synchrotron monochromators with some success, but all have been found to suffer some damage with time. Of the three beryl seems to be most rugged. However, beryl and α-SiO$_2$ are useless near the silicon edge since both contain silicon, resulting in a sharp drop in reflectivity. ADP and mica are common in laboratory mono-

Table 4.4. Properties of Some Silicon and
Germanium Reflections

	h	k	l	$2d$ (Å)	$\Delta E/E$
Si	1	1	1	6.2708	1.3×10^{-4}
	2	2	0	3.8400	5.6×10^{-5}
	4	0	0	2.7154	2.3×10^{-5}
Ge	1	1	1	6.5328	3.4×10^{-4}
	2	2	0	4.0004	1.5×10^{-4}

chromators, but neither has been proven for synchrotron use. InSb has similar properties to silicon and germanium, but only offers a slight addition to the energy range. YB_{66} and Naβ-alumina are listed as future possibilities. YB_{66} has good radiation resistant properties and has been grown in reasonable sizes (~ 2 cm), but crystal quality is still low. Sodium β-alumina has been shown to be reasonably radiation resistant (95), but again the crystal quality is low.

4.5.3. Slits

An important part of any optical design is the location of slits or apertures in the system. For synchrotron radiation they used to limit the size of the beam, exclude scattered radiation, or to improve the energy resolution of a monochromator. As seen in Fig. 4.8, if the full divergence of the source is allowed to strike the monochromator crystals, the energy resolution will be much poorer than the intrinsic crystal value. Therefore, in the absence of collimating optics a slit must be used. Just as for a collimating mirror a single slit is limited by

Table 4.5. Some Candidate Crystals for Low-Energy
Synchrotron Radiation Monochromators

	$2d$ (Å)	Energy Range $(\theta_B = 8-80°)$ (eV)
InSb (111)	7.481	1683–11910
α-SiO$_2$ (10$\bar{1}$0)	8.512	1479–10470
ADP (101)	10.640	1183– 8370
YB_{66} (400)	11.720	1074– 7600
Beryl (10$\bar{1}$0)	15.954	789– 5590
Mica (002)	19.84	635– 4490
Na β-Alumina (0002)	22.49	560– 3962
KAP (10$\bar{1}$0)	26.632	473– 3346

the source size in the amount of collimation it can produce. For an infinitesimal slit opening the divergence is still S_v/F where S_v is the source size and F the source to slit distance. To obtain greater collimation two or more slits are required. The optimum case would have one slit at the source to define a smaller source size, and the second slit operating as a normal single slit collimation. However, this is impossible at a synchrotron source where the nearest slit position is usually several meters form the source. Thus, a two slit collimating system usually results in serious intensity losses.

When focusing optics are used the use of slits becomes more complicated. A slit preceding a focusing mirror has little effect on the image size unless the mirror is set to image the slit rather than the source. Again there is the same problem with intensity losses because the slit cannot be placed at the source position. Another point to remember is that a focusing mirror mixes the horizontal and vertical divergence of the beam. This was pointed out in Section 4.5.2.2 as the cause of a loss in resolution when a monochromator is preceded by a focusing mirror. In this case a horizontal slit preceding the monochromator will have little effect on the resolution. However, a vertical slit to restrict ω_h preceding the mirror can reduce the mixing for improved resolution. When ω_h is small enough, then the vertical divergence is again determined by any horizontal slits present, and as far as the slits are concerned, the system behaves as if the mirror is not present.

This general discussion is only meant as a guideline. To do a good job of calculating the optimum slit settings the phase space properties of the source must be taken into account. This can be done using the methods described in Section 4.3.3 or by using a ray-tracing procedure (63).

4.6. EXAFS BEAMLINES

When all of the components discussed in the preceding sections are combined to make a complete EXAFS beamline, various compromises must be made. As yet, no one has constructed a beamline that can be considered optimized. This section summarizes the characteristics of existing beamlines, and possible improved designs, which may be implemented in the near future.

A basic schematic of an EXAFS line is shown in Fig. 4.21. The optical system consists of up to three elements of which only the monochromator is absolutely necessary. Indeed, the earliest beamlines consisted of a simple double crystal monochromator preceded by slits, and this design still is used for much of the present work. However, experiments requiring the highest intensities would benefit greatly from focusing optics.

In the simplest focusing case the monochromator is preceded by a focusing

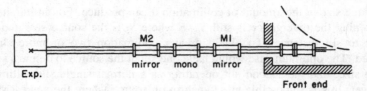

Figure 4.21. Schematic of the major components in a x-ray beamline.

mirror (60), usually a bent cylindrical approximation to an ellipsoid. In Section 4.5.1.1 it was seen that this could lead to a serious reduction in energy resolution if a significant horizontal divergence is collected. The seriousness of this effect depends on the source characteristics. If the source is large, then the energy resolution is already degraded and the addition of the mirror is of little consequence. On the other hand, the newer synchrotron facilities are designed for high brightness for which a high-energy resolution is easy to achieve. For them an option is to put the focusing mirror after the monochromator in position M2. This has the further advantage of a low heat load on the mirror reducing the constraints on the choice of mirror material.

The addition of a collimating mirror in position M1 can increase the monochromator throughput by matching the source to the crystal acceptance. Again the usefulness of the mirror depends on the source properties. If the source is large the collimating mirror has little effect on the beam divergence. A ray tracing study of various collimating cases for the NSLS (Upton, USA) source is described by Heald and Hastings (63). In that study significant gains were found with a collimating mirror. Probably the best design for a source such as the NSLS at energies less than 10 keV would use paraboloids for M1 and M2. Since the sagittal radius of a set of paraboloids is a factor of two larger than for an ellipsoid, the same length mirror can collect more horizontal divergence. Paraboloids have the additional advantage that a simple two crystal monochromator would provide a fixed output beam since beam movement in between the paraboloids is not transmitted to the focus. Such an optical design has been applied to the UV (96) and one such monochromator is also being developed for the x-ray regime (97).

The problem with paraboloids is the high cost for the sizes needed for x-ray beamlines. This has led to the use of less expensive approximate optics. Two options are available for the collimating mirror. It can be a plate bent into a cylindrical approximation of parabola or it can be a spherically ground surface whose meridian radius approximates a parabola. Neither does any collimation of the horizontal and an ellipsoidal type mirror is required for focusing. Of course, the ellipsoid can be approximated by a bent cylinder. This system has somewhat poorer performance, but has the distinct advantage of using readily available mirrors. As mirror fabrication technology improves through the use of diamond turning (65) or other techniques the cost of more ideal optics will come down, and fewer approximate mirrors will be used. This will reduce the need for complicated mirror bending apparatus, although some bent mirrors will continue to be used to allow adjustment of energy cutoffs.

At higher energies mirror glancing angles become so small that their usefulness is limited. In this regime sagittally bent crystals are likely to be used extensively. With existing technology it is possible that 10 mrad or more can be collected with high efficiency. This performance comes at the expense, however, of a more complicated scan procedure. The crystal must be bent or translated as the scan progresses. Thus, although bent crystals are also potentially useful at lower energies, mirror optics will probably remain dominant below 10 keV particularly for EXAFS applications.

The final component to be optimized for EXAFS experiments is the source. The standard bending magnet source is probably approaching its limits in the new generation of dedicated facilities. Major improvement in the source is still possible using undulators and wigglers. Undulators are especially attractive because they concentrate the radiation both spatially and in energy. Thus, much higher fluxes are possible on the sample without an attendant increase in the power absorbed by the optics or the size of the optics. For example, if an undulator concentrates its radiation to a horizontal divergence of ~ 1 mrad then any of these optical solutions can be used with relatively small components. In fact, an unfocused beam would be small enough for many applications.

It seems that these optimizations will be applied in the next few years. When this comes about the intensity improvement will be comparable to the change from laboratory to synchrotron sources made less than 10 years ago. With this in mind, it is safe to predict that the coming years will see the continued growth of EXAFS applications.

REFERENCES

1. G. A. Schott, *Philos. Mag.*, **13**, 189 (1907).

2. G. A. Schott, *Electromagnetic Radiation*, Cambridge University Press, Cambridge, 1912.

3. D. D. Ivanenko and I. Pomeranchuk, *Phys. Rev.*, **65**, 343 (1944).

4. J. Schwinger, *Phys. Rev.*, **70**, 798 (1946).

5. J. Schwinger, *Phys. Rev.*, **75**, 1912 (1949).

6. J. P. Blewett, *Phys. Rev.*, **69**, 87 (1946).

7. F. R. Elder, A. M. Gurewitsch, R. V. Langmuir, and H. L. Pollock, *J. Appl. Phys.*, **18**, 810 (1947).

8. P. L. Hartman, *Nucl. Instrum. Methods*, **195**, 1 (1982).

9. D. H. Tomboulian and P. L. Hartman, *Phys. Rev.*, **102**, 1423 (1956).

10. D. H. Tomboulian and D. E. Bedo, *J. Appl. Phys.*, **29**, 804 (1958).

11. R. P. Madden and K. Codling, *Phys. Rev. Lett.*, **10**, 516 (1963); **12**, 106 (1964); *J. Opt. Soc. Am.*, **54**, 268 (1964).

12. H. Winick, *IEEE Trans. Nucl. Sci.*, **20**, 984 (1973).

13. C. Kunz, in *Synchrotron Radiation Techniques and Applications*, C. Kunz (Ed.), Springer-Verlag, Berlin, 1979, p. 1.

14. H. Winick, in *Synchrotron Radiation Research*, H. Winick and S. Doniach (Eds.), Plenum, New York, 1980, p. 27.

15. M. Howells, in *Reflecting Optics for Synchrotron Radiation*, M. Howells (Ed.), *SPIE Proc.*, **315**, 13, 1982.

16. H. Winick, in *Synchrotron Radiation Research*, H. Winick and S. Doniach (Eds.), Plenum, New York, 1980, p. 11.

17. G. K. Green, BNL Report 50522, 1977; BNL Report 50595, Vol. II, 1977.

18. J. D. Jackson, *Classical Electrodynamics*, Wiley, New York, (1962), Chap. 14.

19. A. A. Sokolav and I. M. Tenkov, *Synchrotron Radiation*, Pergamon Press, New York, 1968.

20. I. Munro and A. Sabersky, in *Synchrotron Radiation Research*, H. Winick and S. Doniach (Eds.), Plenum, New York, 1980, p. 323.

21. J. B. Hastings, *J. Appl. Phys.*, **48**, 1576 (1977).

22. P. Pianetta and I. Lindau, *J. Electron Spectros. Relat. Phenom.*, **11**, 13 (1977).

23. T. Matsushita and U. Kaminaga, *J. Appl. Cryst.*, **13**, 465 (1980).

24. T. Matsushita and U. Kaminaga, *J. Appl. Cryst.*, **13**, 472 (1980).

25. H. Motz, *J. Appl. Phys.*, **22**, 527 (1951).

26. J. E. Spencer and H. Winick, in *Synchrotron Radiation Research*, H. Winick and S. Doniach (Eds.), Plenum, New York, 1980, Chap. 21.

27. B. M. Kincaid, *J. Appl. Phys.*, **38**, 2684 (1977).

28. Y. Farge, *Appl. Optics*, **19**, 4021 (1980).

29. S. Krinsky, W. Thomlinson, and A. Van Steenbergen, BNL Report 31989, 1982.

30. D. Mills, D. Bilderback, and B. W. Batterman, *IEEE Trans. Nucl. Sci.*, **26**, 3854, 1979.

31. C. Jako, N. Hower, and T. Simons, *IEEE Trans. Nucl. Sci.*, **26**, 3851, 1979.

32. T. Oversluizen, *Nucl. Instrum. Methods*, **195**, 399 (1982).

33. H. Betz, P. Hofbauer, and A. Heuberger, *J. Vac. Sci. Tech.*, **16**, (1979).

34. N. M. Ceglio, in *Low Energy X-ray Diagnostics*, D. T. Attwood and B. L. Henke (Eds.), American Institute of Physics, New York, 1981, p. 210. See also Chap. 5 in *High Resolution Soft X-ray Optics*, E. Spiller (Ed.), *SPIE Proc.* **316**, 1981, p. 99.

35. A. Franks, *Sci. Prog. London*, **64**, 371 (1971).

36. A. H. Compton and R. L. Doan, *Proc. Natl. Acad. Sci. USA*, **11**, 498 (1925).

37. L. G. Parratt, *Phys. Rev.*, **95**, 359 (1954).

38. J. H. Underwood, T. W. Barbee, and D. C. Keith, *SPIE Proc.*, **184**, 123, 1979.

39. E. Spiller, A. Segmuller, J. Rife, and R. P. Haelbich, *Appl. Phys. Lett.*, **37**, 1048 (1980).

40. J. P. Henry, E. Spiller, and M. Weisskopf, *SPIE Proc.*, **316**, 166, 1981.

41. T. W. Barbee, in *Low Energy X-Ray Diagnostics*, D. T. Attwood and B. L. Henke (Eds.), American Institute of Physics, New York, 1981, p. 131.

42. A. H. Compton and S. K. Allison, *X-rays in Theory and Experiment*, Van Nostrand, New York, 1935.

43. R. W. James, *The Optical Principles of the Diffraction of X-rays*, Cornell University Press, New York, 1948.

44. B. L. Henke, P. Lee, T. J. Tanaka, R. L. Shimabukuro, and B. J. Fujikawa, *At. Data Nucl. Data Tables*. **27**, 1982.

45. C. MacGillavry, G. Rieck, and K. Lonsdale (Eds.), *International Tables for X-ray Crystallography*, Vol. III, 2nd ed., Kynoch Press, Birmingham, 1968.

46. W. H. McMaster, N. Kerr Del Grande, J. H. Mallett, and J. H. Hubbell, Compilation of X-ray Cross Sections, UCRL-50174.

47. D. H. Bilderback and S. Hubbard, *Nucl. Instrum. Method*, **195**, 85 (1982).

48. P. Beckmann and A. Spizzichino, *The Scattering of Electromagnetic Waves from Rough Surfaces*, Pergamon Press, Oxford, 1963.

49. W. T. Welford, *Opt. Quantum Electron.*, **9**, 269 (1977).

50. E. L. Church, *SPIE Proc.*, **184**, 196 (1979).

51. J. M. Elson and J. M. Bennett, *Opt. Eng.*, **18**, 116 (1979).

52. L. Nevot and P. Croce, *J. Appl. Cryst.*, **8**, 304 (1975).

53. L. Nevot and P. Croce, *Rev. Phys. Appl.*, **15**, 761 (1980).

54. J. K. Silk and P. Burstein, American Science and Engineering Report ASE-4104, Cambridge, MA, 1977.

55. P. Kirkpatrick and A. V. Baez, *J. Opt. Soc. Am.*, **38**, 766 (1948).

56. H. Wolter, *Ann. Phys.*, **10**, 94 and 286 (1952).

57. R. Giacconi, W. P. Reidy, G. S. Vaiana, L. P. Van Speybroeck, and T. F. Zehnpfenning, *Space Sci. Rev.*, **9**, 3 (1969).

58. M. Weisskopf, Ed., *SPIE Proc.*, **184**, entire vol. (1979).

59. J. A. Howell and P. Horowitz, *Nucl. Instrum. Methods*, **125**, 225 (1975).

60. J. B. Hastings, B. M. Kincaid, and P. Eisenberger, *Nucl. Instrum. Methods*, **152**, 167 (1978).

61. J. H. Underwood and D. Turner, *SPIE Proc.*, **106**, 125 (1977).

62. M. R. Howells, *Appl. Opt.*, **19**, 4027 (1980).

63. S. M. Heald and J. B. Hastings, *Nucl. Instrum. Methods*, **187**, 553 (1981).

64. V. Rehn and R. O. Jones, *Opt. Eng.*, **17**, 504 (1978).

65. M. R. Howells and P. Z. Takacs, *Nucl. Instrum. Methods*, **195**, 251 (1982).

66. K. Lindsey, R. Morrell, and M. J. Hanney, in *Reflecting Optics for Synchrotron Radiation*, M. Howells (Ed.), *SPIE Proc.*, **315**, 140, 1981.

67. R. L. Gentilman and E. A. Maguire, in *Reflecting Optics for Synchrotron Radiation*, M. Howells (Ed.), *SPIE Proc.*, **315**, 131, 1981.

68. R. E. Engdahl, in *Reflecting Optics for Synchrotron Radiation*, M. Howells (Ed.), *SPIE Proc.*, **315**, 123, 1981.

69. M. M. Kelly and J. B. West, in *Reflecting Optics for Synchrotron Radiation*, M. Howells (Ed.), *SPIE Proc.*, **315**, 135, 1981.

70. P. Z. Takacs, *Nucl. Instrum. Methods*, **195**, 259 (1982).

71. G. M. Miles, in *Reflecting Optics for Synchrotron Radiation*, M. Howells (Ed.), *SPIE Proc.*, **315**, 65, 1981.

72. W. Gudat and C. Junz, in *Synchrotron Radiation Techniques and Applications*, C. Kunz (Ed.), Springer-Verlag, Berlin, 1979, p. 55.

73. A. Franks, K. Lindsey, and P. R. Stuart, in *Workshop on X-ray Instrumentation for Synchrotron Radiation Research*, H. Winick and G. Brown (Eds.), SSRL Rep. 78/04, p. VII-117 (1978).

74. W. H. Zachariasen, *Theory of X-ray Diffraction in Crystals*, Wiley, New York, 1945.

75. R. W. James, *The Optical Principles of the Diffraction of X-rays*, G. Bell and Sons, London.

76. M. Von Laue, *Röntgenstrahl—Interferenzen*, Akademische Verlag, Frankfurt.

77. B. W. Batterman and H. Cole, *Rev. Mod. Phys.*, **36**, 681 (1964).

78. H. Beaumont and M. Hart, *J. Phys. E*, **7**, 823 (1974).

79. U. Bonse, G. Materlik, and W. Schröder, *J. Appl. Cryst.*, **9**, 223 (1976).

80. K. Kohra, M. Ando, T. Matsushita, and H. Hashizume, *Nucl. Instrum. Methods*, **152**, 161 (1978).

81. G. Materlik and V. Kostroun, *Rev. Sci. Instrum.*, **51**, 86 (1980).

82. J. Cerino, J. Stöhr, N. Hower, and R. Z. Bachrach, *Nucl. Instrum. Methods*, **172**, 227 (1980).

83. M. Lemonnier, O. Collet, C. Depantex, J. M. Estena, and D. Raoux, *Nucl. Instrum. Methods*, **152**, 109 (1978).

84. D. Mills and V. Pollock, *Rev. Sci. Instrum.*, **51**, 1664 (1980).

85. H. H. Johann, *Z. Phys.*, **69**, 185 (1931).

86. J. W. M. Dumond and H. A. Kirkpatrick, *Rev. Sci. Instrum.*, **1**, 88 (1930).

87. T. Johansson, *Z. Phys.*, **82**, 507 (1933).

88. C. J. Sparks, G. E. Ice, J. Wong, and B. W. Batterman, *Nucl. Instrum. Methods*, **195**, 73 (1982).

89. M. Lemonnier, R. Fourme, F. Rousseaux, and R. Kahn, *Nucl. Instrum. Methods*, **152**, 173 (1978).

90. D. E. Sayers, S. M. Heald, M. A. Pick, J. I. Budnick, E. A. Stern, and J. Wong, *Nucl. Instrum. Methods,* **208**, 631 (1983).

91. B. Batterman, private communication.

92. M. Berland, A. Burek, P. Dhez, J. M. Esteva, B. Gauthe, R. C. Karnatek, and R. E. LaVilla, *High Resolution Soft X-ray Optics, SPIE Proc.,* **316**, 169 (1981).

93. Z. Hussain, E. Umbach, D. A. Shirley, J. Stöhr, and J. Feldhaus, *Nucl. Instrum. Methods,* **195**, 115 (1982).

94. A. Bienenstock and H. Winick, SSRL Activity Report 4/1/81–3/31/82, SSRL Report 82/01, III-4 (1982).

95. J. Wong, W. L. Roth, B. W. Batterman, L. E. Berman, D. M. Pease, S. Heald, and T. Barbee, *Nucl. Instrum. Methods,* **195**, 133 (1982).

96. M. R. Howells, *Nucl. Instrum. Methods,* **177**, 127 (1980).

97. W. R. Hunter, R. T. Williams, J. C. Rife, J. P. Kirkland, and M. N. Kalber, *Nucl. Instrum. Methods,* **195**, 141 (1982).

CHAPTER
5

LABORATORY EXAFS FACILITIES

D. C. KONINGSBERGER

Laboratory for Inorganic Chemistry and Catalysis
Department of Chemical Technology
Eindhoven University of Technology
Eindhoven, The Netherlands

5.1. INTRODUCTION

Chapter 4 describes how EXAFS measurements can be carried out using a synchrotron as source for x-ray radiation. The photon density (photons s^{-1} eV^{-1} $mrad^{-1}$) as calculated for an EXAFS beamline of a storage ring is about 10^5–10^6 times higher than the brightness produced by the bremsstrahlung of a rotating anode x-ray source (1). The actual photon flux (photons s^{-1}) available at the experimental EXAFS stations of existing synchrotron beamlines is in most cases lower than the figures calculated for the tangent point of the corresponding beamlines. The concrete shielding of the storage ring and the splitting of the beam port after the front end in several beamlines make it impossible to bring the experimental stations closer than 5–10 m. Therefore, optical elements like mirrors (see Chapter 4) have to be used to collect most of the horizontal and vertical divergence of the beam. X-ray optics is not always present in EXAFS beamlines and mostly slits are used to optimize the resolution leading to a decrease of the available intensity. This means that, in practice, the differences in photon flux between synchrotron and laboratory EXAFS facilities are less drastic. Moreover, for EXAFS measurements carried out in the transmission mode (see Fig. 5.1a) the maximum useful photon flux is limited. It has been measured (2) that for fluxes of about 5×10^8 photons s^{-1} the experimental noise is not determined anymore by the photon statistics but by the noise produced by the electronics of the detection stage. Therefore, one should not be too pessimistic beforehand about the possibility of performing EXAFS experiments with a laboratory facility.

In principle, EXAFS wiggles can be measured by making use of a simple bent LiF monochromator crystal, a standard x-ray tube, and a commercially available Θ, 2Θ goniometer. However, with this setup measuring times for undiluted materials amount to days, with resolutions not better than 20 eV. The use of cryostats or sample cells is very difficult if not impossible. The obtained signal-to-noise ratio and the actual resolution of this type of EXAFS spectrometer make a reliable analysis of the structural parameters of the first-shell difficult, whereas higher coordination shells are hardly resolved in the Fourier transform of the experimental data.

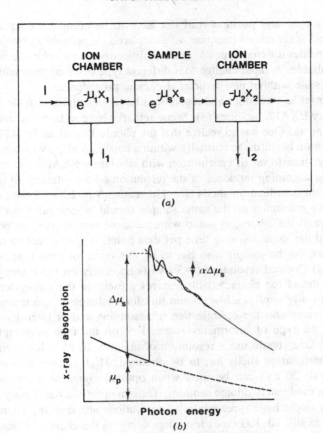

Figure 5.1. (*a*) Detection system for EXAFS in transmission mode. (*b*) An x-ray absorption spectrum ($\Delta\mu_s$ = absorption step due to sample; μ_p = pre-edge absorption, $\mu_s = \Delta\mu_s + \mu_p$).

The aim of this chapter is to demonstrate that good quality EXAFS data can be collected with in-house EXAFS facilities if these systems fulfill the demands that nowadays have to be made with regard to resolution, photon flux, feasibility, and foolproofness. It will be shown that with high quality curved monochromator crystals resolutions of 4–10 eV can be obtained, which give good quality EXAFS data. Intensities can be high enough to realize a signal-to-noise ratio better than 20–1 within scan times in the order of 1 to 20 h. The moving parts of the EXAFS spectrometer have to be especially designed so that the Rowland circle focusing geometry will maintain a positioning accuracy with a resettability in the order of 5 μm, and still allowing a 20–25-kg weight on the sample stage.

Motor positioning and position read out have to be computer controlled. To allow a broad spectrum of energies with emphasis on optimal resolution or on optimal intensities different types of good quality curved monochromator crystals must be available. A rapid change to a different type of monochromator crystal must be possible without new tedious alignment procedures.

Synchrotronlike intensities and resolutions have been claimed in the literature for laboratory EXAFS facilities (3). Some remarks have to be made concerning this optimism: (a) One has to realize that the photon flux of an EXAFS station on a synchrotron beamline is normally within a bandpass of 3 eV. A comparison of the Fourier transforms of synchrotron with laboratory EXAFS data will readily reveal the degrading influence of the resolution on the intensity of the peaks of the different coordination shells. (b) The statistics of EXAFS data obtained with the same resolution on the same sample should be compared for both the synchrotron and the laboratory setup with the same number of data points per spectrum and the same counting time per data point. (c) For a real comparison, the x-ray spot on the sample also has to be the same for both types of spectrometers. (d) Optimal resolution (~ 3 eV) is necessary on laboratory EXAFS systems, to detect the characteristic features present in the absorption edges. This automatically implies a low photon flux for a laboratory spectrometer. For EXAFS measurements the optimization of resolution and or intensity depends strongly on the type of information wanted or on the type of sample to be investigated. One should use a resolution of about 5–10 eV when information of higher coordination shells has to be obtained. High intensities with lower resolutions (10–20 eV) can be used when optimal signal to noise ratios are demanded on moderately diluted samples. Thus an optimized laboratory EXAFS spectrometer might have synchrotronlike resolutions although the intensities in most cases are still 10–100 times less, depending on the characteristics and the type of the synchrotron being used. The higher fluxes available at the storage rings make EXAFS experiments on highly diluted (<25 mM) systems possible, which cannot be realized with a laboratory facility.

The demand for performing EXAFS measurements grew rapidly from 1975 to 1980 and is still growing. The synchrotron EXAFS facilities, which became available, stimulated the EXAFS community to develop techniques that made the laboratory EXAFS spectrometer no longer a curiosity but a scientific instrument. The laboratory EXAFS spectrometer is in principle capable of performing a large class of experiments. To discuss and review all the possibilities a workshop on laboratory EXAFS facilities and their relation to Synchrotron Radiation Sources was organized by Prof. Stern, Washington State University, Seattle in April 1980 (4). Since then, new techniques and technologies have been developed and new laboratory EXAFS facilities have been built and published in the literature.

The aim of this chapter is to give a review of the latest developments. The possibilities, as well as the limitations, of laboratory EXAFS spectrometers will be described. These spectrometers can be constructed using the techniques and technologies nowadays available in well-equipped workshops of universities and industries. A detailed discussion will be given (Section 5.2) about the specifications of laboratory EXAFS facilities. The different components of the laboratory EXAFS system, such as x-ray source, crystals, and detectors, are discussed in Section 5.3, with an emphasis on the quality of the mechanism that has to hold the Rowland circle configuration. Section 5.4 deals with the resolution as calculated for two different types of monochromator crystals (Johann, Johansson) including the important contribution caused by the finite dimension of the focal spot. A survey of a selected number of laboratory made EXAFS spectrometers as described in recent literature will be given in Section 5.5. Section 5.6 compares the laboratory with the synchrotron EXAFS facilities. A summary and conclusions are given in Section 5.7.

5.2. SPECIFICATIONS

5.2.1. Signal-to-Noise Ratio, Photonflux

To obtain structural information on a large class of different samples, laboratory EXAFS spectrometers have to produce a minimum useful photon flux. The relation between the photon flux and the signal-to-noise ratio for a transmission EXAFS experiment will be derived in the following.

The detection system for a transmission EXAFS experiment is given in Fig. 5.1(a). The ionization chambers have absorption coefficients μ_1 and μ_2, with chamber lengths x_1 and x_2, respectively. The sample has a thickness x_s and a linear absorption coefficient μ_s. I is the total number of photons collected per data point entering the first ionization chamber, while I_1 and I_2 represent the signals arising from the ionizations in the chambers. The absorption coefficient of the sample can be calculated from $I_2/I_1 = \exp[-(\mu_s x_s)']$. Due to the absorption coefficients of the ionization chambers, which enter in the measured signal I_1 and I_2, the determined absorption coefficient $(\mu_s x_s)'$ is not exactly equal to $\mu_s x_s$. The noise N in $(\mu_s x_s)'$ is given by

$$N = \sigma(\mu_s x_s)' = [(I_1)^{-1} + (I_2)^{-1}]^{1/2} \tag{1}$$

Assuming that the amplitude of the EXAFS signal is equal to a fraction α (the normal range for α is $0.01 < \alpha < 0.1$) of the edge jump $\Delta\mu_s x_s$ (see Fig. 5.1b) the signal S can be represented by $S = \alpha \, \Delta\mu_s x_s$. The relation between the signal-to-noise ratio (S/N) and I can now be derived:

$$\frac{S}{N} = \frac{(I)^{1/2} \, \alpha \Delta \mu_s x_s}{\left\{ \left(1 - \exp\left(-\mu_1 x_1\right)\right)^{-1} + \exp\left(\mu_s x_s + \mu_1 x_1\right)\left(1 - \exp\left(-\mu_2 x_2\right)\right)^{-1} \right\}^{1/2}} \tag{2}$$

An optimum S/N is obtained with $\mu_1 x_1 = 0.25$, $\mu_2 x_2 = \infty$, and $\mu_s x_s = 2.55$. Using these values, we obtain the total number of photons I necessary to get a certain signal-to-noise ratio (S/N) after t seconds collection time per data point:

$$I = 3.2 \left(\frac{S}{N}\right)^2 \left(\frac{\mu_s}{\alpha \, \Delta \mu_s}\right)^2 \tag{3}$$

Figure 5.2 gives the total number of photons (I) as a function of $\mu_s / \alpha \, \Delta \mu_s$ calculated for S/N values of $20/1$ and $100/1$. To derive reliable structural parameters from EXAFS data a minimum signal-to-noise ratio of about $20/1$ is necessary (5). Maximum useful scan times per spectrum for laboratory EXAFS facilities are in the order of 20 h. Typical EXAFS spectra contain about

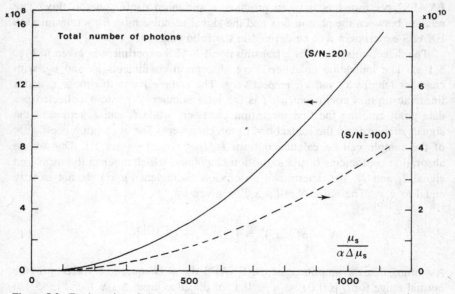

Figure 5.2. Total number of photons collected per data point $I = I_0 t$ as a function of $\mu_s / \alpha \Delta \mu_s$.

600 data points up to $k_{max} = 15$, leading to a maximum acceptable counting time per data point of 100 s. A large class of experiments can be carried out for $\mu_s/\alpha \, \Delta\mu_s \sim 500$. For a laboratory EXAFS spectrometer this implies a minimum useful photon flux I_0 of about 3×10^6 photons s^{-1}, accepting a maximum counting time of 100 s per data point. Typical values for bulk samples are $\alpha = 0.1$ and $\mu_s/\Delta\mu_s = 1$, which gives for $I_0 = 3.2 \times 10^6$ photons s^{-1} and 1 collection time per data point a S/N value of 100/1 in a total counting time of 10 min.

5.2.2. Resolution Versus Intensity

5.2.2.1. Edge and Near-Edge Spectra

Optimum resolution is needed when all characteristic features of the absorption edge are of interest. This automatically implies the use of high index diffraction planes of the monochromator crystal (see Section 5.3.2), which have, however, lower reflectivities (6). Moreover, the smaller d values of these planes prescribe larger distances from the x-ray source (see Section 5.3.2), which in turn lead to lower photon density on the sample position. Figure 5.3 gives the x-ray absorption spectrum of the copper K-edge (8980.3 eV) of copper foil measured at room temperature at SSRL (Stanford University, USA) with a resolution of 2 eV. Measurements on a laboratory EXAFS spectrometer described in Section 5.5.2 equipped with a Johann-type curved ($R = 0.5$ m) Si(400) monochromator crystal resulted in a copper K-edge spectra, which are given in Fig. 5.3 as a function of the resolution of the spectrometer. The resolution can be degraded by increasing the spot size (in horizontal sense) on the Johann monochromator. The mentioned resolutions are calculated from linewidth measurements on the characteristic tungsten impurity lines present on the bremsstrahlung of the molybdenum anode. The results presented in Fig. 5.3 show that a resolution of 2 to 4 eV is necessary to measure edge and near-edge spectra. Resolutions of 2 eV at the copper-edge have been claimed in the literature for laboratory EXAFS spectrometers, using third-order reflections of Ge(111) (7) or Si(111) (8) and fourth order of Si(220) (9). With these higher-order reflections the available intensities are relatively low ($I_0 \sim 10^4$ photons s^{-1}). However, scan times for these types of edge spectra are still within 1–2 h (7–9), which is quite acceptable.

5.2.2.2. High Intensity and Reliable EXAFS Information

The photon density N_d (number of photons s^{-1} cm^{-2}) is inversely proportional to ℓ^2 where ℓ is the x-ray beam path length from focal spot to the sample. Using the formulas given in Section 5.3.2, the relation between the photon density N_d

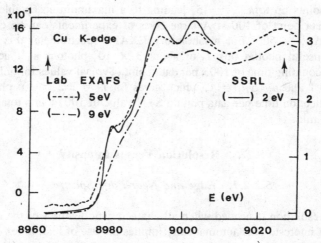

Figure 5.3. Copper K-edge spectra of copper foil (8.3 μm) as a function of the experimental resolution (rotating anode 17 kV, 180 mA).

and the d spacing of the monochromator crystal placed on the Rowland circle can be expressed as:

$$N_d \approx \frac{1}{\ell^2} \approx \frac{E^2 d^2}{R^2} \tag{4}$$

High intensities resulting in short collection times can thus be obtained with a monochromator crystal having a high d value mounted on a Rowland circle configuration with a short radius R. As will be shown in Section 5.4 such a choice (high d value, short radius) automatically leads to a low-energy resolution of the spectrometer. Therefore, a compromise must be made between intensity and resolution, which immediately puts forward the question, which kind of resolutions are acceptable for deriving reliable structural EXAFS information.

The influence of the experimental resolution on the peaks present in the radial structure function derived from the EXAFS data has been investigated in the following. Model EXAFS functions containing Pt–Pt ($R_1 = 2.77$ Å, $R_2 = 5$ Å) and Pt–O ($R_1 = 2.05$ Å, $R_2 = 4$ Å) coordinations have been calculated applying Eq. (4) of Chapter 6, Section 6.3.2. The phase and backscattering amplitude of the Pt–Pt and Pt–O absorber–scatterer combinations were obtained from EXAFS measurements on Pt-foil and Na$_2$Pt(OH)$_6$ (10).

These measurements were carried out at SSRL with a resolution of about 2–

3 eV. The influence of the experimental resolution on the actual EXAFS signal $\chi(E)$ can be simulated by

$$\chi'(E') = \int_{-\infty}^{+\infty} g(E', E)\chi(E)\, dE \qquad (5)$$

with $\chi'(E')$ the simulated "measured" signal. A Gaussian type function was chosen to represent the experimental resolution function:

$$g(E', E) = \frac{1}{(2\pi\tau^2)^{1/2}} \exp\left(-\frac{(E - E')}{2\tau^2}\right) \qquad (6)$$

The full width at half-maximum (FWHM) of this function is $2\sqrt{2 \ln 2}\,\tau$. Figure 5.4 gives the EXAFS functions and the corresponding Fourier transforms for the Pt–Pt and Pt–O coordinations as a function of $\Delta E = [(\Delta E)^2_{\text{SSRL}} + (\Delta E)^2_{\text{conv}}]^{1/2}$. The quantitative results as summarized in Table 5.1 make clear that an experimental resolution of 10 eV is a maximum value that can be allowed in order to derive reliable information of higher coordination shells. With a resolution of about 20 eV the peak of the first neighbor shell degrades to about 20%. If only information about the first coordination shell is desired the highest possible intensities can be used, since a relatively low resolution is sufficient to obtain structural information. For diluted systems a low-energy resolution has to be accepted in order to get a reasonable signal-to-noise ratio, which inevitably leads to a loss of structural information of higher coordination shells.

A decrease of the amplitude of a particular shell can be harmful when theoretical backscattering amplitudes and phases are used for the data analysis. However, amplitudes and phases can also be obtained from EXAFS data on model compounds with similar distances measured under the same conditions as the unknown material. Structural parameters obtained from these type of measurements on laboratory EXAFS spectrometers are reliable, although one still has to investigate possible errors.

5.2.3. Avoidance of Higher Harmonics

Higher harmonics in the x-ray beam form a very important class of thickness effects as discussed in Chapter 3. Higher harmonics have to be avoided as much as possible since they strongly distort the absorption edge and the EXAFS data leading to unreliable results.

When the x-ray absorption coefficient of a sample placed in a photon beam

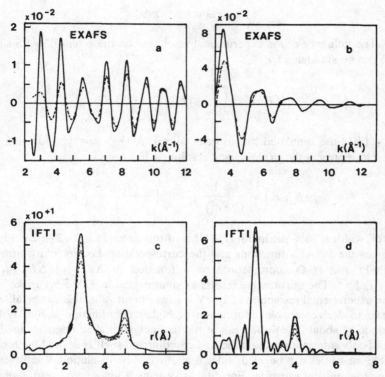

Figure 5.4. Influence of resolution on different coordination shells with different type of neighbors. (Pt—Pt: $N = 4$, $R = 2.77$ Å and $N = 4$, $R = 5$ Å; Pt—O: $N = 4$, $R = 2.05$ Å and $N = 4$, $R = 4$ Å) (a) Calculated Pt—Pt EXAFS (SSRL data: solid line; convolution with $\Delta E = 19.8$ eV: dotted line). (b) Calculated Pt—O EXAFS (SSRL data: solid line; convolution with $\Delta E = 19.8$ eV: dotted line). (c) Magnitude of Fourier transform ($|FT|$) of Pt—Pt EXAFS (SSRL data: solid line; convolution with $\Delta E = 9.5$, 14.7 and 19.8 eV: dotted lines). (d) Magnitude of Fourier transform ($|FT|$) of Pt—O EXAFS (SSRL data: solid line; convolution with $\Delta E = 9.5$, 14.7 and 19.8 eV: dotted lines).

Table 5.1. Influence of Resolution on EXAFS Data[a]

Total Resolution	Convolution		Pt—Pt		Pt—O	
			2.77(Å)	5(Å)	2.05(Å)	4(Å)
ΔE (eV)	ΔE_c (eV)	τ (eV)	D (%)	D (%)	D (%)	D (%)
3	0	0	—	—	—	—
10	9.5	4	7.5	18	4.2	17
15	14.7	6.3	15	32	8.5	29
20	19.8	8.4	23	48	15.5	44

[a]D is the relative decrease of the amplitude of the corresponding peak in the Fourier transform.

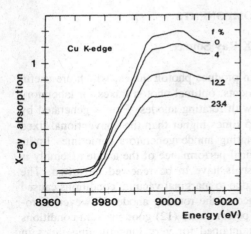

Figure 5.5. Influence of the presence of higher harmonics (fraction f) on the copper K-edge.

that contains higher harmonics is measured with ionization chambers (which are not energy discriminative), the energy response of the detection system is altered. Higher harmonics are hardly absorbed in the sample but contribute to the ionization currents in both chambers.

To investigate the effect of higher harmonics on the x-ray absorption spectrum a copper foil with a thickness of 8.3 μm was measured with a Si(400) monochromator crystal that allows the reflection of higher harmonic radiation. By varying the excitation voltage of the x-ray generator, the harmonic content can be changed. Figure 5.5 shows the influence of higher harmonics on the edge and near-edge spectrum of copper foil. The fraction of higher harmonics has been determined as described in ref. (11). Table 5.2 summarizes the results. Already a fraction of 10% higher harmonics has a large influence on the experimental results. The consequences of avoiding higher harmonics on the attainable photon flux, the choice of the type of monochromator crystal, and the performance of the rotating anode system will be discussed in Sections 5.3.1 and 5.3.3.

Table 5.2. Influence of the Fraction (f) of Higher Harmonics on the Copper Edge

kV	$f(\%)$	$\Delta\mu x$	$\Delta\mu x / \Delta\mu x_{(f=0)} \times 100\%$
17	0	1.73	100
18	4	1.60	92
19	12.2	1.32	76
20	23.4	1.11	64

5.3. COMPONENTS

5.3.1. X-Ray Source

Although fixed sealed x-ray tubes can produce photon fluxes, which are useful for a certain class of EXAFS experiments, optimal photon fluxes for laboratory EXAFS spectrometers are obtained with rotating anodes. Fluxes generated by these type of x-ray sources are 10–15 times higher than the conventional fixed sealed tubes. The complexity of rotating anode generators sometimes limits reliable operation. When the mechanical performance of the anode assembly is not optimal, bearings and vacuum seals have to be renewed too often. The lifetime of the filaments will be short due to marginal vacuum conditions caused by the limited mechanical performance of the rotating anode. However, if rotating anode sources have an optimal performance (12) good vacuum conditions even at high tube powers can be maintained for very long running times in continuous operation (1–2 months using standard vacuum seals, 1–2 years with magnetic fluid type seals). Good vacuum conditions also lead to long lifetimes of the filaments. Choosing the rotating anode as x-ray source several specific points have to be considered, which will be discussed below.

5.3.1.1. High Currents at Low Voltages

The intensity of the bremsstrahlung spectrum intergrated over the whole energy range can be expressed (13) as

$$\text{Int} \approx ZV^2I \tag{7}$$

with Z the atomic number of the target and V and I the tube voltage and current, respectively. To generate high photon fluxes the rotating anode system must be able to provide high currents (200–300 mA), even when the excitation voltage has to be kept low (10–15 kV) to avoid the generation of higher harmonics. At low voltages the maximum current is severely reduced by the space charge in the electron beam. To operate the x-ray generator at high currents with low voltages the space-charge limit can be decreased by increasing the electric field between anode and cathode. This can be done by lowering the distance s (see Fig. 5.6) between cathode and anode. Figure 5.7 gives the $V-I$ characteristic of a GX-21 (Elliot) rotating anode as obtained with a distance s of 6 mm.

5.3.1.2. Optimization of the Focal Spot

For an optimum resolution it is necessary to minimize the horizontal dimension of the focal spot on the rotating anode. The focusing of the electron beam is

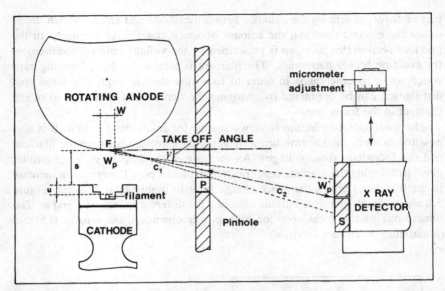

Figure 5.6. Configuration of the GX-21 (Elliot) rotating anode. (W = horizontal spot width, W_p = projected horizontal spot width, s = distance between cathode and focus cup, u = depth of filament in focus cup).

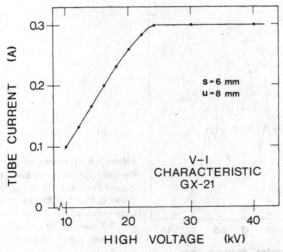

Figure 5.7. V-I characteristic of the GX-21 (Elliot) rotating anode. The maximum current limitation is caused by space charge between anode and cathode.

175

first of all dependent on the distance between cathode and anode, which determines the electric field and the amount of space charge. As discussed in the previous section this distance is prescribed by the voltage hold off requirement for avoiding higher harmonics. The filament is surrounded by a focusing cup, which has a specific shape in order to focus the electric field. The focal spot distribution can be optimized by changing the depth (u) (see Fig. 5.6) of the filament in the focus cup.

The focal spot distribution is very sensitive for small changes in u. It is also possible to focus the electron beam by applying a voltage between the filament and the focus cup (bias voltage). An increase in the bias level has a similar effect as bringing the filament deeper inside the focus cup. However, an increase in the bias level brings the space charge limit to higher tube voltages. Figure 5.8 shows focal spot distributions obtained for different values of s and u. The dimensions of the focal spot for high power operation are usually 0.5 mm (horizontal) × 10 mm (vertical).

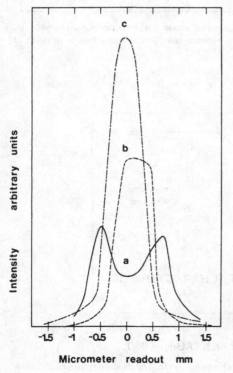

Figure 5.8. The intensity distribution of the x-ray focal spot as a function of the micrometer readout measured with the experimental setup as shown in Fig. 5.6. The following generator settings were used: (a) 20 kV, 200 mA, $s = 9$ mm, $u = 7.5$ mm; (b) 20 kV, 200 mA, $s = 6$ mm, $u = 8$ mm; and (c) 20 kV, 250 mA, $s = 6$ mm, $u = 8$ mm.

5.3.1.3. Choice of Target Material

The target material determines to a great extent the ranges of energies that are useful for EXAFS experiments. The energy regions of the high intense K_α lines and the other intense L_α lines cannot be used for taking good quality EXAFS data. Typical bremsstrahlung spectra for copper, molybdenum, and gold calculated (14) for an excitation voltage of 20 kV are given in Fig. 5.9. A good choice for the energy regions $4 < E < 10$ and $15 < E < 25$ keV is the high Z element gold, which cannot be used for $11 < E < 15$ keV where intense excitation L lines are present. The Mo anode is not useful at energies higher than the 17.5 keV (region of K_α lines and K-absorption edge). The molybdenum

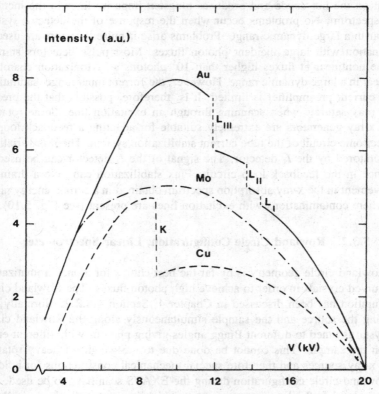

Figure 5.9. Bremsstrahlung spectra for copper molybdenum and gold calculated for an excitation voltage of 20 kV. The positions of the copper K edge and the gold L_I, L_{II}, and L_{III} absorption edges are indicated.

anode might replace the gold anode for $11 < E < 15$ keV. Another important advantage of using high Z elements as target material is the relatively high intensity produced by these elements at low take-off angles. A low take-off angle makes the effective horizontal focal spot size small, which in turn leads to better resolutions (see Section 5.4).

5.3.1.4. Flux Stabilization

Large variations of the incident flux are produced by intense excitation lines in the bremsstrahlung spectra of the x-ray tube. Even with an optimum choice of the target material spurious excitation lines will be present due to impurities in the target material. Also, slow evaporation of the tungsten filament gives a deposit on the hot anode and produces tungsten impurity lines in the incident x-ray spectrum. No problems occur when the response of the detector system is linear in a large dynamic range. Problems arise if pulse detectors are used in combination with large incident photon fluxes. Most pulse detectors start to become nonlinear at fluxes higher than 10^5 photons s^{-1}. Ionization chambers are linear in a large dynamic range. However, the current input range (saturation) of the current preamplifier is limited. It is, therefore, possible that the preamplifier may saturate when scanning through an excitation line. Some rotating anode x-ray generators are extremely suitable for adapting a feedback loop in the electronic circuit of the tube current stabilization system. The flux variations are monitored by the I_1 detector. The signal of the I_1 detector can be used as reference in the feedback loop circuit. Flux stabilization can give a dramatic improvement in the x-ray absorption spectrum obtained in a certain energy range (15) where contaminations with excitation lines are present (see Fig. 5.10).

5.3.2. Rowland Circle Configuration, Linear Spectrometer

The Rowland circle geometry is by far the best choice for monochromatization with curved crystals in order to achieve high photon fluxes. The Rowland circle configuration has been discussed in Chapter 4, Section 4.5.2.2. Normally, by changing the source and the sample simultaneously along the Rowland circle the crystal is tuned to different Bragg angles giving photons with different energies on the sample. This cannot be done due to the weight of heavy rotating anode x-ray sources and therefore special mechanical constructions for holding the Rowland circle configuration during the EXAFS scan have to be used.

Very useful is the linear spectrometer, which is presented schematically in Fig. 5.11. The situations for three different energies are drawn. The x-ray focal

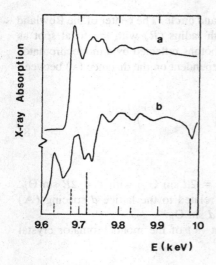

Figure 5.10. X-ray absorption spectrum of ZnO with (curve *a*) and without (curve *b*) x-ray flux stabilization. The positions of the major tungsten excitation lines are given.

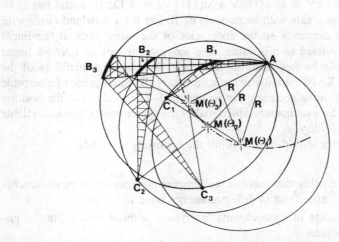

Figure 5.11. The linear spectrometer with positions for three different energies (for angles θ_1, θ_2, and θ_3). *A*: fixed focal point; *B*: center of monochromator crystal; *C*: sample position and *M*: center of Rowland circle.

spot (A) is the only fixed point on the Rowland circle. The center of the Rowland circle (M) itself moves along a circle with radius (R) with the focal spot as central point. The wavelength (λ) of the photons reflected by a monochromator crystal (with lattice spacing d) is linearly dependent on the distance (x) between focal spot and crystal:

$$\lambda = \frac{dx}{R} \qquad (8)$$

This follows from the Bragg condition ($\lambda = 2d \sin \Theta_B$) with $x = 2R \sin \Theta_B$. The energy $E(kV)$ of the spectrometer is related to the lattice d spacing (Å) and the Bragg angle Θ_B by $E = 12.396/2d \sin \Theta_B$.

The accuracy (Δx) of the displacement (x) of the monochromator crystal along the x axis is then given by

$$\Delta x = \left(\frac{\Delta E}{E}\right) \frac{12.396\ R}{Ed} \qquad (9)$$

To make steps of 1 eV at 15.000 eV a Si(111) ($d = 3.1353$) crystal has to be displaced along the x axis with increments of 10 μm for a Rowland circle with $R = 0.5$ m. The demands on the accuracies of the other parts of the linear spectrometer are related to this value. The moving parts of an EXAFS linear spectrometer should be free from any backlash. A good reproducibility of the energy scale for EXAFS measurements is necessary. By making use of electronic rulers a high positioning accuracy can be obtained. The read out of the position can be controlled by a computer, which makes the spectrometer fast and reliable with a good resettability.

The spectrometer should further fulfill the following demands:

1. An easy tunability that makes a realignment unnecessary if measurements have to be carried out in different energy ranges.
2. A rapid change of monochromator crystals without new tedious alignment procedures.
3. Easy and rapid change of the radius of the Rowland circle. This makes it possible to match the system to the curvature of the bent monochromator crystal. It also allows us to adapt the spectrometer to commercially available monochromator crystals.

4. A minimum load on the sample stage of about 25–30 kg weight without any relevant influence on the positioning accuracy.
5. Easy alignment with respect to:
   a. Take-off angle (see Fig. 5.6)
   b. Positioning of the focal point of the x-ray source onto the Rowland circle perpendicular to the used vertical focus.

5.3.3. Monochromator Crystals

The most critical part of the laboratory EXAFS facility is a curved monochromator crystal of high quality. For the Rowland circle geometry (radius R) two types of curved crystals are suitable (a) Johann: diffraction planes bent with radius $2R$, (b) Johansson: idem as (a) but also ground with radius R (see Chapter 4, Fig. 4.16). Optimal resolution with a high intensity can be obtained with Johansson type crystals. However, it is difficult to manufacture (bend, grind, and polish) an undistorted Johansson crystal, which makes the Johann type crystal worthwhile as an alternative. Single crystalline silicon and germanium material of large dimensions and of high quality are commercially available. Commercial sources for silicon and germanium Johansson type crystals are mentioned in recent literature (7, 8, 16). It has to be noted that the EXAFS monochromator crystals must be cut very carefully from the bulk material, in order to have no difference in angle (α) between the surface normal and the normal of the diffraction planes ($\alpha < 0.05°$) (15).

A value of 0.5 m for the radius of the Rowland circle takes care of a good compromise between the demands on resolution, high quality curved crystals and dimensions of the linear spectrometer. Using this R value, the type of index planes and the d value of these planes determine which type of monochromator crystal is most suitable to fulfill the requirements of harmonic rejection, resolution, intensity, or both. Ge(111) crystals are suitable for high intensities and medium resolutions for $4 < E < 7$ keV, whereas Ge(311) is extremely useful for $7 < E < 11$ keV. A Si(311) monochromator is the best choice for high intensities and medium resolution for $11 < E < 15$ keV.

Different methods for bending crystals have been described in the literature (17–22). Figure 5.12 shows the crystal bender as developed by Maas and Brinkgreve (12). A Si(400) crystal (thickness 400 μm) has been glued onto a U-shaped backing block of a well defined thickness and made of a special selected material. By pushing the ends of the U-shaped backing block in opposite directions, the monochromator crystal is bent along a very defined cylindrical surface (measured deviation from $2R = 1$ m is less than 5 μm). The copper-absorption edge spectra as given in Fig. 5.2 were measured with this type of crystal and bender.

5.3.4. Detectors

An extensive review of detectors suitable for laboratory EXAFS spectrometers has been given in ref. (23). Only the most relevant information of the gas proportional counter and the ionization chamber will be briefly discussed here.

Normal gas proportional counters can handle count rates up to 10^6 s^{-1} in the transmission mode. An important advantage of proportional counters is their energy discrimination. Higher harmonics can be rejected, which permits operation of the x-ray tube at higher voltages leading to higher photon fluxes. Multiwire proportional counters with separate electronics for each wire might be used to deal with count rates higher than 10^7 s^{-1}. Great care should be taken to the position and the maximum photon flux of each independent wire system.

Ionization chambers cannot discriminate against higher harmonics but their linearity makes them useful in a large range of photon fluxes (10^5–10^8 s^{-1}). Good, rigid, noise free ionization chambers can easily be built in a standard equipped workshop. Power supplies should have a voltage stability of 1 on 10^4, whereas current amplifiers should be used with extremely low-noise characteristics ($< 10^{-14}$ A). The best operation conditions are guaranteed when the current preamplifier is mounted directly on the ionization chamber with a very short connection wire between collection plate and preamplifier input. Low noise current preamplifiers can be commercially obtained. Special attention should be paid to the material used for the resistance (and capacitor), which determines the amplification factor of the preamplifier.

Figure 5.12. The Johann crystal bender developed by Maas and Brinkgreve (12).

5.3.5. Automating

Automating of EXAFS experiments is absolutely necessary for obtaining data that can be processed and analyzed at a later stage. A preliminary on line analysis should be possible (background subtraction and fast Fourier transformation) in order to investigate the quality of the EXAFS data for making a rapid decision for more and better scans. The reader is further referred to an excellent review of hardware and software considerations of EXAFS spectrometers given in the proceedings of the University of Washington workshop (4).

5.4. CRYSTAL OPTICS AND FOCUSING

5.4.1. Introduction

With an optimally mechanically constructed linear spectrometer, the curved monochromator crystal determines the attainable resolution and intensity. In this section the different contributions to energy resolution will be analyzed and the results of a Monte Carlo ray-tracing computer analysis (24) will be given. With a ray-tracing computer analysis it is possible to calculate the influence of different types of misalignments and to determine the tolerancies of the different components of the linear spectrometer. Both the Johann and the Johansson type of arrangements will be investigated. Based upon the calculations it is possible to select a particular crystal for a certain energy and to predict the intensity that belongs to a certain resolution.

The following contributions to the energy resolution of monochromator crystals can be distinguished:

1. Rocking curve ($\Delta\Theta_c$).
2. Dimensions of the crystal diffraction surface.
3. Finite dimensions of the source.

Berreman (25) computed the reflectance of symmetrically curved ideally perfect crystals in several wavelength ranges. He found that the rocking curves of Si(220) and Ge(220) are hardly affected by bending the crystals with a radius of 1 m. The rocking curves of Si(400) and Ge(400) are broadened by a factor 2 and 1.4, respectively. The influence of the crystal rocking curve ($\Delta\Theta_c$) will still be small in comparison to the contributions 2 and 3 [$\Delta\Theta_c \approx 0.4$ eV for Si(400) at 9000 eV]. The x-rays striking the monochromator crystal have a spread in incident angles $\Delta\Theta_g$ due to the geometry, which is determined by the dimensions of the crystal diffraction surface and the finite dimensions of the x-ray spot. This spread in incident angles determines the resolution if $\Delta\Theta_g \gg \Delta\Theta_c$. For maximum efficiency the rocking curve width $\Delta\Theta_c$ should be equal to $\Delta\Theta_g$. When geometrical effects dominate, the monochromator will reflect only a fraction of the photons incident in the bandwidth determined by $\Delta\Theta_c$. Diffraction efficiency can be increased by increasing the rocking curve of the crystal. By using, for instance, Ge(400) instead of Si(400), the intensity of the diffracted beam is higher due to the 2.5 times larger rocking curve width of Ge(400).

Some calculations of the energy resolution given in the literature (26) were based upon crystals with mosaic-type structures with large rocking curves (27). An analytical analysis of the Johann and Johansson focusing arrangements has been given by Lu and Stern (28) to show the practical usefulness of the Johann arrangement. For this analysis the finite dimensions of the source was not taken into account. The following calculations will include the finite dimension of the source assuming perfect crystals with $\Delta\Theta_c < \Delta\Theta_g$.

5.4.2. Analysis of Factors Determining the Energy Resolution

5.4.2.1. Horizontal Divergence

The main difference between the Johann and Johansson configuration is the contribution of the horizontal divergence. For a Johansson monochromator crystal all photons emitted by an x-ray point focus will have exactly the same incident angle with the crystal planes, which means that there are no energy aberrations in the horizontal plane. For the Johann configuration the horizontal aberrations are given by

$$\left(\frac{\Delta\Theta}{\Theta}\right)_{hor} = \frac{(-\cos\Theta)(1 - \cos\phi)}{(\sin\Theta)(\tan\Theta - \sin\phi)} \tag{10}$$

Θ is the angle between the crystal planes and the direction of the rays from the source to the origin of the crystal. ϕ is a parameter used to describe the position on the crystal surface ($\phi \approx w/2R$, with w the distance between a point on the crystal surface in the plane of the Rowland circle and the center of the crystal). By substituting the Bragg relation [E (keV) $= 12.396/2d \sin\Theta$] the relative energy resolution $(\Delta E/E)_{hor}$ can be given as a function of E, d, w, and R:

$$\left(\frac{\Delta E}{E}\right)_{hor} \approx \frac{-w^2[(Ed)^2 - 38]}{307R^2} \tag{11}$$

By taking crystals with a smaller d spacing or a larger bending radius the contribution of $(\Delta E/E)_{hor}$ can be diminished. For a normal experimental setup $(Ed)^2$ is larger than 38. Since $(\Delta E/E)_{hor}$ is a quadratic function of w the energy resolution function will therefore only contain aberrations to lower energy values. Equation (11) can be integrated along the crystal surface in order to obtain the mean and the standard deviation of the horizontal broadening factor for a certain value of w. More accurate results are obtained with the ray-tracing program. Figure 5.13 shows the energy resolution calculated with the ray-tracing program as a function of w at 10 and 15 keV for a Si(400) ($d = 1.358$ Å) Johann monochromator with $R = 0.5$ m. It can be seen that for small irradiated areas on the crystal the Johann crystal is comparable with the Johansson crystal. With a spectrometer with $R = 0.5$ m, the difference in horizontal energy broadening is only 2 eV at 10 keV when the Si(400) crystal is irradiated over 20 mm.

5.4.2.2. Vertical Divergence

Vertical divergence arises from x-rays that are reflected outside the (horizontal) plane of the Rowland circle. Normally, a (vertical) line focus in combination with a cylindrically bent crystal is used. The effect of vertical divergence, which is almost equal for the Johann and Johansson configuration is given by:

$$\left(\frac{\Delta\Theta}{\Theta}\right)_{vert} = \frac{(h - f)^2}{8R^2 \sin^2\Theta} \tag{12}$$

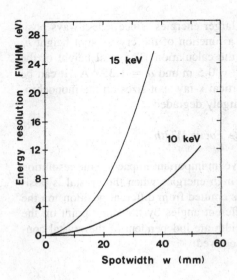

Figure 5.13. The influence of horizontal beam divergence. The energy resolution of a Johann monochromator has been given as a function of the distance (w) between the center of the crystal and a point on the crystal surface in the plane of the Rowland circle.

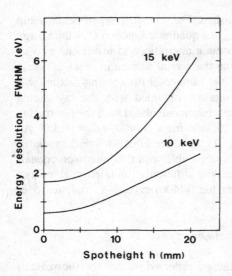

Figure 5.14. The influence of vertical beam divergence. The energy resolution of a Johann- or Johansson-type crystal monochromator has been calculated as a function of the vertical spot height (taken as the vertical distance along the crystal surface with respect to the plane of the Rowland circle).

with h the height of the crystal and f the height of the focal spot, both measured with respect to the origin of the crystal and focal spot, respectively. Using the Bragg relation, $(\Delta E/E)_{vert}$ can be calculated:

$$\left(\frac{\Delta E}{E}\right)_{vert} = \frac{(h-f)^2(Ed)^2}{307R^2} \tag{13}$$

$(\Delta E/E)_{vert}$ causes only aberrations to larger energies, since it is always positive. Figure 5.14 gives $(\Delta E/E)_{vert}$ as a function of the crystal spot height h. The ray-tracing program was used for the calculation. The total height of the x-ray source was taken 10 mm, with $R = 0.5$ m and $d = 1.353$ Å. It can be seen that for high energies and large vertical x-ray spot-sizes on the monochromator crystal the energy resolution is largely degraded.

5.4.2.3. Finite Focus Width

The width of the x-ray focal spot can have an important impact on the resolution function. This is especially the case at high energies when the crystal is positioned close to the x-ray source. Photons emitted from different positions on the finite source will be reflected under different angles by the same point on the crystal. Since the effects of the focus width are independent of the crystal configuration, a general expression can be derived:

$$\left(\frac{\Delta\Theta}{\Theta}\right)_{foc} = \frac{-\Delta\,\sin\,(\gamma + \frac{1}{2}\phi)\,\cos\,\Theta}{2R\,\sin\,(\Theta + \frac{1}{2}\phi)\,\sin\,\Theta} \tag{14}$$

with γ the take-off angle and Δ the half-width of the focal spot. The finite focus width causes a symmetric energy aberration. It can be seen from Eq. (14) that this aberration can be decreased by

1. Decreasing the width of the focal spot.
2. Decreasing the take-off angle.
3. Choosing the optimal side of the crystal, that is, the side of the crystal that has the largest distance to the focal spot.

Figure 5.15 gives the results of the ray-tracing program [$\Delta = 0.25$ mm, $R = 500$ mm, Si(400) as monochromator], showing that at high energies ($E > 15$ keV) a much better resolution can be obtained with a lower take-off angle.

5.4.2.4. Imperfect Grinding of the Crystal

As mentioned in Section 5.3.3, the normal vector of the diffraction planes does not necessarily have to coincide with the normal vector of the mechanical crystal surface. A mismatch between the normal vectors causes an additional deviation

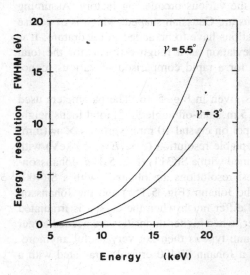

Figure 5.15. The influence of the finite focus width of the x-ray tube. The energy resolution has been determined as a function of the photon energy with the take-off angle as parameter ($\gamma = 5.5°$ and $3°$).

from the Rowland circle and thus a degradation in energy resolution. It can be shown that the deviation α between the normal vectors has to be lower than 0.2°. Larger deviations result in a tailing of the energy resolution function.

5.4.2.5. Angular Broadening

An effect that is of minor importance is the angular broadening (29). Due to the penetration of the photons in the monochromator crystal, photons are reflected from inner crystal planes. These planes are not exactly on the Rowland circle and have, therefore, different incident angles. The broadening is given by:

$$\left(\frac{\Delta\Theta}{\Theta}\right)_{ang} = \frac{-\cos\Theta}{\sin\Theta}\left(2\mu R\cos\Theta\right)^{1/2} \tag{15}$$

with μ the x-ray absorption coefficient of the crystal. The effect of angular broadening can normally (low value of μ) be neglected with respect to the effects mentioned previously.

5.4.3. Resolution as a Function of Energy

To get an impression of the total energy resolution of a particular Rowland circle configuration equipped with a specific type of crystal, Eqs. (10), (12), and (14) can be used to derive a general equation. This can be done by calculating the mean and the standard deviation of the various broadening factors. Assuming that all the energy resolution functions are Gaussian shaped, which is of course an approximation, the standard deviations have to be added in quadrature. It is possible to use the total standard deviation as a rough estimate for the total energy broadening. This may serve for a rapid comparison of various monochromator crystals.

The result of such a calculation is given in Fig. 5.16. The parameters used for the experimental setup are $R = 0.5$ m, take-off angle 5.5°, total focus height 10 mm, total focus width 0.5 mm, spot on crystal 10 mm (vertical) $\times$ w (mm) (horizontal) with w as variable. Acceptable resolution ($8 < E < 15$ keV) with an optimal curved crystal can be obtained with a Si(311) (Fig. 5.16a) Johansson-type monochromator crystal. The best resolutions are obtained with a Si(400) monochromator. The resolution of the Johann (Fig. 5.15b) and the Johansson (Fig. 5.16c) monochromators do not differ much when the crystal is irradiated with a spot of 20-mm width. However, when higher intensities are needed larger spot sizes are necessary and the Johann type is then not very useful anymore. Figure 5.16d gives the resolution of a Johann Si(400 crystal) irradiated with a

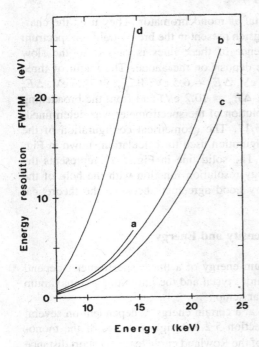

Figure 5.16. The total energy broadening as a function of photon energy for various monochromator crystals: (*a*) Johansson Si(311), $w = 20$ mm; (*b*) Johann Si(400), $w = 20$ mm; (*c*) Johansson Si(400), $w = 20$ mm and (*d*) Johann Si(400), $w = 50$ mm.

horizontal spot size of 50 mm. This curve may be compared with the results obtained for the Johansson crystal irradiated with horizontal spot size of 20 mm, since the resolution for a Johansson monochromator is independent of the horizontal spot size.

In the calculations mentioned previously it is assumed that all parts of the spectrometer are perfectly aligned. Due to a lack of vertical parallelism of source and crystal and also due to imperfectly ground or bent crystals an additional degrading of the resolution may be present. The experimental resolution can then be improved by using a receiving slit that is located on the Rowland circle where the reflected x-ray beam is converging. The width of this slit has to enter the calculation of the experimental energy resolution function. Such calculations were performed by Thulke et al. (8). They determined the actual resolution of their spectrometer equipped with a Ge(311) crystal, by measuring the FWHM of the copper $K_{\alpha 1}$ emission line. This line has a natural linewidth of about 2 eV. The actual energy resolution was found to be 4.5 eV, which was in excellent agreement with their calculations. Brinkgreve et al. (30) reported on measurements of the energy resolution of a laboratory EXAFS spectrometer with a special (31) Si(111) Johansson crystal as monochromator. They used the characteristic x-ray emission lines of tungsten present in the bremsstrahlung spectrum of their x-ray generator. The presence of these lines is caused by the slow evaporation of the filament and the deposit on the anode. The width of three lines were measured ($WL_{\alpha1}$: 8397.6 eV, ΔE_n = 6.5 eV; $WL_{\beta1}$: 9672.3 eV, ΔE_n = 6.9 eV, and $WL_{\gamma1}$: 11,285.9 eV, ΔE_n = 10.2 eV) and from the broadening of these lines the actual energy resolution of the spectrometer was determined. The results are presented in Fig. 5.17. The geometrical configuration of the spectrometer is identical to the configuration used for calculation shown in Fig. 5.16. No entrance slits were used. The solid line in Fig. 5.17 represents the result of the calculation of the energy resolution function with the help of the ray-tracing program. There is a very good agreement between the theoretical and experimental results.

5.4.4. Intensity and Energy

The range of minimum and maximum energy of a linear spectrometer depend on the d-spacing of the monochromator crystal and the minimum and maximum anode-crystal distances of a particular setup.

The intensity that can be attained at a certain energy is dependent on several factors. As already mentioned in Section 5.2.2.2 high d values of the monochromator crystal and a short radius of the Rowland circle lead to a short distance between the anode-focal spot and the crystal. A short distance increases the photon density at the sample position. It also implies that by using a certain crystal the photon flux decreases going from the minimum to the maximum source-crystal distance. For the spectrometer described in (30) the photon flux changes then by a factor of 16. The intensity at a certain energy is further determined by the x-ray spot size on the crystal. The choice of these parameters (d-spacing, R-value, and spot size) is determined by a compromise between intensity and resolution.

The structure factor of the diffraction planes of a particular type of monochromator crystal is also important for the intensity that can be obtained at a certain energy. The reflectivity and thus the rocking curve depend on the structure factor of the index plane that has been used. As mentioned already in Section 5.4.1 the intensity of the diffracted beam can be increased by making use of germanium instead of silicon crystals due to the larger rocking curves of germanium. The structure factor also prescribes the maximum excitation voltage of the x-ray tube. This voltage has to be kept below the value, which starts to generate higher harmonics. For measuring copper EXAFS (~9 keV) 18 kV is a maximum value for a Si(400) crystal, whereas 27 kV can be used for Si(111) or Si(311). The structure factor for the (222) or (622) reflection is zero, implying

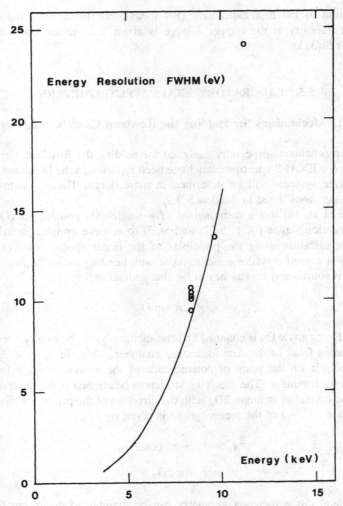

Figure 5.17. The energy resolution of the laboratory spectrometer (30) measured (open circles) as a function of photon energy using a special Si(111) Johansson monochromator. The solid line represents the calculated energy resolution.

no reflection of the first harmonic. This means that for Si(400) the optimal diffracted intensity at the copper K-edge is about 2.25 times lower than for Si(111) or Si(311).

5.5. LABORATORY EXAFS SPECTROMETERS

5.5.1. Mechanisms for Holding the Rowland Circle Configuration

Different mechanisms especially designed for holding the Rowland circle configuration for EXAFS spectrometers have been reported in the literature. In this section three systems will be described in more detail. These systems might fulfill the demands listed in Section 5.3.2.

Thulke et al. (8) use a combination of a single-axis goniometer (GO) and three translation stages (SC1, SC2, and SC3) to achieve optimal focusing and monochromatization using the principles of the linear spectrometer (see Fig. 5.18). For a curved crystal monochromator with bending radius $2R$ the distance x_c between source and crystal has to be changed according to:

$$x_c = 2R \sin \Theta_B \qquad (16)$$

when the Bragg angle Θ_B is changed by rotating the crystal. Normally a receiving slit is used in front of the first ionization chamber. This slit is located on the Rowland circle on the point of convergence of the x-rays reflected from the crystal monochromator. The receiving slit has to be aligned at the same distance x_c from the crystal at an angle $2\Theta_B$ with the direction of the primary x-ray beam. The position (x_s, y_s) of the receiving slit is given by

$$x_s = x_c + x_c \cos 2\Theta_B \qquad (17)$$

$$y_s = x_c \sin 2\Theta_B \qquad (18)$$

The Rowland circle focusing geometry can be maintained during an EXAFS scan by changing x_c, Θ_B, and x_s, y_s according to Eqs. (16)–(18). This requires two motions of the monochromator crystal (x_c, Θ_B) and two of the receiving slit (x_s, y_s). The crystal (CR) is rotated (Θ_B) by the goniometer (GO) placed on the lowest translation stage (SC1), which moves according to Eq. (16). This translation stage moves at the same time the other two stages SC2 and SC3 with the detector platform according to the first term of the sum in Eq. (17). The second term of this sum $(x_c \cos 2\Theta_B)$ is realized by a movement of the upper

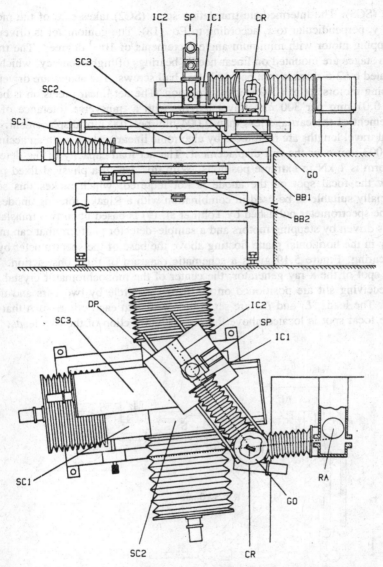

Figure 5.18. Side view (upper half) and top view (lower half) of the EXAFS spectrometer built by Thulke et al. (8). RA: rotating anode; CR: monochromator crystal; IC1 and IC2: ionization chambers; SP: sample cell; GO: single axis goniometer; SC1, SC2, and SC3: lower, intermediate, and upper sliding carriage, respectively; BB1 and BB2: axial ball bearings; DP: detector platform.

193

stage (SC3). The intermediate translation stage (SC2) takes care of the movement y_s perpendicular to x_s according to Eq. (18). The goniometer is driven by a stepping motor with minimum angle increments of 10^{-3} degrees. The translation stages are mounted on linear needle bearings of high accuracy, which are operated by 2.5-threads/cm backlash-free ball screws. The stages are driven by stepping motors set at 1000 steps/revolution. The net linear precision is better than 0.01 mm per 300 mm travel length, with a transverse tolerance of the movement of the carriage less than 0.004 mm per 100 mm displacement. The actual travel lengths are monitored by electronic linear scalars, with an accuracy of 0.003 mm over the full displacement. The net load capacity of the detector platform is 1 kN in extreme positions. With this design a physical fixed point below the focal spot on the anode is not required, which makes this setup especially suitable to be used in combination with a Rigaku rotating anode.

The spectrometer published by Tohji et al. (9) is based upon two translation stages driven by stepping motors and a sample–detector platform that can move freely in the horizontal plane floating above the base of the spectrometer by an air bearing. Figure 5.19 gives a schematic diagram of this construction. The focal spot of the x-ray generator, the center of the monochromator crystal and the receiving slit are positioned on the Rowland circle by two bars and three leads. The leads, L_1 and L_2, are set perpendicular to each other, such that the anode focal spot is located above the virtual intersection of the two leads. The

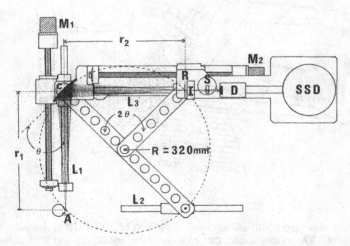

Figure 5.19. The EXAFS spectrometer published by Tohji et al. (9). A: rotating anode; C: monochromator crystal; R: receiving slit; I: ionization chamber; S = sample positioner; D: detector; M_1 and M_2: stepping motors; L_1, L_2, and L_3: leads.

monochromator is positioned at one end of the bar with fixed length $2R$, such that the normal on the crystal surface is parallel to the bar and goes always through the center of the Rowland circle. Both ends of this bar slide along the two leads. The end of this bar, where the crystal is located, is displaced by means of stepping motor M_1. This displacement causes the variation of the Bragg angle Θ_B. The other bar with fixed length R rotates at one end around the center of the Rowland circle, while the other end is connected to the sample–detector platform. This end is able to rotate around the axis determined by the position of the receiving slit. The sample–detector platform with receiving slit slides along lead L_3, which is driven by stepping motor M_2. This platform can move freely above the base of the spectrometer by means of the air bearing. When stepping motors M_1 and M_2 are scanning through an x-ray absorption spectrum the receiving slit follows on the Rowland circle the point of convergence of the x-rays reflected by the monochromator crystal. A smooth translation motion without clearance has been obtained by making use of a circulating ball linear bearing.

The linear spectrometer as described by Brinkgreve et al. (30) is shown in Fig. 5.20. The spectrometer is based upon a mechanism consisting of the following elements:

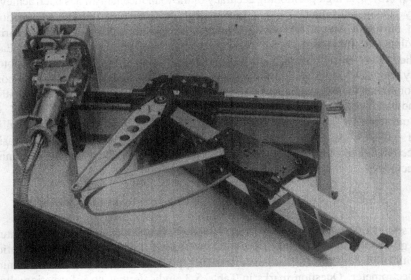

Figure 5.20. Photographic view on the EXAFS linear spectrometer designed and constructed by Brinkgreve et al. (30).

1. The main slide that moves along the main guide in order to change the distance between the x-ray focal spot and the monochromator crystal.

2. The monochromator support, which is attached to the main slide by means of a rotating axis.

3. The sample detector slide, which moves along a guide connected to the main slide by means of the same rotating axis as used for the monochromator support.

4. Three arms of equal length with one common central axis (center of the Rowland circle):

   —One arm from the anode source point.

   —The monochromator arm rigidly attached to the monochromator support.

   —The sample detector arm from the sample focal spot.

The elements described maintain the Rowland circle configuration as long as the distances from the x-ray source to the monochromator axis and from this axis to the sample focal spot are kept equal within the design specifications. The angle between the monochromator crystal and the x-ray beam changes during the displacement of the main slide along the main guide according to Eq. (16). The high positioning accuracy of the total mechanism has been achieved by paying full attention to the kinematical and statistical design specifications of the total layout and its components. In addition, all bearings and bearing points are preloaded in order to eliminate all virtual and actual clearance. Each slide is driven by transmission friction wheels that are coupled to a DC motor. By using DC motors and ruler systems of high mechanical resolution the positioning accuracy is realized by a computer electronic feedback system. With a radius $R = 0.5$ m of the Rowland circle and a Si(400) crystal ($d = 1.3567$ Å) "mechanical" incremental steps of 0.2 eV can be obtained with a resetability of ± 0.1 eV at the copper-edge (9 keV). The sample detector slide can easily accommodate a load capacity of 2.5 kN, without a relevant change in positioning accuracy.

5.5.2. A Survey of Spectrometers Known from the Literature

A survey of laboratory EXAFS spectrometers as known from the recent literature is given in Table 5.3. All systems are based upon a Rowland circle configuration with curved crystals (Johann or Johansson type) to be used as an energy scanning spectrometer. Question marks in Table 5.3. indicate that no information about a particular subject can be found in the corresponding paper.

Table 5.3. A Comparison of Laboratory EXAFS Facilities

	Cohen et al. (33)	Georgopoulos and Knapp (32)	Stern et al. (4), (24)	Khalid et al. (16)	Thulke et al. (8)	Williams (7)	Tohji et al. (9)
X-Ray source							
Standard (S)/Rot. anod (RA)	RA	RA	S	RA	RA	S	RA
Type	Rig. RU 200	Elliot GX-21	Picker	Rig. RU 200	Rig. RU 200	Rigaku	Rig. RU 200
Maximum power (kW)	12	15	1.2	12	12	2	12
Target material	Mo	Au	W, Ag	Mo	Au, Ag	Mo	Mo
Focal spot ($f \times W$ mm^2)	10 × 0.5	10 × 0.5	10 × 0.75	10 × 0.5	10 × 0.5	?	10 × 0.5
Take-off angle (°), Wp (μ)	6, 50	6, 50	4, 52	3, 30	6, 60	?, 40	6, 50
Rowland circle							
Radius (mm)	350	200	510	500	350	400	320
Fixed (F)/Adaptable (A)	F	F	?	A	A	A	?
Linear Spectrometer (LS)/ moving anode (MA)	LS	LS	MA	LS	LS	MA	LS
Number of motors	1	1	1	4	4	1	2
Position readout (Y/N)	N	N	N	N	Y	N	N
Load on detector stage (KN)	?	?	1.5	?	1	?	?
Monochromator crystals							
Johann (J), Johansson (JS)	JS	J	J	Flat/JS	JS	JS	JS
Material	LiF (220)	Si (400)	Si (400)	Si (311)	Si (111), Si (311), Ge (311)	Ge (111)	Si (220), LiF (220)
Commercial (Y/N)	?	?	Y	Y	Y	Y	Y
Dimensions ($h_c \times w_c$ mm^2)	25 × 25	?	?	30 × 50	10 × 40	20 × 127	20 × 50

Table 5.3. (Continued)

	Cohen et al. (33)	Georgopoulos and Knapp (32)	Stern et al. (4), (24)	Khalid et al. (16)	Thulke et al. (8)	Williams (7)	Tohji et al. (9)
Receiving slit							
Dimensions ($h_r \times w$, mm^2)	?	10×0.05	?	$? \times 0.01$	10×0.15	12×0.076	$? \times 0.1$
Focus on slit (Y/N)	Y	Y	Y	?	Y	Y	Y
Resolution							
A. EXAFS							
ΔE at E (eV)	9.5/8600	6/8980	5/9980	6.2/8397	4.6/8200	10.8/8200	?
Crystal	LiF(220)/JS	Si(400)/J	Si(400)/J	Si(311)/JS	Ge(311)/JS	Ge(111)/JS	LiF/JS
Measured	Zn $K_{\alpha 1}$ $K_{\alpha 2}$	?	?	W$L_{\alpha 1}$	Cu $K_{\alpha 1}$	Cu $K_{\alpha 1}$	?
Calculated	8.4	?	?	4.7	4.5	5.5	?
B. XANES							
ΔE at E (eV)	—	—	—	2.1/8980	2/8980	1.5/8200	2/8980
Crystal	—	—	—	Si(311)/2xFlat	Si(333)/JS	Ge(333)/JS	Si(440)/JS
Measured (M)/Estimated (E)	—	—	—	E	E	M	E
Figure copper-edge (Y/N)	—	—	—	Y	Y	Y (Cu $K_{\alpha 1}$)	Y

198

Intensity					
A. EXAFS					
Photons s⁻¹/energy (eV)	5×10^7 / 8980	4×10^7 / 7-10(keV)	5×10^5 / 8980	5×10^6 / 8200	6×10^6 / 8980
Crystal	LiF(200)	Si(400)	Si(400)	Ge(311)	Ge(111)
Calculated (C)/Estimated (E)	E	E	C	C	E
Target/kV⁻¹/mA⁻¹	Mo/50/200	Au/?/?	W/?/?	Au/24/200	Mo/23/30
B. XANES					
Photons s⁻¹/energy (eV)	—	—	—	5×10^4 / 8980	3.7×10^5 / 8980
Crystal	—	—	—	Si(333)/JS	Ge(333)/JS
Scantime (h)	—	—	—	2-3 h	1 h
Detectors					
Gas proportional counter (G)	G	G	—	—	—
Scintillation (S)	—	—	—	S (NaI)	—
Ionization chamber (I)	—	I	I	I	I

	3×10^7 / 8980
	LiF
	C
	?/?/?
	?
	Si(440)/JS
	5 h
	—
	S
	—

Values for the resolution at a certain energy are classified into three categories: (a) determined via measurements of a particular excitation line, (b) calculated via different theoretical approaches, and (c) estimated values. Some authors calculated the photon flux at a certain energy from the measurement of the current produced by the ionization chambers. Others estimated the photon flux probably from the statistics of the experimental data. Sometimes authors indicated that components of their EXAFS apparatus were commercially obtained. This is also mentioned in Table 5.3. In some papers laboratory EXAFS data have been compared with synchrotron EXAFS data. These results will be discussed in Section 5.6.1.

Table 5.3 makes clear that a standard x-ray tube can be used as a source for an EXAFS spectrometer with curved crystals. Stern (34) showed that a photon flux of about 5×10^5 s^{-1} can be obtained at the copper-edge with a resolution of 5 eV using a Si(400) crystal. Measurements were carried out with ionization chambers as detectors. Williams (7) claimed fluxes of 6×10^6 s^{-1} with a resolution of 10 eV at the copper-edge by utilizing a Ge(111) Johansson monochromator (127 mm horizontal spotwidth) and a NaI scintillation detector. His x-ray source is a standard molybdenum-tube (2 kW).

Most spectrometers utilize a rotating anode as an x-ray source. Good resolutions with relatively high intensities were obtained by Thulke et al. (8). With a Rigaku RU 200 x-ray source and a Ge(311) Johansson monochromator crystal (40-mm horizontal spot width) they measured 5×10^6 photons s^{-1} within a bandpass of 4.5 eV at an energy of 8.2 keV. Very high intensities with a moderate resolution were reported by Brinkgreve et al. (30). Using the linear spectrometer especially designed for the Elliot type GX 21 rotating anode (see also Section 5.5.1) and a Si(111) Johansson monochromator crystal (70-mm horizontal spot width) a resolution of 11 eV was measured with an intensity of 3×10^7 photons s^{-1} at the copper K-edge.

5.5.3. Optimization

Table 5.3 makes clear that a large range of intensities (10^4–10^6 photons s^{-1}) and energy resolutions (2–15 eV) have been accomplished by the laboratory EXAFS spectrometers described in recent literature. Some spectrometers are able to produce photon fluxes in the order of 5×10^6 photons s^{-1} with resolutions of about 5 eV. It was demonstrated in Section 5.2.1 that a large class of experiments ($\mu_s/\alpha \, \Delta\mu_s \approx 500$, S/N $\approx 100/1$) can be carried out with a minimum useful photon flux of about 3×10^6 photons s^{-1} accepting a total scan time of 20 h (counting time per data point 100 s). Williams (7) has shown that acceptable results can be obtained with a Ge(111) Johansson crystal that could be fabricated with a horizontal size of 127 mm. Therefore, the spectrometer described by Thulke et al. (8) could be further optimized by using a Ge(311) Johansson crystal with a larger horizontal spot width (40 → 120 mm) resulting in a flux of 1.5×10^7 photons s^{-1} at 8.2 keV (resolution 4.5 eV). The spectrometer of Brinkgreve et al. (30) might produce at 9 keV an intensity of about 2×10^8 photons s^{-1} (resolution 11 eV) by using a gold anode (Mo → Au, increase ~1.6×) and a Ge(111) monochromator crystal (rocking curve, increase ~2.5×) with a horizontal spot width of 120 mm (70 → 120 mm, increase ~1.7×). By changing the type of germanium crystal [Ge(311) → Ge(111)] the optimized spectrometer of Thulke et al. (8) can give a higher number of photons (~5.3×10^7 s^{-1}) with a lower resolution. Conversely the optimized spectrometer of Brinkgreve et al. (30) will produce a lower intensity (~5.7×10^7 s^{-1}) with a higher resolution when using a Ge (311) crystal. The calculations are based upon data, which are derived from real experiments and show that optimized laboratory EXAFS facilities can give photon fluxes that are 10–15 times larger than the fluxes mentioned in the literature until now. Resolutions will then be in the order of 10 eV, which are acceptable for deriving reliable structural information (see also Sections 5.2.1 and 5.6.1).

Not only the photon flux, but also the reliability and the feasibility of laboratory EXAFS spectrometer have to be improved further. Such improvements, together with an optimization of the actual photon flux may lead to a situation where the laboratory EXAFS facility is competitive for a large class of experiments with an EXAFS station at a synchrotron beamline. In the following, attention will be paid to the different components of the laboratory spectrometer, which have to be improved further.

A high quality Johansson-type curved monochromator crystal of large (horizontal) dimension (100–150 mm) forms the key of such an improvement. As already mentioned, normally cylindrical-type Johansson crystals with horizontal dimension of 127 mm have been fabricated (7). Very promising results have been obtained with double curved crystals (31), which can focus more photons in the vertical plane leading to higher x-ray intensities of the sample. The best material for EXAFS monochromator crystals is germanium due to its larger rocking curves. However, germanium is only useful for energies lower than 11 keV, due to the position of its K-edge at 11.103 keV.

A point of further concern is the rotating anode, which should be used as an x-ray source in an optimized laboratory instrument. At low excitation voltages high currents have to be realized with a current regulation circuit of high quality to allow optimum short term stability. The high voltage circuit must be protected against an electronic discharge in the vacuum chamber of the anode. The high voltage has to be switched on again automatically after such an event, allowing long scan times also during nighttime. The housing of the rotating anode must be constructed in such a way that long lifetimes of the bearings are normal.

Magnetic fluid vacuum seals are recommended at present, because they guarantee good vacuum conditions. Since the anode focal spot has to be aligned together with the Rowland circle mechanism, the repositioning of a new filament and its focal cup and also the exchange of different types of anodes must be possible without new alignment procedures.

5.6. LABORATORY VERSUS SYNCHROTRON EXAFS FACILITIES

5.6.1. Comparison of Results on Both Types of Facilities

Information, which might be of great help to get an objective opinion about the actual capabilities of laboratory EXAFS spectrometers consists of a comparison of data obtained on the same sample with a laboratory as well as a synchrotron EXAFS facility. Khalid et al. (16) compared the Fourier transforms of EXAFS data obtained for nickel foil measured on their laboratory spectrometer and on EXAFS station I-5 at SSRL (ring conditions: 2.66 GeV, 60 mA). The Fourier transforms measured on both facilities have the coordination peaks at the same distances (Fig. 5.21). However, the magnitude of the peaks of the Fourier transform of the data collected on the laboratory spectrometer are lower than for the synchrotron EXAFS data. The peak corresponding to the first coordination distance is about 20% lower, whereas the peak of the third shell decreased by about 27%. The SSRL data were taken with a resolution of about 1–2 eV. The authors measured the resolution of their laboratory EXAFS spectrometer to be 6.2 eV at the tungsten $L_{\alpha 1}$ line (8397.6 eV). The convolution calculations of Section 5.2.2.2 show that resolutions of about 20 eV will degrade the first-shell peak by about 20%. Peaks for higher coordination distances decrease much more than 27%, as found for the third shell of the nickel foil data measured on the Delaware spectrometer. This problem has been mentioned already by Stern (4) in his comments on these results. He alluded thickness effects (34) as possible origins for these amplitude differences.

A more detailed comparison of EXAFS data measured on laboratory and synchrotron facilities has been made by van Zon (11). Figure 5.22(a) and 5.22(b) show the x-ray absorption spectrum and the EXAFS of platinum metal foil, measured on beamline I-5 (SSRL), [Si(220) channel cut, resolution 2–3 eV, dedicated operation 3 GeV and 40 mA]. The absorption spectrum was recorded in one scan of 35 min. The actual photon flux at the sample is unknown. In Fig. 5.22c and d the measurements, performed with the Eindhoven EXAFS spectrometer (30), are presented. The measurements were carried out using a Si(400) monochromator with the x-ray generator operating at 240 mA and 21 kV, thus avoiding higher harmonics. The measured resolution was 11.2 eV,

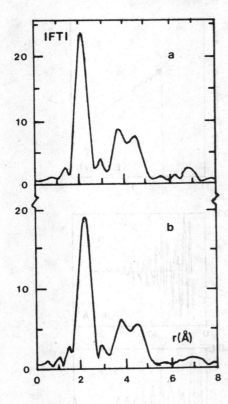

Figure 5.21. Fourier transforms (magnitude |FT|) of Ni foil (E = 8.3 keV) measured both at SSRL (a) and on the laboratory spectrometer of the University of Delaware (b).

while the intensity was estimated to be 3.10^5 photons s^{-1} (2-mm spot width). To get a good signal-to-noise ratio a scan of 15 h was made. There exists a difference between the two absorption spectra in intensity of the white line and in amplitude of the EXAFS signal at low-k-values. To compare the quality of the EXAFS data, a k^2-weighted Fourier transform was performed on both EXAFS functions. By using a high weight factor the difference at low k values are suppressed and thus the influence of the resolution if minimized.

In Figure 5.22(e) the k^2-weighted Fourier transforms on both data sets are displayed. The transforms look very similar and the amplitude of the peaks, for the higher coordination shell as well, hardly differ in amplitude. A more quantitative comparison was obtained by convolution of the SSRL data with a Gaussian energy distribution function. A full width at half-maximum (FWHM) value of 7 eV of the Gaussian resolution function applied to the SSRL data resulted in a good agreement with the laboratory EXAFS data (see Fig. 5.22f). This is a strong indication that no artifacts caused by the laboratory spectrometer are

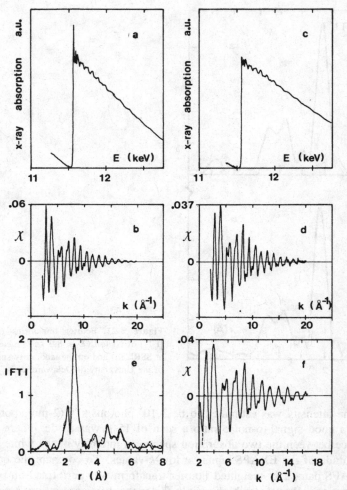

Figure 5.22. Data of Pt–foil collected at the L_{III}-edge ($E = 11.564$ keV): (*a*) X-ray absorption spectrum, measured at SSRL; (*b*) EXAFS spectrum (SSRL); (*c*) X-ray absorption spectrum, measured on the Eindhoven EXAFS facility; (*d*) EXAFS spectrum (lab. EXAFS); (*e*) Fourier transforms (k^2, $\Delta k = 2.6$–19 Å^{-1}) of the EXAFS spectra shown in (*b*) (solid line) and d (dashed line); and (*f*) EXAFS data (laboratory EXAFS solid line, convoluted SSRL data dotted line).

present in the EXAFS data. According to the convolution procedure, the SSRL data should have been measured with a resolution of 8.7 eV. This value is quite large for the channel-cut Si(220) monochromator (vertical divergence is 5 × 10^{-5} rad). However, the SSRL spectrum was measured with a ring energy of 3 GeV and at this energy there is a significant contribution of higher harmonics radiation that is not rejected by the channel-cut Si(220) monochromator. This could explain the differences between the SSRL calculated apparent and the real energy resolution. The results presented in this paragraph make clear that high quality EXAFS data can be obtained on laboratory EXAFS spectrometers. As already discussed in Section 5.2.2.2 a resolution of about 10 eV can be taken as a maximum value to derive reliable information about higher coordination shells.

5.6.2. Analysis of Laboratory EXAFS Data

A full analysis of high quality EXAFS data collected with laboratory EXAFS spectrometers has been presented in the recent literature by Thulke et al. (35, 36), Tohji et al. (37), and Kampers et al. (38). As an example, the EXAFS study of Kampers et al. will be discussed in more detail. They studied with the EXAFS spectrometer described in (30) the influence of the method of preparation on the size of platinum metal crystallites supported on ZSM-5 zeolite. Details of the preparation are given in ref. (38). EXAFS measurements were carried out at room temperature on the platinum L_{III}-edge after passivation (cooling down in flowing H_2 to room temperature and subsequently slowly introducing a mixture of dioxygen and helium). The resolution of the Si(400) monochromator at the platinum-edge was 14 eV with an x-ray spot size on the crystal of 10 (vertical) × 20 mm (horizontal). The x-ray generator operated at 21 kV and 240 mA producing a photon flux of about 3 × 10^6 photon s^{-1} at the sample. Collection times were in the order of 22 h. The EXAFS spectra of a 4.7 wt.% (preparation via impregnation) and a 9.4 wt.% (preparation via ion exchange) Pt/ZSM-5 catalyst are presented in Fig. 5.23a and b, respectively. Due to the passivation procedure it is likely that Pt–Pt and Pt–O bonds are present in the EXAFS spectra. A k^3-weighted Fourier transform (corrected for the Pt–Pt phase shift and backscattering amplitude [see also ref. (10)] gives the best separation between the Pt–Pt and the Pt–O bonds. In a Pt–Pt phase corrected Fourier transform the Pt–Pt contribution peaks at the actual coordination distance (in the transforms of both catalysts at 2.77 Å) (see Fig. 5.22c, d, e, and f). Pt foil and $Na_2Pt(OH)_6$ were measured under the same experimental conditions in order to obtain phase shift and backscattering amplitude functions for the analysis of the Pt–Pt and Pt–O absorber–scatterer pairs. By applying the "difference

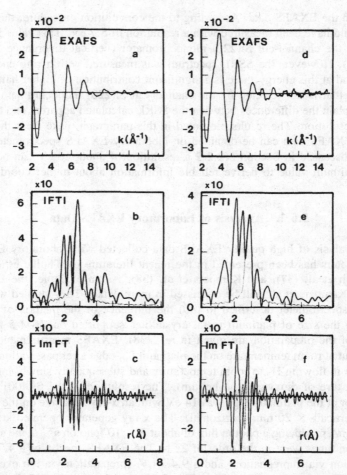

Figure 5.23. EXAFS data measured at the L_{III}-edge of Pt/ZSM-5 catalysts on the Eindhoven EXAFS facility. EXAFS spectrum of (a) 4.7 wt.% Pt/ZSM-5 (impregnation) and (d) 9.4 wt.% Pt/ZSM-5 (ion exchange). Magnitudes (|FT|) (b and e) and Imaginary parts (Im FT) (c and f) of Fourier transforms (k^3, $\Delta k = 2.9$–14 Å^{-1}, Pt—Pt phase and amplitude corrected) of EXAFS spectra (a) and (b), respectively.

file'' technique (11) the following Pt-Pt coordination parameters were obtained: $N = 5.5$, $R = 2.7$ Å, $\Delta 6^2 = 0.003$ Å^2 (sample prepared via impregnation); $N = 2.9$, $R = 2.77$ Å, $\Delta 6^2 = 0.003$ Å^2 (sample prepared via ion exchange). The Fourier transform (Pt-Pt phase and amplitude corrected) of an EXAFS function calculated with these parameters (dotted lines in Fig. 5.23c, d, e, and

f) gives for both catalysts a Pt–Pt peak that superimposes the Pt–Pt peak in the Fourier transform of the experimental data. From this EXAFS study the authors were able to conclude that the impregnation method leads to much larger platinum crystallites on the *ZSM-5* support than the ion exchange method. These results and the results of the analysis of the other laboratory EXAFS data published in the literature (35–37) clearly show that reliable structural information can be obtained from laboratory EXAFS data if the quality of the EXAFS data is comparable with good quality EXAFS data obtained with a synchroton EXAFS facility.

5.6.3. Optimal Use and Advantages of Laboratory and Synchrotron EXAFS Facilities

The useful energy range of in-house EXAFS spectrometers is limited to $4 \leq E \leq 20$ keV, while the lowest concentration that can be measured is in the order of 25–50 mM. There are also limitations on the type of experiments, which can be performed on laboratory EXAFS systems. EXAFS fluorescence experiments on very dilute samples (1–10 mM), polarization studies, time-dependent measurements (using dispersive EXAFS), and experiments, which need highly intense collimated x-ray beams (surface studies) can only reliably be carried out on dedicated synchrotron EXAFS stations. However, for all other type of experiments that are not limited by the energy and concentration ranges mentioned previously, the laboratory EXAFS facility might be fully competitive with a synchrotron EXAFS station. This is only the case if the laboratory facility is optimized according to the points discussed in Section 5.5.3.

A lot of EXAFS data that nowadays are collected on EXAFS stations of synchrotron beamlines could be measured by experimentalists with their own EXAFS facility. Most users of synchrotron facilities will have the experience that 20–40% of the allocated beamtime is lost by injection procedures, machine failures, and ring instabilities. Already today the demand for EXAFS beamtime on synchrotron facilities can hardly be satisfied and there is still a growing interest in doing EXAFS experiments. It should therefore be advantageous for the whole EXAFS community if only those EXAFS experiments that really need the broad energy range, the polarization, and the high intensity or synchrotron radiation, are carried out at synchrotron EXAFS stations, thus relieving this type of EXAFS facility.

The roles of in-house and synchrotron EXAFS facilities have been extensively discussed during the workshop organized by Professor Stern (4). Since 1980 new technologies diminished the limitations of laboratory EXAFS facilities and new prospectives for these facilities have become clear. Most of these prospectives have been discussed in the previous sections. It seems worthwhile

to mention here for both types of facilities those advantages, which are timeless:
Synchrotron EXAFS facility:

- Broad photon energy range.
- High intensities together with high resolutions.
- Education of students by interactions with outstanding scientists, who are doing sophisticated experiments also on other types of synchrotron stations.

Laboratory EXAFS facilities:

- Setting of priority by experimenters.
- Much faster and better interaction between experiment and follow-up leading to new ideas and new experiments.
- Much easier to carry out experiments that need complicated facilities for sample preparation and *in situ* experiments.
- Rapid implementation of ideas.
- Excellent teaching and training possibilities for students.
- Avoidance of travel expenses.

For industries the possibilities for proprietary research are unlimited and no problems will arise in obtaining high ratings for beamtime from the proposal panel review for more routinelike research.

5.7. SUMMARY AND CONCLUSIONS

This chapter has demonstrated that high quality XANES and EXAFS spectra can be collected with laboratory EXAFS facilities. The limitations of these facilities have been discussed and it has been shown that synchrotronlike resolutions, which are sometimes claimed by enthusiastic supporters of synchrotron facilities to be essential for deriving reliable structural EXAFS information, are not necessary. Resolutions of 10–15 eV can still be expected. XANES spectra can be measured with synchrotronlike resolutions within an acceptable time of 1–2 h.

At this moment only a few papers dealing with the analysis of high quality laboratory EXAFS data have been published. Laboratory EXAFS spectrometers, fully optimized according to the points discussed in Section 5.5.3, have not yet been reported in the literature. A key point in this optimization process is the production of high quality monochromator crystals of large (horizontal) dimensions. No large scale applications for laboratory EXAFS facilities are foreseen, if this optimization does not take place and if intensities impinging on the sample remain lower than 10^7 photons s^{-1} (bandpass 5–15 eV). However, the technologies available nowadays can bring such an optimization to reality and many experimentalists may then make fruitful use of their laboratory EXAFS facility.

ACKNOWLEDGMENTS

The author gratefully acknowledges the many fruitful discussions with Dr. J. B. A. D. van Zon, Ir. P. Brinkgreve, T. J. M. Maas, and Dr. M. P. A. Viegers. The author is also thankful for the encouragement and stimulating discussions with Professor E. A. Stern and Professor D. E. Sayers.

REFERENCES

1. P. A. Lee, P. H. Citrin, P. Eisenberger, and B. M. Kincaid, *Rev. Mod. Phys.*, **53**, 805 (1981).

2. R. Cortes, private communication.

3. J. A. Del Cueto and N. J. Shevchik, *J. Phys. E*, **11**, 1 (1978).

4. Laboratory EXAFS Facilities, *AIP Conference Proceedings No. 64*, E. A. Stern (Ed.), American Institute of Physics, New York, 1980.

5. L. Incoccia and S. Mobilio, in *EXAFS and Near Edge Structure*, A. Bianconi, L. Incoccia, and S. Stipcich (Eds.), Springer-Verlag, Berlin, 1983, p. 87.

6. W. H. Zachariasen, *Theory of X-Ray Diffraction in Crystals*, 2nd ed., Wiley, New York, 1967.

7. A. Williams, *Rev. Sci. Instrum.*, **54**, 193 (1983).

8. W. Thulke, R. Haensel, and P. Rabe, *Rev. Sci. Instrum.*, **54**, 277 (1983).

9. K. Tohji, Y. Udagawa, T. Kawasaki, and K. Masuda, *Rev. Sci. Instrum.*, **54**, 1482 (1983).

10. D. C. Koningsberger and D. E. Sayers, *Solid State Ionics*, **16**, 23 (1985).

11. J. B. A. D. van Zon, thesis, Eindhoven University of Technology, Eindhoven, The Netherlands, 1984.

12. T. M. J. Maas and P. Brinkgreve, Internal Report, Eindhoven University of Technology, Eindhoven, The Netherlands, 1984.

13. G. R. Fischer, in Laboratory EXAFS Facilities, *AIP Conference Proceedings No. 64*, E. A. Stern (Ed.), American Institute of Physics, New York, 1980, p. 21.

14. M. Prins, Internal Report, Eindhoven University of Technology, The Netherlands, 1984.

15. G. S. Knapp and P. Georgopoulos, in Laboratory EXAFS Facilities, *AIP Confer-*

ence Proceedings No. 64, E. A. Stern (Ed.), American Institute of Physics, New York, 1980, p. 2.

16. S. Khalid, R. Emrich, R. Dujari, J. Schultz, and J. R. Katzer, *Rev. Sci. Instrum.*, **53**, 22 (1983).

17. D. W. Berreman, *Rev. Sci. Instrum.*, **26**, 1048 (1955).

18. J. W. M. Dumond, *Rev. Sci. Instrum.*, **18**, 626 (1947).

19. J. W. M. Dumond, D. A. Lind, and R. E. R. Cohen, *Rev. Sci. Instrum.*, **18**, 617 (1947).

20. N. G. Webb, *Rev. Sci. Instrum.*, **47**, 545 (1976).

21. A. Franks, *J. Appl. Phys.*, **9**, 349 (1958).

22. P. Georgopoulos and C. H. Tang, in *Advances in X-Ray Analysis*, Vol. 27, J. B. Cohen, J. C. Russ, D. E. Leyden, C. S. Barrett, and P. K. Predecki, (Eds.), Plenum, New York, 1984, p. 299.

23. E. A. Stern, in Laboratory EXAFS Facilities, *AIP Conference Proceedings No. 64*, E. A. Stern (Ed.), American Institute of Physics, New York, 1980, p. 39.

24. J. B. A. D. van Zon, Internal Report, Eindhoven University of Technology, The Netherlands, 1984.

25. D. W. Berreman, *Phys. Rev. B*, **19**, 560 (1979).

26. G. S. Knapp, H. Chen, and T. E. Klippert, *Rev. Sci. Instrum.*, **49**, 1658 (1978).

27. B. E. Warren, in *X-Ray Diffraction*, Addison-Wesley, Reading, MA, 1969, p. 46.

28. Kun-quan Lu and E. A. Stern, in Laboratory EXAFS Facilities, *AIP Conference Proceedings No. 64*, E. A. Stern (Ed.), American Institute of Physics, New York, 1980, p. 104.

29. J. E. White, *J. Appl. Phys.*, **21**, 855 (1950).

30. P. Brinkgreve, T. M. J. Maas, D. C. Koningsberger, J. B. A. D. van Zon, M. H. C. Janssen, A. C. M. E. van Kalmthout, and M. P. A. Viegers, in *EXAFS and Near Edge Structures III*, K. O. Hodgson, B. Hedman, and J. E. Penner-Hahn (Eds.), Springer-Verlag, Berlin, 1984, p. 517.

31. M. P. A. Viegers, Internal Report, Philips Research Laboratories, Eindhoven, The Netherlands, 1984.

32. P. Georgopoulos and G. S. Knapp, *J. Appl. Cryst.*, **14**, 9 (1981).

33. G. G. Cohen, D. A. Fisher, J. Colbert, and N. J. Shevchik, *Rev. Sci. Instrum.*, **51**, 273 (1980).

34. E. A. Stern and K. Kim, *Phys. Rev. B*, **23**, 3781, (1981).

35. W. Thulke, R. Frahm, R. Haensel, and P. Rabe, *Phys. Status Solidi*, **75**, 501 (1983).

36. W. Thulke, R. Haensel, and P. Rabe, *Phys. Status Solidi*, **78**, 539 (1983).

37. K. Tohji, Y. Udagawa, S. Tanabe, and A. Ueno, *J. Am. Chem. Soc.*, **106**, 612 (1984).

38. F. W. H. Kampers, F. B. M. Duivenvoorden, J. B. A. D. van Zon, P. Brinkgreve, M. P. A. Viegers, and D. C. Koningsberger, *Solid State Ionics*, **16**, 55 (1985).

CHAPTER

6

DATA ANALYSIS

D. E. SAYERS

Department of Physics
North Carolina State University
Raleigh, North Carolina

B. A. BUNKER

Department of Physics
University of Notre Dame
Notre Dame, Indiana

6.1. INTRODUCTION

One of the problems that confronts experimenters in EXAFS is the complexity of the acquisition and analysis of their data. Often these complexities are not recognized, even by experienced investigators, at the beginning of their experiments. This situation can arise because at a basic level, EXAFS is conceptually simple. However, for most experiments in order to extract information, the most optimal experimental conditions must be realized and a complete theoretical treatment must be considered. Therefore, a theoretical understanding of the methods of data processing is the basis for obtaining meaningful structural parameters. In this chapter we discuss the analysis of EXAFS data. We presume that optimal experimental conditions as described in Chapters 3, 4, and 5 have been realized so that the data to be analyzed are free of significant distortion. Experimental problems, such as glitches, are discussed only as they relate to the analysis of typical spectra. The theoretical expressions that are used to analyze the data have been developed in Chapter 1. For the most part the standard single-scattering expression is used but where appropriate more complete expressions involving non-Gaussian disorder or multiple scattering are applied.

Analysis of EXAFS data may roughly be separated into three categories. First are those experiments that simply seek to "fingerprint" the spectrum of the sample being studied and to compare it with the spectra of a series of reference compounds. These experiments are usually performed when an investigator starts with the study of a new class of compounds. The primary purpose of these experiments is to develop a general picture of the local structure in terms of a similar reference. In general, this type of experiment is not as sensitive to the experimental conditions and the data are not subject to any significant amount of processing. The second class of experiment seeks more basic information. Usually, these experiments are performed on biological or disordered systems that have an isolated first shell and their purpose is to identify the bonding configuration or valence state of the absorbing atom. The distance is often the most important information being sought in these experiments since the type of nearest neighbors is known from the composition or other experimental techniques. While the data require some processing to extract the frequency of the oscillations, the amplitude information is not extensively analyzed or even considered. The third type of experiment attempts to extract the maximum information possible from the EXAFS data. Sometimes this will involve only the determination of parameters from a single isolated shell with the highest precision possible, while in other cases it may be necessary to analyze two or more shells simultaneously because the contributions from the adjacent shells cannot be effectively isolated. In this type of analysis great care must be taken in the processing of the data to ensure that correct results are obtained. Since the other types of analysis mentioned may be considered as subsets of this more complete treatment, this chapter only attempts to disucss the steps required in a more complete analysis.

The procedures are divided into two parts. First is the processing of the data that includes those steps used to change the raw data into the normalized EXAFS data that can then be interpreted according to the usual theoretical formulations. Some of these steps in the processing include pre-edge subtraction, normalization, deglitching, alignment of the energy scale, EXAFS background removal, and transformation to k space. Second, the data are analyzed to extract the parameters of interest. Some of the procedures discussed are the Fourier transform of the data, including effects of weighting and windows, filtering in r space, single- and multiple-shell fitting, and direct k space analysis (ratio method) including recent generalizations.

Finally, it should be emphasized that the intent of this chapter is to present one possible consistent method of analysis, since during the course of data analysis there may be several ways to perform a given step. This chapter does not provide an exhaustive or critical review of all procedures used in the literature. Instead it concentrates on one basic scheme taking care to point out which elements are critical and which elements are not. For critical steps various options are discussed, particularly when they are in common use by several groups or when they may be the source of some controversy.

6.2. DATA PROCESSING

6.2.1. Introduction

In this section we discuss the steps needed to reduce the raw experimental data to properly normalized EXAFS data for analysis. The primary steps presented include determining the energy scale, deglitching, pre-edge background removal, normalization, and extraction of the EXAFS data. The order of the steps taken is not critical and, in fact, not all the steps described are necessary for the final analysis of the data. However, they are useful for the systematic analysis and comparison of several files at one time as is typically done in EXAFS.

6.2.2. Energy Scale

The EXAFS technique is insensitive to the absolute calibration of the monochromator. A slight misalignment of the monochromator zero position, for instance, will be offset to a good approximation by a simple change in edge energy in later analysis. Larger misalignments, however, will induce errors in energy differences, which will enter into the later interpretation of the EXAFS. Knowledge of the edge energy of a compound relative to the pure element is often useful, with the information from the "chemical shift" (1) being used to extract chemical information about the system. Because of these considerations, it is common to calibrate the monochromator using the pure material as a known reference immediately before the sample of interest. It is also possible to calibrate the edge position simultaneously with the measurement of the EXAFS spectra by introducing a reference compound in front of a third ion chamber placed in tandem with the I_0 and I chambers. Typically the energy scale is chosen so that the binding energy of the core level whose spectrum is being studied is set to some feature on the edge of the reference material such as the inflection point (or first inflection point if there is more than one) or to some characteristic peak in the near-edge structure.

Most x-ray monochromators at synchrotron sources now use two independently adjustable crystals. There are two primary advantages of this design: first, the heating of the first crystal by the x-ray beam may be large enough to distort the crystal and detune the two crystals. The capability of separately moving the two crystals enables the experimentalist to retune the system. The second advantage is that it is often desirable to *deliberately* detune the two crystals to selectively suppress harmonics of the beam. The disadvantage of this flexibility is that the detuning is a change in the average, as well as relative, Bragg angle and therefore the energy calibration may be lost. This is not a crippling problem, but it must be kept in mind when both acquiring and analyzing the data.

6.2.3. Deglitching

Absorption and fluorescence data taken at synchrotron sources often include sharp structure due to spurious reflections from the crystal monochromator; these "glitches" may be many times larger than the EXAFS signal. Because of their small energy width, it may seem that glitches are unimportant in further processing and may be neglected. Much later analysis, however, involves Fourier transform methods. Due to finite data range, the Fourier techniques may cause "spectral leakage," a broadening of spectral features that can magnify the effects of the glitch. Because of this possible later smearing, it is often desirable to remove the worst of the glitches before further processing.

The common technique is to fit the data on both sides of the glitch with a polynomial that is then used to interpolate through the glitch region. For this technique to work, of course, the glitch must be narrow compared with any feature of interest in the data. For data far above the edge, this condition is almost always satisfied because the oscillations are periodic in $k \propto (E - E_0)^{1/2}$. For data within about 100 eV of the edge, however, the width of the glitch may be comparable to some features of the data. In this case, deglitching is rarely satisfactory; the only real solution is to measure the data under different conditions.

Another approach to deglitching is to delete the offending data. Because some sort of interpolation is usually made later in the analysis, however, the replacement of data in the glitch region is merely deferred.

6.2.4. Pre-Edge Background Removal

In analyzing EXAFS data, one is usually interested only in the region above the absorption edge. The region *below* the absorption edge is affected by lower-energy absorption edges, Compton scattering, and other processes that are not of immediate interest. Subtracting out this pre-edge background is useful because it simplifies later data normalization. Otherwise, this step is largely cosmetic because later isolation of the EXAFS oscillations involves another, more important, background subtraction. The removal of the pre-edge background generally involves a fitting of the data before the edge to some functional form and extrapolation of this function into the data region. A traditional function to use for this fit is the "Victoreen" empirical form, $C\lambda^3 - D\lambda^4$, where λ is the x-ray wavelength.

For an in-laboratory EXAFS apparatus where the total I and I_0 may be measured independently (2) and for energies far from other characteristic edges in the sample, the Victoreen formula is known to be a good approximation for the x-ray absorption coefficient (3). This extrapolation can lead to a reliable separation of the partial cross section (e.g., K-edge and L_3-edge) of interest from the total absorption spectrum. However, this is no longer true for the types of apparatus typically found at a synchrotron radiation source and on many newer laboratory EXAFS spectrometers that use tandem ion chambers, since only a fraction of I_0 (and usually I) is measured. Generally, the energy responses of the detectors are different (even if the same gases are used), and also because of harmonic contamination in the beam or nonlinearities in the electronics, the quantity $\ln(S_0/S)$ where S_0 and S are the signals from the detectors (before and after the sample, respectively) contain not only the absorption spectrum but also an energy-dependent term that represents the nonideal response of the apparatus.

Because the Victoreen function is no longer a valid representation of the cross section, if it *is* used, the experimenters must keep in mind that the resulting edge spectrum is distorted from the true absorption coefficient. This does not affect subsequent analysis since the EXAFS background removal and normalization procedures to be discussed in Sections 6.2.5 and 6.2.6 do not need the true cross section in order to extract normalized data. The Victoreen formula may still be used but so can any other function that would model the general shape of the pre-edge region. Generally, a simple polynomial routine is most convenient where either a linear or quadratic polynomial in energy can be used for absorption data and usually a linear polynomial is sufficient for fluorescence data.

6.2.5. Data Normalization

There are two aspects to normalization of EXAFS data: energy dependent and energy independent. Energy-independent normalization basically removes the effects of sample thickness so that different samples may be directly compared. For convenience it is usually carried out before the isolation of the oscillations. On the other hand, energy-dependent normalization is necessary for transmission data only if comparing to theoretical calculations or data from other absorption edges. This is done after the oscillations have been isolated.

In general, the oscillatory part of the x-ray absorption coefficient is normalized by the smooth monotonically decreasing cross section of the edge of interest, $\mu_0(E)$, according to the equation that defines the EXAFS function:

$$\chi(E) = \frac{\mu(E) - \mu_0(E)}{\mu_0(E)} \tag{1}$$

A serious difficulty with this, however, is that $\mu_0(E)$ in general is neither measurable independently (what is measured may be called $\mu_b(E)$ but has the problems discussed in the previous section) nor can it be calculated accurately, especially near an absorption edge. It also may be distorted by the system response as discussed in the previous section and by the difficulty in removing the pre-edge background to sufficient accuracy. By normalization to the measured background all these *additive* effects become *multiplicative* effects and large errors in the energy-dependent EXAFS amplitude may result.

If data analysis is to proceed using model compounds then this definition can be replaced with

$$\chi(E) = \frac{\mu(E) - \mu_b(E)}{\mu(E_i)} \tag{2}$$

where $\mu(E_i)$ is the absorption coefficient chosen systematically at a point near the edge so that Eq. (2) represents data normalized to the edge step. It is assumed in the complete analysis that the energy dependence of $\mu_0(E)$ in the unknown and reference compound is the same since μ_0 should depend primarily on the absorbing atom and not on its chemical state. The normalization point E_i should be chosen far enough above the edge to be away from any near-edge structure yet not so far above the edge that the cross section has fallen appreciably from its value at threshold. Typically this point is in the range of 20 to 50 eV above the edge. To be self-consistent with the definition [Eq. (1)] the point E_i should be on the μ_b (background) curve described in Section 6.2.6. Initially, if the derivative of the μx versus E data is examined this point can be chosen to lie at a node of the derivative $d(\mu x)/dE$ and its location at a node verified after the $\mu_b(E)$ curve is calculated. Finally, the normalization point should be chosen as consistently as possible among the unknown and standard sets used in a particular experiment. Examples of normalization points will be shown later in this section. Alternatively, the background curve calculated in Section 6.2.6 may be extrapolated to $E = 0$, and this extrapolated $\mu_b(E = 0)$ may be used in Eq. (2). However, this procedure is more sensitive to the background curve particularly near the threshold where distortions may occur.

If theoretical values of amplitudes and phases are to be used for analysis then the complete energy-dependent normalization must be performed. This may be done most easily after the step normalization and isolation of the EXAFS oscillations by dividing the results by an appropriate energy dependent cross section [e.g., those of McMaster et al. (4)]. It should be emphasized again that this energy-dependent background should not be done using the smooth background curve that is determined when isolating the EXAFS oscillations since this will generally be distorted from the true cross section.

Finally, it should be stressed that data obtained using fluorescence or electron yield is complicated by a number of slowly varying "gain" factors that increase the apparent absorption above the edge. These multiplicative factors depend on the particular material being studied and therefore do *not* cancel out in most analysis. In this case, it is desirable to compensate for these gain factors in later analysis. Further discussion of this problem may be found in Chapter 10.

6.2.6. Isolation of the Fine-Structure Oscillations

This step consists of subtracting out the smooth background from the absorption data. There are several commonly used methods for this procedure all of which are dependent on the background being slowly varying compared with the EXAFS oscillations. These methods include the subtraction of:

1. A cubic least-squares-spline approximant to the data (5).
2. A simple polynomial background, followed by digital high-pass filtering (6).
3. A cubic smoothing-spline approximant to the data (7).

The first technique is probably the most foolproof and is recommended for most applications. A cubic spline is a curve constructed out of linked cubic polynomials; the function value and first derivative are matched at the "knots" where the polynomials meet. The number of knots is variable and depends on the particular data; typical values range from 2 to 6. The obvious danger of using too many knots is that the spline may have enough degrees of freedom to follow the EXAFS oscillations. The use of too few knots can make for poor background removal that will interfere with later analysis.

The second technique is basically a Fourier-filtering operation to remove the low-frequency background. The initial polynomial background subtraction mainly serves to minimize data termination effects. In practice, this technique is limited by the care needed to minimize spectral leakage and is now rarely used. It should be pointed out that the filtering of low r components (or contributions from the background) or high r components (noise) is essentially cosmetic since only the contributions of the background or noise in the r range of interest will affect the analysis and these cannot be separated from the signal by filtering.

The third technique differs from the first in that the abscissa of every data point is considered a knot. This algorithm uses a set of data Y_j and a set of inverse weighting factors W_j to construct a set of spline coefficients S_j using

$$\text{SM} \geq \sum_j \left(\frac{(Y_j - S_j)}{W_j}\right)^2 \tag{3}$$

where SM is a user-specified parameter that determines the "tightness" of the fit and the sum is over the pertinent data points. Within this constraint, the algorithm constructs the curve with the lowest average curvature. Best results appear when the weights W_j roughly follow the amplitude of the first-shell EXAFS, but the results are rather insensitive to this. An exception to this is the first (lowest energy) point of the data range; it is often useful to use a small value for the inverse weight W_1 to force the spline to go through the first point. This affords a high degree of control over the resulting curve that is sometimes necessary if the data range is short. This method is desirable for special cases involving difficult background with large curvature.

In general, the smoothing spline is more flexible but in its simple form is difficult to use: Great care is needed not to reduce the resulting EXAFS amplitudes. This technique has been extended by including the spline fit as part of an iterative background removal scheme. The basic idea of this method is to (a) make a spline fit with an initial guess for the smoothing parameter, (b) subtract the data from the spline fit, (c) Fourier transform the resulting χ data, and (d) examine the low-frequency (background) components and the first-shell peak of the data transform to evaluate whether the background curve has frequency components high enough to modify the data. If this is true, the process is repeated with a higher value for the smoothing parameter SM.

Extremely important and helpful for optimum background removal is the monitoring of the derivative of the generated background curve since it is sensitive to small changes in the shape of the curve. This is useful for any type of background removal to make sure that sufficient curvature is included but that there are no oscillations in the background whose frequencies are comparable to those of the EXAFS oscillations.

The major problem with inadequate background removal comes about later in the analysis. When Fourier transforming the data, the slowly varying background contributes spectral components at low r. Because of the finite data range, truncation effects will cause leakage of the low-r background peak to higher frequencies, possibly contaminating the first- or second-shell data. When the data is weighted with k^3, especially, the ripple induced by the background can cause large distortions in the r-space spectrum.

Serious complications in background removal arise whenever the curvature of the background approaches the curvature due to the EXAFS oscillations. This is a particular problem at the low-energy end of the spectrum. Systems with low-Z backscatterers are especially difficult, because the EXAFS amplitude is large in the lower-k region and the background curve may itself be a strong function of k. In these cases, it is often useful to "anchor" the background curve through some point near the edge. It is important to stress that one must be systematic in the selection of this point for all samples and reference compounds. If the low-energy endpoint of the background curve is allowed total freedom, severe distortion of the resulting EXAFS data is possible at low-k values. By being systematic and using model compounds in the data analysis, background distortions tend to cancel and a reliable analysis can be obtained. This is a much greater problem if theoretical amplitude and phase functions are used for analysis since a background distortion will affect the EXAFS amplitude in an unknown way. In fact, this problem is one of the most important reasons why, if possible, it is better to use experimentally determined and not theoretically calculated amplitude and phase functions. This complication has been ameliorated somewhat by the suggestion of Cox (8) to add an energy dependent function to the fitting equation (to be discussed in Section 6.3).

Sometimes one might get the impression that background removal is something of an art. For the worst cases this is probably true. However, for most spectra background removal is straightforward as long as experimental problems did not cause complications. Any background removal technique has the capability of distorting the data. Whenever the curvature of the background curve is comparable to that of the EXAFS, there will be difficulties in isolating the oscillations. Because of this, it is essential to be systematic and consistent in the choice of parameters for background removal. Often it is desirable to make several attempts at background removal with different choices of parameters, follow the different sets of EXAFS data through the entire chain of analysis and test the results for consistency.

6.2.7. Examples of Data Processing

To show how the processing described previously proceeds, two examples are given in this section. The first is the data from a rhodium foil taken in transmission at a temperature of about 90 K. The second example is a fluorescence spectrum at the iron iron K-edge taken for a 10-mM solution of the protein ferritin. Both samples were measured at SSRL (Stanford University, USA) on line I-5 under dedicated operation. In Fig. 6.1a and b the raw data are shown plotted as a function of the photoelectron energy. The origin of energy was taken to be the primary inflection point in the edge for rhodium foil ($E_K = 23{,}220$ eV). In the case of ferritin the origin of energy was found by calibrating the spectrometer with an iron foil ($E_K = 7112$ eV). For the transmission data (Fig. 6.1a) $\mu x = \ln(I_0/I)$ is plotted while for the fluorescence data (Fig. 6.1b) I_f/I_0 is plotted. For the pre-edge fit, a quadratic function was used for the rhodium foil and a linear curve for the ferritin. In Fig. 6.1c and d the step-normalized data are shown for rhodium and ferritin, respectively. In Figs. 6.2 and 6.3 the background removal process is shown for rhodium and ferritin, respectively. In Figs. 6.2a and 6.3a is shown an expanded version of the data curves over which the background is removed. Also shown as a dotted line is the background curve determined using the smoothing spline technique described in Section (6.2.6). In Figs. 6.2b and 6.3b the derivative of the background curves are shown. The nonmonotonic behavior of these curves is an indication of the distortions in the cross section discussed previously. Finally, the step normalized EXAFS data are shown in Figs. 6.2c and 6.3c.

6.3. GENERAL PRINCIPLES OF ANALYSIS

6.3.1. Introduction

In this section the techniques for analyzing the data obtained by the processing described in the previous section are described. Since new analytical procedures are constantly being developed in the context of analyzing particular problems, this section does not attempt to provide a comprehensive discussion of all possible techniques. Instead, it represents the personal opinions of the authors of one possible complete set of techniques. After a review of the equations that are used to analyze the data, this section discusses Fourier transforms and their inverses including problems of ranges, windows and filtering, interpretation of

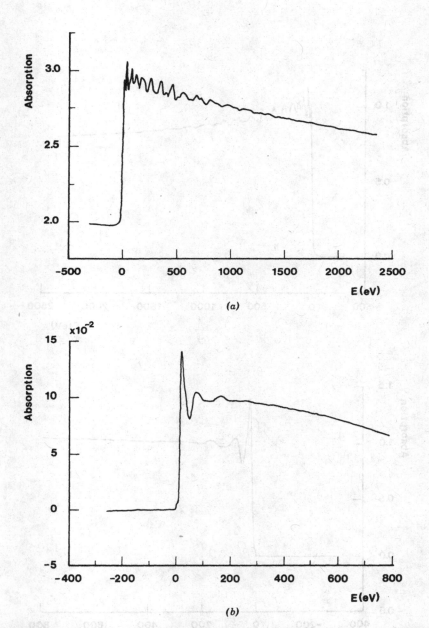

Figure 6.1. Raw absorption versus photoelectron energy for (a) the rhodium K-edge (23220 eV) in rhodium metal foil at 80 K and (b) the iron K-edge (7112 eV) in the protein ferritin taken in the fluorescence model. (c) The normalized absorption spectrum for the rhodium foil data in (a). A Victoreen pre-edge function has been used and the spectrum normalized at 107 eV above the edge. (d) The normalized fluorescence spectrum for the ferritin data of (b). A linear pre-edge function was removed and the spectrum normalized at 95 eV above the edge.

221

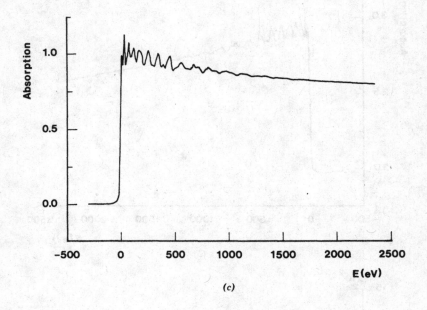

(c)

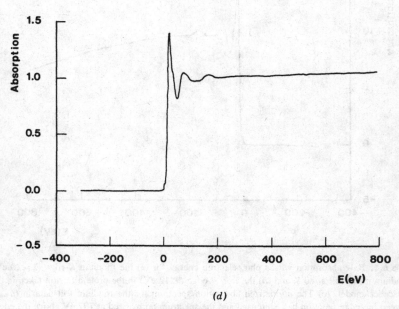

(d)

Figure 6.1. (*Continued*)

222

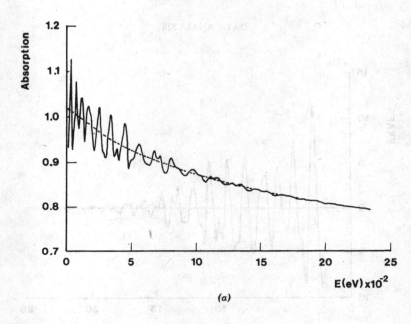

(a)

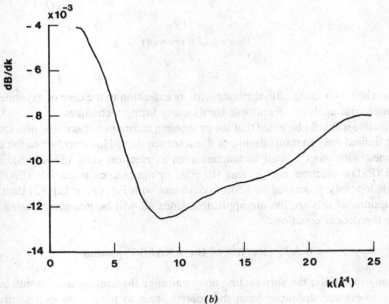

(b)

Figure 6.2. (*a*) The normalized data of Fig. 6.1*c* for rhodium metal showing on an expanded scale for the data above the edge. The calculated background curve used to extract the EXAFS is shown as a dashed line. (*b*) The derivative of this background curve versus the photoelectron wave vector *k*. (*c*) The normalized EXAFS for rhodium metal plotted versus *k*.

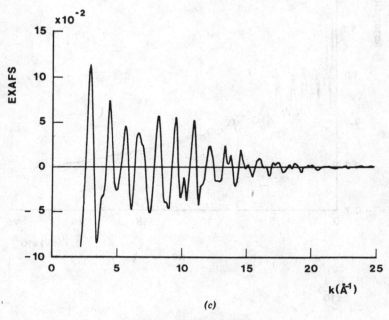

(c)

Figure 6.2. *(Continued)*

single-shell data using ratio methods with its extension to the case of asymmetric disorder, and analysis of multiple shells using fitting techniques.

Finally, it should be noted that the processing techniques discussed previously were limited only to transmission or fluorescence data. However, when the data obtained with other related techniques such as electron yield (SEXAFS), ion-yield (PSD), electron energy loss (ELFS), or appearance potentials (EAPFS) are analogously processed to a form consistent with Eq. (1) or Eq. (2) then the discussions of this section are applicable since all will be interpreted using the same theoretical equations.

6.3.2. Review of the EXAFS Equation

In many materials, the surrounding atoms arrange themselves into coordination shells where the distances from the central atom to the atoms in a shell are approximately equal and there is only one atomic species in each shell. In this case we use for an unoriented sample with small or approximately Gaussian disorder the following EXAFS equation [see Eq. (61), Chapter 1]

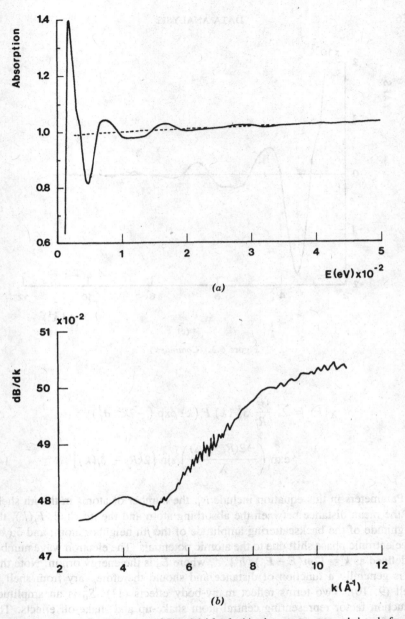

Figure 6.3. (a) The normalized data of Fig. 6.1d for ferritin shown on an expanded scale for the data above the edge. The calculated background curve used to extract the EXAFS is shown as a dashed line. (b) The derivative of this background curve versus the photoelectron wave vector k. (c) The normalized EXAFS for ferritin plotted versus k.

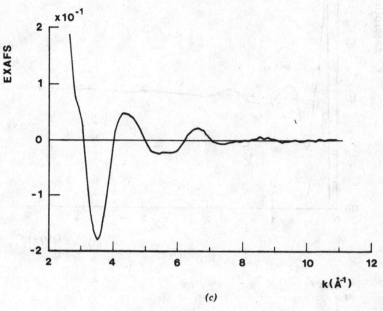

(c)

Figure 6.3. *(Continued)*

$$\chi(k) = \sum_j \frac{N_j}{kR_j^2} S_0^2(k) \, F_j(k) \, \exp\left(-2k^2 \, \sigma_j^2\right)$$

$$\exp\left(\frac{-2(R_j - \Delta)}{\lambda}\right) \sin\left[2kR_j + \delta_j(k)\right] \tag{4}$$

Parameters in this equation include N_j, the number of atoms in the jth shell; R_j, the mean distance between the absorbing atom and the jth shell; $F_j(k)$, the magnitude of the backscattering amplitude of the jth neighbor atom; and $\delta_j(k)$, the electronic phase shift due to the atomic potentials. The electron wave number is defined as $k = [2m(E - E_0)/\hbar^2]^{1/2}$, where E_0 is the energy origin. Note that E_0 is generally a function of distance and should therefore vary from shell to shell (9, 10). Two terms reflect many-body effects (11): S_0^2 is an amplitude reduction factor representing central atom shake-up and shake-off effects. The mean free path (λ) of the electron is due to the finite core hole lifetime and interactions with the valence electrons. The parameter Δ is a ''core radius'' of the central atom that eliminates double counting of inelastic processes in the core region. The nearest-neighbor distance R_1 appears to be a reasonable approximation for Δ. The relative mean squared disorder between the absorbing atom and an atom in the jth shell including static and thermal contributions is σ_j^2. Expressing the spectrum in this way is valid only if the disorder is small; for example, if $2k^2\sigma_j^2 \ll 1$. More general expressions involving sums or integrals over distribution functions have been presented in ref. (12) and in Chapter 1, Section 1.3.2 and will be discussed in Section 6.3.7.

A more complicated case arises when (a) several different atomic species constitute a coordination shell; (b) the disorder in the radial distances is large so that the Debye–Waller approximation is invalid; or (c) two shells are not totally resolved so that they cannot be spectrally isolated. There is no *fundamental* difficulty in handling these situations, but analysis is more complicated. Most of the discussion in this section involves analyzing EXAFS data according to Eq. (4) but generalizations of this equation particularly in the case of large or anisotropic disorder will also be considered.

Multiple-scattering processes may greatly complicate the EXAFS spectrum. Although the scattering amplitude is relatively small for a particular backscattering atom, the multiple scattering may be large if enough backscatterers exist at specific distances. These effects may be particularly large when an atom shadows a more distant atom (13). This tends to "focus" the outgoing electron wave and greatly enhance the EXAFS amplitude of the higher shell. A fortunate aspect of most of these multiple-scattering processes is that they have a large characteristic path length. By the techniques discussed in the next section, this greatly simplifies data interpretation.

6.3.3. Spectral Isolation: Fourier Transforming

6.3.3.1. General Principles, Justification

Equation (4) obviously contains a great deal of structural information about the local environment. The challenge is the isolation of the parameters of interest. With the data having contributions from as many as 10 coordination shells, it is very important to get rid of superfluous degrees of freedom. A great simplification in analysis is realized through the use of spectral isolation techniques. Because the fine structure corresponding to different coordination shells oscillate at different frequencies, spectrally isolating these different contributions could greatly simplify the analysis. In addition, the contributions from multiple scattering may be isolated from the low-r peaks. In practice, first-shell data are *never* contaminated by multiple scattering and second-shell data are very rarely contaminated. Data from third and higher shells, on the other hand, may be very difficult to interpret because of multiple scattering. Isolating the data corresponding to different distances is feasible because the primary k dependence in Eq. (4) is the sine term. A number of other parameters in Eq. (4) are energy dependent, but usually vary slowly enough that they act as a small correction to the frequency components of the spectra.

To further justify the use of the Fourier transform, we may rewrite Eq. (4) as

$$\chi(k) = \sum_j \left| A_j(k) \right| \sin \left[2kR_j + \delta_j(k) \right]$$

$$= \sum_j \text{Im} \left[A_j(k) \exp \left(2ikR_j \right) \right] \qquad (5)$$

with

$$A_j(k) = \frac{N_j}{kR_j^2} S_0^2(k) F_j(k) \exp \left(-2k^2 \sigma_j^2 \right) \exp \left[i\delta_j(k) \right] \exp \left(\frac{-2(R_j - \Delta)}{\lambda} \right)$$

The Fourier transform of this expression is (for positive distance)

$$\tilde{x}(r) = \frac{1}{2i} \sum_j \tilde{A}_j(r) * \delta(r - R_j) \qquad (6)$$

where "*" denotes convolution and $\tilde{A}$ the Fourier transform of A. As a trivial example, we may take the case of a Gaussian amplitude of width Δk centered at k_0 and a phase shift that is linear in k:

$$A(k) = \exp \left[\left(\frac{k - k_0}{\Delta k} \right)^2 + i(\delta_0 + \delta_1 k) \right] \qquad (7)$$

The Fourier transform of this is

$$\tilde{A}(r) = \Delta k \exp \left[-\frac{\Delta k^2}{4} (\delta_1 + 2r)^2 + i \left(\delta_0 + \frac{k_0}{2} (\delta_1 + 2r) \right) \right] \qquad (8)$$

where the Fourier transform used here is defined as

$$\tilde{A}(r) = \frac{1}{\sqrt{\pi}} \int A(k) \, e^{2ikr} \, dk$$

Combining Eqs. (6) and (8), we have the transform of the idealized EXAFS

$$\tilde{\chi}(r) = \frac{\Delta k}{2i} \exp \left(i\delta_0 + ik_0\delta_1 + 2ik_0r - \frac{\Delta k^2}{4} (2r - 2R - \delta_1)^2 \right) \qquad (9)$$

The magnitude of this transform is

$$\left|\tilde{\chi}(r)\right| = \frac{\Delta k}{2} \exp\left[-\Delta k^2(r - R - \delta_1/2)^2\right] \tag{10}$$

so it may be seen that the result is simply a Gaussian centered at $R + \delta_1/2$.

To realistically evaluate this approach, it is important to know the actual variation of the amplitude and phase with k. Figure 6.4 shows that the calculated backscattering amplitude $F(k)$ for platinum, germanium, and carbon (14). For backscattering atoms of $Z < 40$ or 50, the amplitude is relatively smooth over the EXAFS data range. For higher Z atoms, $F(k)$ is more complicated. This additional structure in the amplitude gives rise to sidebands about the central frequency in the transform. In many cases, this causes no problem: If the different shells are far enough apart in radial distance, the complicated peak—sidebands and all—may still be well isolated from data of other shells.

Similarly, Fig. 6.5 shows the contributions to the electronic phase shift $\delta(k)$ due to the central and backscattering atoms for platinum, germanium, and carbon. Because the phase shift is not absolutely linear in k, there will also be some broadening of the transform peak: There is no longer a "sharp" frequency

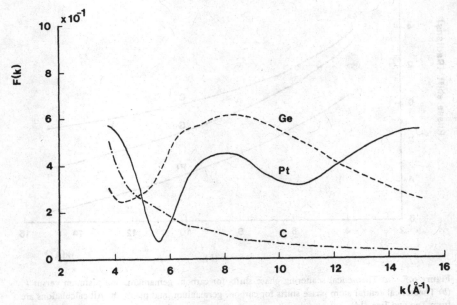

Figure 6.4. Theoretical backscattering amplitudes versus k for C($Z = 6$), Ge($Z = 32$), and Pt($Z = 78$) after Teo and Lee (14).

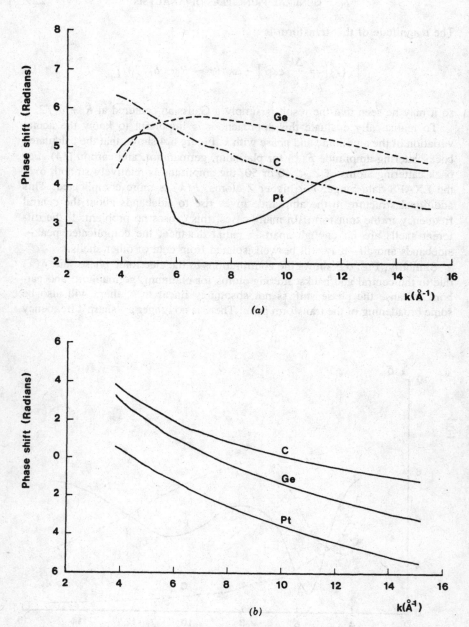

Figure 6.5. (a) Theoretical scattering phase shifts for carbon, gemanium, and platinum versus k. (b) Theoretical central atom phase shifts for carbon, germanium, and platinum. All calculations are from Teo and Lee (14).

associated with the EXAFS. For the vast majority of systems of interest, however, this broadening due to phase nonlinearity is comparable to other contributions to peak width, such as finite data range.

Other energy-dependent parameters include S_0^2 and λ. The central factor S_0^2 appears to exhibit only very weak k dependence (11) and can probably be neglected in the discussion. The mean free path λ, on the other hand, may change significantly through the data range. As opposed to the other k-dependent parameters, the energy dependence of the mean free path may greatly complicate the analysis. Luckily, first-shell data are essentially independent of the mean free path and are, therefore, relatively simple. Analysis of higher shell data may be difficult to interpret unless a suitable known material of similar local chemical environment is available for comparison.

6.3.3.2. Fourier Transform Techniques

Spectral isolation is useful in decoupling EXAFS parameters, but the Fourier transform is not necessarily the end of the analysis chain. In most cases, it is merely an intermediate step used to isolate the data from different shells.

One of the practical problems associated with Fourier-transforming data is due to the finite data range. Because of truncation effects, the transform may be contaminated with spurious peaks—often interfering with physical features of interest. This truncation ripple is easy to understand in terms of the convolution theorem when transforming over a finite range in k space, the "true" infinite range transform is convoluted with the transform of the window function when the data are sharply truncated. The data transform is convolved with

$$\tilde{W}(r) = \exp{(ik_0 r)} \frac{\sin{(\Delta k r)}}{r} \tag{11}$$

where k_0 is the center of the transform range and Δk is the width. If a smoother window such as a $\cos^2{[a(k - k_0)]}$ is used, the truncation ripple will be minimized at the expense of some peak broadening. This $\cos^2$ window is often referred to as a Hanning function (15). Explicitly:

$$W(k) = \cos^2{\left(\frac{k - k_0}{\Delta k}\right)} \quad \text{for } |k - k_0| < \Delta k$$

$$= 0 \quad \text{for } |k - k_0| > \Delta k \tag{12}$$

where k_0 and Δk are as defined before. The transform of this window function is

$$\tilde{W}(r) = \exp{(2ik_0r)} \frac{\sin{(\Delta kr)}}{r} \frac{\pi^2}{2(\pi^2 - \Delta k^2 r^2)} \qquad (13)$$

Comparing this result with Eq. (11), it is apparent that the oscillations damp out much faster at large $r - r_0$ using the smooth window, but the central peak is significantly broader.

It is often useful to combine a flat region over the middle of the data range with a Hanning function at the ends, that is:

$$W(k) = \sin^2{\left(\frac{\pi(k - k_1)}{2(k_2 - k_1)}\right)} \qquad k_1 < k < k_2$$

$$= 1 \qquad\qquad\qquad\qquad k_2 < k < k_3$$

$$= \cos^2{\left(\frac{\pi(k - k_3)}{2(k_4 - k_3)}\right)} \qquad k_3 < k < k_4 \qquad (14)$$

where k_1, k_2, k_3, and k_4 delimit three different data ranges corresponding to the two end regions and the flat central region. Figure 6.6 shows the effect of the modified Hanning window for rhodium metal. In this figure, k_2 and k_3 are constrained to be

$$k_2 = k_1 + \frac{F(k_4 - k_1)}{2}$$

$$k_3 = k_4 - \frac{F(k_4 - k_1)}{2} \qquad (15)$$

where F is the "Hanning fraction": The relative range over which the Hanning function is acting. For $F = 0$, there is no smooth truncation; for $F = 1.00$, a full Hanning function is used over the entire data range.

In summary, window functions have several purposes:

1. To weight different portions of the data. For instance, low-k data tend to be less reliable because of errors in chemical transferability and other effects. Greater weighting of the high-k data deemphasizes this problem.

2. To truncate the data smoothly at the ends of the transform range so as to minimize truncation ripple at the expense of some peak broadening.

3. To make data from two different samples appear as close as possible to one another. For instance, in comparing data from a sample measured

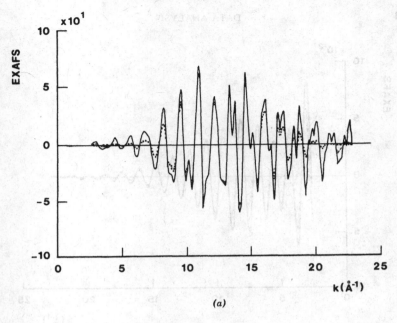

(a)

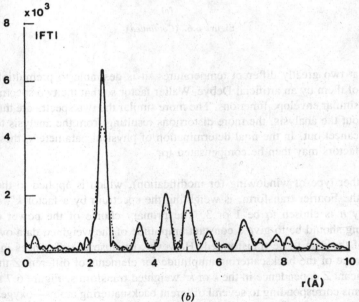

(b)

Figure 6.6. (a) $k^3 \chi (k)$ for rhenium metal with $F = 0$ (———), $F = 0.2$ (----), and $F = 1.00$ (. . .).
(b) The Fourier transforms of the curves in (a), over the range 2.6–22.7 Å^{-1}. (c) The first-shell
filtered data with the respective Hanning windows included. An inverse transform range of 1.71–
2.72 Å was used in all cases.

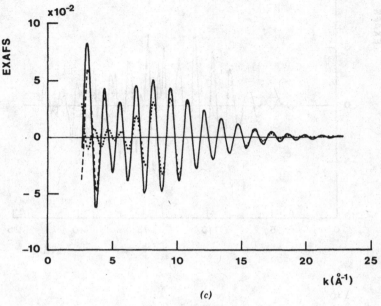

(c)

Figure 6.6. (*Continued*)

at two greatly different temperatures, it is desirable to premultiply one of them by an artificial Debye–Waller factor so that the two spectra have similar envelope functions. The more similar the two spectra are throughout the analysis, the more distortions resulting from the analysis tend to cancel out. In the final determination of physical parameters, these prefactors may then be compensated for.

Another type of windowing (or modification), which is applied to the data before the Fourier transform, is weighting the spectrum by a factor k^n, where typically n is chosen to be 1 or 3. The primary choice of the power of the weighting should be to give a constant amplitude of the weighted data over the range of the data to be transformed. However, because of differences in the k dependence of the backscattering amplitude for elements of different Z there is a significant Z dependence in the k or k^3 weighted transforms. Figure 6.7 shows transforms corresponding to several different backscattering atoms—oxygen, cobalt, and rhodium; all have the same (k^1) weighting and use no Hanning window. Note that as the atomic number of the backscatterer changes, the transform peak also changes significantly. The transform peak corresponding to oxygen backscattering is very broad and featureless. This is because the backscattering amplitude drops quickly with increasing electron energy. A much better choice

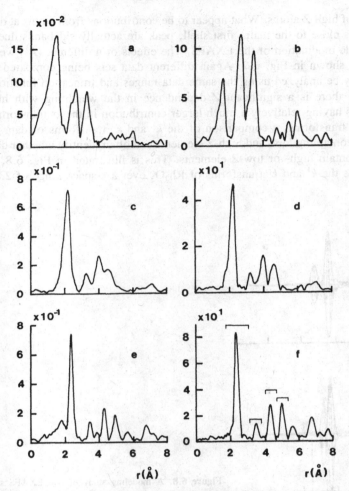

Figure 6.7. A comparison of k^1 and k^3 weighted Fourier transforms for Rh_2O_3 (2.8–18.2 Å^{-1}) (*a*), (*b*); cobalt foil (2.9–16.8 Å^{-1}) (*c*), (*d*) and rhodium foil (2.6–21.6 Å^{-1}) (*e*), (*f*). Only the magnitude of the transform is shown. Note the differences in peak shape of the first shell as the Z of the backscatterer is increased from $O(Z = 8)$ to $Co(Z = 27)$ to $Rh(Z = 45)$. Note also the growth of the Rh—Rh peak at about 2.7 Å relative to the Rh—O peak at 1.8 Å.

of weighting would be k^3 so as to emphasize the higher energy part of the spectrum. This artificial weighting of the spectrum is quite justifiable as long as other materials being compared with this are treated the same way. The transform peak for cobalt, on the other hand, is well suited to k^1 weighting. The transform peak corresponding to rhodium appears quite complicated, due to the energy-dependent structure in the backscattering amplitude and phase characteristic of high-Z atoms. What appear to be contributions from atoms at different distances close to the main, first shell, peak are actually sidebands due to the amplitude modulation of the EXAFS. The effects of a different (k^3) weighting are also shown in Fig. 6.7. Again different data sets being compared should generally be analyzed using the same data ranges and truncation functions.

Since there is a significant Z dependence in the weighting with higher-Z elements having relatively a much larger contribution in the k^3 transforms than in the k transforms, a comparison of the k^1 and k^3 transforms of data from a sample containing low- and high-Z elements can often identify whether different shells contain high- or low-Z elements. This is illustrated in Fig. 6.8, which compares the k^1 and k^3 transforms of Rh_2O_3 over a k-space range of 2.8–18.2

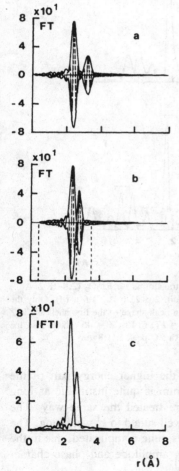

Figure 6.8. A modeling study of two EXAFS shells to show the effect of interference. In (a) the magnitude (solid) and imaginary part (dashed) of the modeled data is shown for two shells separated by 0.8 Å where very little interference occurs. In (b) the separation was reduced to 0.38 Å with all other parameters having the same values as (a). The separation of the peaks with the sharp minimum at 2.7 Å is clearly seen to be due to significant interference between the side lobes. Also shown are inverse transform ranges to separate the first shell (0.5–2.7 Å), the second shell (2.7–3.5 Å), and both shells together (0.5–3.5 Å). In (c) the magnitude of the Fourier transform of each shell used in Fig. 6.9b is shown separately. The overlap of the curves shows the amount of interference that occurs and is a measure of the errors introduced if each peak in Fig. 6.9b is filtered separately.

$Å^{-1}$. The growth of the second shell relative to the first in the k^3 transform versus the k^1 transform clearly shows that the second shell contains a significant contribution from rhodium.

Finally, a few other general comments about the windowing of the EXAFS data should be emphasized. The data naturally has a sharp window at the lower and upper ends of the spectrum because of the finite range of the data. Any additional weighing and windowing should be done bearing in mind the effects that a particular modification will have on the transform. Usually it is a good rule to compare the effects of applying a window since in the transform there is always a compromise between localizing the side lobes and reducing and broadening the peak because of loss of data. One helpful procedure at this stage of the analysis is to graphically inspect the weighted and windowed data since it is this function that will be input into the Fourier transform. Choices of windows, effect of noise, glitches, and distortions of the data, which may occur during the background removal particularly at high k, will be apparent. Generally one can decide on the proper k weighting to give a uniform amplitude of the weighted data across the k-space range subject to the limitations of the signal to noise. If the transform limits are chosen carefully and the upper limit is chosen before the weighted noise dominates the signal it is often found that other windowing is not necessary.

6.3.3.3. *Interpretation of Transforms*

Although the Fourier transform is usually an intermediate step in the analysis, a great deal of structural information may be learned from direct inspection. For systems with small disorder (e.g., $2k^2\sigma^2 \ll 1$) with a single atomic species in a shell, the radial distances may be determined with relatively high precision. In particular, if a sample and a suitable standard compound that is similar in local structure and chemical state are compared, then distance *difference* determinations of about 0.01-Å accuracy are readily obtainable, particularly for high-Z backscattering atoms. Inspection of the transform is also very useful as an overview of the data, giving information about various shell radii and amplitudes, disorder, noise level, and so on. As mentioned previously, it is also a critical part of careful background removal.

Highly accurate structural determinations are possible using r-space techniques such as multiparameter fitting of the transform function (16). Another approach is the use of an "optical" transform that explicitly includes the atomic phase shift (19):

$$\tilde{\chi}(r) = \int \chi(k)\, e^{2ikr + i\delta(k)}\, dk \qquad (16)$$

This approach has two advantages: (a) the phase shift is removed so that distances may be read off immediately from the transform peak positions, and (b) because nonlinearities of the phase shift are removed, the peaks are in principle less broadened and more symmetric. However, this is only useful in single component systems or can be done shell by shell if the shells are isolated and contain only one type of atom. Even then there may be a shell to shell variation of the phase shift (9). Phase corrected transforms have recently been used in catalytic studies to separate the contributions of different elements (17–19). To determine amplitude factors such as coordination number and Debye–Waller factors, however, some fitting of the peak shape is generally necessary. The energy origin E_0 is also critical when including the phase shift. Choice of the E_0 will be discussed in Section 6.3.6.

Often only the magnitude of the Fourier transform is examined. Since it is a complex Fourier transform, the entire transform contains not only information in the magnitude but in the phase as well. Examination of the imaginary (or real) part of the transform, as well as the magnitude, can be useful since due to the phase information its shape is more characteristic of the types of atoms that make up a given shell. By suitable comparison with reference compounds or theoretically modeled data it is often possible to tell the types of atoms contributing to a given shell. If a given shell may actually have contributions from more than one type of atom, it is the real or imaginary part of the transform that can give conclusive information. This is due to the fact that the phases of different atoms are linearly combined in the real and imaginary part but not in the magnitude of the Fourier transform.

Viewing the imaginary part of the transform can be particularly helpful when applying the phase corrected transform discussed previously. The imaginary part of the transform should peak at the peak of the magnitude of the transform when a proper phase correction has been made (13) (i.e., a proper choice of the type of backscatterer).

6.3.4. Inverse Transforms

To isolate the single-shell data, it is necessary to inverse transform the Fourier transformed data over the r-space range corresponding to that shell. As when performing the forward transform, it is often desirable to multiply the data by a smooth truncation function to minimize truncation ripple. A Hanning function may also be useful for smoothly windowing the particular shell of interest. A problem with this window function (and also the commonly used Gaussian window) is that the transform is very sensitive to the choice of window position. Because the window is rounded at the position of the transform peak, a shift in window position will effect a shift in apparent peak position with the resultant error in distance determination. To minimize this problem it is useful to combine a flat region over the transform peak with a Hanning function at the ends, analogous to Eq. (14).

It is imperative not to back transform over too restricted a range in r space. If too small a range is used, the amplitude and phase of the data will be artificially constrained: The phase will appear more linear than is correct, and the amplitude will be smoothed from the true value, leading to incorrect determination of Debye–Waller factors and coordination numbers. Figure 6.7f shows typical back-transformation ranges for the first-four shells of rhodium metal foil.

There are several other pitfalls in choosing back-transforming ranges. When two shells are not truly isolated, they often interfere with one another. This interference may actually reduce the apparent amplitude where the two peaks are overlapping often appearing as a sharp minimum and making it appear that they *are* isolated at first. The major indication that interference is occurring is that the peaks are generally quite asymmetrical with the interference region dropping much faster in amplitude than the opposite side of the peak. Very large errors may ensue if an attempt is made to separately back transform the interfering peaks: It is highly preferable to transform the two peaks together and use a fitting approach in the k-space analysis. In Fig. 6.8 an example of the effects of interference is shown in a modeling study as two peaks are brought closer to each other. Figure 6.8c shows the position and shape of each peak separately. In Fig. 6.9 the inverse transforms performed using the ranges shown in Fig. 6.8 are compared to the input data. As can be seen the attempts to isolate single shells from the data produces significant distortions that would effect subsequent analysis. However, the inverse covering both shells, adequately reproduces the input data and can, in principal, be accurately analyzed using the multiple-shell fitting procedure described in Section 6.3.8.

6.3.5. Interpretation of Single-Shell Data

Once the data have been Fourier filtered to isolate the single-shell EXAFS, the analysis may become more specialized. Because the amplitude and phase may be separately obtained from the real and imaginary parts of the transform, detailed physical information is more directly determined than in r-space analysis.

Analysis may proceed with fitting procedures using theoretical parameters (usually the phase and backscattering amplitude) to account for some of the parameters in Eq. (4). It is also possible to determine phase and backscattering amplitude from model compounds and analyze the data with either fitting or the ratio method discussed later. Because of the distortions that can be introduced in the background removal and the difficulty in performing the energy-dependent normalization required when using theoretical parameters, it is usually more reliable to analyze the data using model compounds provided that suitable model compounds can be found. A more complete discussion of fitting procedures and the relative merits of theoretical empirical parameters will be discussed.

The ratio technique determines Debye–Waller factors and coordination num-

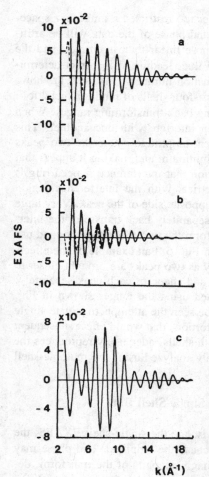

Figure 6.9. The effect of filtering on the two-shell modeling study. (*a*) A comparison of the first-shell modeled data (solid) with the first-shell filtered data (dashed) using a range of 0.5–2.7 Å. (*b*) A similar comparison of the second-shell data using the filter range of 2.7–3.5 Å. (*c*) A comparison of the two-shell model data with the two shell inverse data (Δ*r* = 0.5–3.5 Å). The large differences in (*a*) and (*b*) will lead to significant errors in determination of N and σ^2 for these shells.

bers by plotting the log of the ratio of the amplitudes of sample and standard versus k^2. Writing $\chi(k) = A(k) \sin [\phi(k)]$, we may take the ratio

$$\frac{A_1}{A_2} = \frac{(N_1/R_1^2)\, S_{01}^2(k)\, F_1(k)\, \exp\left[-2k^2\sigma_1^2 - 2(R_1 - \Delta_1)/\lambda_1\right]}{(N_2/R_2^2)\, S_{02}^2(k)\, F_2(k)\, \exp\left[-2k^2\sigma_2^2 - 2(R_2 - \Delta_2)/\lambda_2\right]} \quad (17)$$

which may be rewritten

$$\ln\left(\frac{A_1}{A_2}\right) = \ln\left(\frac{N_1 R_2^2}{N_2 R_1^2}\right) + 2k^2(\sigma_2^2 - \sigma_1^2) - 2[(R_1 - \Delta_1) - (R_2 - \Delta_2)]/\lambda$$

(18)

if the sample and standard are similar: that is, if $S_{01}^2;(k) = S_{02}^2(k)$, $F_1 = F_2$ and $\lambda = \lambda_1 = \lambda_2$, the last factor involving mean free path terms may be neglected to a good approximation for first-shell data because the core radii Δ tend to be approximately equal to the nearest-neighbor distance. Therefore, within these approximations, this log(ratio) as plotted versus k^2 will be a straight line of slope $-2\Delta\sigma^2$ with an intercept of $\ln(R_2^2 N_1 / R_1^2 N_2)$ (6). Any significant deviation from a straight line is indicative of (a) poor data, (b) errors in analysis, (c) nontransferability of backscattering amplitude or mean free path, (d) large, non-Gaussian disorder, or (e) multiple atomic constitutents in the shell being studied.

Similarly, the phase difference may be written

$$\phi_1 - \phi_2 = 2k(R_1 - R_2) + \delta_1(k) - \delta_2(k)$$

(19)

For a suitable standard and choice of energy origin, $\delta_1(k) = \delta_2(k)$, so this phase difference is a linear function of k with zero intercept and a slope of $2(R_1 - R_2)$. Presumably, the distance R_2 of the standard is known, so that R_1 is obtained directly in a model-independent way. As has been shown by Martens et al. (20) some problems can be encountered if the energy origin (E_0) is different for the two samples. Usually this causes the graph of $\phi_1 - \phi_2$ versus k to be nonlinear. The relative energy shift between the two samples can be adjusted until linearity is achieved. This is one method for correcting for differences in the energy origin between samples as will be discussed in the next section.

It should be noted that use of the ratio method is not restricted to systems with small disorder. The simple interpretation discussed previously is not appropriate, but more detailed analysis of the amplitude ratios and phase differences is often undertaken. This will be discussed in more detail in Section 6.3.7.

6.3.6. Determination of Energy Origin

The energy origin, E_0, compensates for many of the approximations in the simple EXAFS formula such as the small-atom approximation (2). Also, varying E_0 allows for a much more general concept of phase-shift transferability: Many small changes in chemical environment may be compensated for by changing E_0.

For systems with a single type of backscattering atom in a shell and small ($2k^2\sigma^2 \ll 1$) disorder, the energy origin may be determined self-consistently from the EXAFS data. In particular, as noted previously, the phase difference between a sample and standard should be a linear function of k with zero intercept if E_0 is chosen correctly and the standard is suitable. Two criteria present themselves: (a) choose E_0 such that the extrapolated $k = 0$ intercept of the phase difference is zero, or (b) vary E_0 to obtain the best fit to a straight line. In practice, criterion (a) appears the more reproducible. Note that the absolute value of E_0 is relatively unimportant: Varying E_0 by the same amount for sample and standard causes rather small changes. Therefore, it is common to fix the energy origin for one of the data sets and vary E_0 for the other to satisfy the above criteria.

This discussion is concerned with analysis in k-space. Similar methods are possible in r-space analysis. For instance, when using the optical transform method [Eq. (16)], the magnitude and imaginary parts of the transform will peak at the same distances if E_0 is chosen correctly and a suitable standard is used (13).

It is also possible to determine E_0 self-consistently in the context of multi-parameter model-dependent fitting. This will be discussed in more detail in the section on fitting. If theoretical modeling is used, the difference in energy origin between the theoretical parameters and the experimental data is even more important since normally the origin of the experimental data is at or near the Fermi level, while the origin of the theoretical data is at the "muffin-tin" zero, which can be significantly different from the experimental origin.

6.3.7. Large Disorder

As discussed in Chapter 1, Section 1.3.2 large disorder, either thermal structural or site, modifies both the amplitude and phase of the EXAFS. For a distribution of distances within the jth coordination shell with only one type of neighbor atoms, Eq. (4) may be rewritten (21)

$$\chi_j(k) = \operatorname{Im} \frac{N_j}{kR_j^2} S_0^2(k) \, F_j(k) \exp\left(\frac{-2(R_j - \Delta)}{\lambda}\right) \int P_j(r)$$
$$\cdot \exp\left\{i[2kr + \delta_j(k)]\right\} dr \tag{20}$$

letting $r = R_j + \Delta_r$ and performing the integration over Δ_r

$$\chi_j(k) = \frac{N_j}{kR_j^2} S_0^2(k) F_j(k) Q_j(k) \exp\left(\frac{-2(R_j - \Delta)}{\lambda}\right)$$

$$\cdot \sin\left[2kR_j + \delta_j(k) + \phi_j(k)\right] \tag{21}$$

where

$$\tilde{P}(k, R_j) = Q_j \exp\{i\phi_j(k)\} = \int P_j(r) \exp\{i2k\Delta r\} d(\Delta r) \tag{22}$$

[See also Chapter 1, Eqs. (29), (30), (31), and (61).] It can be seen that the magnitude $Q_j(k)$ of the effective distance distribution $\tilde{P}(k, R_j)$ changes the amplitude and the argument $\phi_j(k)$ changes the phase of EXAFS.

Three techniques are commonly used to analyze data with large disorder:

1. Analysis of beats: If two (or in principle, more) distances comprise a coordination shell, than the interference between their contributions will give rise to "beats" in the EXAFS: Large oscillations in the amplitude accompanied by large phase changes (22). For systems with (a) a single kind of atom in the shell, and (b) only two distances involved, the data are particularly easy to interpret. If the first minimum in the EXAFS amplitude occurs at the point k_0 the difference in distances is $\pi/2k_0$. In addition, the amplitude at k_0 reveals information about coordination numbers and Debye–Waller factors. For more complicated systems than some sort of model-dependent fitting is generally necessary.

2. Method of cumulants: This method (23, 24) is basically a generalization of the ratio method. As noted previously, the distribution $P(r)$ may be expanded in a moment series. In principle, these moments could provide a model-independent way of determining the r-space distribution. In practice, however, this series is very slow to converge and is, therefore, rather ill-suited to the problem. A superior method is to expand the logarithm of the transform of the distribution (see Chapter 1, Section 1.3.2)

$$\ln\left[\tilde{P}(k, R_j)\right] = \sum_{n=0}^{\infty} \frac{(2ik)^n}{n!} C_n(R_j) \tag{23}$$

where the C_n are called cumulants. The high-order cumulants represent the deviation of the effective distribution $P(r)$ from a Gaussian; therefore, the cumulants tend to converge much faster than a simple moment expansion. Equations (18) and (19) may therefore be generalized (temporarily ignoring disorder in the "standard"):

$$\ln\left(\frac{A_1}{A_2}\right) = \ln\left(\frac{N_1}{N_2}\right) + C_0 - C_2\frac{(2k)^2}{2!} + C_4\frac{(2k)^4}{4!} - \cdots$$

$$\phi_1(k) - \phi_2(k) = 2k(C_1 + R_j) - C_3\frac{(2k)^3}{3!} + \cdots \tag{24}$$

Therefore, the cumulant expansion presents a generalized ratio method that can handle problems of large disorder. By a model-*independent* fitting of the amplitude ratio and phase difference, the cumulants may be determined and the actual radial distance distribution reconstructed.

This technique becomes much more difficult, however, if (a) more than one type of atom constitutes the shell, or (b) the mean free paths of the sample and standard exhibit much different k dependence. This technique has been applied to CuBr by Tranquada and Ingalls (25).

3. r-Space Analysis: Hayes and co-workers (16) have applied an r-space deconvolution procedure to extract asymmetric distribution functions. Using an amplitude function in r-space that has no disorder or Gaussian disorder, they deconvolve the r-space distribution function, $P(r)$, using the transform of Eq. (20) and the convolution theorem of Fourier analysis. This procedure has been applied to the determination of distribution functions in ionic conductors.

In systems where more than one shell or shells containing different types of atoms are involved the procedures described previously cannot be applied since all depend on an interpretation of a single term of Eqs. (4) or (20). Instead, a fitting procedure must be used, which will be described in the next section.

6.3.8. Fitting Procedures

A common alternative approach to the analysis of single-shell EXAFS data and the only approach if more than one shell at a time must be analyzed is the use of numerical fitting procedures using the EXAFS parameterization function for the fitting routine. However, the form of the EXAFS equation requires that nonlinear fitting algorithms be used even for single-shell analysis. Generally for each shell N, $F(k)$, σ, λ, R, E_0, $\delta(k)$, Δ, and S_0^2, must be parameterized or entered as input data for each shell, if the single scattering Eq. (4) is used. Typically, up to four parameters are varied for each shell (N, σ, R, and E_0), fixing the other parameters using either model compounds or theoretical values. If theoretical values for the amplitude and phases are used then explicit corrections must also be made for S_0^2, λ, and Δ. If model compounds are used then the amplitude of the model compound also contains the contributions of S_0^2, λ, and Δ provided that the bond distance of the model compound is close to ($<$ 0.1 Å) that of the unknown. Since these parameters are generally insensitive to the chemical environment of the absorbing atom, no further correction for them need to be made. The empirical amplitude also contains the disorder contribution (σ^2) for the model compound and for the temperature at which the data is taken. Thus the disorder parameter that is determined is the relative mean square displacement $\Delta\sigma^2 = \sigma^2 - \sigma_m^2$ and not the absolute displacement. The fixed values that are used may be input either as tabulated values or as parameterized values. Care must be taken when using parameterized functions that the functional form that is used is valid for the particular case being studied and does not introduce errors from the parameterization itself. Similarly, when using tabulated theoretical functions such as those of Teo and Lee (14) care must be taken in interpolating the values for elements that are not explicitly calculated.

The fitting problem now may be reduced to minimizing the sum of the squares of the residuals

$$S^2 = \sum S_i^2 = \sum (\chi_i^T - \chi_i^E)^2 \, W_i \qquad (25)$$

relative to the parameters that are being varied.

The sum is taken over the data points i being used for the analysis, W_i is a weighting function that may be used to account for the changing statistical accuracy of the data points, and χ^T and χ^E are the theoretically calculated and experimental EXAFS functions, respectively. The use of the minimization of the sum of the squares has not been universal; other groups have used related functions such as χ^2 or the crystallographic R factor. There does not appear to be any significant difference in the results of programs using these different requirements. This does point to one of the problems in using curve fitting techniques in EXAFS analysis: There is still not a universally recognized formalism for the type of analysis as in crystallography. This was emphasized recently by Pendry (26) who sought to identify a function analogous to the R factor for optimal minimization of EXAFS data. Finally, the theoretically calculated and experimentally determined EXAFS functions, χ^T and χ^E in Eq. (25) represent one or more shells of data depending on the range of r-space data included in the inverse transform and the number of theoretical shells used to interpret the data.

The numerical considerations for the minimization of the function, S^2 of Eq. (25) have been discussed by Cox (8) who also proposed adding an additional term to the calculated EXAFS, χ^T, to account for inaccuracies in the determination of μ_0. This appears to be useful particularly when theoretical parameters are being used. In general, standard mathematical procedures for finding the minimum of a nonlinear function are used.

Algorithms for doing these calculations are found in most large numerical analysis packages available commercially or at large computer centers. The advantage of using these routines is that they often provide statistical results as well as the solution, particularly the correlation matrix showing the relationship between the parameters. In general, as is well known, R and E_0 and N and σ^2 are highly correlated with each other for a given shell with only weak correlations between the parameters of that shell describing the phase and those describing the amplitude. This generalization must be used cautiously, however, since for some problems, particularly involving weak, or similar shells the intershell correlation may become significant thus complicating the analysis.

If only a single shell is to be analyzed using numerical fitting then the problem may be simplified by explicitly calculating the amplitude and phase of the sample data from the inverse transform using

$$A(k)_{\text{exp}} = \left[R^2(k) + I^2(k) \right]^{1/2} \tag{26}$$

$$\phi(k)_{\text{exp}} = \tan^{-1} \left(\frac{I(k)}{R(k)} \right) \tag{27}$$

with $I(k)$ and $R(k)$ the imaginary and real part of the inverse transform, respectively. Then the problem is reduced to two single problems where only two parameters $(N, \Delta\sigma^2)$ and (R, E_0) need to be varied for A_{exp} and ϕ_{exp}, respectively. Since, as discussed previously the correlation between the amplitude and phase parameters is small, well-defined solutions can usually be found. A comparison between ratio and first single-shell fitting results for room temperature copper metal using the data at 80 K as a reference has been given (7), which shows both techniques are comparable in their accuracies finding coordination numbers within 5% of their known values and σ^2 values within 10% of the theoretical values (27).

For multiple-shell fitting typically up to four parameters per shell can be allowed to vary. Generally the range of the data limits the number of independent parameters, which can be determined to about six. Therefore, to fit two or more shells of data some way must be found to restrict the number of parameters that are varied. Fortunately, the correlation between the parameters can be used to help. Generally the largest errors are introduced by errors in the phase and since these are not strongly correlated between shells a first fit can be done either with one dominant shell or with approximate amplitude parameters fixed to get distance information. The distance parameters can then be fixed, particularly for a dominant shell ot refine the remaining parameters. The values of E_0 obtained from this fit represent another way of determining the proper E_0 for each shell. If necessary this process can be iterated to get a self-consistency.

As pointed out by Cox (8) and others the minimization procedures used do not guarantee that an absolute minimum has been found. With any fitting (particularly involving a large number of parameters) some care must be taken to see that the solution found is the correct one and not a local minimum. There is yet no rigorous procedure to do this, but there are a number of techniques that may be used alone or in combination to help test the validity of the solution. One is to start the fit at different initial values of the parameters to be varied to see if the same solution is found. A criterion may be that the solution chosen has the lowest residuals. One may also plot the EXAFS function calculated with the best function not only against the inverse data used in the fitting but also against the unfiltered EXAFS itself, particularly if a dominant shell was analyzed. This enables us to see if the solution is consistent with the data. Finally, one may Fourier transform the fitted function and compare it with the transform of the experimental data. The imaginary part of the transform is very useful in this comparison since as mentioned previously, it contains both phase and amplitude information. In our experience a multiple-shell system that has good agreement with the data in k space and with the imaginary part of the FT in r space results in the best physically acceptable solution to the analysis. Two examples of multiple-shell fitting are shown in Figs. 6.10 and 6.11. In Fig. 6.10 a supported rhodium catalyst reduced at 473 K is analyzed using the first shell of Rh—Rh metal and Rh—O in Rh_2O_3 as standards. The dominant main peak in the transform was inverse transformed and analyzed first as a single Rh—Rh shell and then as a combination of Rh—Rh and Rh—O. While the

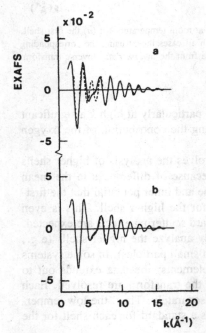

Figure 6.10. Inverse Fourier transform of the first coordination shell for the experimental data (solid) for 3 wt. % Rh/MgO reduced at 473 K compared with calculated spectra (dashed), obtained using parameters from a single Rh—Rh fit (a) and a double Rh—Rh—Rh—O fit (b). Parameters were obtained by minimizing residuals relative to the shell coordination number, distance, relative disorder, and inner potential.

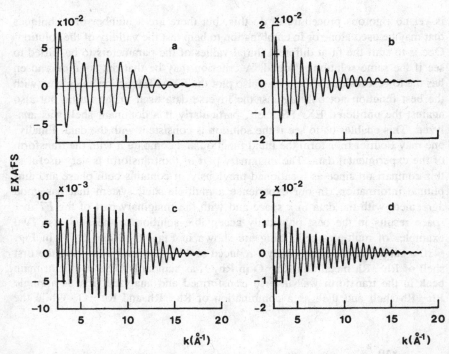

Figure 6.11. Isolated single-shell data for rhodium foil at room temperature for (*a*) the first shell, (*b*) second shell, (*c*) third shell, and (*d*) fourth shell. In all cases the fit using the corresponding shell from the 80 K data as a model is indistinguishable from the inverse data. Inverse transform ranges and structural parameters are shown in Table 6.1.

Rh—Rh accounts for the majority of the data particularly at high k a significant systematic difference at low k remains. Adding the contribution of the oxygen shell gives an excellent fit in k space.

The other example of fitting results involves the analysis of higher shells in bulk materials. It has been known that because of differences in the mean free path and shell-to-shell differences in phase and inner potential that the first-shell phase and amplitude are poor models for the higher shell. This is even true for those shells where no significant forward scattering effects are expected. In many materials it is desirable to accurately analyze the higher shells (e.g., to gain information about the size and shape of small particles). In some systems (e.g., rhodium, platinum and other high-Z elements) the data extends out to such a high k range that the higher peaks in the transform are resolved. Each shell can be filtered and, therefore, analyzed separately. Thus, the low temperature data from the metal foil may be used as a standard for each shell for the analysis at higher-temperature data from the foil or for the analysis of higher shells in metal clusters. In this way even shells that have strong focusing effects [such as the fourth shell in face center cubic (fcc) systems] can be analyzed. The results of such an analysis of the room temperature EXAFS data for rhodium metal foil is shown in Fig. 6.11. The results of single-shell fitting are shown for the first, second, third, and fourth shells using the 80 K temperature data as a reference. The parameters obtained from the fitting are given in Table 6.1. Clearly coordination numbers that are accurate to within 10–20% can be obtained for data of good quality. This is sufficient to analyze higher shells of small particles (<2.0 nm) to obtain information about their average size and shape.

6.3.9. Empirical Versus Theoretical Parameters

The preceding discussion of the analytical procedures primarily concentrated on the use of empirically determined phase and amplitude functions as part of the analysis. However, as alluded to in the discussion, theoretically derived parameters are often used by many groups. It has often been a point of discussion about which method is preferable.

The primary advantage of using empirically derived parameters from model compounds is that if they are analyzed in the same fashion as the unknown sample then factors such as errors in the background removal, normalization, multielectron effects (S_0^2 and Δ), and the inelastic mean free path are automatically compensated for by the direct comparison of the model compound and

Table 6.1. **Structural Parameters for the First-Four Shells of Rhodium Metal at Room Temperature Using Rhodium Foil at 80 K As Reference**[a]

Shell Inverse Range (Å)	Crystallographic		EXAFS Fitting			
	N	R(Å)	N	R(Å)	$\Delta\sigma^2 \times 10^3$ (Å^2)	E_0(eV)
1 1.76–3.10	12	2.69	12	2.69	2.3	−0.2
2 3.10–3.76	6	3.80	4.8	3.80	2.0	−0.1
3 4.02–4.70	24	4.66	24.7	4.67	3.1	−0.4
4 4.70–5.40	12	5.38	12.1	5.38	3.1	0.4

[a]The parameters were obtained from a k^3 transform (2.6–18.4 Å^{-1}) and a fitting range of 5–17 Å^{-1}.

the unknown. However, this method has the disadvantage that it may not be possible to find a suitable model compound or, even more seriously, that an unsuitable compound may be used in the analysis. In addition, the determination of the empirical parameters of the model compound may introduce errors that do not cancel in the analysis introducing systematic errors into the final results. Generally these problems are not serious and may be checked through the analysis particularly using modeling programs. A discussion of one approach using empirical parameters has been given by Cramer (28). Stern (29) discussed the criteria for a good reference or standard compound arguing that it must not only contain the same central atom–scattering atom pair but also have about the same distance and coordination geometry. As he points out the XANES is a good way to monitor this.

The advantage of theoretical parameters is that they can be calculated, without reference to model compounds and be tabulated for use for all elements, without the necessity of finding a standard compound meeting these requirements. Some disadvantages include the difficulties of doing a proper energy dependent normalization, accounting for multielectron effects and performing corrections for inadequacies in background distortion. The theoretical values most often used are the tabulated values of Teo and Lee (14). These have been found to give reliable results to better than 0.02 Å in distance and about 10% in coordination number particularly for elements in the middle of the periodic table. However, they may not be reliable for the low-Z elements where chemical effects have a more significant effect on the potential and for high-Z elements where an inadequate number of phase shifts were used in the calculations. A more comprehensive approach has been reported by Pendry and co-workers (17), which includes a more complete calculation of the theoretical parameters in a specific coordination environment. However, to completely calculate a complex spectrum a very long and complicated calculation must be done.

Is there any inherent advantage in either approach? Clearly both approaches have been successfully used in a wide variety of systems so that any advantage cannot be significant. Ultimately, both approaches seem to use the same criterion. In order to most accurately analyze EXAFS data the local environment (and therefore the potential) must be known, in one case to find a suitable reference and in the other to construct the most accurate potential. It seems that one must know the answer to find the answer but that is not precisely correct. In a complex system where little is known about the local structure, all information including the XANES and the chemical composition must be used along with the measurement of a number of possible standards in order to self-consistently arrive at an answer. In the hands of competent investigators either technique seems to be capable of the same limits of accuracy. The only advantage, at this time, for the use of empirical parameters is that effects of data processing can be more easily minimized than by using theoretical parameters.

6.4. CONCLUSIONS

The previous sections have presented one approach to the systematic processing and analysis of EXAFS data. As mentioned at the outset, the discussion was not intended to comprehensively discuss all possible approaches. The intention was to present one possible complete and logical sequence of procedures and to discuss important points or sources of errors that may occur. In addition, specialized analytical procedures such as bond angle determination (30, 31), use of polarization dependence, particularly in SEXAFS measurement (32) or the use of phase and amplitude corrected transforms to separate the contributions of two closely spaced shells (18,19) have not been discussed here. However, the latter two methods will be discussed in Chapters 10 and 8, respectively.

The procedures outlined here are fairly long and complex compared to many spectroscopies. All of the procedures are not needed for simpler experiments that may seek simple distance or coordination information particularly from a single shell. However, most problems studied at present require a more complete analysis. This complexity of analysis shows why EXAFS will not become a routine "mail order" spectroscopy for most problems. On the other hand, it is a sign of maturity of the field that the most critical elements of analysis are now understood and utilized by most groups. Differences in analytical approaches reflect more the personality of the investigators and the history of the group or individual rather than substantive differences in procedures.

Since not all of the results come from rigorous and well-defined procedures a significant element of judgment based on experience is required particularly studying complex systems. New groups should test their procedures carefully with model or known systems or collaborate with experienced groups. In any case it is wise not to routinely accept the output of the computer as a definitive result. Two steps that appear to be underutilized in studying EXAFS are to plot any final curve based on analytically determined parameters to see if the results are consistent with the raw data. In particular, it is important to see whether small differences that may be indicated in the analysis can really be detected in the original data. The other approach is to use modeling programs to model the results and take their transforms. One must be able to adjust any parameter to see if it has a significant effect on the results when compared to the data, either visually or using some numerical criteria.

While the ultimate analysis of EXAFS data is complex, the intelligent and systematic application of sound analytical procedures allows a considerable amount of reliable structural information to be obtained from systems for which, in some cases no other technique is available. The EXAFS technique has already demonstrated itself to be a powerful tool in studying a wide variety of systems and with the advent of more powerful sources and new generations of instrumentation its range of applicability will be considerably extended. Undoubtedly this will lead to new and refined analytical procedures since they are generally developed in the context of specific applications. However, the procedures outlined here will remain at the core of any complete data analysis.

ACKNOWLEDGMENTS

We would like to thank Dr. A. N. Mansour and Dr. D. C. Koningsberger for their assistance in the preparation of this chapter.

REFERENCES

1. B. K. Agarwal and L. P. Varma, *J. Phys. C*, **3**, 535 (1970).
2. F. W. Lytle, D. E. Sayers, and E. A. Stern, *Phys. Rev. B*, **11**, 4825 (1975).
3. *International Tables of Crystallography*, Vol. III, Lonsdale (Ed.), Kynoch, Birmingham, 1962, Section 3.2.
4. W. H. McMaster, N. Kerr Del Grande, J. H. Mallet, and J. H. Hubell, *Compilation of X-ray Cross Sections*, National Technical Information Service, Springfield, 1969.
5. T. K. Eccles, Ph.D. thesis, Stanford University, 1977. (Available as SSRL Report No 78/01.)
6. E. A. Stern, D. E. Sayers, and F. W. Lytle, *Phys. Rev. B*, II, 4836 (1975).
7. J. W. Cook, Jr. and D. E. Sayers, *J. Appl. Phys.*, **52**, 5024 (1981).
8. A. D. Cox, in *EXAFS for Inorganic Systems*, C. D. Garner and S. S. Hasnain (Eds.), DL/SCI/R17, Daresbury Laboratory, Daresbury, 1981, p. 51.
9. J. J. Rehr and E. A. Stern, *Phys. Rev. B*, **11**, 4413 (1976).
10. W. L. Schaich, *Phys. Rev. B*, **14**, 4413 (1976).
11. E. A. Stern, B. A. Bunker, and S. M. Heald, *Phys. Rev. B*, **21**, 5521 (1980).
12. G. Beni and P. M. Platzman, *Phys. Rev. B*, **14**, 9514 (1976).
13. P. A. Lee and J. B. Pendry, *Phys. Rev. B*, **11**, 2795 (1975).
14. B. K. Teo and P. A. Lee, *J. Am. Chem. Soc.*, **101**, 2815 (1979).
15. E. O. Brigham, The Fast Fourier Transform, Prentice–Hall, Englewood Cliffs, NJ, 1974.
16. T. M. Hayes and P. N. Sen, *Phys. Rev. Lett.*, **34**, 956 (1975).
17. J. B. Pendry, in *EXAFS for Inorganic Systems*, C. D. Garner and S. S. Hasnain (Eds.), DL/SCI/R17, Daresbury Laboratory, Daresbury, 1981, p. 5.
18. F. W. Lytle, R. B. Greegor, E. C. Marques, D. R. Sandstrom, G. H. Via, and J. H. Sinfelt, *Proceedings of Advances in Catalytic Chemistry II*, Salt Lake City, 1982.
19. J. B. A. D. van Zon, D. C. Koningsberger, H. F. J. van 't Blik, R. Prins, and D. E. Sayers, *J. Chem. Phys.*, **80**, 3914 (1984).

20. G. Martens, P. Rabe, N. Swentner, and A. Werner, *Phys. Rev. B*, **17**, 1481 (1978).

21. P. Eisenberger and G. S. Brown, *Solid State Commun.*, **29**, 481 (1979).

22. G. Martens, P. Rabe, N. Swentner, and A. Werner, *Phys. Rev. Lett.*, **39**, 1411 (1977).

23. J. J. Rehr, private communication.

24. G. Bunker, *Nucl. Instrum. Methods*, **207**, 437 (1983).

25. J. M. Tranquada and R. L. Ingalls, *Phys. Rev. B*, **28**, 3520 (1983).

26. J. B. Pendry, private communication.

27. E. Sevillano, H. Meuth, and J. J. Rehr, *Phys. Rev. B*, **20**, 4908 (1979).

28. S. Cramer, in *EXAFS for Inorganic Systems*, C. D. Garner and S. S. Hasnain (Eds.), DL/SCI/R17, Daresbury Laboratory, Daresbury, 1981, p. 47.

29. E. A. Stern, in *EXAFS for Inorganic Systems*, C. D. Garner and s. S. Hasnain (Eds.), DL/SCI/R17, Daresbury Laboratory, Daresbury, 1981, p. 40.

30. B. K. Teo, *J. Am. Chem. Soc.*, **13**, 3992 (1981).

31. J. J. Boland and J. D. Baldeschwieler, *J. Chem. Phys.*, **80** 3005 (1984).

32. J. Stöhr, L. I. Johansson, S. Brennan, M. Hecht, and J. N. Miller, *Phys. Rev. B*, **22**, 4052 (1980).

PART
C

APPLICATIONS

CHAPTER

7

BIOCHEMICAL APPLICATION OF X-RAY ABSORPTION SPECTROSCOPY

STEPHEN P. CRAMER

Schlumberger-Doll Research
Ridgefield, Connecticut

7.1. INTRODUCTION

7.1.1. Historical Prologue

Working barely 50-m apart on opposite sides of an electron storage ring, scientists have made major experimental breakthroughs in two vastly different fields, high-energy physics and bioinorganic chemistry. The instrument that made these experiments possible is the particle storage ring SPEAR, which was originally designed for electron–positron collision experiments. The successful discovery of the ψ particle in 1974 using the electron–positron colliding beams within the SPEAR ring led to a Nobel prize for Burton Richter and paved the way for bigger rings such as PEP (Stanford, USA), PETRA (Hamburg, Germany), CESR (Ithaca, USA), and LEP (Geneva, Switzerland). While these high-energy experiments were being conducted, a group of physicists, chemists, and biologists were exploring the utility of synchrotron radiation for spectroscopy and diffraction experiments. The achievements of these original "parasites" at the Stanford Synchrotron Radiation Laboratory, using the x-rays generated by synchrotron radiation, helped them gain partial control of SPEAR. Realization of the potential applications also prompted the construction of rings designed expressly for the production of synchrotron radiation, such as the National Synchrotron Light Source at the Brookhaven National Laboratory, the Synchrotron Radiation Source at the Daresbury Laboratory, and the Photon Factory in Japan.

X-ray absorption spectroscopy is one field that experienced a renaissance with the development of synchrotron radiation sources. The initial excitement with the *E*xtended *X*-Ray *A*bsorption *F*ine Structure, EXAFS, was such that virtually any spectrum with a wiggle could be published as a breakthrough. As with other new techniques, a good amount of exaggeration and overinterpretation accompanied this explosive phase. Now, as the field matures and more beam time becomes available, the average quality of work is improving dramatically. EXAFS is inexorably becoming an accessible standard technique, with all its peculiar strengths and weaknesses. In the future, the focus of interest will gradually revert from the technique itself to the samples being measured.

The experimental methodology and data analysis procedures for EXAFS emerged at the time when the rapidly expanding horizons in bioinorganic chemistry posed a wealth of important structural questions. Fortunately, this was also a period when synthetic models for metalloprotein active sites were becoming available. For example, synthetic analogs for the iron–sulfur electron transfer proteins were novel compounds at the time they were being used as EXAFS standards to answer questions about strain and distortion in iron–sulfur proteins. Similarly, "picket-fence" porphyrin compounds were used within months of their crystallographic characterization to investigate hemoglobin cooperativity. In other cases, model compound development has been driven by the EXAFS results. The development of molybdenum, iron, and sulfur cluster syntheses subsequent to the nitrogenase EXAFS studies is an example of such a chronology. The convergence of physical theory, experimental techniques, and synthetic inorganic chemistry to solve biological problems has made this a truly interdisciplinary field.

7.1.2. Scope and Structure of This Chapter

The intent of this chapter is to summarize and where appropriate to criticize the structural results obtained from EXAFS studies of biochemical systems. The barest outlines of current EXAFS methodology are treated in Section 7.2. Section 7.3 summarizes the literature in detail through 1983, organized according to the specific element under study. For 1984–1986, only the most significant papers are included. A few topics of major importance, such as nitrogenase and cytochrome oxidase, have been treated in significant detail. Cursory treatment of other areas no doubt reflects the prejudices of the author. Finally, Section 7.4 treats some of the areas in which new developments in instrumentation or data analysis might be expected, and Section 7.5 presents a brief summary.

7.1.3. Previous Reviews and Other Literature

The other chapters in this volume will present a more complete treatment of the theory and nonbiochemical applications of EXAFS. Previous reviews of biochemical uses of EXAFS include those by Eisenberger and Kincaid (1), Shulman et al. (2), Cramer and Hodgson (3), Powers (4), Scott (5), Chan and Gamble (6), Teo (7), and Bordas (8), as well as material in the volume edited by Teo and Joy (9).

7.2. EXAFS METHODOLOGY

7.2.1. Data Collection

Almost all biochemical EXAFS experiments involve spectroscopically dilute samples. This means that absorption by the species of interest, such as a metal active site, is small compared to absorption by the matrix, such as protein and water solvent. Since EXAFS itself is a small modulation, it is almost impossible to measure EXAFS accurately in the presence of a large matrix background absorption. To overcome this problem, the majority of biochemical experiments have been done using fluorescence detection (10).

A fluorescence excitation spectrum is equivalent to an absorption spectrum when the sample is dilute or spectroscopically thin. Fluorescence detection has the advantage of stripping away the background absorption by solvent and protein or other matrix, since the photons absorbed by the matrix usually do not give rise to fluorescence. A certain fraction of incident photons are scattered, however, and for very dilute samples, some means for rejecting this background signal resulting from scattered radiation is required. Electronic rejection using solid-state detectors is often limited by the low count rate capabilities of these detectors (10). A 13-element intrinsic Ge detector has recently been developed to overcome this restriction (11). Crystal analyzers have also been used (12, 13), but they are limited to a small solid angle acceptance. For most applications, the use of a filter and slit combination gives the best results (14, 15). A comprehensive description of these techniques is given in Chapters 3 and 10, Sections 10.2 and 10.3.

7.2.2. Sample Handling and the Radiation Damage Issue

Free radicals are produced when x-rays strike biological samples (16). Although direct photoreaction of biological metal centers may occur, the more important damage mechanism is reaction with migrating free radicals and hydrated electrons. Chance et al. reported that room temperature irradiation of cytochrome oxidase caused reduction by hydrated electrons at the rate of 2 μM/s (17). At lower temperatures, Powers et al. found that a radical concentration of 500 μM existed in their cytochrome oxidase samples (18). Furthermore, this radical decayed by 40% after removal from the beam for 10 min. However, little change in the EPR or optical spectra of the metal centers themselves was observed, as long as the samples were maintained at $-130°C$. Thus, as long as the sample remained frozen at low temperature, the radicals appeared to recombine with themselves rather than migrating to and reacting with the enzyme active sites.

The photoreduction results of Powers and co-workers have been challenged by Brudvig et al. (19). This group exposed cytochrome oxidase solutions to monochromatized synchrotron radiation as well as to a broad-band tungsten anode source. Although they clearly observed the creation of free radicals, they observed no reduction of the oxidase metal centers as measured by EPR or optical methods.

Measurable photoreduction of cytochrome c has been reported by Korzun et al. (20). This group found about 3–5% photoreduction of 2-mM cytochrome c using 3×10^{10} photons s^{-1} over a period of about 45 min. When starting with reduced samples, no photooxidation was observed. Radical production in cytochrome oxidase has also been observed by Scott (21), and photoreduction of iron in superoxide dismutase has been reported by Roe et al. (22).

The physics of radiation damage have recently been discussed (23); X-ray damage does not appear to be an insurmountable problem for biochemical EXAFS, as long as appropriate precautions are taken. These may include the use of low temperature, frequent sample changes, radical traps, or excess oxidant or reductant. Every responsible EXAFS study should include some control over sample integrity.

7.2.3. Data Analysis

The appropriate methods for interpretation of x-ray absorption edge and EXAFS spectra have been the subject of heated debate for more than a decade. This has led to a healthy skepticism about the precision and accuracy of many of the reported results. Although the situation may appear chaotic to those outside the field, there has actually been a slow convergence of the disparate analysis methods, to the point where a de facto consensus exists, even though there are many details to be worked out. This section will briefly summarize the basic principles of data analysis, details of which are discussed in Chapter 6.

7.2.3.1. *Edge Interpretation*

Many biochemical problems can be posed in such a way that the data in the absorption edge region is sufficient for an answer. To satisfy the hunger for acronyms, edge structure is sometimes called XANES, X-Ray Absorption Near-Edge Structure. Edges are in many ways more complicated than EXAFS and their physics has been discussed in Chapters 2 and 11 and in ref. (24). For current purposes, it need only be mentioned that there are two extreme approaches to edge analysis, a purely atomic bound-state viewpoint and a multiple-scattering formalism. The former approach considers edge features to be derived from ligand field perturbations of atomic states, and a representative application is the interpretation of perovskite edges by Shulman et al. (25). The alternate viewpoint regards the excited states involved in edge excitations as scattering states in which the excited electron is strongly scattered off the neighboring atoms. Examples are the edge interpretations by Kutzler et al. (26, 27) and the calculations of Bianconi (28) and Pendry (29, 30).

In view of the complexities of numerical simulation, it is fortunate that there are many cases in which qualitative interpretation of edges is sufficient to yield unambiguous results. One case involves oxidation state changes that open up previously filled subshells and thereby new transitions. For example, upon oxidation of Cu(I) to Cu(II), the filled $3d^{10}$ shell becomes $3d^9$, and a weak forbidden $1s \rightarrow 3d$ transition may appear. More dramatic is the change from Au(I) to Au(III), where oxidation of the gold opens up intense, fully allowed $2p \rightarrow 5d$ transitions. Enhancement of bound state transitions by terminal ligands, as in the case of V=O or Mo=O bonding, has also been observed.

7.2.3.2. *EXAFS Analysis*

Details of EXAFS analysis procedures are presented in Chapter 6, and the purpose of this discussion is merely to provide a framework in which the current state of EXAFS analysis can be viewed. The expression representing the EXAFS formula for an unoriented sample with Gaussian disorder (not including corrections for many-body effects) can be given by [Chapter 1, Eq. (47)]:

$$\chi(k) = \left(\frac{N_b}{kR_{ab}^2}\right) F(k) \exp\left(-2k^2\sigma_{ab}^2\right) \exp\left(-\frac{2R_{ab}}{\lambda}\right) \sin\left[2kR_{ab} + \delta_{ab}(k)\right]$$

In this expression N_b is the number of b-type neighbors at distance R_{ab} with Gaussian deviation σ_{ab}^2, and k is the photoelectron wave number. $F(k)$ is the amplitude function and $\delta_{ab}(k)$ is the phase shift function that depends on a and b. The divergences in technique often reflect different approaches to obtaining phase shift and amplitude functions.

1. Qualitative procedures: It can be shown that the Fourier transform of the EXAFS from k space to frequency space has peaks related to the absorber–scatterer distances, R_{ab}. For various reasons, however, the Fourier transform is often dramatically distorted from the true radial distribution function. An alternative method for interpretation of EXAFS spectra is simple comparison of the spectra of model compounds with that of the unknown (18). This method has some intuitive appeal, since if a model and unknown have identical spectra, the unknown *may* have the same structure.

2. Quantitative methods: If it is known that the EXAFS represents only a single type of absorber–scatterer interaction, then analysis is straightforward. The simplest procedure involves plotting the logarithm of the EXAFS amplitude ratio for the unknown versus a model compound versus k^2 (31). The coordination number is obtained from the ordinate intercept and the $\Delta\sigma^2$ is determined from the slope. Furthermore, if the phase and amplitude are extracted separately by Fourier filtering (32), then an appropriate plot of the phase difference will give the difference in distances (33). These single-shell analysis methods clearly give the most unambiguous EXAFS results, with the only room for controversy being the suitability of the model compounds or theoretical functions chosen for the analysis.

The presence of overlapping shells in the radial distribution function requires an iterative nonlinear approach to a structural solution, since the correct N, R, and σ can no longer be expressed as independent quantities in a linear equation. One problem with all curve-fitting methods is that there is a *correlation* between the numerical values obtained for the different variables. Fortunately for EXAFS interpretation, the highly correlated quantities can be divided into two groups. The correlated phase terms are the total phase shift $\delta(k)$, the distance R, and if used, the E_0 shift ΔE_0, while the most important amplitude terms are the amplitude function $A(k)$, the coordination number N, and the Debye–Waller factor

$\exp(-2\sigma^2 k^2)$. Since the phase and amplitude quantities are relatively uncorrelated, accurate distances can often be obtained despite ambiguous coordination numbers.

The main problem that persists with curve-fitting is assessing the significance of additional components. Extra degrees of freedom will always improve a fit, and distinguishing minor components from numerical artifacts remains an art.

7.2.4. Limitations of EXAFS

The strengths of EXAFS for biological applications have been widely advertised, and most authors have tried to balance their enthusiasm by defining the inherent limitations of the technique. Apart from experimental difficulties, there are intrinsic constraints that restrict the conclusions of any investigation.

7.2.4.1. Resolution

The resolution of distances achievable by EXAFS is determined by the range of data available. Although EXAFS oscillations have no intrinsic energy cutoff, they are experimentally observable only over a finite range. The minimum useful k value is in the range of 4–5 Å^{-1}, or 60–100 eV above E_0. Below this low cutoff, multiple-scattering effects invalidate the simple EXAFS formalism. At the other extreme, the maximum k value may be determined by appearance of another edge, or by disappearance of the damped EXAFS signal into the noise. For biological samples k_{max} is typically 10–15 Å^{-1}, and differences in distances of 0.1–0.15 Å cannot be unambiguously resolved.

A second type of resolution pertains to identification of the elemental nature of the scattering neighbor. These distinctions are made on the basis of characteristic phase shift and amplitude functions. For example, it is often stated that oxygen can be distinguished from sulfur by the fact that the two phase shifts are different by nearly π, whereas oxygen and nitrogen are indistinguishable because of similar phase shifts. However, it is rarely appreciated that Z resolution is also related to the range of data. Although a detailed analysis is beyond the scope of this chapter, it should be remembered that on a narrow range of data, incorporation of the wrong a–b interaction at the wrong distance might still improve the quality of the fit.

7.2.4.2. Accuracy

For single-shell systems with Gaussian pair distribution functions, the distance accuracy of EXAFS is determined by the degree of transferability of phase shifts. Numerous tests of this chemical transferability have been conducted using experimental (34, 35) or theoretical (36) phase shifts. The transferability is usually excellent, leading to an accuracy of ± 0.01 Å.

The distance accuracy achievable for complex unknown structures is limited not by phase shift transferability, but by intercomponent correlations and a lack of knowledge about the underlying pair distribution function. The latter difficulty is rarely appreciated. The total EXAFS in a given shell represents a weighted average of contributions from all interactions over a certain R interval. As Eisenberger and Brown have pointed out, if the pair distribution is asymmetric, the EXAFS average distance will not be the true average (37). In fact, even a Gaussian pair distribution requires a small correction factor. Although the arguments were derived for continuous distributions in amorphous materials, they are equally valid for the discrete distributions expected in a biological active site. Thus, although 0.03 Å accuracy for individual distances may be reasonable, caution should prevail when the average pertains to three or more interactions.

The accuracy of coordination numbers derived from EXAFS is limited by two factors, the low-k cutoff and multiple-scattering effects. It is often stated that coordination numbers are ambiguous because of the correlation with the Debye–Waller factor. However, if the data could be used down to $k = 0$, this exponential term, or any other distribution dependent amplitude effect, would become negligible. Therefore, the main uncertainty in first coordination sphere coordination numbers ultimately derives from the low-k limit. It is conceivable that further progress in XANES calculations, which use the low-k data, will dramatically improve first coordination sphere amplitude analysis.

The second source of uncertainty in coordination numbers is multiple scattering, for atoms beyond the first coordination sphere. It is well appreciated that a substantial amplitude enhancement occurs when a second shell scatterer lies directly behind an intervening atom (38, 39). Less appreciated is the fact that amplitudes can be reduced by intervening atoms when the angles are more acute (40, 41). Since the geometry of such interactions is rarely known in an EXAFS study, a considerable uncertainty should be ascribed to the overall amplitude.

7.2.4.3. Sensitivity

A final limitation on the prediction of structure by EXAFS is the lack of sensitivity to certain weak or nonbonding interactions. Ligands that are disordered or rapidly exchanging, such as water molecules in aqueous solutions (42), are often not observed. At room temperature, loosely bound ligands such as thioethers are also difficult to observe. Weakly scattering nitrogen interactions are sometimes buried under stronger sulfur contributions. Finally, long distance metal–metal interactions are frequently obscured in the absence of a bridging ligand. In short, predictions of unknown structures must always take into account the possibility that certain chemically important components are not observed in the EXAFS.

7.2.5. Types of Structural Problems

In biological applications there are three different types of structural problems. In some cases, there is virtually no information about the species being investigated. The initial EXAFS work on the molybdenum in nitrogenase is an example of such a study. The opposite extreme concerns sites that have been extensively studied by other techniques, and in which extremely well-defined structural questions can be posed. The EXAFS studies of hemoglobin and rubredoxin fall into this latter category. Finally, the most common case in EXAFS applications involves sites whose structure has been partially revealed by other techniques. In such cases EXAFS is used to test an existing hypothesis and/or to complete a proposed structure. This is perhaps the most dangerous application of the technique, because there is a tendency to use EXAFS to confirm preexisting notions, rather than presenting all of the structural possibilities that are consistent with the data. Currently, the art involved in the use of EXAFS involves discerning which structural features are required by the data and which features are merely consistent with the data.

7.3. RESULTS

7.3.1. Iron

7.3.1.1. Fe/S Environments

Fe—S cluster sites have been a fruitful area of study for EXAFS. Their high symmetry and strong sulfur electron backscattering give rise to large and readily interpretable EXAFS signals. Some of the types of sites and the corresponding EXAFS Fourier transforms are illustrated in Fig. 7.1.

1. 1-Fe sites: Rubredoxin is an electron transfer protein that contains a single iron atom per 6000 daltons, coordinated by four cysteine thiolate ligands. Because of its low molecular weight and the four sulfur atoms, this protein is ideally suited for x-ray absorption spectroscopy, and it was among the very first metalloproteins to be studied by EXAFS. The early x-ray diffraction crystal structure predicted four Fe—S distances of 2.31, 2.33, 2.39, and 1.97 Å (43), later revised to 2.24, 2.32, 2.34, and 2.05 Å (44). The apparent distortion of

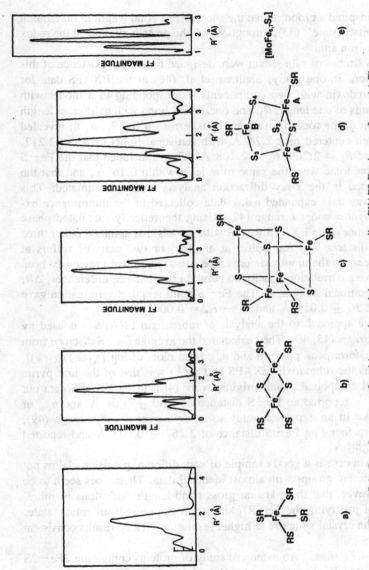

Figure 7.1. Fe_x-S_y structures found in proteins and the corresponding EXAFS Fourier transforms. (a) The rubredoxin type $[Fe(SR)_4]^{-,2-}$ site: Fourier transform of oxidized form, modified from Fig. 4 of ref. (49). (b) The plant ferredoxin type $[Fe_2S_2(SR)_4]^{-,2-}$ site: Fourier transform of oxidized form, modified from Fig. 3 of ref. (51). (c) The $[Fe_4S_4(SR)_4]^n$ site: Fourier transform of oxidized HIPIP, modified from Fig. 3 of ref. (51). (d) Proposed $[Fe_3S_4(SR)_4]^n$ site: Fourier transform of oxidized form of aconitase, modified from figure of ref. (54). (e) Fourier transform of FeMo-co iron EXAFS, modified from figure of ref. (57).

the Fe—S site inspired a proposal that the short Fe—S bond length in rubredoxin reflected an "entatic state" (45), through which the protein modulated the redox potential of the iron site.

The EXAFS studies of rubredoxin were designed to test the existence of this unusual distortion. In one study, Shulman et al. (46) fit the EXAFS data for lyophilized rubredoxin with two components, corresponding to a model with three Fe—S bonds of one length R_3, and one Fe—S bond with a different length R_1. Examination of the root mean square (rms) error for the R_3, R_1 fit revealed a symmetric well centered near 2.24 Å with statistical limits of $R_3 = 2.217$, $R_1 = 2.389$, and $R_3 = 2.268$, $R_1 = 2.108$ Å. They concluded that the Fe—S distances in rubredoxin were the same to at least within 0.16 Å, and that the distortion reported in the x-ray diffraction analysis was overestimated. This original work was later expanded using data collected in the fluorescence excitation mode over a wider k range (47). Using theoretically calculated phase shifts and amplitudes, the EXAFS was fit with models that assumed either three sulfurs at one distance and one sulfur at another, or two pairs of sulfurs at different distances. In the amplitude envelope data, which had previously been discarded, a strong correlation was found between the distance differences, ΔR, and the rms vibrational motion, σ. The Fe—S bond length determination gave an average of 2.26 ± 0.015 Å and $R_3 - R_1 = 0.00 \pm 0.1$ Å.

An alternative approach to the analysis of rubredoxin EXAFS was used by Stern and co-workers (48, 49). They calculated the average Fe—S distance from the Fourier transform peak position and σ_{stat} from plots of log $[\chi_1(k)/\chi_2(k)]$, where $\chi_1(k)$ was the rubredoxin EXAFS and $\chi_2(k)$ was that of the iron pyrite used as a model compound. Their original work (48), using EXAFS data out to only 6.4 Å^{-1}, reported an Fe—S distance of 2.30 ± 0.04 Å and σ_{stat} of 0.06 ± 0.04 Å. In an expanded study using data over a wider range (49), Bunker and Stern found an Fe—S distance of 2.267 ± 0.003 Å and reported $\sigma_{stat} = 0.032 \, ^{+0.013}_{-0.032}$ Å.

The rubredoxin case is a good example of very different precision claims put forth by two different groups with almost identical data. There does seem to be agreement, however, that there are no gross bond-length distortions in rubredoxin worthy of justifying the Fe(SR)$_4$ site as an example of an entatic state. Refinement of the crystal structure to higher resolution yielded results consistent with the EXAFS (50).

2. 2-Fe and 4-Fe sites: An extensive study of proteins containing 2Fe—2S and 4Fe—4S clusters has been reported (51), and the reported Fe—S and Fe—Fe distances are summarized in Table 7.1. No significant structural differences were found between the protein clusters and the corresponding small molecule model compounds. The most serious discrepancy between EXAFS and crystallography occurred in the case of HIPIP (high potential iron protein), where Teo et al.

Table 7.1. Summary of EXAFS Results on Biological Fe–S Clusters

Sample	Conditions	Fe–S N[a]	Fe–S R (Å)	Fe–S σ (Å)	Fe–Fe N	Fe–Fe R (Å)	Fe–Fe σ (Å)	Fe–X N	Fe–X R (Å)	Fe–X Type	References
P. aerogenes rubredoxin	Lyophilized oxidized	4	2.24	—	—	—	—	—	—	—	46
P. aerogenes rubredoxin	Solution oxidized	4	2.245(16)	0.047(18)	—	—	—	—	—	—	47
P. aerogenes rubredoxin	Solution reduced	4	2.32(2)	0.057(25)	—	—	—	—	—	—	47
C. pasteurianum rubredoxin	Lyophilized oxidized	4	2.30(4)	—	—	—	—	—	—	—	48
C. pasteurianum rubredoxin	Lyophilized oxidized	4	2.267(3)	—	—	—	—	—	—	—	49
rhubarb ferredoxin	Solution oxidized	4	2.233(22)	0.063(19)	1	2.726(40)	0.057(18)	—	—	—	51
rhubarb ferredoxin	Solution reduced	4	2.241(28)	0.059(22)	1	2.762(48)	0.076(31)	—	—	—	51
HIPIP	Solution oxidized	4	2.262(13)	0.060(11)	3	2.705(26)	0.088(9)	—	—	—	51
HIPIP	Solution reduced	4	2.251(13)	0.001(27)	3	2.659(50)	0.088(17)	—	—	—	51
C. pasteurianum ferredoxin	Solution oxidized	4	2.249(16)	0.063(15)	3	2.727(35)	0.092(13)	—	—	—	51
C. pasteurianum ferredoxin	Solution reduced	4	2.262(14)	0.062(11)	3	2.744(32)	0.098(13)	—	—	—	51
D. gigas ferredoxin II	Solution oxidized	4.4(12)	2.250(13)	0.043[b]	2.5(12)	2.698(20)	0.067[b]	1.0(2)	1.850(17)	O	52
D. gigas ferredoxin II	Solution reduced	4.6(10)	2.266(15)	0.043[b]	2.3(11)	2.673(22)	0.067[b]	1.7(3)	1.796(15)	O	52
Beef heart aconitase	Frozen solution oxidized	—	2.24	—	—	2.71	—	—	—	—	54
A. vinelandii FeMo-co	Frozen NMF	3.4(16)	2.247(20)	—	2.3(9)	2.656(27)	—	0.41(1)	2.760(32)	Mo	57
	solution as isolated	—	—	—	—	—	—	1.2(10)	1.814(65)	O	

[a]Coordination numbers reported as integers were held fixed in the analysis.
[b]σ value held fixed during optimization.

(51) determined a 0.05 Å contraction in the Fe—Fe distances upon reduction, whereas crystallography predicted a 0.1 Å expansion.

3. *3-Fe sites:* The 3-Fe cluster site of *Desulfovibrio gigas* ferredoxin II has been examined by EXAFS (52) and shown to have a dramatically different structure from that of the 3-Fe site predicted by crystallography for *Azotobacter vinelandii* ferredoxin I (53). The Fourier transforms of both oxidized and reduced forms of the *D. gigas* protein clearly show evidence of Fe—S and Fe—Fe interactions. However, an additional Fe—O interaction was required at 1.85 (or 2.06 Å) and 1.80 (or 2.00 Å), respectively, for the oxidized and reduced forms. Antonio et al. (52) state that the number of oxygen neighbors increases upon reduction of the protein. However, the fact that Fe—O interactions could be included at two different distances suggests that the Fe—O bonds seen by EXAFS may be an artifact.

The 2.7 Å Fe—Fe distances reported in the *D. gigas* EXAFS study are 1.5 Å shorter than the 4.2 Å separations observed in the *A. vinelandii* ferredoxin I crystal structure. It would take a major revision of the crystal structure to achieve compatibility with the EXAFS data. An alternate explanation is that two distinct types of 3-Fe cluster conformations are possible, an extended "twisted-boat" form as observed crystallographically and a more compact form as observed by EXAFS. If such forms were interconvertible, it could be of great importance to the biosynthesis and catalytic chemistry of these and related clusters.

An EXAFS study by Beinert et al. of beef heart aconitase also found a short Fe—Fe distance (54). The authors propose a model for interconversion of Fe_3S_4 and Fe_4S_4 clusters, in which the 3-Fe cluster still has only sulfur donor ligands.

4. *Iron–molybdenum cofactor:* Interpretation of the iron EXAFS of the nitrogenase iron–molybdenum cofactor is complicated, since recent Electron Nuclear Double Resonance, ENDOR, experiments on ^{57}Fe-substituted MoFe protein indicate that FeMo-co contains six spectroscopically distinct iron sites (55). Nevertheless, both ENDOR and Mössbauer (56) experiments suggest that the iron in FeMo-co can be divided into two subgroups each containing three roughly equivalent species. One group appears to have the nearly isotropic hyperfine tensor expected for high-spin ferric ions, while the other three irons show a greater anisotropy that is at least consistent with high-spin ferrous ions.

Three structurally significant peaks are observed in the Fourier transform of the iron EXAFS of FeMo-co (57, 58), as shown in Fig. 7.1e. Comparison with the spectra of model compounds lead to assignments as representing Fe—O(N), Fe—S(Cl), and Fe—Fe(Mo) interactions, and an iterative curve-fitting procedure led to the structural predictions in Table 7.1. The authors summarize their conclusions by saying that "the iron atoms in FeMo-co have an average of 3.4 ± 1.6 S (Cl) at 2.25 (2) Å, 2.3 ± 0.9 Fe atoms at 2.66 (3) Å, 0.4 ± 0.1 Mo atoms at 2.76 (3) Å, and 1.2 ± 1.0 O (N) atoms at 1.81 (7) Å as nearest neighbors." Unfortunately, no final activity data was reported for the FeMo-co used in these experiments.

7.3.1.2. Hemoproteins

The local symmetry provided by a porphyrin ring structure results in strong EXAFS signals from hemoproteins. Contributions are observed not only from the porphyrinic nitrogens and axial ligands, but also from distant shells of carbon atoms in the porphyrin macrocycle. As illustrated in Fig. 7.2, the power of EXAFS for hemoprotein studies stems from its sensitivity to these numerous distances. Unfortunately, when questions about doming or puckering of the porphyrin arise, the distances provided by EXAFS may not be sufficient to distinguish between surprisingly different interpretations of the data.

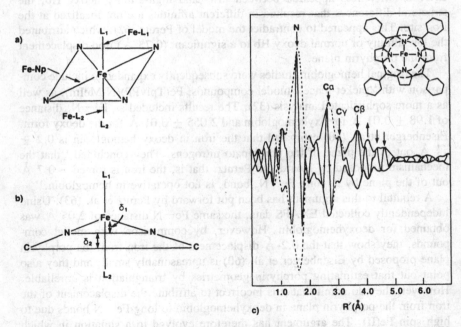

Figure 7.2. (*a*) Schematic illustration of structural results derived in most hemoprotein EXAFS studies. (*b*) Doming effects (δ_1 and δ_2), which complicate interpretation of hemoprotein EXAFS. (*c*) EXAFS Fourier transform for Fe(II)TPP at 77 K, from ref. (59). Contributions from as far as the phenyl groups can be identified.

1. Hemoglobin: Hemoglobin is the iron-containing protein that binds oxygen in red blood cells. It is composed of two α subunits and two β subunits, each of which contains an Fe–protoporphyrin IX prosthetic group, which can bind O_2. Hemoglobin works because it binds oxygen cooperatively, that is, the binding of an oxygen molecule at one subunit affects the affinity of the other subunits for oxygen. The cooperativity is positive, meaning that the oxygen affinity increases as more oxygen is bound. The central question in hemoglobin chemistry is the precise molecular nature of this cooperativity.

EXAFS studies of hemoglobin have attempted to probe the structural changes that occur at the iron site during conversion from low-affinity to high-affinity forms, as well as between oxy and deoxy forms. The first work reported an average Fe—N distance of 1.99 Å (60). Furthermore, no significant differences were found between the spectra of normal deoxyhemoglobin A and that of deoxy hemoglobin Kempsey. The latter protein has a mutation ($\alpha_2\beta_2^{99}$Asp $\rightarrow$ Asn), which causes it under certain conditions to remain in the high-affinity form even though deoxygenated (61). Eisenberger et al. (60) reasoned that since negligible EXAFS differences appeared between low- and high-affinity deoxy Hb, the structural differences that resulted in different affinities are not localized at the iron site. This appeared to contradict the model of Perutz (62), which attributed the low affinity of normal deoxy Hb to a significant (0.75 Å) iron displacement from the porphyrin plane.

The original hemoglobin studies were subsequently expanded to include comparison with "picket-fence" model compounds, Fe(TpivPP)(N-MeIm), as well as a more sophisticated analysis (32). The results included an Fe—N_p distance of 1.98 ± 0.01 Å for oxyhemoglobin and 2.055 ± 0.01 Å for the deoxy form. Eisenberger et al. (60) estimated that the iron in deoxy hemoglobin is $0.2 \pm \substack{0.1 \\ 0.2}$ Å out of the plane of the porphinato nitrogens. They concluded "that the mechanism proposed by Hoard and Perutz, that is, the iron is forced ~ 0.7 Å out of the plane by its long Fe—N_p bond, is not operative in hemoglobin."

A rebuttal to this argument has been put forward by Perutz et al. (63). Using independently collected EXAFS data, the same Fe—N distance of 2.05 Å was obtained for deoxyhemoglobin. However, by comparison with model compounds, they show that the 0.2-Å displacement of the iron from the porphyrin plane proposed by Eisenberger et al. (60) is unreasonably small, and they also point out that estimating porphyrin geometries by triangulation is unreliable. However, they do admit that it is incorrect to attribute the displacement of the iron from the porphyrin plane in deoxyhemoglobin to long Fe—N bonds due to high spin Fe(II). The argument has therefore evolved to a situation in which there is agreement over the EXAFS results to 0.01 Å, yet there is still no consensus concerning the implications for cooperativity.

2. Cytochrome P-450 and chloroperoxidase: Cytochrome P-450 is the name given to a class of Fe–protoporphyrin IX containing enzymes that act as mono-oxygenases (64–66):

$$\overset{P\text{-}450}{2H^+ + 2e^- + RH + O_2 \rightarrow ROH + H_2O}$$

The name derives from the characteristic 450-nm absorption band observed upon reduction of the enzyme in the presence of carbon monoxide. A proposed reaction scheme for the enzyme is illustrated in Fig. 7.3. The main goal of the EXAFS studies of this enzyme has been to characterize the nature of the axial ligand(s) X as well as the Fe—N_p and Fe—X bond lengths throughout the catalytic cycle.

The first EXAFS study of this system (67) was confined to a comparison of oxidized rabbit liver microsomal protein, P-450 LM-2, and chloroperoxidase, a related protein from the fungus *Caldariomyces fumago*. Average Fe—N_p and Fe—S distances of 2.00 and 2.19 Å, respectively, were found for the LM-2 protein, which was predominantly in the low-spin state. High-spin chloroperoxidase, yielded longer distances, Fe—N = 2.05 Å and Fe—S = 2.30 Å. Just as important, in both cases it was not possible to obtain a reasonable fit without an axial sulfur ligand.

A more extensive study of bacterial P-450 from *Pseudomonas putida* was subsequently reported by Hahn et al. (68). High-spin ferric and ferrous, low-

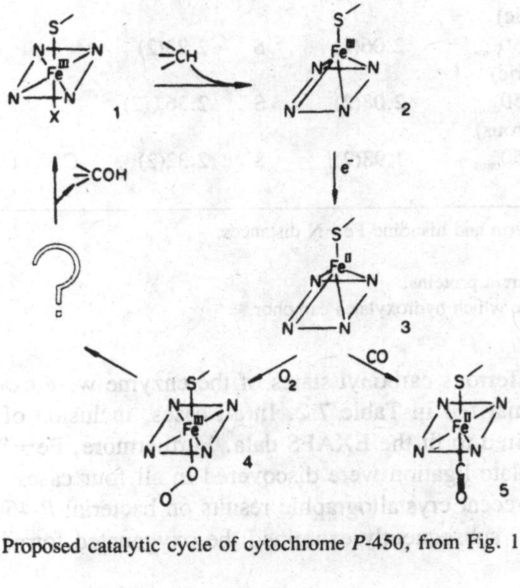

Figure 7.3. Proposed catalytic cycle of cytochrome P-450, from Fig. 1 of ref. (68).

Table 7.2.   EXAFS Results on Heme Iron Proteins

Sample	Fe—N$_p$ R (Å)	Fe—L$_1$ Type	R (Å)	Fe—L$_2$ Type	R (Å)	References
Oxyhemoglobin	1.99^a(2)					60
Oxyhemoglobin	1.986(10)	N$_{his}$	2.07^b	O	1.75^b	32
Deoxyhemoglobin	2.055(10)	N$_{his}$	2.12^b			32
Deoxyhemoglobin (Kempsey)	2.055(10)	N$_{his}$	2.12^b			32
Deoxyhemoglobin	2.05	N$_{his}$	2.12			63
Cytochrome c^c	1.98	S$_{met}$	2.30			20
Cytochrome oxidase (oxidized, Fe$_{a3}^{3+}$)	2.01	S	2.60	N	2.14	18
Cytochrome oxidase (oxidized, Fe$_a^{3+}$)	1.99	N	1.99	N	1.99	18
Cytochrome oxidase (reduced, Fe$_{a3}^{2+}$)	1.99	N	2.07	O	1.75	18
Cytochrome oxidase (reduced, Fe$_a^{2+}$)	1.99	N	1.99	N	1.99	18
Chloroperoxidase (high-spin, ferric)	2.05(3)	S	2.30(3)			67
Cytochrome P-450-LM-2 (low-spin, ferric)	2.00(3)	S	2.19(3)			67
Cytochrome P-450$_{cam}^d$ (low-spin, ferric)	2.00(2)	S	2.22(2)			68
Cytochrome P-450$_{cam}$ (high-spin, ferric)	2.06(2)	S	2.23(2)			68
Cytochrome P-450$_{cam}$ (high-spin, ferrous)	2.08(2)	S	2.36$_{av}$(2)			68
Cytochrome P-450$_{cam}$ (ferrous-CO)	1.98(2)	S	2.32(2)	C	1.72(2)	68

^{a}Average of porphyrin and histidine Fe—N distances.
^{b}Held fixed.
^{c}Average over different proteins.
^{d}The monoxygenase which hydroxylates camphor.

spin ferric, and ferrous carbonyl states of the enzyme were examined, and the results are summarized in Table 7.2. In all cases, inclusion of an axial sulfur ligand was required to fit the EXAFS data. Furthermore, Fe—S distances consistent with thiolate ligation were discovered in all four cases. The results are consistent with recent crystallographic results on bacterial P-450 (69).

Dawson et al. subsequently examined the oxygenated forms of bacterial P-

Table 7.3. Crystallographic and EXAFS Results on Hemerythrins

Sample	Technique	Fe—O$_{bridge}$ R (Å)	Fe—O—Fe R (Å)	Fe—O—Fe Θ (deg)	Fe—X Number	Fe—X R (Å)	References
Phascolopsis gouldii azidomethemerythrin	EXAFS	1.735(25)	3.38(5)	165 ± 8	2-3 N / 2-3 O	2.07 / 2.13	80
Phascolopsis gouldii azidomethemerythrin	EXAFS	1.80(5)	3.49(8)	152^{+28}_{-13}	3 N, 2 O	2.15(5)	79
Themiste dyscritum azidomethemerythrin	X-ray diffraction		3.25-3.30				208
Phascolopsis gouldii methydroxohemerythrin	EXAFS	1.82(5)	3.54(8)	153^{+27}_{-13}	3 N, 2 O	2.15(5)	79
Themiste dyscritum methydroxohemerythrin	X-ray diffraction		3.05				77
Phascolopsis gouldii oxyhemerythrin	EXAFS	1.83(5)	3.57(8)	155^{+25}_{-55}	3 N, 2 O	2.16(5)	79
Phascolopsis gouldii deoxyhemerythrin	EXAFS		3.13(3)		2.5 N, / 3.0 O	2.14(5) / 2.02(5)	81

450 as well as chloroperoxidase (70). An Fe–S distance of 2.37 Å was found in both cases, and it is argued that the sulfur is in the thiolate form.

3. *Cytochrome c:* The c-type cytochromes are important electron transfer proteins containing an iron porphyrin with axial histidine and methionine ligands. The absorption edge region has been studied by Shulman et al. (21) as well as by Labhardt and Yuen (71). Shulman found little change in the edge upon reduction, whereas the latter group found a 1.5-eV shift to lower energy. Labhardt and Yuen also studied the EXAFS using an unusual maximum entropy transform. They concluded that no significant geometrical changes occurred upon reduction, but they did not quote any absolute values for Fe—X bond lengths.

The c-type cytochromes have a wide range of redox potentials, and it has been proposed that the potentials might be influenced by the length of the axial Fe—S_{met} bond (72). Korzun et al. (20) examined a series of c-type cytochromes with potentials from +34 to +365 mV. An average Fe—N bond length of 1.98 Å was found in all cases, and a Fe—S bond length of 2.30 ± 0.02 Å for all except flavocytochrome c-555 where a value of 2.34 ± 0.02 Å was found. It thus appears that cytochrome-c Fe—X bond lengths are relatively independent of both redox potential and oxidation state.

4. *Cytochrome oxidase:* This enzyme is a complex polymer that contains two a-type cytochromes as well as two distinct copper species. The copper results are discussed in detail in Section 7.3.2.3. Powers et al. made very detailed proposals concerning the iron environments in cytochromes a and a_3 (18). These predictions are summarized in Table 7.2 and Fig. 7.6(c).

5. *Peroxidase:* Horseradish peroxidase catalyzes the oxidation of organic substrates with concomitant reduction of peroxide to water (73). Treatment of the enzyme with H_2O_2 can produce two different intermediates, bright green compound I and red compound II. In a comparative study of peroxidases and catalases, Chance and co-workers proposed that compound I possesses a 1.64-Å Fe—O bond, whereas the compound II Fe–O distance is 1.93 Å (74). This distinction differs from the results of Penner-Hahn et al., who found the same short 1.64-Å Fe—O bond in both compounds I and II (75, 76).

7.3.1.3. *Hemerythrin and Related Proteins*

Hemerythrin is an oxygen-binding protein containing iron that is found in certain invertebrate animals. The oxygen-binding site contains two iron atoms, and x-ray crystallography investigations by two different groups have shown the coordinating ligands to be histidine imidazoles and carboxylates from glutamate and asparagine residues (77, 78). However, the crystallography has only been done on inactive met-azide or met-hydroxide forms of the enzyme, illustrated

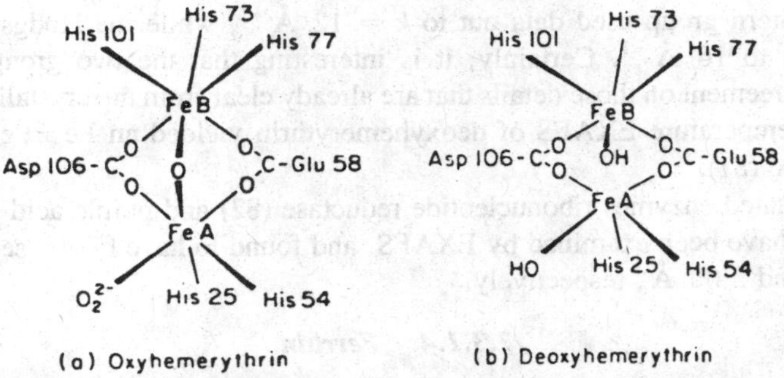

Figure 7.4. Proposed structures of the binding sites in oxy- and deoxyhemerythrin, from Fig. 6 of ref. (81).

in Fig. 7.4 and there is some ambiguity about the active oxy and deoxy structures. EXAFS studies have provided useful information about the latter forms, as well as more precise details about the nature of the oxo bridge. The EXAFS results are compared with the crystallography in Table 7.3.

The results of the Stern group (79) and the Hodgson group (80) are basically in agreement, although the analysis methods and error estimates differ substantially. The Stern group analyzed the iron first coordination sphere as a single component and derived an average Fe—O,N distance of 2.15 ± 0.05 Å. The Hodgson group divided the first sphere into two subgroups, which yielded 2.1 oxygens at 2.13 Å and 2.5 nitrogens at 2.08 Å.

Since the basic coordination environment is known, the most important feature for the EXAFS analysis of the hemerythrin binuclear iron site is the Fe—O—Fe geometry. The intervening oxygen complicates interpretation of the Fe—Fe EXAFS because multiple-scattering processes have to be accounted for. Using the trinuclear complex $[Fe_3O(gly)_6(H_2O)_3]^{+7}$ as a model compound for the Fe—Fe interaction, the Stern group found no amplitude enhancement in the Fe—Fe interaction and an Fe—Fe distance of 3.49 ± 0.08 Å. In contrast, the Hodgson group found a twofold enhancement in the Fe—Fe intensity, which they attribute empirically to a bridging angle of 165°, and they compute an Fe—Fe distance of 3.38 ± 0.05 Å.

There is a discrepancy in the estimations of amplitude enhancement effects, since the Stern group found only a twofold amplitude enhancement for the $[Fe_2O \cdot (4$-chloro-2,6 pyridine-dicarboxylate$)_2(H_2O)_4] \cdot 4H_2O$ dimer, which is known to have a linear Fe—O—Fe bridge, whereas the Hodgson group finds that at 180° the amplitude enhancement is a factor of 4. The theoretical calculations of Teo predict an even larger effect, a 10-fold amplitude enhancement for linear A—B—C arrangements (39). Part of the discrepancy may stem from the fact that the Stern group used data out to $k = 12$ Å^{-1}, while the Hodgson group used data to 16 Å^{-1}. Certainly, it is interesting that the two groups are in closest agreement on those details that are already clear from the crystallography.

Low-temperature EXAFS of deoxyhemerythrin yielded an Fe—Fe distance of 3.13 Å (81).

The related enzymes ribonucleotide reductase (82) and purple acid phosphatase (83) have been examined by EXAFS, and found to have Fe–Fe separations of 3.22 and 3.00 Å, respectively.

7.3.1.4. Ferritin

Ferritins are iron storage proteins that are often described as a core of ferric oxide surrounded by a protein coat (84). EXAFS analysis has been used to compare the structures of ferritin with iron–dextran, iron nitrate hydrolysate, and iron–glycine (85). The greatest resemblance was with iron dextran, and quantitative analysis (86) yielded 5.8 ± 0.4 oxygens at 1.95 ± 0.01 Å, with a second shell of 6.0 ± 0.5 irons at 3.34 ± 0.02 Å. The determination of the number of irons in the second shell, and the associated error, is somewhat arbitrary since it depends on the model compound chosen and estimation of multiple-scattering and mean free path effects.

The Fe(III) complex with apoferritin has been found to have 5.2 ± 1.0 oxygen neighbors at 2.09 Å, substantially longer than ferritin itself (87).

7.3.1.5. Iron-Tyrosinate Proteins

Iron-tyrosinate proteins are characterized by a distinctive visible absorption from Tyr→Fe(III) charge transfer bonds (88). Protocatechuate 3,4 dioxygenase catalyzes the intradiol cleavage of protocatechuic acid to β-carboxy-cis, cis-muconic acid. The iron site is unusual, being neither a heme nor an Fe—S cluster. The iron site had been proposed to be a tetrahedral iron-sulfur center, (89, 90) but this idea has since been rejected in favor of a site with histidine and tyrosine ligation (91). A combined EXAFS and Raman study yielded a broad Fourier transform first-shell peak (92). The data are consistent with combined histidine and tyrosine ligation, but more quantitative interpretation of the data is hindered by the limited resolution of the restricted range EXAFS data.

Somewhat better spectra were reported by Scheider et al. in a comparative study (93) of carbonate and thioglycolate complexes with the iron-binding protein ovotransferrin. The carbonate complex was simulated by an octahedral complex with a 1.96-Å average Fe-O distance, and evidence for 1–2 sulfur neighbors at 2.32 Å was observed in the thioglycolate spectrum. Edge spectra for iron-tyrosinate proteins have also been analyzed (94). Lindley et al. also examined transferrin EXAFS, but they found two shells of 1.85 and 2.04 Å (95).

7.3.1.6. Iron in the Photosynthetic Reaction Center

Photosynthetic "reaction center" is a membrane-bound bacteriochlorophyll–protein complex in which the primary photochemical events occur (96, 97). The complex contains protein, bacteriochlorophyll, bacteriopheophytin, two quinones, and a single iron atom. There is some evidence that the iron facilitates electron transfer between the two quinones (98). Two independent groups have used x-ray absorption to probe the structure of this site.

Eisenberger et al. compared low-temperature reaction center iron EXAFS with that of model compounds (99). They concluded that the coordination sphere was octahedral with oxygen and/or nitrogen ligands at 2.10 ± 0.02 Å with a large, ~ 0.1 Å, amount of static disorder. A long-distance interaction at 4.12 ± 0.03 Å was also observed.

Bunker used the amplitude ratio method to analyze a similar iron reaction center EXAFS (100). This group obtained an average Fe—(N, O) distance of 2.14 ± 0.02 Å with a static σ of 0.07 Å. They also postulated that the long distance transform feature arose from three third-shell histidine interactions.

Since the completion of the EXAFS work, the reaction center has been crystallized (101).

7.3.2. Copper Proteins

7.3.2.1. "Blue" Copper Proteins

The "blue" copper proteins contain a distorted tetrahedral copper site with a cysteine thiolate ligand (102), and it is the sulfur to copper charge transfer transition that gives these proteins their intense blue color (103). Azurin from *Pseudomonas aeruginosa* was the first blue copper protein studied by EXAFS (104), and an unusually short Cu—S bond length of 2.10 ± 0.02 Å was reported. Further comparative study of azurin, stellacyanin, and plastocyanin yielded Cu—S_{cys} bond lengths of 2.11 Å for all the oxidized forms, as well as 2.22 Å for reduced plastocyanin and 2.25 Å for reduced stellacyanin (105).

These results have been criticized by Peisach et al. (106) and Blumberg and Powers (107). Using "analytical methods developed at Bell Laboratories, as well as the more primitive method originated by Hodgson and Doniach", they found a Cu—S distance of 2.15 ± 0.025 Å in both azurin and stellacyanin. Crystallographic information subsequently became available for plastocyanin (108) and azurin (109), and the reported Cu—S_{cys} bond length for plastocyanin was 2.13 ± 0.05 Å. This is within experimental error of the values reported by most groups.

Single crystal EXAFS studies of plastocyanin failed to observe a Cu—S_{met} interaction (110), even when the polarization was oriented along the Cu—S_{met}

axis. The authors propose that the copper and methionine sulfur move independently, indicating the absence of significant bonding between the two atoms.

More recently, azurin EXAFS was reexamined by Groenveld et al. (111). They report a 2.23-Å Cu–S distance in oxidized high pH azurine, longer than any other group has reported. A study by Sano and Namura (112) of the blue copper site in nitrite reductase gave a Cu–S distance of 2.17 Å, while Chapman et al. found a Cu–S distance of 2.13 Å in oxidized umecyanin, lengthening to 2.21 Å upon reduction (113).

7.3.2.2. Hemocyanin

Hemocyanin is an oxygen-binding protein found in molluscs and arthropods. It is found in many polymeric forms, but the smallest functional subunits have molecular weights of 50,000 to 75,000 and contain two copper atoms that bind a single molecule of dioxygen (114). Hemocyanin can exist in a variety of chemical forms, and the structures proposed for the oxy and deoxy species are summarized in Fig. 7.5. The key information sought in most hemocyanin EXAFS studies is the Cu—Cu distance, although the number and type of ligands in the copper first coordination sphere are also of interest.

The earliest EXAFS study of giant keyhole limpet, *Megathura crenulata*, hemocyanin did not yield an unambiguous Cu—Cu distance (115). Average first-shell distances of 1.96, 1.99, and 2.01 Å were reported for oxygenated, deoxygenated, and nitric oxide-treated forms, with estimated errors of ±0.04 Å. The copper coordination number was predicted as three or four in the deoxy form and five or six in both oxygenated and nitric oxide-treated samples, with coordination solely by nitrogen and/or oxygen donors in all three cases.

Another study of arthropod hemocyanin confirmed the previous first coordination sphere findings (116, 117). It also predicted Cu—Cu distances of 3.68 and 3.39 Å in oxy and deoxy forms, respectively.

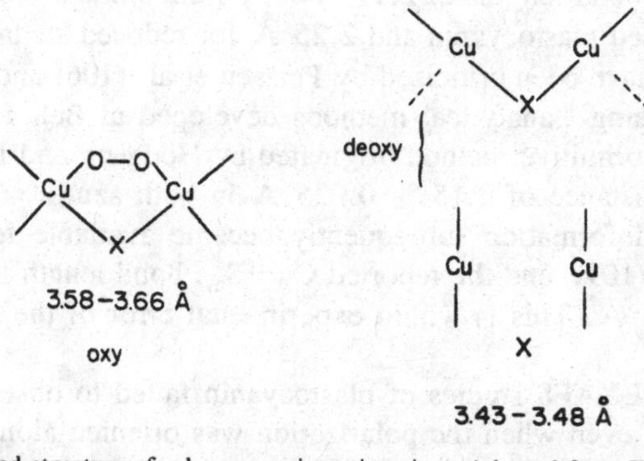

Figure 7.5. Proposed structures for hemocyanin active sites. Adapted from Fig. 14 of ref. (120).

The interpretation of hemocyanin EXAFS was further refined by application of a group-fitting procedure that optimized all the copper imidazole interactions by assuming a rigid ring and fixed bonding angles (118, 119). The resultant analysis gave a Cu—Cu distance of 3.55 Å in oxyhemocyanin. However, the Hodgson group did not observe a Cu—Cu distance in deoxyhemocyanin. Spiro and co-workers (120) later applied a similar group-fitting analysis to hemocyanin, but in searching for Cu—Cu interactions they first filtered out contributions to the EXAFS from the first coordination sphere. They found a 3.66-Å Cu—Cu distance in oxyhemocyanin and a 3.43-Å distance in the deoxy protein by this procedure.

A study of hemocyanin copper absorption edges found that the x-ray spectra did not change in parallel with UV absorption changes (121). Two intermediates in the oxygenation process are invoked to explain the results. However, these authors apparently normalized the spectra to the same peak height at the edge. Since this peak height changes during oxygenation, the normalization process could have introduced nonlinearity. Normalization to a point in the EXAFS region would have been more appropriate.

7.3.2.3. Cytochrome Oxidase

Cytochrome oxidase is a complex enzyme containing two distinct types of heme iron, heme a and heme a_3, and two distinct types of copper, Cu_A, and Cu_B. The enzyme catalyzes the reduction of dioxygen to water, in parallel with the oxidation of four molecules of cytochrome c and the translocation of protons across the mitochondrial membrane (122, 123).

$$O_2 + 4 \text{ cyt } c^{red} + 4H^+ \rightarrow 2H_2O + 4 \text{ cyt } c^{ox}$$

The first x-ray absorption study by Chan and co-workers of cytochrome oxidase involved only the edge region, and yielded the controversial conclusion that the Cu_A site in the oxidized enzyme was actually Cu(I) bound to a sulfur radical (124). A subsequent study reported edge data on the CN^- form of the enzyme (125). These reports were challenged by Powers et al. (126), who ascribed the Chan results to photoreduced enzyme, a claim subsequently rejected by the Chan group (19). Although there are real spectral differences, the photoreduction criticism does not appear justified, however, since the low energy feature attributed to photoreduction is also observed in the Yonetani "pulsed" state spectrum (127). EPR (128) and ENDOR (129) work on the $g_{av} = 2.06$ signal from resting cytochrome oxidase appear consistent with a delocalized Cu(II) assignment.

In their studies of cytochrome oxidase x-ray absorption edges, Powers et al. concluded (126) that the Cu_B site in cytochrome oxidase resembles the blue copper site in stellacyanin. They showed that the higher energy half of the oxidized oxidase spectrum resembles that of oxidized stellacyanin, while the lower energy half of the reduced edge resembles reduced stellacyanin. They then generated the Cu_A edge by subtracting the appropriate stellacyanin edge from the total oxidase spectrum. The results were tested for consistency by comparing observed mixed valence edges with those calculated from sums of stellacyanin and calculated Cu_A edges. This test for consistency does not prove uniqueness, however, since any species that had the same edge *change* upon reduction as Cu_B would yield results consistent with the data.

The first EXAFS study of cytochrome oxidase (130) showed a double peak Fourier transform for the resting enzyme. This was interpreted as an indication of not only $Cu-N$ or $Cu-O$ ligation, but also two or three $Cu-S$ bonds (per two coppers) with an average $Cu-S$ bond length of 2.27 Å. Assuming that the Cu_A site is more covalent than that of Cu_B, these results led to the conclusion that the Cu_A site had at least two sulfur ligands.

Expanded studies by this group (131, 132) of four different states of the enzyme yielded the results summarized in Table 7.4. The principal change, which can be unambiguously interpreted, is a 0.05 Å increase in the average $Cu-S$ distance upon reduction of Cu_A. This result stems from a comparison of the resting state and mixed valence-formate data. Ascribing most or all of the sulfur ligands to Cu_A is consistent with the edge studies.

Powers and co-workers have also published on the EXAFS of the oxidase metal centers (18, 133). The first paper presented an extremely detailed model for both the iron and copper sites, and the various competing models are compared in Fig. 7.6. Powers and co-workers stated that their analysis gives a "unique solution within experimental error to the contribution and identity of each site". Furthermore, they criticize curve fitting as "futile, because the first shell of the two copper sites may contain eight or so neighbors with different distances and/or different chemical types," and conclude that the "only alternative is to model each site with a single compound."

Despite the distinction drawn between analysis using model compounds and curve fitting, the underlying logic remains the same, namely, that changes in EXAFS reflect changes in structure. The same basic limitations in the resolution and sensitivity of EXAFS persist no matter how the data are analyzed. The proposed structure is not a unique solution, since there are variations that could be made, which would make negligible changes in the EXAFS.

Given this fundamental disagreement over interpretation, it is useful to compare features of the Powers–Chance model with the Scott et al. results, tabulated in Table 7.4. Apart from the fact that the Powers model has more detail, the most serious discrepancies between the two groups concerning the first coordination spheres are the average $Cu-S$ distance in the oxidized enzyme and the average $Cu-N$ distance in the reduced enzyme. Part of this difference may stem from the use of different enzyme preparations, the Yonetani preparation (134) and the Hartzell and Beinert method, respectively (135).

It is worth noting that there is other information available that raises questions about assigning Cu_B as a "blue" copper. The proposed sequence homology with the blue copper proteins is not convincing (136), and an alternative interpretation based on ceruloplasmin homology has been proposed (137). Assignment of the near-infrared band at 830 nm to Cu_B by Chance et al. has been questioned by other work (138–140). Experiments with Ag^+ and Hg^{2+} show that Cu_B does not behave chemically like other blue copper sites (141). EPR signals observed upon decoupling Cu_B from cytochrome a_3 show a rhombic g tensor and hyperfine coupling similar to half-met hemocyanin rather than blue copper (139). Finally, ^{14}N ENDOR of the $Cu_B(II)$ site reveals three distinct nitrogeneous ligands with couplings similar to the type 3 $Cu(II)$ site in laccase (140). All of these data suggest a Cu_B site more akin to type 3 copper than to "blue" copper.

Scott et al. examined cytochrome oxidase Cu EXAFS from a number of sources in order to resolve disputes about sample preparation (142). For all samples they found an average coordination (per 2 coppers) of 6 ± 1 nitrogens and/or oxygens at 1.99 ± 0.03 Å and 2 ± 1 sulfurs at 2.28 ± 0.02 Å. No evidence was found for the short Cu—S distance expected for a blue copper site, even in the Yonetani preparation. Analysis of longer distance features indicated a Cu-Fe interaction at 3.00 ± 0.03 Å in all of their samples except the enzyme prepared by the Yonetani method. They argue that the latter preparation, generally used for the studies by Chance and co-workers represents an inactive form, and that structures illustrated in Figure 7.6 are more physiologically relevant.

Chance and Powers have examined the EXAFS of cytochrome oxidase depleted of Cu by extensive dialysis against cyanide (143). From the iron point of view, the peak they had assigned to the 3.8-Å Fe-Cu interaction is diminished as expected. Furthermore, they no longer see a well-resolved Fe—S peak, and they conclude that the heme a_3 axial Fe—S distance has contracted from 2.6 to 2.3 Å. They examined the "pulsed" form of oxidase obtained by dithionite reduction and reoxidation of the enzyme. The main Fe EXAFS Fourier transform peak is again unsplit, while the Cu EXAFS now shows a split transform peak. They conclude that the bridging sulfur is now bound to the Cu_B site. They also examined the Hartzell–Beinert preparation of the enzyme used in most of the Scott et al. studies. They conclude that in this form the S atom has moved closer to Fe and further from Cu, compared to the Yonetani form. However, their kinetic data seem to indicate that the Hartzell–Beinert should more closely resemble the "pulsed" enzyme in which the Fe—S bond is broken. Regardless of the interpretation, their work further emphasizes the difficulty in comparing data on different preparations of this enzyme. The Cu spectra of membrane-bound oxidase has been reported to resemble oxidase prepared by the Yonetani method (127).

Scott et al. have examined a cytochrome oxidase sample with Cu_A removed

Table 7.4. Summary of Copper Protein EXAFS and Diffraction Results

Sample	Cu–N(O)		Cu–S		Cu–X		References
	N	R (Å)	N	R (Å)	N	R (Å)	
Lyophilized oxidized *P. aeruginosa* azurin	3.0	1.97(2)	1.1	2.12(2)			104
	2.6	1.97(2)	1.3	2.10(2)	0.6 S'	2.24(5)	104
Solution oxidized azurin	3.3	1.97	0.9	2.11	0.7 S'	2.56	105
		1.975(25)		2.175(25)			107
	2	1.95	1	2.23			111
Solution oxidized plastocyanin	2.3	1.97	1.1	2.11	0.7 S'	2.66	105
Crystalline oxidized plastocyanin	1	2.04(5)	1	2.13(5)	1 S'	2.90(5)	108
Solution reduced plastocyanin	2.2	2.05	0.8	2.22			104
Solution oxidized stellacyanin	2.9	2.04	1.1	2.11	0.7 S'	2.72	105
	2	1.96(3)	1	2.20(3)	1 S'	2.84(3)	106
		1.975(25)		2.175(25)			107
Solution reduced stellacyanin	2.7	2.07	0.7	2.25			105
Pellet oxygenated *M. crenulata* hemocyanin	5–6	1.96(4)					115

Sample	N	R (Å)	N	R (Å)		R (Å)	Ref
Solution, oxygenated *B. canaliculatum* hemocyanin	3.5(1.5)	1.96(2)			1 Cu	3.68(5)	117
Pellet, deoxygenated *M. crenulata* hemocyanin	3–4	1.99(4)			—	—	115
Frozen solution, deoxygenated *B. canaliculatum* hemocyanin	2.0(1.5)	1.95(2)			1 Cu	3.39(5)	117
Frozen solution, oxygenated *M. crenulata* hemocyanin	2.0(5)	2.01(2)			1.1 Cu	3.55(2)	118
	2.0(5)	1.92(2)					
Frozen solution, oxygenated *B. canaliculatum* hemocyanin	2.4(5)	1.90(2)			1 Cu	3.66(5)	120
	2.04	2.00(2)					
Frozen solution, deoxygenated *M. crenulata* hemocyanin	0.5(2)	1.95(2)			—	—	118
	1.6(4)	1.95(2)					
Frozen solution, deoxygenated *B. canaliculatum* hemocyanin	1.3(3)	1.92(2)			1 Cu	3.43(5)	120
	1.0(2)	1.95(2)					
Oxidized (Beinert preparation) cytochrome oxidase		1.96		2.28			131,132
Oxidized (Yonetani preparation) cytochrome oxidase	$\tfrac{3}{2}$	1.97	2	2.27	$\tfrac{1}{2}$S	2.82	18
			2	2.18	$\tfrac{1}{2}$S(N, O)	?	
			$\tfrac{1}{2}$		$\tfrac{1}{2}$Fe	3.75	
Reduced (Beinert preparation) cytochrome oxidase		1.98		2.34			131,132
Reduced (Yonetani preparation) cytochrome oxidase	$\tfrac{1}{2}$	2.00	2	2.35	$\tfrac{1}{2}$S(N, O)	?	18
	$\tfrac{1}{2}$	2.07	2	2.25			
			$\tfrac{1}{2}$	2.22			
Copper metallothionein			2	2.36			166

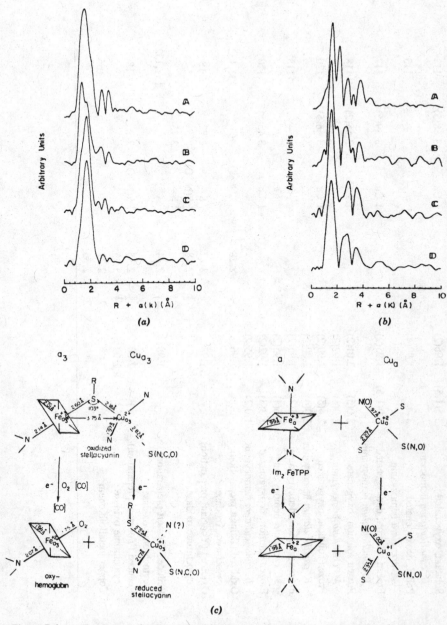

Figure 7.6. (*a*) Fourier-transformed copper EXAFS data of the A, fully oxidized; B, mixed valence formate; C, mixed valence CO; and D, reduced CO states. From Fig. 7 of ref. (18). (*b*) Fourier-transformed iron EXAFS data of the A, fully oxidized; B, mixed valence formate; C, mixed valence CO; and D, reduced CO states. From Fig. 8 of ref. (18). (*c*) Structures proposed by Powers et al. for the cytochrome oxidase active site. From Fig. 25 of ref. (18). (*d*) orientation-dependent copper EXAFS, from ref. (145). (*e*) orientation dependent Fe EXAFS, from ref. (145).

286

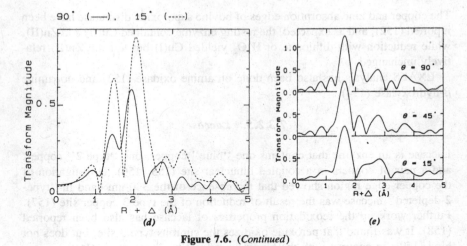

Figure 7.6. *(Continued)*

by treatment with *p*-(hydroxymercuribenzoate) and subsequent dialysis (144), instead of the Cu_B-depleted enzyme originally used by Chance et al. They find that the Cu_B site can be simulated by an environment with 3N or O ligands at 1.98 Å, and 1 S or Cl ligand at 2.30 Å.

A Cu_A-depleted form of oxidase, prepared by brief expose to bathocupreine sulfate, has been examined by Powers et al. (145). They see evidence for 1 or 2 S ligands and report that the Cu_B is "less tetrahedral than is the oxidized 'resting' state of the untreated enzyme."

George et al. have recently reported the orientation-dependent iron and copper XAS of cytochrome oxidase in hydrated membrane multilayers (146). These samples, prepared by partial dessication of membraneous oxidase on mylar films, are known from EPR and optical studies to be 1-dimensionally ordered with the heme axes perpendicular to the membrane normal (147–149). The copper edge region is highly anisotropic, and George et al. conclude that Cu_B is a tetragonal species with the long axis nearly perpendicular to the membrane normal. They find average Cu—N,O and Cu—S distances of 1.94 and 2.28 Å, respectively. The iron EXAFS was consistent with the presence of an axial sulfur or chlorine ligand at 2.33 ± 0.04 Å.

7.3.2.4. "Type 2" Copper

Superoxide dismutase enzymes catalyze the decomposition of the superoxide radical (150, 151). The cyclic mechanism involved oxidation of one superoxide to dioxygen and reduction of the next superoxide to hydrogen peroxide:

$$2O_2^{\cdot -} + 2H^+ \rightarrow O_2 + H_2O_2$$

The copper and zinc absorption edges of bovine superoxide dismutase have been reported (152), and as expected, the resting enzyme contained Cu(II) and Zn(II), while reduction with dithionite or H_2O_2 yielded Cu(I) but left the Zn(II) relatively unchanged.

EXAFS analysis has also been done on amine oxidase (153) and dopamine β-hydroxylase (154).

7.3.2.5. Laccase

Laccase is an enzyme that contains one "blue" copper, one "type 2" copper, and a pair of coppers in a coupled binuclear site (155, 156). Investigation of the copper-edge region showed that the absence of the 330-nm band in "type-2-depleted" laccase was the result of reduction of the type-3 copper site (157). Further work on the reoxidation properties of laccase has also been reported (158). It was found that peroxide oxidizes the cuprous type 3 site, but does not bind (159), in contrast with the binding proposal by Frank et al. (160).

An EXAFS comparison of native versus type-2-depleted laccase was reported by Woolery et al. (161). By comparison of Fourier transforms with k^2 and k^3 weighting, they claim to see a Cu-Cu interaction at 3.4 Å in native enzyme which disappears on removal of the type II Cu. Their data contains wide gaps near 11 Å^{-1}, presumably due to glitches, and the sensible type-2-depleted EXAFS only extends to 10 Å^{-1}. Given the complexity of the analysis, their evidence for the loss of a Cu-Cu interaction must be viewed *con granulo salis*.

7.3.2.6 Copper Metallothionein

Metallothioneins are small cysteine-rich proteins that bind copper and numerous other metals (162–164). They are thought to be involved in the detoxification of heavy metals, metal storage, and perhaps in the regulation of available copper and zinc levels in the organism. The cadmium-zinc protein has been crystallized (165), and is found to have two separate domains. The β domain contains a $Cd_1Zn_2Cys_9$ cluster, whereas the α domain contains a Cd_4Cys_{11} cluster. Both clusters feature tetrahedral metal coordination.

Although the Cd,Zn protein structure is well understood, Cu(I) binds to metallothionein with a different stoichiometry, and hence must have a different structure. The earliest EXAFS report by Bordas et al. (166, 167) assumed a stoichiometry of 4 Cu(I):8 Cys for the yeast enzyme. Using data extending to 11 Å, this group found an average of 2 Cu—S bonds at 2.22 Å and 2 longer Cu—S bonds at 2.36 Å. Combining this result with the 1Cu:2S stoichiometry led them to propose a cubane-type Cu-S cluster for the yeast enzyme.

Freedman et al. have reported EXAFS of the canine liver metallothionein (168). They claim to find 4 sulfur neighbors per copper at an average distance of 2.27 ± 0.02 Å. However, their analysis assumed 4 coordinate copper and fit various models with x sulfur and $(4 - x)$ nitrogen neighbors. Thus, the alternatives of simply 3-coordinate copper, or nonintegral average Cu-S coordination, were never considered. Furthermore, although their data extended to only 11 Å^{-1}, they inferred the presence of Cu—Cu interactions at 2.74, 3.32, and 3.88 Å. From the presence of these longer Cu—Cu interactions they dismiss the cubane model of Bordas et al. and suggest adamantane-like structures with 4 or 5 coppers per cluster.

The copper metallothionein from *Neurospora crassa* was studied by Smith et al. (169). They found an average Cu environment with 3–4 sulfur neighbors at 2.20 Å and 1–2 Cu neighbors at 2.71 Å. They suggest a compact cluster and rule out previously proposed chain structures with 2-coordinate Cu.

Since the presence of multiple domains complicates the EXAFS interpretation, George et al. investigated the rat liver β domain, obtained from subtilisin proteolysis, which binds 6 Cu(I) per molecule (170). Comparison of the EXAFS spectrum with the model compound *tris*-tetramethyl thiourea Cu(I) yielded a Cu—S coordination number of approximately 3. Based on known Cu—S bond lengths for 2-, 3-, and 4-coordinate copper(I), they concluded that the 3 Cu—S bonds per cluster was the most chemically reasonable interpretation. A trigonal prismatic Cu_6S_9 cluster was therefore proposed for the β domain.

Because the early results of Bordas et al. on the yeast enzyme conflicted with all subsequent analyses of the mammalian enzyme, the EXAFS of the yeast enzyme has been reexamined by George et al. (171). EXAFS from both the copper and sulfur edges was obtained. The copper EXAFS yielded an averge Cu—S distance of 2.23 Å, with an average coordination number between 2 and 3. The XANES and EXAFS were quite similar to previous mammalian hepatic data. The sulfur EXAFS suggested an average S—Cu coordination of 1–2 with an average distance of 2.22 Å. A Cu_8S_{12} cluster is proposed to explain the combined copper and sulfur XANES and EXAFS.

An average 3-coordinate copper was also found in Cu, Zn metallothionein (172).

7.3.3. Molybdenum and Tungsten

Molybdenum is present in a number of diverse enzymes (173–176). One of the unique aspects of these enzymes is the presence of "cofactors," either the iron–molybdenum cofactor, "FeMo-co," of nitrogenase (177, 178) or the molybdenum cofactor, "Mo-co," apparently common to the remainder of these enzymes (179, 180). These cofactors are small molecules containing molybdenum, which can be released from one protein and then inserted into a second apoprotein structure to form an active enzyme. The EXAFS work on these enzymes (181) has attempted to elucidate how the catalytic properties of the cofactors are modified by environmental and structural changes (see Fig. 7.7).

Tungsten enzymology is not yet well developed. In sulfite oxidase and nitrate

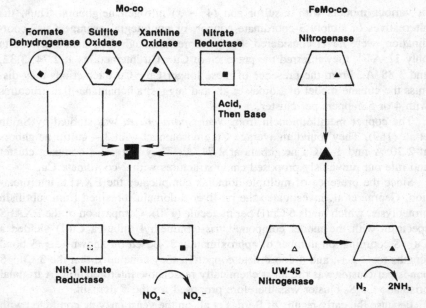

Figure 7.7. Schematic illustration of propeties of molybdenum cofactor and iron–molybdenum cofactor, from ref. (181).

reductase, tungsten can replace molybdenum to form inactive proteins, while in certain formate dehydrogenases tungsten appears to be the biologically active metal.

7.3.3.1.   Nitrogenase and the Iron–Molybdenum Cofactor

Nitrogenase catalyzes the reduction of dinitrogen to ammonia (182–184):

$$(6 + x)e^- + y\text{MgATP} + (6 + x)\text{H}^+ + \text{N}_2 \rightarrow$$

$$y\text{MgADP} + y\text{P}_i + (x/2)\text{H}_2 + 2\text{NH}_3$$

The number of electrons consumed in hydrogen production, x, and the number of ATP molecules hydrolyzed, y, depend on the reaction conditions. Two different proteins are involved in the reaction, an iron protein sometimes referred to as component II or azoferredoxin, and a molybdenum–iron protein alternatively called component I or molybdoferredoxin.

The first absorption edge results on nitrogenase (185) were invalid because of the use of inactive protein. However, the subsequent edge and EXAFS data using the transmission mode on lyophilized protein (186) have since been reproduced on enzyme in solution by both transmission (187) and fluorescence detection techniques (188).

Examination of the nitrogenase absorption edge region revealed some qualitative information about the nitrogenase molybdenum environment. The edge *position* is near the low end of the range observed for molybdenum, indicating a low oxidation state and/or, a covalent environment. The edge *shape* shows only a single inflection point, whereas all terminal oxomolybdenum model compound edges had an additional low-energy inflection arising from a bound state transition. Thus, it was concluded from the edge data alone that the nitrogenase molybdenum was devoid of terminal oxo groups when active, and perhaps existed in a high sulfur environment.

The split Fourier transform, Fig. 7.8a and the beat pattern in the EXAFS itself led to the conclusion that two major components are required to explain the nitrogenase EXAFS. Based on comparison with model compounds as well

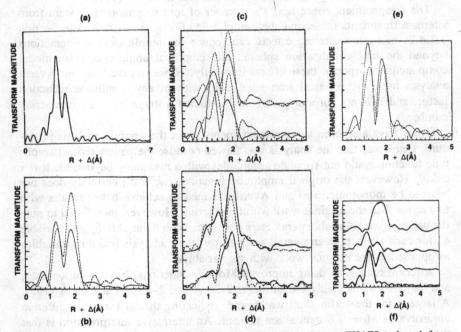

Figure 7.8. Fourier transforms of molybdenum and tungsten enzyme EXAFS, adapted from ref. (130). (*a*) Nitrogenase Mo—Fe protein. (*b*) Sulfite oxidase: oxidized (——) versus reduced (———). (*c*) Xanthine dehydrogenase: top, oxidized active (——) versus reduced active (———); bottom, oxidized desulfo (——) versus reduced desulfo (———). (*d*) Nitrate reductase, top, oxidized *Chlorella* (———) versus reduced *Chlorella* (——); bottom, oxidized *E. coli* (———) versus reduced *E. coli* (——). (*e*) *Desulfovibrio gigas* Mo—(2Fe—2S) protein; oxidized (———) versus reduced (———). (*f*) Formate dehydrogenases and models. Top to bottom: *C. thermoaceticum* W enzyme; WO_2 (ox)$_2$; WO_2(tox)$_2$; *C. pasteurianum* Mo enzyme.

as theoretically calculated EXAFS, the data were interpreted as the sum of Mo—S and Mo—Fe interactions. This led to the essential conclusion that nitrogenase contains a molybdenum, iron, sulfur cluster, as in Figure 7.9, and that the molybdenum–iron distance was 2.72 ± 0.05 Å.

Two alternative structural models consistent with the EXAFS were originally suggested, one with a linear Fe-(S)$_2$-Mo-(S)$_2$-Fe cluster and another model based on an MoFe$_3$S$_4$ cube. The subsequent synthesis of a double-cubane model compound and the similarity of its EXAFS to that of nitrogenase led Hodgson and co-workers to conclude "the Mo—Fe protein possesses a structural fragment similar to the MoFe$_3$S$_4$ cube that constitutes one half of the complex [Mo$_2$Fe$_6$S$_9$(SEt)$_8$]$^{3-}$ (189). This conclusion was challenged by Teo and Averill (190), who reinterpreted the original Hodgson group EXAFS data and found two rather than three iron neighbors. They proposed models with roughly linear Fe-(S)$_x$-Mo-(S)$_x$-Fe local environments (190, 191). Other proposals include "string-bag" models (192, 193) and more extended arrays (194).

The disagreement concerning the number of iron neighbors may stem from alternate treatments of second-shell amplitude effects. As discussed in Section 7.2.4.2. multiple-scattering effects can reduce the amplitudes of interactions beyond the first coordination sphere. The empirical amplitudes from model compounds incorporate these effects implicitly. However, the Teo and Averill analysis used a theoretical iron amplitude, without any amplitude reduction factor, and it is not surprising that they obtained a lower Mo—Fe interaction number.

There have also been some misstatements about the original analysis procedure. Lee stated that the analysis was flawed because the parameterized amplitude function could not simulate amplitudes with a maximum beyond the low k cutoff. However, the original amplitude function, $c_0 k^{c_1} e^{c_2 k^2}$, certainly does not have to be monotonic. Teo and Averill claimed to achieve better results with two terms than the original work with three terms. However, they failed to note the use of more adjustable parameters per term. With a variable E_0 and variable σ^2 for each shell, plus an overall scale factor, their analysis had nine variables as opposed to the original work with six variables.

A recent reanalysis using improved Mo—Fe functions, better solution data, and only two shells yields 4–6 sulfurs at 2.36 Å and 2–4 irons at 2.68 Å (195). A large σ for the sulfur shell was found, indicating that at least two different unresolvable Mo—S distances are present. An alternative interpretation is that there is destructive interference with an unresolved Mo-O, N component. In either case, the length of the long Mo—S bonds reported in the original studies may well be an overestimate.

Single crystal EXAFS studies of nitrogenase have been conducted (195), in order to distinguish between linear and cubane cluster models. The x-rays derived from synchrotron radiation sources are highly polarized, and the EXAFS of a linear Mo—Fe cluster will show a larger orientation dependence than a

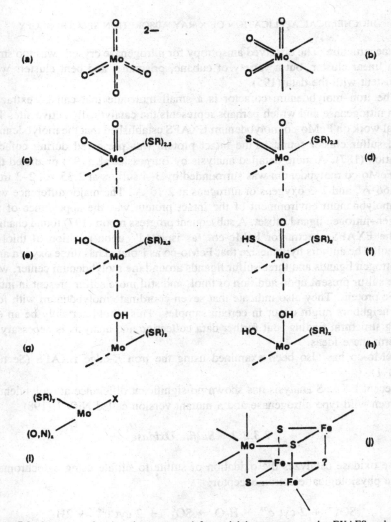

Figure 7.9. Summary of active sites suggested for molybdenum enzymes by EXAFS and other techniques, adapted from ref. (181) and citations therein. (*a*) Free MoO_4^{2-}, the form of molybdenum typically found in aqueous solution under physiological conditions. (*b*) Trioxo species, established for *C. pasteurianum* formate dehydrogenase. (*c*) Dioxo, thiolate species, found in oxidized sulfite oxidase, oxidized *Chlorella* nitrate reductase, oxidized desulfo xanthine oxidase, oxidized *D. gigas* Mo—(2Fe—2S) protein. (*d*) Oxo, sulfido, thiolate species proposed for oxidized, intact xanthine oxidase. (*e*) Oxo, hydroxo, thiolate environment proposed for reduced desulfo xanthine oxidase, reduced *D. gigas* Mo—(2Fe—2S) protein, reduced sulfite oxidase. (*f*) Oxo, thiol, thiolate environment proposed for reduced intact xanthine oxidase. (*g*) Model for oxidized, low pH, *E. coli* nitrate reductase. (*h*) Plausible model for reduced *M. formicicum* formate dehydrogenase. (*i*) Model for reduced *E. coli* nitrate reductase. (*j*) One of several models for local molybdenum environment in nitrogenase.

293

cubane structure. The observed anisotropy for nitrogenase crystals was too small for a linear cluster, but a variety of cubane, prismane, and bent clusters were consistent with the data (195).

The iron–molybdenum cofactor is a small molecule that can be extracted from nitrogenase and which perhaps represents the catalytically active site. The initial work on FeMo-co molybdenum EXAFS established that the molybdenum, iron, sulfur cluster found in the intact protein was preserved during cofactor isolation (187). A more detailed analysis by Burgess et al. (196) predicted that the FeMo-co molybdenum was surrounded by 3–4 sulfurs at 2.35 Å, 2–3 irons at 2.66 Å, and 2–3 oxygens or nitrogens at 2.10 Å. The major difference with the molybdenum environment of the intact protein was the appearance of the oxygen–nitrogen ligand subset. A subsequent progress report (197) found changes on the EXAFS spectra for FeMo-co "as isolated" upon addition of thiol or selenol. The authors hypothesize that FeMo-co as isolated has three oxygen and/or nitrogen ligands and three sulfur ligands around the molybdenum center, with more sulfur present upon addition of thiol, and still more sulfur present in intact MoFe protein. They also indicate that seven-coordinate molybdenum with four iron neighbors might occur in certain samples. This would certainly be an exciting structural finding, but further data collection and analysis is necessary to confirm these ideas.

FeMo-co has also been examined using the iron K-edge EXAFS (Section 7.3.1.1).

Recent EXAFS analysis has shown no significant difference at molybdenum between wild-type nitrogenase and a mutant version called NifV$^-$ (198).

7.3.3.2. Sulfite Oxidase

Sulfite oxidase catalyzes the oxidation of sulfite to sulfate using cytochrome c as the physiological electron acceptor:

$$SO_3^{2-} + 2 \text{ cyt } c^{ox} + H_2O \rightarrow SO_4^{2-} + 2 \text{ cyt } c^{red} + 2H^+$$

Structurally, is the simplest molybdenum enzyme, and it typically exists as a dimer of equivalent 55–60-kdalton subunits, each containing molybdenum and cytochrome b_5-like heme (199).

Sulfite oxidase x-ray absorption studies (200, 201) reveal a dramatic difference in the absorption edges and EXAFS Fourier transforms for the oxidized and dithionite-reduced species, as illustrated in Fig. 7.8b. Curve-fitting analysis of the EXAFS data suggests that there are two oxo groups in the oxidized form of the enzyme, but only a single terminal oxo in the reduced enzyme. One interpretation is that an oxo group is protonated upon reduction of the enzyme at low pH, and that the proton on this hydroxyl is the source of the observed hyperfine splittings.

There is also reported a Mo—S interaction corresponding to 2 or 3 Mo—S bonds (201). Some reinterpretation may be required now that chloride binding is better appreciated (202).

7.3.3.3. Xanthine Oxidase

Xanthine oxidase (X.O.) belongs to a class of enzymes collectively known as molybdenum iron–sulfur flavin hydroxylases, which also includes aldehyde oxidase and xanthine dehydrogenase (203, 204). The reactions catalyzed by these enzymes are quite numerous, involving a wide range of purines, pteridines, pyrimidines, and aldehydes:

$$RH + H_2O \rightarrow ROH + 2H^+ + 2e^-$$

An important aspect of molybdenum hydroxylase chemistry is the conversion from active or "intact" to inactive "desulfo" forms, either gradually during storage or rapidly with cyanide (205).

$$\underset{\text{active}}{X.O.^{ox}} + CN^- \rightarrow X.O.^{red} + SCN^-$$

The earliest EXAFS studies of xanthine oxidase (206, 207) produced contradictory results. The analysis by Tullius et al. (206) found 1.5 terminal oxygens at 1.71 Å, as well as 2.1 sulfurs at 2.54 Å. In contrast, Bordas et al. (207) found 1 terminal oxygen at 1.75 Å, 2 sulfurs at 2.46 Å, and another Mo—S interaction at 2.25 Å. Both groups also found a long Mo—S at about 2.85 Å.

A subsequent study by Cramer et al. (201) of chicken liver xanthine dehydrogenase yielded the values summarized in Table 7.5. This study confirmed the presence of a terminal Mo—S, interaction at 2.15 Å in the oxidized active enzyme. Furthermore, this short bond disappeared upon dithionite reduction, whereas the terminal Mo=O stayed in place. As expected, cyanolyzed xanthine dehydrogenase showed no evidence for a short Mo=S bond. Rather, this species appeared to have two terminal Mo=O bonds in its oxidized form, and only a single oxo when reduced. Both forms of the enzyme had two sulfurs at about 2.47 Å in the oxidized forms, which shortened by about 0.1 Å upon reduction. It was also shown that long methionine bonds, while perhaps present, would be difficult to observe. The EXAFS Fourier transforms are shown in Fig. 7.8c.

Cramer and Hille have also investigated arsenite-inhibited (208) and alloxanthine-inhibited (209) forms of the enzyme, and George et al. have examined a xanthine oxidase–lumazine complex prepared by rapid freezing (210). In every one of these experiments, a Mo=O bond remains in place in the reduced enzyme.

Table 7.5. Summary of EXAFS Results on Molybdenum Enzymes and Proteins

Enzyme	Conditions	Mo=O N	Mo=O R (Å)	Mo—S N	Mo—S R (Å)	Type	Mo—X N	Mo—X R (Å)	References
Nitrogenase (C. pasteurianum)	Lyophilized "semireduced"	0	—	3–4	2.35	Mo—Fe Mo—S'	2–3 1–2	2.72 2.49	186
Nitrogenase (A. vinelandii)	25 mM Tris HCl pH 7.4 "semireduced"	0		3–4	2.35	Mo—Fe Mo—S'	2–3 1–2	2.73 2.46	187
Sulfite oxidase (Chicken liver)	50 mM KP, pH 7.8 oxidized	2	1.68	2–3	2.41				201
Sulfite oxidase (Chicken liver)	50 mM KP, pH 7.8 dithionite-reduced	1	1.69	3	2.38				201
Xanthine dehydrogenase (Chicken liver)	5 mM KP, pH 7.8 oxidized, active	1	1.70	2	2.47	Mo=S	1	2.15	201
Xanthine dehydrogenase (Chicken liver)	5 mM KP, pH 7.8 reduced, active	1	1.68	3	2.38				201
Xanthine dehydrogenase (Chicken liver)	5 mM KP, pH 7.8 oxidized, desulfo	2	1.67	2	2.46				201
Xanthine dehydrogenase (Chicken liver)	5 mM KP, pH 7.8 reduced, desulfo	1	1.66	2–3	2.33				201
Xanthine oxidase (Bovine milk)	Lyophilized oxidized, active	1–2	1.75	2	2.49	Mo=S Mo—S'	1 1	2.25 2.89	207
Xanthine oxidase (Bovine milk)	Lyophilized oxidized, desulfo	2	1.74	2	2.49	Mo—S"	1	2.91	207
Xanthine oxidase (Bovine milk)	100 mM Tris-HOAc pH 8.5 oxidized	1.5	1.71	2.1	2.54	Mo—S'	1.1	2.84	206
Nitrate reductase (E. coli)	50 mM KP, pH 7.0 reduced	2	1.72	2	2.36	Mo—Fe,P,Mo Mo—O,N	1 2	2.79 2.10	218
Nitrate reductase (Chlorella)	80 mM KP, pH 7.6 oxidized	2	1.72	2–3	2.44				218
Nitrate reductase (Chlorella)	80 mM KP, pH 7.6 reduced	1	1.67	3	2.38	Mo—O,N	1	2.07	218
Mo—(2Fe—2S) protein (D. gigas)	100 mM Tris HCl pH 7.6 oxidized	2	1.68	2	2.47	Mo—O,N	1	1.90	213
Mo—(2Fe—2S) protein (D. gigas)	100 mM Tris HCl pH 7.6 dithionite reduced	1	1.68	2	2.38				213
Formate dehydrogenase (C. pasteurianum)	50 mM Tris HCl pH 8 as isolated	3	1.74	0					223

7.3.3.4. Desulfovibrio gigas Mo-(2Fe—2S) Protein

No enzymatic function was first known for this molybdenum-containing protein isolated from *Desulfovibrio gigas*. It has a molecular weight of about 120,000, and contains one molybdenum and six irons, the latter as 2Fe—2S clusters (211, 212). As in all its other properties, the EXAFS of the *D. gigas* protein strongly resembles that of desulfo xanthine oxidase (213). The Fourier transforms shown in Fig. 7.8d are consistent with a primarily dioxo-Mo(VI) species as isolated, which is converted to a monoxo species upon reduction with dithionite. Two sulfur ligands are clearly present, and additional oxygen or nitrogen ligands presumably complete the coordination sphere. Recent biochemical work indicates that the protein is in fact an aldehyde oxidase (214), and that the EXAFS experiments were done on the inactive desulfo form.

7.3.3.5. Nitrate Reductase

There are two known types of nitrate reductase enzymes, both of which reduce nitrate to nitrite (215):

$$NO_3^- + 2e^- + 2H^+ \rightarrow NO_2^- + H_2O$$

Assimilatory nitrate reductases reduce nitrate to nitrite so that the nitrogen can ultimately be reduced to ammonia and then incorporated into amino acids. These enzymes contain molybdenum, heme, and FAD. A representative case is the enzyme from *Chlorella vulgaris*, which is a 360,000-dalton homotetramer (216) containing one of each prosthetic group per subunit. The *dissimilatory* or *respiratory* nitrate reductases use nitrate merely as terminal electron acceptor in the absence of dioxygen. An example is the *E. coli* enzyme, which has a molecular weight of 200,000 and contains one molybdenum and about 16 iron atoms (217).

X-ray absorption has been used to compare the molybdenum sites of the *Chlorella* and *E. coli* enzymes (218). As illustrated in Fig. 7.8e, the EXAFS Fourier transforms for the two enzymes are quite different. The curve-fitting results indicate that the *Chlorella* enzyme molybdenum site is nearly identical to that of hepatic sulfite oxidase, shuttling between monoxo and dioxo forms. However, the *E. coli* enzyme exhibited different behavior, perhaps indicating the presence of a form without Mo=O bonds. In both cases at least 2 sulfur ligands are also present.

Since EPR has shown significant pH and anion effects on the *E. coli* enzyme structure (219), the EXAFS has been reexamined in buffers designed to yield a single form (220). The results confirm that the oxidized high pH form is lacking a terminal oxo group, while the oxidized low pH, chloride form appears to be a monooxo Mo(VI) species unique among Mo enzymes.

7.3.3.6. Formate Dehydrogenase

Depending on the organism, the physiological role of formate dehydrogenase enzymes may be to oxidize formate or to reduce CO_2 (221, 222). In the latter case, the CO_2 reductase activity is generally the first step in carbon fixation, since the formate produced can be further metabolized via the tetrahydrofolate pathway to one-carbon precursors for cell materials.

$$HCO_2^- \rightleftarrows CO_2 + H^+ + 2e^-$$

Preliminary EXAFS results are available for the tungsten enzyme from *Clostridium thermoaceticum* and the molybdenum enzyme from *Clostridium pasteurianum* (223). The Fourier transform of the tungsten L_{III}-edge EXAFS shows clearly that the tungsten site is not the dioxo type of species found in many molybdenum enzymes, nor is it similar to the cluster-type site proposed for nitrogenase. Curve-fitting analysis indicates the presence of at least two sulfur ligands at about 2.39 Å as well as at least two oxygen or nitrogen ligands at about 2.10 Å.

In contrast with the *C. thermoaceticum* tungsten enzyme, the molybdenum site in *C. pasteurianum* formate dehydrogenase appears to have three oxo groups. This is indicated by the unique absorption edge and the Fourier transform, Fig. 7.8*f*, which has a strong peak in the region characteristic of oxo ligation. Surprisingly, there is no evidence for coordination by sulfur ligands, which have been found in all other molybdenum enzymes examined to date. The unusual EXAFS spectrum of the *C. pasteurianum* enzyme is complemented by an anomalous EPR spectrum which shows no Mo hyperfine splittings and is suggestive of a pterin radical (224).

7.3.4. Zinc

7.3.4.1. Aspartate Transcarbamylase

Aspartate transcarbamylase catalyzes the conversion of aspartate to carbamyl aspartate (225). The zinc EXAFS of this protein has been compared with Zn (dimethyldithiocarbamate)$_2$ (226). The enzyme spectra were found to be consistent with ligation by four sulfur atoms at a mean distance of 2.34 ± 0.03 Å.

7.3.4.2. Zinc Metallothionein

Metallothioneins are small, cysteine-rich proteins that have important roles in zinc and copper metabolism and are also involved in the detoxification of heavy metals such as cadmium and mercury (227). The zinc EXAFS of sheep metallothionein has been reported (228), and an average coordination sphere with four sulfurs at 2.29 Å was proposed. An environment with 4 sulfurs at 2.33 Å was preposed for rabbit liver zinc metallothionein (172).

7.3.4.3. Zinc Superoxide Dismutase

The zinc absorption edge of superoxide dismutase has been reported (113). Not surprisingly, the zinc was redox inactive. Zinc and copper EXAFS of the Cu,Zn enzyme have been interpreted by assuming that the "bridging" histidine is actually statistically distributed between zinc-only and copper-only binding (229).

7.3.4.4. Aminolaevulate Dehydratase

EXAFS analysis of zinc in this enzyme yielded a native structure with 3 sulfurs at 2.28 Å and a single O or N ligand (230).

7.3.5. Calcium

Calcium is an element of immense biological importance (231), but it is a difficult element for the application of EXAFS. This is because calcium tends to have complicated low symmetry coordination spheres. In addition, under biological conditions calcium almost invariably has oxygen coordination, so ligand identification is not a significant piece of new information. Nevertheless, by comparison with model compounds some qualitative information has been obtained from calcium edge and EXAFS spectra. A detailed review of calcium EXAFS has been prepared by Hasnain (232).

7.3.5.1. Phospholipids and Miscellaneous Proteins

Powers et al. reported edge and EXAFS data for a number of calcium model compounds (233). A correlation between the edge position and the coordination number was observed, with a roughly 3-eV shift to higher energy going from six-coordinate $CaCO_3$ to nine-coordinate $CaNO_3 \cdot 4H_2O$. Using these trends, the phosphilipids DPPC (dipalmitoyl phosphatidylcholine) and DPPE (dipalmitoyl phosphatidylethanolamine) were studied in 0.85 M $CaCl_2$ solution as a function of temperature and lipid concentration. It was concluded that the calcium is six-coordinate, although the same -1-eV edge shift is later assigned to seven-coordination in the proteins conconavalin A and thermolysin. The EXAFS bond length reported for Con A, 2.32 $\pm$ 0.05 Å, is considerably shorter than the crystal structure average of 2.7 $\pm$ 0.3 Å. Less drastic but systematic shortenings were observed in the thermolysin and muscle calcium-binding pervalbumin.

There is some question as to whether EXAFS gives a reliable average bond length for highly distorted calcium environments. As mentioned in Section 7.2.4.2, Brown and Eisenberger have described how asymmetric radial distribution functions in amorphous materials can lead to inaccurate EXAFS determinations (37). Apart from the asymmetry effect, the longer Ca—O bonds will tend to have larger Debye–Waller factors, which will diminish their contribution to the EXAFS. In view of these problems, the results should be viewed with some caution.

Bianconi et al. have investigated troponin (234). Similarities in the spectra of bovine milk calcium phosphate with that of the $CaHPO_4 \cdot 2H_2O$ have also been reported (235).

7.3.5.2. Bone

Although the x-ray diffraction patterns of bone mineral closely resemble those of disordered hydroxyapatite (236), some differences do exist. This has stimulated several comparative studies of rat bone mineral with synthetic calcium compounds (237, 238).

7.3.5.3. ATP

Bianconi et al. (239) compared the K-edge of CaATP with several model compounds. The similarity with $CaHPO_4$ suggested similar bonding. For a further discussion compare Chapter 11, Section 5.2.

7.3.6. Manganese

7.3.6.1. Chloroplasts

The photooxidation of H_2O to O_2 in chloroplasts is known to require manganese (240, 241), yet the chemical and structural basis of this catalysis is not well understood. Oxygen evolution is known from kinetic data to involve at least five intermediate states, S_0–S_4 (242). A complex set of EPR signals associated with the S_2 state (243–245) is strong evidence that these S-state transitions involve the 4 Mn ions that are present in the oxygen-evolving complex (OEC) of photosystem II (PSII).

The earliest XAS studies by Klein's group involved whole chloroplast preparations—a heroic effort given the low, submillimolar concentrations of Mn involved. The experiments were further complicated by the presence of 2 pools of Mn, a "loosely bound" pool essential for O_2 evolution, and a second pool that remains after treatment with Tris buffer and osmotic shock. The original edge difference spectrum (246), presumed to represent just the active Mn fraction, was relatively broad and at a position suggestive of a mixture of Mn(II) and Mn(III). The EXAFS difference spectrum (247) yielded a complex Fourier transform with large peaks out to 5 Å, some of which are clearly noise artifacts. Still, there is a major feature that is consistent with a Mn–Mn or Mn–Fe interaction at about 2.72 Å. The results were interpreted in terms of an oxo-bridged dinuclear Mn site, but many other interpretations are possible.

Subsequent experiments by Goodin et al. (248) used PSII particle preparations to obtain a more concentrated preparation without spurious Mn. The edge position was observed to shift from 6549.2 to 6551.7 eV upon conversion from the resting S_1 state to the EPR-active S_2 state. EXAFS of the PSII complex was interpreted in terms of a binuclear species with a metal—metal distance of 2.7 Å, as well as Mn—N,O interactions at 1.75 and 1.98 Å (249).

More recently, the Klein group has reported (250) an edge shift to lower energy (6550.2 eV) for a sample treated with hydroxylamine and subsequently illuminated at 195 K, a hydroxylamine-induced S_0 state. This was compared to positions of 6551.4 eV for S_1 and 6552.4 eV for both S_2 and S_3. The relative shifts are interesting, although some change in calibration seems to have occurred between this report and previous work. By averaging two years worth of spectra obtained on various PSII preparations, they have also found a new feature, which may represent a second Mn-Mn interaction at 3.3 Å (251).

7.3.7. Miscellaneous

7.3.7.1. Potassium

Huang et al. compared the absorption edge of frog blood cell potassium with those of aqueous K^+ and other model compounds (252). Although the authors conclude that the edge is different from simply aquated K^+, this may be an artifact of a sloping base line. If the background due to the water absorption were subtracted, most of the reported spectral differences would disappear.

The EXAFS of potassium–valinomycin has also been recorded (253), and an average K—O distance of 2.79 ± 0.02 Å was found.

7.3.7.2. Vanadium

Certain species of the invertebrate tunicate or "sea squirt" concentrate vanadium to an astonishing degree in blood cells called vanadocytes. Within the vanadocytes are vacuoles known as vanadophores, which were claimed to contain greater than molar concentrations of V(III) and sulfuric acid (254). The finding of high sulfuric acid concentrations now appears to be an artifact, however, and a new study indicates a normal intracellular pH (255). Characterization of the chemical form of vanadium in intact vanadocytes has been impeded because the light green blood turns into a red–brown "Henze solution" upon exposure to air. Upon standing this hemolysate eventually turns blue–green.

Tullius et al. compared the vanadium edges of intact cells with Henze solution and the blue hemolysate (256). Surprisingly, only 10% VO^{2+} was found, which would have yielded a strong pre-edge bound state transition. The EXAFS of intact vanadocytes as well as Henze solution was consistent with six oxygen donor ligands with an average V—O distance of 1.99 Å.

Vanadium-containing nitrogenases have recently been obtained from *Azotobacter chroococcum* (259) and *Azotobacter vinelandii* (258). The presence of a V,Fe,S cluster is confirmed, with a V—Fe distance of about 2.8 Å (259, 260).

7.3.7.3. Nickel

Nickel is present in the enzymes urease, hydrogenase, CO dehydrogenase, S-methyl reductase, and in the small molecule called *F*-430, the prosthetic group of S-methyl reductase (261). Hasnain and Piggott have reported spectra for jackbean urease which resemble a model compound with nitrogen and oxygen ligands in an octahedral environment (262). Lindahl et al. presented the EXAFS of the F_{420}-reducing hydrogenase from *Methanobacterium thermoautotrophicum* (263). They derived an environment with about 3 S ligands at 2.25 Å with a large σ of 0.09 Å. Scott et al. reported about 4 Ni—S interactions at 2.20 Å in *Desulfovibrio gigas* hydrogenase (264). Albracht et al. have recently reported evidence from ^{33}S hyperfine splittings in the Ni EPR for perhaps 2 S ligands (265). No N hyperfine splittings were observed, so the remaining ligands are still unknown.

The Ni EXAFS spectrum of isolated *F*-430 was compared to that of intact S-methyl reductase by Scott et al. A 0.1–0.2 Å increase in the Ni-N distance was observed, which was interpreted as the "planarization" of an initially ruffled corphine ligand upon axial coordination of *F*-430 to protein-derived ligands (266).

Cramer et al. have compared the spectra of CO dehydrogenase with the other Ni enzymes, concluding that it contains a unique Ni site (267). The edge features suggest a square pyramidal or distorted square planar environment, while the EXAFS is consistent with a mix of Ni—S bonds at 2.20 Å and Ni—N,O bonds at about 2.0 Å.

7.3.8. Exogenous Metals

Apart from interest in naturally occurring inorganic species, another biologically related application of x-ray absorption spectroscopy is the study of the effects of exogenous metals. These are often of therapeutic interest, and representative examples are platinum antitumor drugs and gold antiarthritics.

7.3.8.1. *Platinum*

Cis-Pt(NH$_3$)$_2$Cl$_2$ is an anticancer drug that inhibits DNA replication (268). A proposed mechanism involves formation of a dimer aquation product before interaction with DNA (269,270). Teo et al. (271) have examined the interactions of *cis*- and *trans*-dichlorodiamine—platinum (II) with calf thymus DNA, using the platinum L_1-edge. No evidence for a Pt—Pt interaction was found, and a planar coordination with four oxygen nitrogen ligands was deduced.

Hitchcock et al. also addressed this issue (272), but starting with the dimer aquation product [(NH$_3$)$_2$Pt(OH)$_2$Pt(NH$_3$)$_2$]$^{2+}$. Using the platinum L_3-edge, they show that a Pt—Pt interaction exists after the platinum dimer reacts with DNA. Thus, the question of drug interaction with DNA must involve the kinetics and equilibria of dimer formation. However, for the dilute platinum concentrations involved in therapeutic applications, it seems unlikely that a significant dimer concentration exists.

7.3.8.2. *Gold*

Gold-based drugs are important in the treatment of arthritis (273), and gold EXAFS has been used to study the structure of the drugs themselves as well as the fate of the gold once inside a living organism. Mazid et al. (274) studied the L_3-edge of disodium thiomalato-*S*-gold(I) (Mycrosin) and thioglucopyranosyl-*S*-gold(I) (Solganol). The authors concluded that the drugs are polymeric with AuS$_2$ coordination and a longish average bond length of 2.37 Å.

Elder et al. (275) used édge and EXAFS studies to examine gold in model compounds and aurosomes, the lysosomal bodies in the kidneys of rats where gold is found to concentrate (276). The edge structure of the models was found to be particularly useful, since a sharp spike characteristic of a $2p \rightarrow 5d$ transition is observed in gold (III) and gold (II) compounds, but not in gold (I) or gold (0), which have a filled $5d$ shell.

Examination of aurosomes induced by chronic or single dose administration revealed the presence of gold (I), even if the gold was administered as gold (III). Curve fitting the EXAFS led to prediction of a AuS$_2$ environment with an average bond length of 2.30 Å. The same 2.30-Å bond length was found for Mychrosin and several other Au—S compounds. This value is significantly shorter than the previously reported value (274). The source of this discrepancy may be the simulation procedure used by Mazid et al. (274), which tends to give longer distances than curve-fitting based on model compounds.

7.4. FUTURE PROSPECTS

7.4.1. New X-Ray Sources

Electron storage rings designed and dedicated for the production of synchrotron radiation are now available at the SRS, Daresbury, Great Britain, the NSLS, Upton, USA, and the Photon Factory, Tsukuba, Japan. These new rings, as well as dedicated operation of existing sources, such as SSRL, CHESS, LURE, and DESY have improved the quality of biological research for several reasons. The increased intensity and beam stability of dedicated sources, especially from insertion devices (277), will lead to better EXAFS spectra and a wider range of data. Just as important, the greater availability and dependability of beam time allows for more controls to be done in biological experiments, and it is no longer justifiable to report incomplete or unreproduced results with the excuse that insufficient beamtime was available.

Further improvements in brightness appear possible if higher energy rings such as PEP can be used for synchrotron radiation. The fundamental energy of an undulator on this ring can be tuned all the way to 20 keV (278).

7.4.2. Improved Detection Systems

Many biological samples are difficult to run at millimolar concentrations, and the sensitivity limits of filtered fluorescence detection systems are close to being reached. With simple filters and Soller slits, it is not possible to eliminate all of the scattered radiation and filter fluorescence background. An additional level of background discrimination can be achieved electronically using high resolution solid-state detectors, but a single detector is usually count-rate limited. Warburton considered the relative merits of energy-dispersive detector (EDD) arrays, filter/slit combinations, and barrel monochromator designs, and concluded that EDD arrays were superior in the range of 3 mM to 30 μM (279).

A prototype mercuric iodide detector for such an EDD area has been demonstrated (280), and a 13-element intrinsic Ge detector has also been tested (11). Mercuric iodide has the advantage that it can be used with thermoelectric cooling, whereas Ge technology requires liquid nitrogen temperatures. As a dedicated instrument in a national facility, one can envisage a computer-controlled area of 100 elements of either type designed to handle the high count rates of the next generation beam lines.

It also appears possible to improve on the early barrel monochromator designs (11) by replacing crystals with multilayer reflectors (281). Such a device would provide scatter rejection without filter fluorescence, while still accepting a reasonable solid angle. If these detector improvements are combined with a new generation of storage rings, EXAFS experiments on samples in the 1–100 μM regime should eventually become feasible.

7.4.3. Kinetics

With conventional techniques, times on the order of minutes are required to obtain an EXAFS spectrum. This has precluded EXAFS analysis of unstable species or reaction kinetics. Several approaches are available to reach the millisecond time scale. For concentrated samples, dispersive mode EXAFS looks promising (282–284). In this experiment a broad band of x-ray energies is passed through the sample, and a bent crystal disperses the transmitted x-rays at different angles. A multichannel position sensitive detector is then used to collect the entire EXAFS spectrum at the same time. Data collection times on the order of milliseconds are feasible by this method.

An alternative procedure for studying time-dependent samples is to synchronize a pulsed laser with the x-ray bursts occurring as the electron bundles pass in front of the beam line (285). Mills and co-workers have used such a procedure to examine photolyzed myoglobin–CO complexes (285).

One technique that might be more viable for biological systems is freeze-quench EXAFS (210). In this procedure the desired sample is shot into liquid isopentane shortly after generation. Typical mixing and freezing times are on the order of a couple milliseconds. If the sample is stable at low temperatures, the spectrum can be collected by transmission or fluorescence techniques on the usual time scale.

Finally, with a sufficiently intense source, it might be possible to obtain a fluorescence excitation spectrum in a short time period by using a rapidly rotating monochromator crystal and a synchronized detection system (286). As x-ray source technology improves, it is becoming clear that rapid EXAFS experiments are detector-limited rather than photon limited.

7.4.4. Oriented Samples

The spectra of single crystals or oriented membranes contain more information than the spectra of nonoriented samples (287). Most synchrotron radiation sources are linearly polarized, and the EXAFS obtained using polarized radiation is sensitive to the projection of the absorber–backscatterer axis along the E vector. In principle, therefore, it is possible to use polarized single-crystal EXAFS to deduce the geometry and orientation of metalloprotein active sites within the unit cell. The methodology should be quite similar to the techniques already developed for single-crystal optical or EPR studies. Representative applications have already been discussed (110, 195).

Single-crystal studies of absorption edges are also of value in understanding the nature of the transitions that are observed (27). For example, investigation of the orientation dependence of a feature in the $[CuCl_4]^{2-}$ edge resulted in an electric quadrupole assignment (288). Further work along these lines will help clarify the complexities of metalloprotein edges.

7.4.5. Merger with XANES

As XANES calculations become more readily available, it appears possible that they may be combined with EXAFS data to put better limits on coordination numbers and to predict geometries. Ideally, every EXAFS structural result should be checked for compatibility with the XANES region. It is conceivable that EXAFS bond lengths would be used as constraints in XANES calculations designed to yield bond angles and coordination numbers.

7.4.6. Circular Dichroism, Magnetic Circular Dichroism, and Magnetic Dichroism

Circular dichroism (CD) and magnetic circular dichroism (MCD) in the x-ray region will soon be possible. Helical wigglers may be one source of circularly polarized x-rays, while crystals operating under the Bormann effect are another. Metalloenzyme active sites are intrinsically chiral, but it remains to be seen how large the CD effects are. Magnetic dichroism has recently been demonstrated (289, 290). There might be a wealth of information in these unexplored areas.

7.4.7. Improved Documentation

An area that will ultimately benefit the EXAFS community as a whole involves improved communication of experimental results. It is gradually becoming standard procedure to include EXAFS spectra as supplementary material in publications, and editors should be encouraged to require this. Furthermore, an EXAFS program library and database have been established at SRS (291). Once the database becomes an accepted repository for published EXAFS spectra, it will be much easier to resolve disputes within the community about experimental artifacts and reproducibility.

7.4.8. Soft X-Ray Applications

Biochemical applications of the 2–3-keV region, which contains the K-edges of phosphorus (2149 eV), sulfur (2472 eV), and chlorine (2822 eV), as well as the molybdenum L-edges, are now being developed. Although biochemical samples rarely contain just a single species of P, S, or Cl, there are often distinctive XANES features which can be used as fingerprints. In such applications, the x-ray absorption analysis using resonance features is similar to what would be done with IR bands or NMR peaks.

Hedman et al. have reported the use of undulator synchrotron radiation for this spectral region (292). Although the flux available is actually less than from vacuum beam lines, running the monochromator in a helium atmosphere provides a more stable I_0 for dilute metalloenzyme spectroscopy. They reported representative spectra for sulfur compounds as well as whole living vanadocyte blood cells from tunicates.

Mo L-edges contain splittings which can be used to deduce the symmetry and strength of the molybdenum ligand field (293). A small splitting of 1.6 eV has been observed in the Mo L-edges of nitrogenase from wild-type as well as nif V⁻ mutant sources (294). In contrast, a large splitting of 2.3 eV is observed in the L_3 edge of oxidized desulfo xanthine oxidase, compared to 3.0 eV for oxidized sulfite oxidase (295). This information can be used to exclude tetrahedral Mo sites for the latter two enzymes. Combining EXAFS coordination numbers and distances with L-edge-derived d-orbital splittings should eventually allow calculation of bond angles within the first coordination sphere.

The next step for biochemical x-ray absorption is to enter the vacuum x-ray region, in order to use the L-edges of the 3-d transition elements as well as the K edges of C,N, and O. Although this region has been used primarily for surface science studies, moderate depth penetration can be achieved using fluorescence instead of electron yield detection. Although problems with the radiation and vacuum stability of samples will have to be addressed, the wealth of high resolution structure in this region will make the struggle worthwhile.

7.5. SUMMARY

X-ray absorption spectrocopy has already been applied to many of the outstanding problems in bioinorganic chemistry, but the field is in no imminent danger of becoming depleted. As the understanding of these systems improves, the questions posed about them become more precise. For example, the original EXAFS studies of nitrogenase revealed the presence of a molybdenum, iron, sulfur cluster. However, rather than solving all structural questions, this result prompted single-crystal studies designed to probe the cluster bond angles and orientation within the protein. It also led to attempts to observe changes in the cluster dimensions after various types of reactions. The same pattern of initial discovery, followed by new questions requiring more difficult experiments, has emerged in virtually every successful biochemical application of x-ray absorption spectroscopy. The growing sophistication of bioinorganic chemistry, along with the increasing power of the EXAFS technique, will keep the two fields profitably intertwined for many years to come.

Note added in proof: A derivative of *A. vinelandii* ferredoxin I which has the Fe_4S_4 cluster removed was found to have a 3-Fe cluster similar in dimensions to those from *D. gigas* ferredoxin II and aconitase, rather than the x-ray crystal structure (296).

The hemoglobin controversy persists. According to Galloway, "early proponents of EXAFS rashly made unfounded claims of its superiority over crystallographic analysis of haemoglobin . . . and got a bloody nose for their pains" (297). Although he kindly concedes that "EXAFS has been rehabilitated," Galloway questions whether XANES work such as recent Mo-CO analyses (298) will also require "rescue operations." Shulman has responded to these comments (299) by pointing out that the disagreement was not about the EXAFS distances but about the validity of triangulation.

New Fe EXAFS data have been reported for FeMo-co by Arber et al. (300). In contrast with previous work (57), no distinct shell of oxygen or nitrogen ligands was observed. Furthermore, evidence for Fe-Fe or Fe-Mo distances at about 3.6 Å was seen in the new spectra, suggesting similarities of FeMo-co with prismane-type structures.

REFERENCES

1. P. Eisenberger and B. M. Kincaid, *Science*, **200**, 1441 (1978).

2. R. G. Shulman, P. M. Eisenberger, and B. M. Kincaid, *Ann. Rev. Biophys. Bioeng.*, **1**, 559 (1978).

3. S. P. Cramer and K. O. Hodgson, *Prog. Inorg. Chem.*, **25**, 1 (1979).

4. L. Powers, *Biochim. Biophys. Acta*, **683**, 1 (1982).

5. R. A. Scott, in *The Biological Chemistry of Iron*, H. B. Dunford, D. H. Dolphin, K. N. Raymond, and L. C. Sieker (Eds.), Reidel, Boston, 1982, p. 475.

6. S. I. Chan and R. C. Gamble, *Methods Enzymol.*, **54**, 323 (1978).

7. B. K. Teo, *Acc. Chem. Res.*, **13**, 412 (1980).

8. J. Bordas, in *Uses of Synchrotron Radiation in Biology*, H. B. Stuhrman (Ed.), Academic Press, New York 1982, p. 107.

9. B. K. Teo and D. C. Joy (Eds.), *EXAFS Spectroscopy*, Plenum, New York, 1981.

10. J. Jaklevic, J. A. Kirby, M. P. Klein, A. S. Robertson, G. S. Brown, and P. Eisenberger, *Solid State Commun.*, **23**, 679 (1977).

11. S. P. Cramer, O. Tansch, M. Yocum, and G. N. George, *Abstracts of the 5th Synchrotron Radiation Instrumentation Conference*, Madison, Wisconsin (1987).

12. J. B. Hastings, P. Eisenberger, B. Lengeler, and M. L. Perlman, *Phys. Rev. Lett.*, **43**, 1807 (1979).

13. M. Marcus, L. S. Powers, A. R. Storm, B. M. Kincaid, and B. Chance, *Rev. Sci. Instrum.*, **51**, 1023 (1980).

14. S. P. Cramer and R. A. Scott, *Rev. Sci. Instrum.*, **52,** 395 (1981).

15. E. A. Stern and S. M. Heald, *Rev. Sci. Instrum.*, **50,** 1579 (1979).

16. A. Bass and H. Broida (Eds.), *Formation and Trapping of Free Radicals*, Academic, New York, 1960.

17. B. Chance, P. Angiolillo, E. K. Yang, and L. Powers, *FEBS Lett.*, **112,** 178 (1980).

18. L. Powers, B. Chance, Y. Ching, and P. Angiolillo, *Biophys. J.*, **34,** 465 (1981).

19. G. W. Brudvig, D. F. Bocian, R. C. Gamble, and S. I. Chan, *Biochim. Biophys. Acta*, **624,** 78 (1980).

20. Z. R. Korzun, K. Moffat, K. Frank, and M. A. Cusanovich, *Biochemistry*, **21,** 2253 (1982).

21. R. A. Scott, personnel communication.

22. A. L. Roe, K. O. Hodgson, R. C. Reem, E. I. Solomon, and J. W. Whittaker, Stanford Synchrotron Radiation Laboratory Annual Activity Report, 1986, Proposal 932 B.

23. H. H. Tawad and D. E. Watt, *Int. J. Radiat. Biol. Relat. Stud. Phys., Chem. Med.*, **50,** 665 (1986).

24. A. Bianconi, L. Incoccia, and S. Stipcich (Eds.), *EXAFS and Near Edge Structure*, Springer-Verlag, Berlin, (1983).

25. R. G. Shulman, Y. Yafet, P. Eisenberger, and W. E. Blumberg, *Proc. Natl. Acad. Sci. USA*, **73,** 1384 (1976).

26. F. W. Kutzler, C. R. Natoli, D. K. Misemer, S. Doniach, and K. O. Hodgson, *J. Chem. Phys.*, **73,** 3274 (1980).

27. F. W. Kutzler, R. A. Scott, J. M. Berg, K. O. Hodgson, S. Doniach, S. P. Cramer, and C. H. Chang, *J. Am. Chem. Soc.*, **103,** 6083 (1981).

28. A. Bianconi, in *EXAFS and Near Edge Structure*, A. Bianconi, L. Incoccia, and S. Stipcich (Eds.) Springer-Verlag, Berlin, 1983, p. 118.

29. J. B. Pendry, in *EXAFS and Near Edge Structure*, A. Bianconi, L. Incoccia, and S. Stipcich (Eds.), Springer-Verlag, Berlin, 1983, p. 4.

30. D. D. Vredensky, D. K. Saldin, and J. B. Pendry, *Comput. Phys. Commun.* **40,** 421 (1986).

31. E. A. Stern, F. W. Lytle, and D. E. Sayers, *Phys. Rev. B.*, **11,** 4836 (1975).

32. P. Eisenberger, R. G. Shulman, B. M. Kincaid, G. S. Brown, and S. Ogawa, *Nature (London)*, **274,** 30 (1978).

33. G. Martens, P. Rabe, N. Schwentzer, and A. Werner, *Phys. Rev. B*, **17,** 1481 (1978).

34. S. P. Cramer, K. O. Hodgson, E. I. Stiefel, and W. E. Newton, *J. Am. Chem. Soc.*, **100,** 2748 (1978).

35. P. H. Citrin, P. Eisenberger, and B. M. Kincaid, *Phys. Rev. Lett.*, **36,** 1346 (1976).

36. B. K. Teo and P. A. Lee, *J. Am. Chem. Soc.*, **101,** 2815 (1979).

37. P. Eisenberger and G. S. Brown, *Solid State Commun.*, **29,** 481 (1979).

38. P. A. Lee and J. B. Pendry, *Phys. Rev. B*, **11**, 2795 (1975).
39. B. K. Teo, *J. Am. Chem. Soc.*, **103**, 3990 (1981).
40. P. Eisenberger and B. Lengeler, *Phys. Rev. B*, **22**, 3551 (1980).
41. S. P. Cramer, K. S. Liang, A. J. Jacobson, C. H. Chang, and R. R. Chianelli, *Inorg. Chem.*, **23**, 1215 (1984).
42. S. P. Cramer, P. K. Eidem, M. T. Paffett, J. R. Winkler, Z. Dori, and H. B. Gray, *J. Am. Chem. Soc.*, **105**, 799 (1983).
43. K. D. Watenpaugh, L. C. Sieker, J. R. Herriott, and L. H. Jensen, *Cold Spring Harbor, Symp. Quant. Biol.*, **36**, 359 (1971).
44. K. D. Watenpaugh, L. C. Sieker, J. R. Herriott, and L. H. Jensen, *Acta Crystallogr. Sect. B*, **29**, 943 (1973).
45. B. L. Valle and R. J. P. Williams, *Proc. Natl. Acad. Sci. USA*, **59**, 498 (1968).
46. R. G. Shulman, P. Eisenberger, W. Blumberg, and N. A. Stombaugh, *Proc. Natl. Acad. Sci. USA*, **72**, 4003 (1975).
47. R. G. Shulman, P. Eisenberger, B. K. Teo, B. M. Kincaid, and G. S. Brown, *J. Mol. Biol.*, **124**, 305 (1978).
48. D. E. Sayers, E. A. Stern, and J. R. Herriott, *J. Chem. Phys.*, **64**, 427 (1976).
49. B. Bunker and E. A. Stern, *Biophys. J.*, **19**, 253 (1977).
50. K. D. Watenpaugh, L. C. Sieker, and L. H. Jensen, *J. Mol. Biol.*, **138**, 615 (1980).
51. B. K. Teo, R. G. Shulman, G. S. Brown, and A. E. Meixner, *J. Am. Chem. Soc.*, **101**, 5624 (1979).
52. M. R. Antonio, B. A. Averill, I. Moura, J. J. G. Moura, W. H. Orme-Johnson, B. K. Teo, and A. V. Xavier, *J. Biol. Chem.*, **257**, 6646 (1982).
53. D. Ghosh, W. Furey, Jr., S. O'Donnell, and C. D. Stout, *J. Biol. Chem.*, **256**, 4185 (1981).
54. H. Beinert, M. H. Emptage, J.-L. Dreyer, R. A. Scott, J. E. Hahn, K. O. Hodgson, and A. J. Thomson, *Proc. Natl. Acad. Sci. USA*, **80**, 393 (1983).
55. B. M. Hoffman, R. A. Venters, J. E. Roberts, M. Nelson, and W. H. Orme-Johnson, *J. Am. Chem. Soc.*, **104**, 4711 (1982).
56. B. H. Huynh, E. Münck, and W. H. Orme-Johnson, *Biochim. Biophys. Acta*, **527**, 192 (1979).
57. M. R. Antonio, B. K. Teo, W. H. Orme-Johnson, M. J. Nelson, S. E. Groh, P. A. Lindahl, S. M. Kauzlarich, and B. A. Averill, *J. Am. Chem. Soc.*, **104**, 4703 (1982).
58. B. K. Teo, M. R. Antonio, R. H. Tieckelman, H. C. Silvis, and B. A. Averill, *J. Am. Chem. Soc.*, **104**, 6126 (1982).
59. S. P. Cramer, Ph.D. Thesis, Stanford University, 1978.
60. P. Eisenberger, R. G. Shulman, G. S. Brown, and S. Ogawa, *Proc. Natl. Acad. Sci. USA*, **73**, 491 (1976).
61. M. F. Perutz, J. E. Ladner, S. R. Simon, and C. Ho, *Biochemistry*, **13**, 2163 (1974).

62. M. F. Perutz, *Nature (London)*, **228**, 726 (1970).

63. M. F. Perutz, S. S. Hasnain, P. J. Duke, J. L. Sessler, and J. E. Hahn, *Nature (London)*, **295**, 535 (1982).

64. R. E. White and M. J. Coon, *Ann. Rev. Biochem.*, **49**, 315 (1980).

65. I. C. Gunsalus and S. G. Sligar, *Adv. Enzymol.*, **47**, 1 (1979).

66. J. T. Groves, *Ann. N.Y. Acad. Sci.*, **471**, 99 (1986).

67. S. P. Cramer, J. H. Dawson, L. P. Hager, and K. O. Hodgson, *J. Am. Chem. Soc.*, **100**, 7282 (1978).

68. J. E. Hahn, K. O. Hodgson, L. A. Andersson, and J. H. Dawson, *J. Biol. Chem.*, **257**, 10934 (1982).

69. T. L. Poulos, B. C. Finzel, and A. J. Howard, *Biochemistry*, **25**, 5314 (1986).

70. J. H. Dawson, L. S. Kau, J. E. Penner-Hahn, M. Sono, K. S. Eble, G. S. Bruce, L. P. Hager, and K. O. Hodgson, *J. Am. Chem. Soc.*, **108**, 8114 (1986).

71. A. Labhardt and C. Yuen, *Nature (London)*, **277**, 150 (1979).

72. Z. R. Korzun, K. Moffat, K. Frank, and M. A. Cusanovich, *Biophys. J.*, **33**, 305a (1981).

73. M. R. Mauk and A. W. Girotti, *Biochemistry*, **13**, 1757 (1974).

74. B. Chance, L. Powers, J. Ching, T. Poulos, G. R. Schonbaum, I. Jamazaki, and K. G. Paul, *Arch. Biochem. Biophys.*, **235**, 596 (1984).

75. J. E. Hahn, T. J. McMurry, M. Renner, L. Latos-Grazynski, K. S. Eble, I. M. Davis, A. L. Balch, J. T. Groves, J. H. Dawson, and K. O. Hodgson, *J. Biol. Chem.*, **258**, 12761-4 (1983).

76. J. E. Penner-Hahn, K. S. Eble, T. J. McMurry, M. Renner, A. L. Balch, J. T. Grovez, J. H. Dawson, and K. O. Hodgson, *J. Am. Chem. Soc.*, **108**, 7819 (1986).

77. R. E. Stenkamp and L. H. Jensen, *Adv. Inorg. Biochem.*, **1**, 219 (1979).

78. W. A. Hendrickson, *Nav. Res. Rev.*, **31**, 1 (1978).

79. W. T. Elam, E. A. Stern, J. D. McCallum, J. Sanders-Loehr, *J. Am. Chem. Soc.*, **105**, 1919 (1983).

80. W. A. Hendrickson, M.-S. Co., J. L. Smith, K. O. Hodgson, and G. L. Klippenstein, *Proc. Natl. Acad. Sci. USA*, **79**, 6255 (1982).

81. W. T. Elam, E. A. Stern, J. D. McCallum, and J. Sanders-Loehr, *J. Am. Chem. Soc.*, **105**, 1919 (1983).

82. R. C. Scarrow, M. J. Maroney, S. M. Palmer, L. Que, Jr., S. P. Salowe, and J. Stubbe, *J. Am. Chem. Soc.*, **108**, 6832 (1986).

83. S. M. Kauzlarich, B. K. Teo, T. Zirimo, S. Burman, J. C. Davis, and B. A. Averill, *Inorg. Chem.*, **25**, 2781 (1986).

84. H. N. Munro and M. C. Linder, *Physiol. Rev.*, **58**, 317 (1978).

85. S. M. Heald, E. A. Stern, B. A. Bunker, E. M. Holt, and S. L. Holt, *J. Am. Chem. Soc.*, **101**, 67 (1979).

86. E. C. Theil, D. E. Sayers, and M. A. Brown, *J. Biol. Chem.*, **254**, 8132 (1979).

87. D. E. Sayers, E. C. Theil, and F. J. Rennick, *J. Biol. Chem.*, **258**, 14076 (1983).

88. L. Que, Jr., *Coord. Chem. Rev.*, **50,** 73 (1983).

89. W. E. Blumberg and J. Peisach, *Ann. N.Y. Acad. Sci.*, **222,** 539 (1973).

90. M. Fujiwara and M. Nozaki, *Biochim. Biophys. Acta*, **327,** 306 (1973).

91. L. Que, Jr. and R. M. Epstein, *Biochemistry*, **20,** 2545 (1981).

92. R. H. Felton, W. L. Barrow, S. W. May, A. L. Sowell, S. Joel, G. Bunker, and E. A. Stern, *J. Am. Chem. Soc.*, **104,** 6132 (1982).

93. D. J. Schneider, A. L. Roe, R. J. Mayer, and L. Que, Jr., *J. Biol. Chem.*, **259,** 9699 (1984).

94. A. L. Roe, D. J. Schneider, R. J. Mayer, J. W. Pyrz, J. Widom, and L. Que, Jr., *J. Am. Chem. Soc.*, **106,** 1676 (1984).

95. P. F. Lindley, S. Bailey, R. W. Evans, R. C. Garratt, B. Gorinsky, S. S. Hasnain, and R. Sara, *Biochem. Soc. Trans.*, **14,** 542 (1986).

96. G. Feher and M. Y. Okamura, in *The Photosynthetic Bacteria*, R. K. Clayton and W. R. Sistram (Eds.), Plenum, New York, 1978, p. 349.

97. R. E. Blankenship and W. W. Parson, *Ann. Rev. Biochem.*, **47,** 635 (1978).

98. R. E. Blankenship and W. W. Parson, *Biochem. Biophys. Acta*, **545,** 424 (1979).

99. P. Eisenberger, M. Y. Okamura, and G. Feher, *Biophys. J.*, **37,** 523 (1982).

100. G. Bunker, E. A. Stern, R. E. Blankenship, and W. W. Parson, *Biophys. J.*, **37,** 539 (1982).

101. J. Diesendorfer, O. Epp, K. Miki, R. Huber, and H. Michel, *Nature (London)*, **318,** 618 (1985).

102. R. Malkin and B. G. Malmstrom, in *Advances in Enzymology*, vol. 33, F. F. Nord (Ed.), Wiley-Interscience, New York, 1970, p. 177.

103. E. I. Solomon, J. W. Hare, and H. B. Gray, *Proc. Natl. Acad. Sci. USA*, **73,** 1389 (1978).

104. T. D. Tullius, P. Frank, and K. O. Hodgson, *Proc. Natl. Acad. Sci. USA*, **75,** 4069 (1978).

105. T. D. Tullius, Ph.D. thesis, Stanford University, 1979.

106. J. Peisach, L. Powers, W. E. Blumberg, and B. Chance, *Biophys. J.*, **38,** 277 (1981).

107. W. E. Blumberg and L. Powers, *Fed. Abstr.*, No. 3462, 862 (1982).

108. H. C. Freeman, in *Coordination Chemistry -21*, J. P. Laurent (Ed.), Pergamon Press, Oxford, 1981, p. 29.

109. E. T. Adman and L. H. Jensen, *Israel J. Chem.*, **21,** 8 (1981).

110. R. A. Scott, J. E. Hahn, S. Doniach, H. C. Freeman, and K. O. Hodgson, *J. Am. Chem. Soc.*, **104,** 5364 (1982).

111. C. M. Groeneveld, M. C. Feiters, S. S. Hasnain, J. van Rijn, J. Reedijk, and G. W. Canter, *Biochim. Biophys. Acta*, **873,** 214 (1986).

112. M. Samo and T. Matsubara, *Chem. Lett.*, 2121 (1984).

113. S. K. Chapman, W. H. Orme-Johnson, J. McGinnie, J. D. Sinclair-Day, A. G. Sykes, P. I. Ohlsson, and K. G. Paul, *J. Chem. Soc., Dalton Trans.*, **10,** 2063 (1986).

114. K. E. van Holde and K. I. Miller, *Q. Rev. Biophys.*, **15**, 1 (1982).

115. T. K. Eccles, Ph.D. thesis, Stanford University, 1977.

†16. J. M. Brown, Ph.D. thesis, Princeton University, 1978.

117. J. M. Brown, L. Powers, B. Kincaid, J. A. Larrabee, and T. G. Spiro, *J. Am. Chem. Soc.*, **102**, 4210 (1980).

118. M.-S. Co, R. A. Scott, and K. O. Hodgson, *J. Am. Chem. Soc.*, **103**, 986 (1981).

119. M.-S. Co and K. O. Hodgson, *J. Am. Chem. Soc.*, **103**, 3200 (1981).

120. G. L. Woolery, L. Powers, M. Winkler, E. I. Solomon, and T. G. Spiro, *J. Am. Chem. Soc.*, **106**, 86 (1984).

121. R. Torensma and J. C. Phillips, *Biochem. J.*, **209**, 373 (1983).

122. A. Azzi, *Biochim. Biophys. Acta*, **594**, 231 (1980).

123. M. Denis, *Biochimie*, **68**, 459 (1986).

124. V. W. Hu, S. I. Chan, and G. S. Brown, *Proc. Natl. Acad. Sci. USA*, **74**, 3821 (1977).

125. V. W. Hu, S. I. Chan, and G. S. Brown, *FEBS Lett.*, **84**, 287 (1977).

126. L. Powers, W. E. Blumberg, B. Chance, C. H. Barlow, J. S. Leigh, J. Smith, T. Yonetani, S. Vik, and J. Peisach, *Biochim. Biophys. Acta*, **546**, 520 (1979).

127. L. Powers, B. Chance, and Y. Ching, *J. Biol. Chem.*, **262**, 3160 (1987).

128. W. Froncisz, C. P. Scholes, J. S. Hyde, Y.-H. Wei, T. E. King, R. W. Shaw, and H. Beinert, *J. Biol. Chem.*, **254**, 7482 (1979).

129. B. M. Hoffman, J. E. Roberts, B. M. Swanson, S. Speck, and E. Margoliash, *Proc. Natl. Acad. Sci. USA*, **77**, 1452 (1980).

130. R. A. Scott, S. P. Cramer, R. W. Shaw, H. Beinert, and H. B. Gray, *Proc. Natl. Acad. Sci. USA*, **78**, 664 (1981).

131. R. A. Scott, in *The Biological Chemistry of Iron*, H. B. Dunford, D. H. Dolphin, K. N. Raymond, and L. C. Sieker (Eds.), Reidel, Boston, 1982, p. 475.

132. S. P. Cramer, R. A. Scott, R. Shaw, and H. Beinert, in *Symposium on Inorganic and Biochemical Perspectives in Copper Coordination Chemistry*, SUNY, Albany, NY Abstract No. 4 (1982).

133. L. Powers, B. Chance, Y. Ching, B. Muhoberac, S. T. Weintraub, and D. C. Wharton, *FEBS Lett.*, **138**, 245 (1982).

134. T. Yonetani, *J. Biol. Chem.*, **235**, 845 (1960).

135. C. R. Hartzell and H. Beinert, *Biochim. Biophys. Acta*, **368**, 318 (1974).

136. G. J. Steffens and G. Buse, in *Cytochrome Oxidase*, T. E. King, Y. Orii, B. Chance, and K. Okunuki (Eds.), Elsevier/North-Holland, New York, 1979, p. 79.

137. B. Reinhammer, *Inorg. Chim. Acta*, **79**, 70 (1983).

138. M. K. Johnson, D. G. Eglinton, P. E. Goodling, C. Greenwood, and A. J. Thomson, *Biochem. J.*, **193**, 699 (1981).

139. H. Beinert, R. W. Shaw, R. E. Hanson, and C. R. Hartzell, *Biochim. Biophys. Acta*, **635**, 73 (1981).

140. R. Boelens and R. Wever, *FEBS Lett.*, **116**, 223 (1980).

141. G. W. Brudvig and S. I. Chan, *FEBS Lett.*, **106,** 139 (1979).

142. R. A. Scott, J. R. Schwartz, and S. P. Cramer, *Biochemistry*, **25,** 556 (1986).

143. B. Chance and L. Powers, *Curr. Top. Bioenergy.*, **14,** 1 (1985).

144. P. M. Li, J. Gelles, S. I. Chan, R. J. Sullivan, and R. A. Scott, *Inorg. Chem.*, **26,** 2091 (1987).

145. L. S. Powers, Y. Ching, B. Chance, and D. C. Wharton, *Biophys. J.*, **51,** 300 (1987).

146. G. N. George, S. P. Cramer, T. G. Frey, and R. C. Prince, in *Advances in Membrane Biochem. and Bioenergetics.*, C. H. Kim (Ed.), Plenum, New York, 1987.

147. M. Erecinska, D. F. Wilson, and J. K. Blasie, *Biochim. Biophys. Acta*, **501,** 53 (1978).

148. M. Erecinska, D. F. Wilson, and J. K. Blasie, *Biochim. Biophys. Acta*, **501,** 63 (1978).

149. J. K. Blasie, M. Erecinska, S. Samuels, and J. S. Leigh, *Biochim. Biophys. Acta*, **501,** 33 (1978).

150. I. Fridovich, *Adv. Enzymol.*, **41,** 35 (1974).

151. S. L. Marklund, *Adv. Clin. Enzymol.*, **4,** 135 (1986).

152. W. E. Blumberg, J. Peisach, P. Eisenberger, and J. A. Fee, *Biochemistry*, **17,** 1842 (1978).

153. R. A. Scott and D. M. Dooley, *J. Am. Chem. Soc.*, **107,** 4348 (1985).

154. S. S. Hasnain, G. P. Diakun, P. F. Knowles, N. Binsted, C. D. Garner, and N. J. Blackburn, *Biochem. J.*, **221,** 545 (1984).

155. B. Reinhammar and B. G. Malmstrom, in *Copper Proteins*, T. G. Spiro (Ed.), Wiley-Interscience, New York, 1981, p. 109.

156. O. Farver and I. Pecht, in *Copper Proteins*, T. G. Spiro (Ed.), Wiley-Interscience, New York, 1981, p. 151.

157. C. D. Lubien, M. E. Winkler, T. J. Thamann, R. A. Scott, M.-S. Co, K. O. Hodgson, and E. I. Solomon, *J. Am. Chem. Soc.*, **103,** 7014 (1981).

158. J. E. Hahn, M.-S. Co, D. J. Spira, K. O. Hodgson, E. I. Solomon, *Biochim. Biophys. Res. Commun.*, **112,** 737 (1983).

159. J. E. Penner-Hahn, B. Hedman, K. O. Hodgson, D. J. Spira, and E.I. Solomon, *Biochem. Biophys. Res. Comm.* **119,** 567 (1984).

160. P. Frank, O. Farver, and I. Pecht, *J. Biol. Chem.*, **258,** 11112 (1983).

161. G. L. Woolery, L. Powers, J. Peisach, and T. G. Spiro, *Biochemistry*, **23,** 3428 (1984).

162. M. Vasak and J. H. R. Kägi, *Met. Ions Biol. Syst.*, **15,** 213 (1983).

163. D. R. Winge, B. L. Geller, and J. Garvey, *Arch. Biochem. Biophys.*, **208,** 160 (1981).

164. U. Weser, H.-J. Hartmann, A. Fretzdorff, and G.-J. Strobel, *Biochim. Biophys. Acta*, **493,** 465 (1977).

165. W. F. Furey, A. H. Robbins, L. L. Clancy, D. R. Winge, and C. D. Stout, *Science*, **234,** 704 (1985).

166. J. Bordas, M. H. J. Koch, H.-J Hartmann, and U. Weser, *FEBS Let.*, **1401**, 19 (1982).

167. J. Bordas, M. H. J. Koch, H.-J. Hartmann, and U. Weser, *Inorg. Chim. Acta*, **78**, 113 (1983).

168. J. H. Friedman, L. Powers, and J. Peisach, *Biochemistry*, **25**, 2342 (1986).

169. T. A. Smith, K. Lerch, and K. O. Hodgson, *Inorg. Chem.*, **25**, 4677 (1986).

170. G. N. George, D. R. Winge, C. D. Stout, and S. P. Cramer, *J. Inorg. Biochem.*, **27**, 213 (1980).

171. G. N. George, D. R. Winge, and S. P. Cramer, to be published.

172. I. L. Abrahams, I. Bremner, G. P. Diakun, C. D. Garner, S. S. Hasnain, I. Ross, and M. Vasak, *Biochem. J.*, **236**, 585 (1986).

173. *Molybdenum Enzymes*, T. Spiro (Ed), Wiley, New York, 1985.

174. *Molybdenum and Molybdenum-Containing Enzymes*, M. P. Coughlan (Ed.), Pergamon Press, New York, 1980.

175. *Molybdenum Chemistry of Biological Significance*, W. E. Newton and S. Otsuka (Eds.), Plenum, New York, 1980.

176. J. T. Spence, *Coord. Chem. Rev.*, **48**, 59 (1983).

177. E. I. Stiefel and S. P. Cramer, in *Molybdenum Enzymes*, T. Spiro (Ed.), Wiley, New York, 1985, p. 89.

178. P. T. Pienkos, V. K. Shah, and W. J. Brill, in *Molybdenum and Molybdenum-Containing Enzymes*, M. P. Coughlan (Ed.), Pergamon Press, New York, 1980, p. 385.

179. S. P. Cramer and E. I. Stiefel, in *Molybdenum Enzymes*, T. Spiro (Ed.), Wiley, New York, 1985, p. 411.

180. J. L. Johnson, in *Molybdenum and Molybdenum-Containing Enzymes*, M. P. Coughlan (Ed.), Pergamon Press, New York, 1980, p. 345.

181. S. P. Cramer, in *Advances in Inorganic and Bioinorganic Mechanisms*, Vol. II, A. G. Sykes (Ed.), Academic, London, 1983, pp. 259.

182. L. E. Mortenson and R. N. F. Thorneley, *Ann. Rev. Biochem.*, **48**, 387 (1975).

183. R. V. Hageman and R. H. Burris, in *Molybdenum and Molybdenum-Containing Enzymes*, M. P. Coughlan (Ed.), Pergamon Press, New York, 1980, p. 403.

184. W. H. Orme-Johnson and E. Munck, in *Molybdenum and Molybdenum-Containing Enzymes*, M. P. Coughlan (Ed.), Pergamon Press, New York, 1980, p. 427.

185. S. P. Cramer, T. K. Eccles, F. Kutzler, K. O. Hodgson, and L. E. Mortenson, *J. Am. Chem. Soc.*, **98**, 1287 (1976).

186. S. P. Cramer, K. O. Hodgson, W. O. Gillum, and L. E. Mortenson, *J. Am. Chem. Soc.*, **100**, 3398 (1978).

187. S. P. Cramer, W. O. Gillum, K. O. Hodgson, L. E. Mortenson, E. I. Stiefel, J. R. Chisnell, W. J. Brill, and V. K. Shah, *J. Am. Chem. Soc.*, **100**, 3814 (1978).

188. S. D. Conradson, K. O. Hodgson, B. K. Burgess, W. E. Newton, M. W. W. Adams, and L. E. Mortenson, Stanford Synchrotron Radiation Laboratory 1981 Activity Report, VIII-26-VIII-27.

189. T. E. Wolff, J. M. Berg, C. Warrick, K. O. Hodgson, R. H. Holm, and R. B. Frankel, *J. Am. Chem. Soc.*, **100**, 4630 (1978).

190. B. K. Teo and B. A. Averill, *Biochem. Biophys. Res. Commun.*, **88**, 1454 (1979).

191. B. K. Teo, in *EXAFS Spectroscopy: Techniques and Applications*, B. K. Teo and D. C. Joy (Eds.), Plenum, New York, 1981, p. 13.

192. J. Lu, in *Nitrogen Fixation*, W. E. Newton and W. H. Orme-Johnson (Eds.), University Park Press, Baltimore, 1980, p. 343.

193. K. R. Tsai, in *Nitrogen Fixation*, W. E. Newton and W. H. Orme-Johnson (Eds.), University Park Press, Baltimore, 1980, p. 373.

194. G. Christou, K. S. Hagen, and R. H. Holm, *J. Am. Chem. Soc.*, **104**, 1744 (1982).

195. A. M. Flank, M. Weininger, L. E. Mortenson, and S. P. Cramer, *J. Am. Chem. Soc.* **108**, 1049 (1986).

196. B. K. Burgess, S. S. Jang, C. B. Jou, J. G. Li, G. D. Friesen, W. H. Pan, E. I. Stiefel, W. E. Newton, S. D. Conradson, and K. O. Hodgson, in *Current Perspectives in Nitrogen Fixation*, A. H. Gibson and W. E. Newton (Eds.), Australian Acad. Sci., Canberra, 1981, p. 71.

197. S. D. Conradson, K. O. Hodgson, B. K. Burgess, W. E. Newton, M. W. W. Adams, and L. E. Mortenson, Stanford Synchrotron Radiation Laboratory Activity Report 82/01, Proposal 485 B.

198. M. K. Eidsness, A. M. Flank, B. E. Smith, A. C. Flood, C. D. Garner and S. P. Cramer, *J. Am. Chem. Soc.*, **108**, 2746 (1986).

199. K. V. Rajagopalan, in *Molybdenum and Molybdenum-Containing Enzymes*, M. P. Coughlan (Ed.), Pergamon Press, New York, 1980, p. 241.

200. S. P. Cramer, H. B. Gray, and K. V. Rajagopalan, *J. Am. Chem. Soc.*, **101**, 2772 (1979).

201. S. P. Cramer, R. Wahl, and K. V. Rajagopalan, *J. Am. Chem. Soc.*, **103**, 7721 (1981).

202. R. C. Bray, *Polyhedron*, **5**, 591 (1986).

203. R. C. Bray, in *Advances in Enzymology and Related Areas of Molecular Biology*, A. Meister (Ed.), Wiley, New York, 1979, p. 107.

204. R. C. Bray, in *The Enzymes*, P. D. Boyer (Ed.), Vol. XII, Academic, New York, 1975, p. 299.

205. V. Massey and D. Edmondson, *J. Biol. Chem.*, **245**, 6595 (1970).

206. T. D. Tullius, D. M. Kurtz, Jr., S. D. Conradson, and K. O. Hodgson, *J. Am. Chem. Soc.*, **101**, 2776 (1979).

207. J. Bordas, R. C. Bray, C. D. Garner, S. Gutteridge, and S. S. Hasnain, *Biochem. J.*, **191**, 499 (1980).

208. S. P. Cramer and R. Hille, *J. Am. Chem. Soc.*, **107**, 8164 (1985).

209. R. Hille, M. K. Eidsness, and S. P. Cramer, to be published.

210. G. N. George, R. C. Bray, and S. P. Cramer, *Biochem. Soc. Trans.*, **14**, 651 (1986).

211. J. G. G. Moura, A. V. Xavier, R. Cammack, D. O. Hall, M. Bruschi, and J. LeGall, *Biochem. J.*, **173**, 419 (1978).

212. J. G. G. Moura, A. V. Xavier, M. Bruschi, J. LeGall, D. O. Hall, and R. Cammack, *Biochim. Biophys. Res. Commun.*, **72**, 782 (1976).

213. S. P. Cramer, J. G. G. Moura, A. V. Xavier, and J. LeGall, *J. Inorg. Biochem.*, **20**, 75 (1985).

214. N. A. Turner, B. Barata, R. C. Bray, J. Deistung, J. Le Gall, and J. J. G. Moura, *Biochem. J.* **243**, 755 (1987).

215. E. J. Hewitt and B. A. Notton, in *Molybdenum and Molybdenum-Containing Enzymes*, M. P. Coughlan (Ed.), Pergamon Press, New York, 1980, p. 273.

216. W. D. Howard and L. P. Solomonson, *J. Biol. Chem.*, **257**, 10243 (1982).

217. M. W. W. Adams and L. E. Mortenson, *J. Biol. Chem.*, **257**, 1791 (1982).

218. S. P. Cramer, L. P. Solomonson, M. W. W. Adams, and L. E. Mortenson, *J. Am. Chem. Soc.*, **106**, 1467 (1984).

219. G. N. George, R. C. Bray, F. F. Morpeth, and D. H. Boxer, *Biochem. J.*, **227**, 925 (1985).

220. G. N. George, D. Boxer, and S. P. Cramer, to be published.

221. L. Ljungdahl, in *Molybdenum and Molybdenum-Containing Enzymes*, M. P. Coughlan (Ed.), Pergamon Press, New York, 1980, p. 463.

222. R. K. Thauer, G. Fuchs, and K. Jungermann, in *Iron-Sulfur Proteins*, Vol. 3, W. Lovenberg (Ed.), Academic, New York, 1977, p. 121.

223. S. P. Cramer, C. L. Liu, L. E. Mortenson, J. T. Spence, S. M. Liu, I. Yamamoto, and L. G. Ljundahl, *J. Inorg. Biochem.*, **23**, 119 (1985).

224. R. C. Prince, C. L. Liu, T. V. Morgan, and L. E. Mortenson, *FEBS Lett.*, **189**, 263 (1985).

225. G. R. Jacobson and G. R. Stark, in *The Enzymes*, P. D. Boyer (Ed.), Vol. IVB 3rd ed., Academic, New York, 1973, p. 225.

226. J. C. Phillips, J. Bordas, A. M. Foote, M. H. J. Koch, and M. F. Moody, *Biochemistry*, **21**, 830 (1982).

227. J. H. R. Kagi and M. Nordberg, *Experentia Suppl.*, **34** (1979).

228. C. D. Garner, S. S. Hasnain, I. Brenner, and J. Bordas, *J. Inorg. Biochem.*, **16**, 353 (1982).

229. J. C. Phillips, R. Bauer, J. Dunbar, and J. T. Johansen, *J. Inorg. Biochem.*, **22**, 179 (1984).

230. S. S. Hasnain, E. M. Wardell, C. D. Garner, M. Schlösser, and D. Beyersmann, *Biochem. J.*, **230**, 625 (1985).

231. R. H. Kretsinger, *Ann. Rev. Biochem.*, **45**, 239 (1976).

232. S. S. Hasnain, in *EXAFS and Near Edge Structure*, A. Bianconi, L. Incoccia, and S. Stipcich (Eds.), Springer-Verlag, Berlin, 1983, p. 330.

233. L. Powers, P. Eisenberger, and J. Stamatoff, *Ann. N.Y. Acad. Sci.*, **307**, 113 (1978).

234. A. Bianconi, S. Alema, L. Castellani, P. Fasella, A. Giovanelli, S. Mobilio, and B. Oesch, *J. Mol. Biol.*, **165**, 125 (1983).

235. C. Holt, S. S. Hasnain, and D. W. L. Hukins, *Biochim. Biophys. Acta*, **719**, 299 (1982).

236. A. S. Posner and F. Betts, *Acc. Chem. Res.*, **8**, 273 (1978).

237. E. D. Eanes, L. Powers, and J. L. Costa, *Cell Calcium*, **2**, 251 (1981).

238. R. M. Miller, D. W. Hukins, S. S. Hasnain, and P. Lagarde, *Biochem. Biophys. Res. Commun.*, **99**, 102 (1981).

239. A. Bianconi, S. Doniach, and D. Lublin, *Chem. Phys. Lett.*, **59**, 121 (1978).

240. R. Radmer and G. Cheniae, in *Topics in Photosynthesis*, Vol. 2, J. Barber (Ed.), Elsevier, Amsterdam, 1977, p. 303.

241. J. Amesz, *Biochem. Biophys. Acta*, **726**, 1 (1983).

242. B. Kok, B. Forbusch, and M. McGloin, *Photochem. Photobiol.*, **11**, 457 (1970).

243. G. C. Dismukes and Y. Siderer, *Proc. Natl. Acad. Sci.* USA, **78**, 274 (1981).

244. Ö. Hansson and L. E. Andreasson, *Biochim. Biophys. Acta*, **679**, 261 (1982).

245. J. C. de Paula, W. F. Beck, and G. W. Brudwig, *J. Am. Chem. Soc.* **108**, 4002 (1986).

246. J. A. Kirby, D. B. Goodin, T. Wydrzynski, A. S. Robertson, and M. P. Klein, *J. Am. Chem. Soc.*, **103**, 5537 (1981).

247. J. A. Kirby, A. S. Robertson, J. L. Smith, A. C. Thompson, S. R. Cooper, and M. P. Klein, *J. Am. Chem. Soc.*, **103**, 5529 (1981).

248. D. B. Goodin, V. K. Yachandra, R. D. Britt, K. Sauer, and M. P. Klein, *Biochem. Biophys. Acta*, **767**, 209 (1984).

249. V. K. Yachandra, R. D. Guiles, A. McDermott, R. D. Britt, S. L. Dexheimer, K. Sauer, and M. P. Klein, *Biochem. Biophys. Acta*, **850**, 324 (1986).

250. R. D. Guiles, V. K. Yachandra, A. E. McDermott, R. D. Britt, S. L. Dexheimer, K. Sauer, and M. P. Klein, Proceedings of 13th Annual Users Conference, Stanford Synchrotron Radiation Laboratory Report 86/02, 1986, p. 29.

251. R. D. Guiles, V. K. Yachandra, A. E. McDermott, R. D. Britt, S. L. Dexheimer, K. Sauer, and M. P. Klein, Proc. *7th Int. Congr. Photosynthesis*, 1987, in press.

252. H. W. Huang, S. H. Hunter, W. K. Warburton, and S. C. Moss, *Science*, **204**, 191 (1979).

253. H. W. Huang and C. R. Williams, *Biophys. J.*, **33**, 269 (1981).

254. R. M. K. Carlson, *Proc. Natl. Acad. Sci. USA*, **72**, 2217 (1975).

255. A. L. Dingley, K. Justin, I. G. Macara, G. C. McCleod, and M. F. Roberts, *Biochim. Biophys. Acta*, **720**, 384 (1982).

256. T. D. Tullius, W. O. Gillum, R. M. K. Carlson, and K. O. Hodgson, *J. Am. Chem. Soc.*, **102**, 5670 (1980).

257. R. Robson, R. R. Eady, T. H. Richardson, R. W. Miller, M. Hawkins, and J. R. Postgate, *Nature (London)*, **322**, 388 (1986).

258. B. J. Hales, E. E. Case, J. E. Morningstar, M. F. Dzeda, and L. A. Mauterer, *Biochemistry*, **25**, 7251 (1986).

259. J. M. Arber, B. R. Dobson, R. R. Eady, P. Stevens, S. S. Hasnain, C. D. Garner, and B. E. Smith, *Nature*, **325**, 372 (1987).

260. S. P. Cramer, L. E. Mortenson, M. Eideness, R. C. Bray, G. N. George, D. Lowe, B. E. Smith, R. Thorneley, and B. Haler, Stanford Synchrotron Radiation Laboratory Activity Report, (1986) Proposal 945B.

261. A. J. Thomson, *Nature (London)*, **298**, 602 (1982).

262. S. S. Hasnain and B. Piggott, *Biochim. Biophys. Res. Commun.*, **112**, 279 (1983).

263. P. A. Lindahl, N. Kojima, R. P. Hausinger, J. A. Fox, B. K. Teo, C. T. Walsk, and W. H. Orme-Johnson, *J. Am. Chem. Soc.*, **106**, 3062 (1984).

264. R. A. Scott, S. A. Wallin, M. Czechowski, D. V. Der Vartarian, J. Le Gall, and H. D. Peck, Jr., I. Moura, *J. Am. Chem. Soc.*, **106**, 6864 (1984).

265. S. P. J. Albracht, A. Kroeger, J. W. van der Zwaan, G. Unden, R. Boecher, H. Mell, and R. D. Fontijn, *Biochim. Biophys. Acta*, **874**, 116 (1986).

266. R. A. Scott, P. L. Hartzell, R. S. Wolfe, J. Le Gall and S. P. Cramer, in *Frontiers in Bioinorganic Chemistry*, A. V. Xavier (Ed.), VCH, Weinheim, 1986, p. 20.

267. S. P. Cramer, M. K. Eidsness, W. H. Pan, T. A. Morton, S. W. Ragsdale, D. V. Der Vartanian, L. G. Ljundahl, and R. A. Scott, *Inorg. Chem.* 1987, in press.

268. B. Rosenberg, *Biochimie*, **60**, 859 (1978).

269. R. Faggiani, B. Lippert, C. J. L. Lock, and B. Rosenberg, *J. Am. Chem. Soc.*, **99**, 77 (1977).

270. B. Lippert, C. J. L. Lock, B. Rosenberg, and M. Zvagulis, *Inorg. Chem.*, **17**, 2971 (1978).

271. B. K. Teo, P. Eisenberger, J. Reed, J. K. Barton, and S. J. Lippard, *J. Am. Chem. Soc.*, **100**, 3225 (1978).

272. A. P. Hitchcock, C. J. L. Lock, and W. M. C. Pratt, *Inorg. Chim. Acta*, **66**, L45 (1982).

273. Empire Rheumatism Council, *Ann. Rheum. Dis.*, **20**, 315 (1961).

274. M. A. Mazid, M. T. Razi, P. J. Sadler, G. N. Greaves, S. J. Gurman, M. H. J. Koch, and J. C. Phillips, *J. Chem. Soc. Chem. Commun.*, 1261 (1980).

275. R. C. Elder, M. K. Eidsness, M. J. Heeg, K. G. Tepperman, C. F. Shaw III, and N. Schaeffer, in *ACS Symposium Series*, S. J. Lippard (Ed.), American Chemical Society, Washington, DC, 1983, p. 385.

276. F. N. Ghadially, *J. Rheumatol. Suppl.*, **5**, 45 (1979).

277. G. Brown, K. Halbach, J. Harris, and H. Winick, *Nucl. Instrum. Methods*, **208**, 65 (1983).

278. G. S. Brown, *Nucl. Inst. Meth.*, **A246**, 149 (1986).

279. W. K. Warburton, *Nucl. Inst. Meth.*, **A246**, (1986).

280. W. K. Warburton, J. S. Iwanczyk, A. J. Dabrowski, B. Hedman, J. Penner-Hahn, A. L. Roe, K. O. Hodgson, and A. Beyerle, *Nucl. Inst. Meth.*, **A246**, 558 (1986).

281. T. W. Barbee, Jr., *Opt. Eng*, **25**, 989 (1986).

282. A. M. Flank, A. Fontaine, A. Jucha, M. Lemonnier, and C. Williams, in *EXAFS*

and Near Edge Structure, A. Bianconi, L. Incoccia, and S. Stipcich (Eds.), Springer-Verlag, Berlin, 1983, p. 405.

283. R. P. Phizackerly, Z. U. Rek, G. B. Stephenson, S. D. Conradson, K. O. Hodgson, T. Matsushita, and H. Oyanagi, *J. Appl. Crystallogr.*, **16**, 220 (1983).

284. E. Dartyge, C. Depautex, J. M. Dubuisson, A. Fontaine, A. Jucha, P. Leboucher, and G. Tourillon, *Nucl. Inst. Meth.*, A246, 452 (1981).

285. D. E. Mills, A. Lewis, A. Harootunian, J. Huang, and B. Smith, *Science*, **223**, 811 (1984).

286. S. M. Heald, personal communication.

287. J. E. Hahn and K. O. Hodgson, in *Inorganic Chemistry Towards the 21st Century*, M. H. Chisholm (Ed.), ACS Symposium Series, vol. 211, American Chemical Society, Washington, DC, 1983, p. 431.

288. J. E. Hahn, R. A. Scott, K. O. Hodgson, S. Doniach, S. R. Desjardins, and E. I. Solomon, *Chem. Phys. Lett.*, **88**, 595 (1982).

289. B. T. Thole, G. van der Laan, and G. A. Sawatsky, *Phys. Rev. Lett.*, **55**, 2086 (1985).

290. G. van der Laan, B. T. Thole, G. A. Sawatsky, J. B. Goedkoop, J. C. Fuggle, G. M. Esteva, R. Karnatak, G. P. Reimeika, and H. A. Dabkowska, *Phys. Rev. B*, **34**, 6529 (1986).

291. E. Pantos, in *EXAFS and New Edge Structure*, A. Bianconi, L. Incoccia, and S. Stipcich (Eds.), Springer-Verlag, Berlin, 1983, p. 110.

292. B. Hedman, P. Frank, J. E. Penner-Hahn, A. L. Roe, K. O. Hodgson, R. M. K. Carlson, G. Brown, J. Cerino, R. Hettel, T. Troxel, H. Winick, and J. Yang, *Nucl. Inst. Meth.*, **A246**, 797 (1986).

293. B. Hedman, J. E. Penner-Hahn, and K. O. Hodgson, in EXAFS and Near Edge Structure III, K. O. Hodgson, B. Hedman, and J. E. Penner-Hahn (Eds.), Springer Verlag, Berlin, 1984, p. 64.

294. S. P. Cramer, B. E. Smith, M. K. Eidness, G. N. George, C. D. Garner and A. Hood, Stanford Synchrotron Radiation Laboratory Activity Report, 1986, Proposal 948B.

295. S. P. Cramer, M. W. W. Adams, G. N. George, E. I. Stiefel, V. Minak, J. Enemark, W. Cleland, B. E. Smith, and L. P. Solomonson, Stanford Synchrotron Radiation Laboratory Activity Report, 1986, Proposal 1030 B.

296. P. J. Stephens, T. V. Morgan, F. Devlin, J. E. Penner-Hahn, K. O. Hodgson, R. A. Scott, C. D. Stout, B. K. Burgess, *Proc. Natl. Acad. Sci. USA*, **82**, 5661 (1985).

297. J. Galloway, *Nature*, **318**, 602 (1987).

298. A. Bianconi, A. Congiu-Castellano, P. J. Durham, S. S. Hasnain, and S. Phillips, *Nature*, **318**, 685 (1985).

299. R. G. Shulman, *Proc. Natl. Acad. Sci. USA*, **84**, 973 (1987).

300. J. W. Arber, A. C. Flood, C. D. Garner, S. S. Hasnain, and B. E. Smith, *J. Phys. Colloq. C8*, **47**, 1159 (1986).

CHAPTER

8

CATALYSIS

R. PRINS and D. C. KONINGSBERGER

Laboratory for Inorganic Chemistry and Catalysis
Department of Chemical Engineering
Eindhoven University of Technology
Eindhoven, The Netherlands

8.1. INTRODUCTION

Catalysts are materials that are used to improve the activity and/or selectivity of chemical reactions. To make an efficient use of their capacities, and also because they often are expensive, catalysts are used in a diluted or well-dispersed form. Thus in homogeneous catalysis, where catalysts, reactants, and products are dissolved in one and the same phase, the catalyst is a coordination complex

321

or organometallic compound dissolved in the reaction medium. In heterogeneous catalysis, where the catalyst is in the solid phase while the reactants are in the gaseous or liquid phase, the catalyst may consist of a solid material with a relatively high surface area and large porosity, such as metal sponge (Raney nickel). In this way a large part of the catalytic material can be reached by the gaseous or liquid reactant molecules. In most cases, however, a heterogeneous catalyst consists of a combination of an inert, porous solid material and the catalytically active material. The inert solid functions as a firm support that supplies a large internal surface area on which the active material is deposited in the form of very small particles. Materials like alumina, silica, zeolites, and carbon, with specific surface areas ranging from 1 to 1000 $m^2 g^{-1}$, are used in large quantities in the petroleum and chemical industries as support materials for catalytically active materials like metals (platinum, rhodium, nickel), oxides (V_2O_5, RuO_2), and sulfides (MoS_2, Co_9S_8). In this way it is possible to reach a very high dispersion of the active material. In most industrial catalysts the dispersion is better than 10% (crystallite size < 15 nm) and in some cases, such as platinum on γ-alumina (denoted as Pt/γ-Al_2O_3) used in the catalytic reforming of naphtha to high-octane quality gasoline, the dispersion may even approach 100% (crystallite size $\approx$ 1 nm). In that case all catalyst atoms are exposed to the gaseous phase and may participate in the catalytic reaction.

It is clear that for spectroscopic studies of such catalysts only techniques can be used in which the radiation can penetrate into and through the support particles and thus reach the interior where the majority of the active material is located. But this also means that the radiation will be able to penetrate the active material.

As a consequence any method, which because of sensitivity reasons has to look at the active material inside the support, is a bulk technique for the active material. This holds for methods like IR, NMR, ESR, Mössbauer, and other spectroscopic techniques, which are used to characterize catalysts on support. The EXAFS technique is also a bulk technique, unless use is made of special, surface-sensitive detection techniques. These techniques are very useful in studies of surfaces of single crystals and films (see Chapter 10) but are not very helpful in the study of material inside the pores of another solid material, like catalysts.

In this chapter the application of the x-ray absorption technique to the study of the structural and electronic state of the catalyst is described. It is shown that by studying the edge and near-edge structure of x-ray absorption spectra information can be obtained about the electronic state of the atoms constituting the catalytic material. This gives valuable information on their oxidation state and sometimes on the symmetry of the site in which the atoms are located. The extended fine structure, on the other hand, gives information about the local structure of the atoms in the catalytic particles, on the number and kind of neighbor atoms, and their distances. The EXAFS technique is especially useful for the structural study of catalysts since, because of their high dispersion, the catalytically active material on the support of heterogeneous catalysts is present in such tiny little crystallites or particles that other techniques, like x-ray diffraction or electron diffraction, cannot be used. In this respect catalysts have much in common with truly amorphous systems. In case the nature of the chemical state of the active material on the support is unknown or under doubt, the EXAFS method may even clarify which chemical compound is present, as we shall see in Section 8.3.3.

For all heterogeneous catalysts the EXAFS technique opens the possibility to determine the dispersion of the active material and the size and shape of its crystallites. This can be done because one can determine from the EXAFS data the number of nearest-neighbor atoms around a certain atom and also, if present, the coordination numbers of higher coordination shells. These data can be used to test possible models for size and shape of the crystallites. For instance, the nonexistence or broadening of the EXAFS peak in the radial structure function due to atoms in the second-neighbor shell points to very small crystallites or a very amorphous material. The observation of many neighbor shells, on the other hand, proves the presence of well-developed crystals (Section 8.4.1.1).

For many catalysts the EXAFS technique is the only technique with which structural information can be obtained at all. Whenever the dispersion of the active material is high and the size of the crystallites is below 3 nm, the x-ray diffraction lines become too broad to be observable. The standard technique used for such catalysts is the chemisorption method in which the amount of gas is measured that can adsorb on the surface of the active material. From this the size of the active crystallites can be calculated, but in order to do so one has to make an assumption on the stoichiometry of the number of gas molecules that on average adsorb on one surface atom. In practice the chemisorption method works well for metal-on-support catalysts but not for metal oxides and sulfides on support. However, at very high dispersions the method also breaks down for metallic catalysts since in that case nearly all metal atoms are exposed to the gas phase and thus the average number of neighboring metal atoms becomes low. Under these circumstances the number of adsorbed molecules per metal atom may increase dramatically above the commonly assumed values at dispersions of 90% or less (1). This invalidates the chemisorption method, at least for the time being, for the determination of high metal dispersions. The EXAFS technique is extremely useful in studying these high metal dispersions but it is also useful in studying dispersions of metal oxides and sulfides. Some examples will be presented in Sections 8.3.2, 8.3.3, and 8.4.1.

For homogeneous catalysts the EXAFS technique gives information about the local structure (number and kind of surrounding atoms) of the atom or ion in question. In this respect an EXAFS study of homogeneous catalysts is very similar to that of other complexes in solution, for example, metal enzymes. For a discussion of these biochemical catalysts the reader is referred to Chapter 7.

In this chapter most of the examples of the application of the EXAFS method to heterogeneous and homogeneous catalysis published thus far are discussed. Applications of SEXAFS are not given here but in Chapter 10, since the majority of examples of this technique lie in the field of surface physics. On the other hand, we present the results of measurements of the absorption edges of some catalysts, although a general treatment of the near-edge structure of materials is presented in Chapter 11. We preferred to discuss the measurements of these absorption edges in combination with the EXAFS results.

Before discussing the applications of EXAFS techniques to catalysis in detail, we begin to discuss in Section 8.2 a few general, experimental points that apply to many catalysts. The first is the preparation of the catalyst and the sample to be used in the EXAFS measurement. The second point concerns the need of *in situ* treatment of the catalyst in a special cell to avoid contamination of the catalyst surface by oxygen or any other gas.

8.2.   EXPERIMENTAL

8.2.1.   Sample Preparation

The high dispersion of the active material on the support of a heterogeneous catalyst makes this material very vulnerable to attack by gaseous compounds. The active material in 1 g of catalyst may easily have a surface area of a few square meters and this amount of surface can, for instance, be covered by all oxygen molecules present as a 1-ppm impurity in 1000 L of inert gas. This means that the contents of an ordinary glovebox, which is filled with an extremely pure inert gas, contains enough oxygen to cover and transform the active material present in a 1-g catalyst sample. Attention should also be paid to the fact that the inside surfaces of gloveboxes continuously give off impurity gases by desorption. From these considerations it becomes obvious that all catalyst handling and treatment should be performed with utmost care. Results obtained on catalysts that have been exposed to air, even for very short periods of time, cannot be trusted to represent the actual catalyst and even results obtained from catalysts handled in good quality gloveboxes should not be taken for granted. Therefore, catalyst preparation has to be performed in *in situ* cells, in which both treatment and measurement can be carried out.

Given the fact that the sample should be treated *in situ* and that, because of the mass absorption coefficients, the sample thickness has to be in the order of a few millimeters, the most appropriate sample form is that of a self-supporting disk (wafer) in open contact to the surrounding atmosphere. Such disks are made in a similar way as disks used in IR investigations, namely, by pressing the catalyst powder in a die at a pressure high enough for the material to form an adherent entity. The pressure should not be raised too high, otherwise the catalyst particles will be squashed and the active material will no longer be attainable by the gas to be used in the catalyst preparation procedure.

Catalyst preparation of disk-shaped samples is in principal not much different from conventional preparation methods for heterogeneous catalysts. Usually the catalytically active material is brought into the support by impregnating the support material with an aqueous solution containing the cations or anions of the elements that are to be the constituent elements of the active material. For instance, for the preparation of a platinum on alumina catalyst (Pt/Al_2O_3) one may either use a solution of $Pt(NH_3)_4(OH)_2$ or of H_2PtCl_6, while the preparation of a MoS_2/Al_2O_3 catalyst may be started with the impregnation of an alumina sample with a solution of $(NH_4)_6Mo_7O_{24}$. After impregnation the catalyst is dried to remove the water, which otherwise would cause sintering of the active material in subsequent high-temperature treatments. The catalyst now consists of a support with patches of salt distributed over its surface. The dispersion of the salt will depend on the interaction between the salt and the support surface. If the dispersion is high, one may directly proceed with the treatment leading to the final catalyst. For a metal-on-support catalyst, this treatment consists of a reduction with hydrogen, for a metal sulfide-on-support catalyst the treatment consists of reduction and sulfidation with a mixture of, for example, H_2 and H_2S. Such treatments are known as "direct reduction method" and "direct sulfidation method," respectively.

If the dispersion of the salt on the support surface is low, one may want to improve it before starting the final preparation step. It is well known that precursor particles are usually transformed as a whole into product particles and bad precursor dispersions therefore never lead to good catalysts. Improvement of the dispersion of the elements in the salt can, amongst others, be done by oxidation of the salt to metal oxides at elevated temperatures. For instance, both $Pt(NH_3)_4(OH)_2$ and H_2PtCl_6 will be oxidized to PtO_2, while $(NH_4)_6Mo_7O_{24}$ will transform into MoO_3. If the resulting metal oxide has a good interaction with the support, its dispersion may be better than that of the preceding salt. A further advantage of this so-called calcination step is that other elements, which are unwanted in the final catalyst, are blown away as gaseous products. In the examples considered, these gases would be NH_3, NO_x, and H_2O for $Pt(NH_3)_4(OH)_2$ and $(NH_4)_6Mo_7O_{24}$ and HCl and H_2O for H_2PtCl_6. After the calcination step the catalyst is brought into its final active state by reduction, sulfidation, or any other high temperature treatment.

The shaping of the catalyst disk can be done at several points in the catalyst preparation scheme. For instance the pressing of disks can be performed with the support material. In that case the impregnation and drying are carried out with the disks and the subsequent preparation steps have to be carried out with the disk in its proper position in the *in situ* EXAFS cell. Another possibility is that the disk is pressed after impregnation and drying. Still another possibility is to first prepare a final batch of catalyst powder and then passivate the catalyst material by slowly admitting air (or a mixture of oxygen and inert gas) to the catalyst at room temperature. Care should be taken to do the passivation rather slowly. Otherwise hot spots may arise due to run-away exothermic oxidation of the active material and as a consequence the resulting catalyst may have a bad dispersion due to sintering of the oxidized active material. The carefully passivated catalyst has oxygen adsorbed on the surface of the active material (partial oxidation of this material may also have taken place) and can now be handled in air. Disks can be prepared from this passivated catalytic material, which then can be put in the EXAFS cell to give the final preparation treatment, for example, reduction.

In our laboratory we prefer to prepare the EXAFS samples by the latter route, via passivation of a batch of catalyst powder and preparing disks from the passivated powder. The reason is that in our experience the final treatment step can be performed under much milder conditions after passivation than after calcination or directly after drying. This means that one can take the passivated catalyst material and use it for the preparation of EXAFS disks, as well as for samples to be used in other characterization studies, for instance, hydrogen chemisorption, XPS, IR, and so on. In all these cases one uses the same material and one has to apply only one, not too difficult, final treatment. The chances that in this way the results obtained from the different studies are indeed comparable, are much better than with other methods.

8.2.2. *In Situ* Cells

Compounds that are insensitive to air can be measured in the open air. Pure metals are normally measured as thin foils and salts or other solid materials in pressed disk form. If the material itself is not adherent enough to be pressed into disks, one can admix some polyethylene and press disks at a somewhat elevated temperature (2). Another possibility is to put the material between adhesive tape, between mylar sheets (3), or make a paste that can be put between mylar (4) or aluminium foil (5). Materials that are slightly sensitive to air (bulk solids, which are reasonably protected by their surface-to-volume ratio) may be handled in gloveboxes. Thus Vlaic et al. have mounted samples into a sample holder inside a glovebag filled with dry nitrogen and transferred the sample holder into a cell equipped with mylar windows (4). The cell could be pumped to 10^{-4} Pa and cooled to liquid nitrogen temperature for the duration of the x-ray absorption measurement.

Air sensitive materials, such as metal and metal sulfide catalysts, have to be measured in *in situ* cells. A few designs for such cells have been described in the literature and all have demonstrated that they are able to give reliable results. The first cell described in the literature was that of Lytle et al. (6) (Fig. 8.1). It consists of two boron nitride parts that are tightly clamped together by screws onto a stainless steel mounting block. The mating surfaces of the boron nitride cell are scrupulously polished, to obtain a good seal. At the junction of the two boron nitride cell parts there is a cavity to hold the catalytic material. The outside surfaces of this cavity, which are in the x-ray path, are machined to a thickness of 1 mm to minimize x-ray absorption by the cell. Different cells can be made with different cavity depths to permit variations in sample thickness. The stainless steel mounting block has connections for inlet and outlet of gases for catalyst treatment and also inlets and outlets for liquid nitrogen to cool the sample for measurements at low temperatures. A heating element surrounds the sample and makes it possible to heat the sample to 773 K. The sample cell and surrounding heating element are housed in a water cooled aluminium block equipped with two mylar windows. The region between cell and aluminium block can be evacuated. By maintaining a gas flow through the sample cell during a measurement and simultaneously evacuating the region surrounding the cell, leaks can only occur outwardly, away from the cell and catalyst. All mono- and bimetallic EXAFS and white line studies of Sinfelt et al. have been performed with this cell (Sections 8.4.1–8.4.3). The sample in Lytle's cell should be in powdered form, so that the gas can flow through it. This has the advantage that

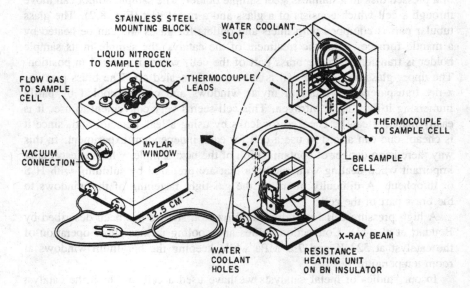

STAINLESS STEEL
MOUNTING BLOCK

WATER COOLANT
SLOT

LIQUID NITROGEN
TO SAMPLE BLOCK

THERMOCOUPLE
LEADS

FLOW GAS
TO SAMPLE
CELL

THERMOCOUPLE
TO SAMPLE CELL

MYLAR
WINDOW

VACUUM
CONNECTION

BN SAMPLE
CELL

~12.5 CM

WATER
COOLANT
HOLES

X-RAY BEAM

RESISTANCE
HEATING UNIT
ON BN INSULATOR

Figure 8.1. *In situ* cell of Lytle et al. (6).

one can use powdered catalytic material without having to press a disk from it. Actually the cell performs as a flow reactor and is therefore very well suited for *in situ* studies. On the other hand, gas absorption studies are difficult to perform, because the cell is not vacuum tight.

Recently, Lytle et al. (7) have added the possibility of fluorescence detection to their cell. They made a window at a right angle to the x-ray beam and followed up on Stern's idea of using an ionization chamber for the detection of the fluorescent x-rays in combination with a $Z - 1$ filter and a Soller slit to suppress secondary fluorescence (8). In this way a good S to N ratio could be obtained even for low catalyst loadings.

A simple cell, resembling cells used in *in situ* IR studies, has been used by Katzer and coworkers (9). It consists of two connecting cylindrical tubes. The double walled vertical cylinder can be filled with liquid nitrogen to cool the frame in which the sample holder is put. The sample consists of a pressed disk in a steel sample holder. The horizontal cylinder is closed at both sides by beryllium windows, which can be removed to put the sample into position at the bottom of the vertical cylinder. Electrical leads for heating the sample and inlets and outlets for gases for catalyst treatment are connected to the horizontal cylinder. The cell is rather easy to construct and to operate, but uses large amounts of liquid nitrogen for cooling, because of poor thermal insulation.

An even simpler cell has been used by Clausen et al. in studies of metal–sulfide hydrodesulfurization catalysts (10). Also, in this case the sample consists of a pressed disk in a stainless steel sample holder. The sample holder can move through a cell which consists of a glass and a metal part (Fig. 8.2). The glass tubular part is equipped with inlets and outlets for gases and can be heated by a mantle furnace. After the treatment of the catalyst the sample in its sample holder is transferred to the brass part of the cell, where it is fixed in position. The upper glass part of the cell is subsequently sealed off. The brass part has x-ray transparent beryllium or mylar windows and can be cooled by partly immersing it into liquid nitrogen. This cell seems very convenient to use. It is easily made, it can be checked for leaks by using a leak detector and, since it is cheap, one can afford to use a cell for each independent experiment. In this way there is no danger of contamination of the next sample, which is especially important when dealing with catalysts that are prepared by sulfiding with H_2S or thiophene. A difficulty might be the gas-tight fastening of the windows to the brass part of the cell.

A high pressure cell for hydrodesulfurization studies has been described by Boudart et al. (11). Convection baffles and cooling jackets allow operation of the catalyst at 523 K and 7.3 MPa while keeping the beryllium windows at room temperature.

In our studies of metal catalysts we have used a cell in which the catalyst

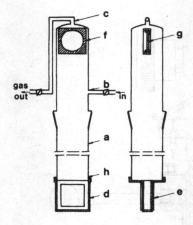

Figure 8.2. Glass sample *in situ* cell used by Clausen et al. (10): (*a*) pyrex glass part, (*b*) inlet of gas, (*c*) outlet of gas, (*d*) brass part, (*e*) beryllium (or Mylar) window, (*f*) stainless steel sample holder, and (*g*) catalyst wafer.

treatment is done with the sample in a position different from that in which it is measured (12) (Fig. 8.3). The pressed-disk-type sample is placed in a sample holder at the bottom of a cylindrical metal tube. Leads for heating and inlets and outlets for cooling the sample with liquid nitrogen run through this inner cylinder. The inner tube and sample holder are positioned in the middle of a wider metal cylinder, which consists of an upper, treatment part and a lower, measuring part. When the inner cylinder is pulled up to a position in which the sample is situated inside the upper part, the upper part is closed off from the lower part by means of a metal flange that is fixed to the bottom of the sample holder. The catalyst is reduced or oxidized in this position by admission of H_2 or O_2 through inlets and outlets through the outer cylinder and by heating the sample holder via the electrical leads through the inner cylinder. The maximum temperature attainable is 773 K. The flange is cooled and this ensures that the lower part of the cell does not come into contact with hot gases. In this way the beryllium windows in the lower part are protected. After catalyst treatment the inner tube with sample is pushed down into the measuring position. During the EXAFS measurement the temperature of the sample can be varied between 77 and 623 K and the cell can be evacuated or filled with gas. In this way the influence of gas adsorption on the catalyst EXAFS spectrum can be easily studied. Recently, we developed a simpler version of this cell in which catalyst treatment (heating) and measurement (cooling) are done with the sample in the same position. The beryllium windows in this cell are water cooled for protection. Because of the strong adsorption of sulfur we use one cell for metal sulfide catalysts and another for other types of catalysts.

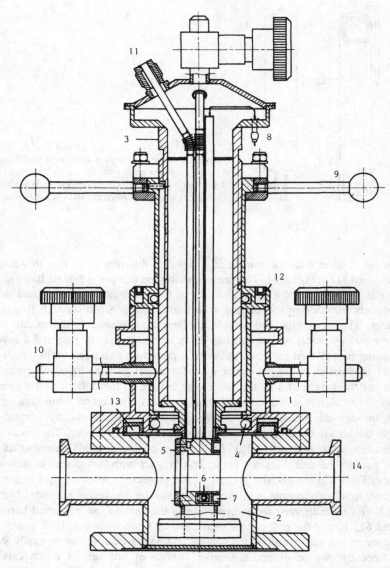

Figure 8.3. Two-compartment cell of Koningsberger and Cook (12): (1) compartment for sample treatment, (2) compartment for measurements, (3) movable inner tube, (4) O-ring seal, (5) sample holder, (6) heater, (7) sample cooling channel (N_2), (8) thermocouple, (9) locking ring, (10) vacuum valve, (11) gas inlet, (12)–(13) cooling channel (H_2O), and (14) beryllium windows.

8.3. EXAFS OF NONMETALLIC CATALYSTS

8.3.1. Homogeneous Catalysts

Up to now only two studies have been reported on real homogeneous catalyst systems, with the catalyst dissolved in solution. All other studies have been concerned with so-called immobilized or heterogenized homogeneous catalysts, in which the actual catalytic complex is attached to a solid support by means of physical or chemical interactions. Immobilization is done to facilitate the separation of catalyst and product in the final solution. In most cases it does not interfere with the chemistry taking place during the catalytic reaction.

In a series of three articles Reed et al. reported transmission EXAFS measurements on Wilkinson's catalyst $RhCl(PPh_3)_3$, the related bromo analogue $RhBr(PPh_3)_3$, and on these catalysts bound to polymers (13–15). These studies are of interest for a better understanding of the structure of the Wilkinson hydrogenation catalyst, especially when supported on polymers. The measurements were performed at SSRL (Stanford University, USA) both on the rhodium and bromine K-edges. No details were given of the sample cell used. The data analysis used was more or less equal to the procedure described in Chapter 6. For the chlorine containing catalysts $Rh(Ph_2P-CH_2-CH_2-PPh_2)_2^+$ and $RhCl_3 \cdot nH_2O$ were used as model compounds to obtain the phase and scattering functions for Rh—P and Rh—Cl bonds. For the bromine containing catalysts theoretical phase shifts and amplitudes were used. In the analysis of the rhodium data, contributions from the neighboring halogen and phosphorus atoms were taken into account. The difference between the halogen and phosphorus phase shifts is large enough to distinguish between the contributions of these atoms. In the fitting procedure the sum of the squares of the fit residuals was plotted for several integral values of the number of ligand X and P atoms around rhodium, always assuming a total coordination of four. Since there are two types of phosphorus atoms in the $RhX(PPh_3)_3$ complex, one in trans and two in cis position relative to the halogen atom, also the number of cis and trans phosphorus atoms was varied. In all cases a best fit could be obtained with the sum of the squares of the residuals at least a factor of two better than for the next best fit, indicating that the final best fit represented the actual structure rather well. A small percentage of contribution from other structures could not be ruled out, however. The best fit for the $RhX(PPh_3)_3$ compounds was obtained for $N_X = 1$ and $N_P = 3$. The interatomic distances obtained were in good agreement with x-ray results (Table 8.1).

For the bromine containing compounds, bromine K-EXAFS was measured and analyzed. Since the bromine atom has only one nearest neighbor, the

Table 8.1. Interatomic Distances in Wilkinson's Catalysts as Determined by EXAFS and by X-Ray Diffraction

| | X = Cl | | | | X = Br | | |
| | I | | II | | I | I* | II |
	X-ray[a]	EXAFS	EXAFS	X-ray[b]	EXAFS	EXAFS	EXAFS
Rh—X	2.376	2.35	2.33	2.587	2.54	2.50	2.50
Rh—P_1	2.214	2.23	2.16	2.176	2.18	2.14	2.16
Rh—P_2	2.326	2.35	2.23		2.31	2.26	2.32

[a]P. B. Hitchcock, M. McPartlin, and R. Mason, *Chem. Commun.*, 1367 (1969); M. J. Bennett and P. B. Donaldson, *Inorg. Chem.*, **16**, 655 (1977).
[b]X-ray results for RhBrP(*o*-vinylphenyl)₃. C. Nave and M. R. Truter, *Chem. Commun.*, 1253 (1971).

EXAFS spectrum contains only a single frequency (Fig. 8.4*a*) and hence a single peak in the Fourier transform. In contrast to this, the rhodium atom has three different neighboring atoms and the rhodium EXAFS spectrum thus shows a clear interference between several frequencies, while the Fourier transform contains three peaks (Fig. 8.4*b*).

For RhX(PPh₃)₃ bound to a weakly cross-linked phosphinated polystyrene polymer the best fit was obtained for $N_X = 2$ and $N_P = 2$. For Rh(I) compounds such a coordination can only be achieved through dimerization. Apparently the starting compound I (Fig. 8.5) has exchanged one of its triphenylphosphine

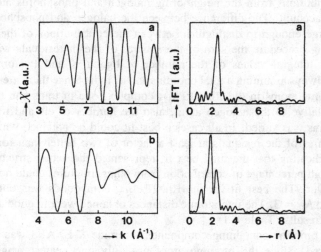

Figure 8.4. (*a*) Bromine *K*-EXAFS and its Fourier transform (magnitude |*FT*|) in RhBr(PPh₃)₃. (*b*) Rhodium *K*-EXAFS and its Fourier transform (magnitude |*FT*|) in RhBr(PPh₃)₃ (14).

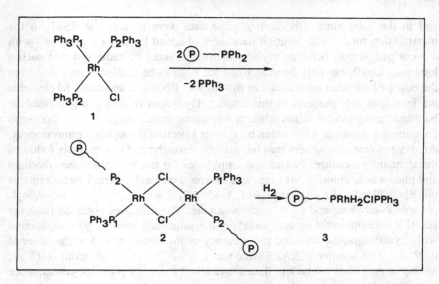

Figure 8.5. Dimerization of RhCl(PPh₃)₃, when attached to a derivatized polymer P (13).

ligands for a PPh₂L ligand, where L = cross-linked polystyrene $-p-C_6H_4-PPH_2$, and has subsequently dimerized to compound II (Fig. 8.5). The suggestion that weakly cross-linked polystyrene–divinylbenzene (PS–DVB) copolymers are swellable and mobile and as a consequence promote dimer formation, was proved by measuring the EXAFS spectra of RhBr(PPh₃)₃ on 20% cross-linked PS–DVB. In this case the best fit was found for $N_X = 1$ and $N_P = 3$, demonstrating that in this polymer, which is less mobile and has a reduced free volume, dimer formation is substantially reduced. The interatomic distances obtained for the compound I* are presented in Table 8.1.

These structural results are very interesting and encouraging, because they could not have been obtained by x-ray crystallography. Furthermore, they are very relevant for catalysis since the catalyst activity of Wilkinson's catalyst has been found to decrease when bound to polymers. The EXAFS results provide the first real evidence that this is due to dimerization and can be influenced by catalyst (polymer) architecture.

An EXAFS study of real homogeneous catalysts has been reported by Stults et al. (16). In this study model compounds for the rhodium catalyst used in the asymmetric hydrogenation of prochiral alkenes were used. The large interest in this class of catalysts stems from their successful use in the industrial synthesis of L-dopa, a pharmaceutical used in the treatment of Parkinson's disease. There-fore, a structural study of several model complexes was done both in solution and in the solid state. Rhodium K-edge data were obtained at SSRL in the transmission mode. The solution data were obtained by using a glass cell with a 5-cm path length between mylar windows. It must be remarked that such a long path length can only be used when the edge to be studied is far away from the edges of all other atoms present in solution. Rhodium and other $4d$ elements are in a favorable position in this respect. Hydrogen or nitrogen gas could be bubbled through the solution while it was being stirred magnetically. Solutions containing a substrate were added by syringe injection through a septum stopper. An oxygen-free atmosphere was maintained throughout. Data analysis followed the standard procedure. Phases and amplitudes for the bonds between rhodium and phosphous, chlorine, oxygen, and carbon were derived from measurements on $RhCl(PPh_3)_3$, Rh_2O_3, and $(RhL_1L_2)^+ \cdot BF_4^-$, with L_1 = cis-bisdiphenyl-phosphino-ethylene and L_2 = cyclooctadiene. The distances obtained from the EXAFS measurements on compounds in the solid state were in good agreement with crystallographic data and the accuracy of the distances was in the order of 0.02 Å. The solution EXAFS data had a lower accuracy of about 0.07 Å, arising in part from the lack of data above $k = 14$ Å^{-1}, a probable consequence of vibrational damping. A general increase of rhodium-to-neighboring atom bond distances was observed on going from a solid to the corresponding solution. This might reflect relaxation of crystal packing forces, or a failure of the fitting parameters (which were derived from solid-state data) to adequately describe the solution data. In the least squared fit for the unknown compounds only N, R, and σ of the different interatomic distances were varied.

Studies of catalyst and catalyst-substrate intermediates concentrated on the solvated Rh^+ species and on the complex formed with α-acylamino-cinnamic acid. When hydrogen was bubbled through a solution of a $Rh(I)(C_8H_{12})$ complex in methanol, with two monodentate phosphine ligands or one bidentate bis-phosphino ligand, two oxygen atoms and two phosphorus atoms were observed in the first coordination shell around rhodium. The two oxygen atoms are most probably from coordinated methanol. Addition of α-acylamino-cinnamic acid to this solution resulted in an EXAFS spectrum that could be analyzed in terms of two different Rh—P distances, a Rh—O, and a Rh—C distance. Apparently a square planar complex is formed with two distinct phosphorus atoms, one trans to the acyl oxygen atom in the substrate and the other trans to the alkene group in the substrate (Fig. 8.6). The Rh—C and Rh—P distances from the *in situ* prepared catalyst–substrate complex agreed favorably with the values for the crystalline compound that was synthesized independently. The Rh—O distance in the *in situ* complex was found to be surprisingly short and may indicate a real difference between the crystalline complex and the active catalytic species in solution. Thus by performing *in situ* EXAFS measurements structural information on two intermediates in the catalytic cycle of asymmetric hydrogenation has been obtained. This structural information is extremely valuable for this

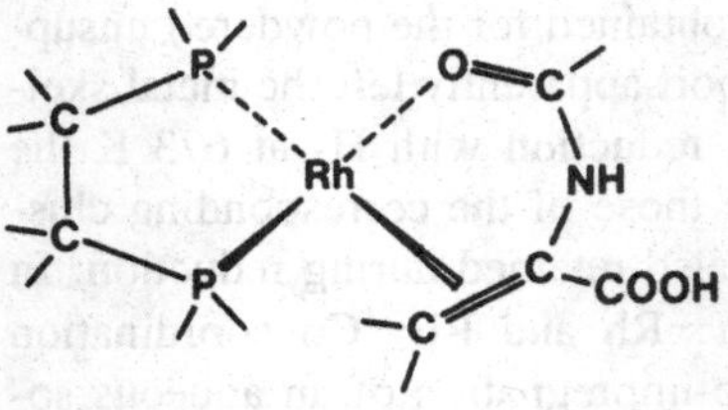

Figure 8.6. Proposed chemical structure for the complex formed when adding α-acylaminocinnamic acid to a solution containing a rhodium(I)diphosphino complex (16).

branch of catalysis, in which the ultimate product of the catalytic reaction is completely determined by the structure of the active catalyst–substrate complex.

Another EXAFS investigation of a real homogeneous catalyst was described by Goulon et al. (17). They studied nickel and cobalt octoate (salts of $C_7H_{15}COOH$) in benzene, before and after addition of triethylaluminium. These Ziegler-type catalysts are used for hydrogenation. Airtight cells with variable optical paths were made of polyethylene or preferably of Delrin and care was taken to fill the cells under perfectly anaerobic conditions. Data analysis was performed by Fourier transformation, corrected for the phase shifts and amplitude factors of the relevant absorber and backscatterer atoms by theoretical estimates, and inverse Fourier transformation. Before reduction the nickel ions were surrounded by six oxygen atoms, at least four of which belong to a monodentate carboxyl group. After addition of triethylaluminium the analysis definitely showed metallic Ni—Ni and Co—Co distances and Ni—X, respectively, Co—X, distances. It was suggested that X might be the carbon atom of an ethyl group attached to a metal atom. Unambiguous fitting of the data with a two-shell model proved impossible. The results suggested that the reduced systems contain small, probably amorphous, metal clusters onto which ethyl groups are attached.

In a short communication Besson et al. showed that the combination of EXAFS with other spectroscopic techniques can be very helpful in the characterization of the structure of complexes (18). They published some results of a combined IR and EXAFS study of the adsorption of osmium carbonyl complexes on the surface of a silica support. When a solution of $Os_3(CO)_{12}$ in CH_2Cl_2 was brought into contact with silica and subsequently dried and heated to 423 K, 2 mol of carbon monoxide evolved and a new IR spectrum was observed. The EXAFS spectrum changed too, but it still contained peaks ascribed to Os—Os bonds, albeit shorter, and no evidence for second neighbor osmium peaks. Therefore, it was concluded that no aggregation to metal particles had occurred and that the cluster framework was still intact.

Similarly, K-EXAFS data from unsupported $Rh_2Co_2(CO)_{12}$ and $RhCo_3(CO)_{12}$ complexes and from these complexes on γ-Al_2O_3 (19) revealed that in the catalysts made by impregnation of the dehydrated γ-Al_2O_3 support with an alkane solution of the complexes, the Rh—Rh and Rh—Co bond lengths and coordination numbers were about the same as those obtained for the powdered unsupported complexes. Putting the clusters on support apparently left the metal skeletons of the clusters intact. After subsequent reduction with H_2 at 673 K the coordination numbers were about the same as those of the corresponding clusters, indicating that the metal skeletons were also retained during reduction. In contrast to these results for the clusters, Rh—Rh and Rh—Co coordination numbers obtained for catalysts prepared by co-impregnation of an aqueous solution of $RhCl_3$ and $CoCl_2$ and subsequent reduction, were much higher and their sums were close to the bulk value of 12. The classic aqueous impregnation method apparently led to rather large bimetallic particles.

Joyner et al. studied the influence of heat treatment of a cobalt–porphyrin catalyst supported on active carbon (2), since such treatments are known to lead to improved oxygen reduction electrocatalysts. Although the data were rather noisy they could see a change in the EXAFS spectrum after a heat treatment of the catalyst. A qualitative analysis of the data indicated that a plausible explanation was that in the cobalt K-EXAFS spectrum of the untreated cobalt porphyrine there are four nitrogen atoms in the first shell and eight carbon atoms in the second shell around cobalt, while after heat treatment only the four nitrogen atoms were retained. This might indicate that during heat treatment the carbon bridges connecting the pyrrole fragments in the porphyrin framework are broken.

8.3.2. Transition Metal Compounds

Studies of several transition metal ions on Al_2O_3 catalysts have been performed. Metal ions in and on oxidic supports can occur in different sites and may form various compounds with the support. Nevertheless the resulting EXAFS spectra will be rather similar: The first-shell interatomic distances will be about the same in all oxidic sites and compounds and it is only in the second and third neighbor shells that different environments will show up. This is clearly demonstrated in the copper K spectra of CuO and $CuAl_2O_4$ (4,20), where the major difference is found in the range $4 < R < 6$ Å. Since this region is less well developed in catalysts with a high dispersion, it is not so easy to distinguish by means of EXAFS between CuO formation or dissolution of Cu^{2+} into Al_2O_3 catalyst. In combination with other spectroscopic techniques, like x-ray diffraction (XRD), x-ray photoelectron spectroscopy (XPS), ESR, and diffuse reflectance spectroscopy, still worthwhile information could be obtained.

The analysis of the cobalt and nickel K-EXAFS spectra of oxidic cobalt and nickel on Al_2O_3 was done by Fourier and inverse Fourier transformation of the first coordination shell, and the log ratio technique was used to determine N and σ for the oxygen ions around Co^{2+} and Ni^{2+} (21). The coordination number of oxygen ions around Co^{2+} was found to increase from 4 to 5.3 when increasing the cobalt loading of the support. This confirms that at low loading the Co^{2+} ions are in tetrahedral sites in the Al_2O_3, while at high loading bulk Co_3O_4 is formed on the catalyst surface. At all loadings the coordination number for Ni^{2+} was higher than that of Co^{2+}, in line with the higher preference of Ni^{2+} for octahedral sites. Near-edge structures confirmed the EXAFS results, a weak pre-edge peak was observed in all cases, but most intense for Co/Al_2O_3 catalysts. In model compounds the $1s$–$3d$ peak has indeed been observed to be more intense for tetrahedral than for octahedral coordination of the absorbing ion (22) (cf. also Chapter 11, Section 11.5.7).

Studies of Ni^{2+} and Co^{2+} ions in oxides and other materials have also been performed by Tohji et al. (23, 24). The aim of their investigations was to study the preparation of nickel and cobalt metal catalysts via the method of hydrolysis of a mixed solution of a metal alkoxide and ethyl orthosilicate or n-propyl orthotitanate. For that reason these investigations will be discussed in Section 8.4.1.2 together with other EXAFS investigations of metal catalyst preparation.

The titanium and vanadium K-EXAFS and XANES spectra of vanadium oxide supported on titanium dioxide, which catalyzes the selective oxidation of o-xylene to phthalic anhydride, has been studied by Kozlowski et al. (25). Less well-resolved EXAFS spectra of pure rutile and anatase have been measured by Vlaic et al. (26) at Frascati. The study of Kozlowski et al. was performed at Daresbury with a channel cut Si(111) crystal monochromator at low energy of the storagering, thus avoiding contamination of the data by second or third harmonics. Well-resolved spectra were obtained for V_2O_5 and for the anatase and rutile phases of TiO_2. High surface area TiO_2 was found to have the same structure as highly crystalline anatase. However, the correlation of atomic positions was lost beyond distances involving more than three linked TiO_6 octahedra. The monolayer V_2O_5 on TiO_2 catalyst had a different structure than that of crystalline V_2O_5. EXAFS combined with XANES analysis indicated that most probably the V^{5+} ions in the catalyst are surrounded by two terminal oxygen ions at short distance and two bridging oxygen ions at larger distances.

Sato et al. have determined the structure of a molybdenum metathesis catalyst with the aid of a laboratory EXAFS apparatus with a rotating anode x-ray generator and a flat Ge(111) crystal monochromator in fourth-order reflection (27). The catalyst was prepared by impregnation of a solution of $Mo_2(C_3H_5)_4$ on γ-Al_2O_3. After subsequent reduction and oxidation the molybdenum K-absorption spectrum was measured. The Fourier transform of χk^3 showed clear peaks due to Mo—O and Mo—Mo distances. Curve fitting was performed by the theoretical method as well as by the empirical method, with Mo and K_2MoO_4 as model compounds. The analysis showed that there are 4–5 oxygen ions at a distance of 2.76 Å and about one molybdenum ion at a distance of 3.2 Å of the absorbing molybdenum ion. This provided proof for the suggestion made in other studies that the catalyst consists of dimeric molybdenum oxide species on the Al_2O_3 surface.

EXAFS studies of zeolites A and Y ion exchanged with Co^{2+} and Mn^{2+} were published by Morrison et al. (28-30). Both sixfold and fivefold coordination by oxygen atoms at distances equal to that of $M(H_2O)^{2+}$ in solution was observed for hydrated zeolites. For dehydrated zeolites a peak due to backscattering from nearest-neighbor oxygen atoms in the zeolite framework six ring was observed, and also a second peak due to backscattering from silicon, aluminium, and remaining oxygen atoms in the six ring. Some of these results were at variance with crystallographic results. According to Morrison et al. this suggests that there exists static and dynamic disorder of the metal cation complexes in the zeolite and that equivalent crystallographic sites may be occupied by inequivalent cations. The EXAFS technique, being atom specific and insensitive to long-range disorder, may then turn out to be a better probe of the coordination environment of the metal cation. Further studies would be of interest.

Titanium K-edge measurements on $TiCl_3$ and $TiCl_2$ Ziegler–Natta catalysts have been published by Reed et al. (3) and by Vlaic et al. (4). The results obtained by these two groups are rather conflicting. One group reports a Ti—Cl distance in $TiCl_3$ of 2.22 Å and the other a distance of 2.46 Å. The reason for this discrepancy is not clear, since both sets of raw data look good. It may be that one of the samples had suffered from oxygen contamination. On the other hand, the most possible product of such a contamination (anatase) has a completely different EXAFS spectrum (25) than observed for the two $TiCl_3$ samples. New measurements under very careful conditions seem to be called for.

8.3.3. Hydrodesulfurization Catalysts

Several studies of Mo, Co—Mo, and Ni—Mo on Al_2O_3 hydrodesulfurization (HDS) catalysts have been published, both of the oxidic precursors to the catalysts and of the sulfidic catalysts themselves (10, 11, 31-36). The samples studied were either commercial catalysts or were prepared by the usual method of impregnation of Al_2O_3 with a solution of $(NH_4)_6Mo_7O_{24}$, followed by drying and calcining. Subsequently $Co(NO_3)_2$ was added by impregnation, drying, and calcining. Clausen et al. (10, 31, 36) and Boudart et al. (11, 32) prepared their EXAFS samples by pressing the catalysts in the oxidic state and sulfiding them *in situ* in the EXAFS cell. Kohatsu et al. (35) and Parham and Merrill (33) first sulfided the catalysts in a reactor and made EXAFS samples with the aid of a glovebag and nitrogen purging. While Parham and Merrill claimed that the molybdenum EXAFS of their samples had not suffered from this handling,

Kohatsu reported that especially nickel could not be kept in the sulfided form in this way.

The most extensive studies of the oxidic precursors to the hydrodesulfurization catalysts have been published by Parham and Merrill (33) and Chiu et al. (34). They not only studied oxidic molybdenum and Co—Mo on Al_2O_3, but also many oxidic molybdenum model compounds. All oxidic catalysts showed one strong peak in the Fourier transform of the molybdenum K-EXAFS data at $R = 1.73$ Å. This distance is typical for Mo—O distances in tetrahedrally coordinated molybdenum and for the shorter Mo—O distances observed in octahedrally coordinated molybdenum. For instance in MoO_3 the six Mo—O distances range from 1.67 to 2.34 Å. Other spectroscopic techniques have indeed indicated that after calcination a substantial part of molybdenum is in the tetrahedral form (37). Smaller peaks in the radial structure function were observed at 2.55, 2.86, and 3.5 Å by Chiu et al. (34). They used a new EXAFS data reduction method, analogous to a method used in gas-phase electron diffraction. By smoothing the data to minimize extraneous noise, deconvoluting the spectra to correct for finite energy width of the beam, and correcting for truncation errors they were able to obtain radial distribution curves with substantially lower background noise and high reproducibility of the full width at half-maximum (FWHM) of the major peaks. It is a pity, however, that the authors did not present the uncertainties in the magnitudes of the resulting peaks. Looking at the two figures in which they present "original" data, it is obvious that some of the smaller peaks are very hard to differentiate from noise and this has to result in a rather high uncertainty in the final result. This might explain why the intensities of the small peaks plotted as a function of molybdenum loading on the support were difficult to rationalize. The relative intensity of the peak at 1.73 Å clearly diminished with increasing molybdenum loading, while after addition of cobalt to a catalyst with constant molybdenum loading, increases as well as decreases in the intensity were observed. The decrease in intensity of the main peak in the radial structure function with increasing molybdenum loading was paralleled by an increase in the relative size of a particular peak in the near-edge structure, which was assigned to the $1s \rightarrow 5p$ transition. This increase in intensity in one part of the spectrum and decrease in another part is not uncommon and might be due to the sum rule, which states that the sum over all final states j of the oscillator strengths f_{ij} of all transitions from initial state i to final state j is a constant and equal to the number of electrons in the atom. The observed variations in intensity were correlated by Chiu et al. (34) with increasing distortions of the oxygen atoms surrounding the molybdenum cations with increasing molybdenum loading. A quantification of these distortions could, however, not be made. A quantitative analysis is difficult because of the spread in distance of the first-shell oxygen neighbors of molybdenum. As Eisenberger and Brown have clearly pointed out, the truncation of the EXAFS data below

$k = 3$ Å^{-1} leads to a loss of correlations between $\rho(r)$ and $\rho(r + \Delta r)$ for Δr > 1 Å (38, 39). This means that in disordered materials such as catalysts the second and further neighbor peaks are easily lost and that even the first neighbor peak may suffer a loss in intensity when the spread in the corresponding metal-ligand distances is large. The work of Chiu et al. shows that this is probably the case for many molybdenum oxides and for molybdenum oxide catalysts. The intensity of the first Mo—O peak of a series of model molybdenum oxide compounds strongly decreased with increasing width. An empirical correction for this effect applied to the catalysts still led to too low a coordination number of 2.5 for catalysts with high molybdenum loadings. Further progress for these catalysts and for materials with similar large degrees of disorder is severely hampered because of these inherent limitations of the EXAFS method at low-k values. Recent studies have indicated, however, that the single scattering formalism might be applicable even in the low-k region (40, 41) and this might be very helpful in future studies of the structure of disordered systems (cf. also Chapter 9, Section 9.3.3).

Analysis of the molybdenum K EXAFS spectra of the sulfided catalysts proved much easier than those of the oxidic catalysts. The Fourier transform of the Mo/Al$_2$O$_3$, as well as Co—Mo/Al$_2$O$_3$ catalyst showed two peaks (Fig. 8.7a) at the same distances as in MoS$_2$ (11, 31–33, 36) and back transformation of the first and second peak showed them to be due to backscattering by sulfur and molybdenum atoms, respectively (31). Whereas Clausen et al. reported a coordination number of six for the coordination of molybdenum by sulfur as in bulk MoS$_2$ (31, 36), Parham and Merrill reported values that increased with increasing sulfidation temperature but which were always below four (33). Boudart et al. published values below six (32), as well as values equal to six (11). A future reliable determination of the Mo—S coordination number is highly desirable because it may give us information about the size of the MoS$_2$ crystallite on the support and of the existence of an excess or shortage of sulfur at the edges of these crystallites. For the Mo—Mo coordination number all authors (11, 31–33, 35, 36) found a value substantially below the MoS$_2$ bulk value of six. This indicates either that some degree of disorder exists outside the first coordination shell or that the MoS$_2$ crystallites in the catalysts are very small, with a substantial fraction of surface molybdenum atoms. The latter conclusion is in accordance with Mössbauer results (42), which showed that the Co—Mo/Al$_2$O$_3$ catalyst contains MoS$_2$-like structures in a highly dispersed state. EXAFS measurements at low temperature might give more quantitative information on the Debye–Waller factor and on the Mo—Mo coordination number. A preliminary measurement at 77 K performed at Cornell High Energy Synchrotron Source CHESS (33) showed very good resolution up to 17.8 Å^{-1} and indicated that the size of the MoS$_2$ crystallites might not be so small after all, because the second Mo—S and second and third Mo—Mo shells could be

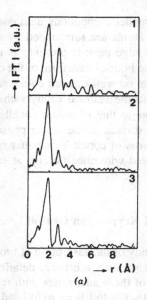 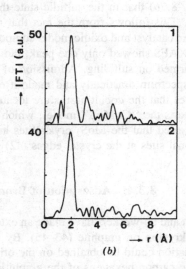

Figure 8.7. (a) Magnitude of (k^3) Fourier transform ($|FT|$) of (1) well crystallized MoS$_2$, (2) sulfided Mo/Al$_2$O$_3$, and (3) sulfided Co-Mo/Al$_2$O$_3$. (b) Magnitude of (k^3) Fourier transform ($|FT|$) of (1) Co$_9$S$_8$ and (2) sulfided Co-Mo/Al$_2$O$_3$ (31).

observed. In any case the EXAFS results have proven beyond any doubt that in the sulfided Mo/Al$_2$O$_3$ catalyst, as well as in the cobalt promoted catalyst, the active material consists of small crystallites of MoS$_2$. That no molybdenum oxysulfides are present in the sulfided catalysts, as suggested several times in the literature (37), was furthermore proven by the absence of a low-energy 1s–4d shoulder in the molybdenum K-edge.

Boudart et al. have studied the molybdenum K-EXAFS as a function of cobalt loading in sulfided Co—Mo/Al$_2$O$_3$ catalysts (32). By measuring the intensity of the Mo—S peak in the radial structure function they demonstrated that there exists a correlation between the number of sulfur neighbors and the HDS activity as a function of the Co/(Co + Mo) atomic ratio. For N_{Mo-S}, as well as for the activity a vulcano type of curve was obtained. At low Co/(Co + Mo) ratios the cobalt is situated at the edge of the small MoS$_2$ crystallites, forming the so-called "CoMoS" phase proposed by Topsøe et al. (42). The additional sulfur atoms that go together with this cobalt also function as ligands for the molybdenum edge atoms, thus increasing the average Mo—S coordination number. At large Co/(Co + Mo) ratios most of the cobalt forms separate Co$_9$S$_8$ particles, thus depleting cobalt—and thus the extra sulfur—from the "CoMoS" phase.

Comparison of the cobalt K-absorption spectra of the Co—Mo/Al$_2$O$_3$ catalyst with those of several oxidic and sulfidic model compounds demonstrated (Fig. 8.7b) that in the sulfidic state the cobalt atoms are surrounded by sulfur (10). This follows from the fact that no $1s-4p$ edge peak is present, as in the oxidic catalyst and oxidic model compounds. The Fourier transform of the cobalt K-EXAFS showed only one peak, indicating that no well-ordered cobalt sulfide is formed on sulfiding. Admission of oxygen to the sulfided catalyst changed the spectrum drastically and made it comparable to that of oxidic cobalt. This proves that the cobalt atoms are located at the surface of the active phase and supports the "CoMoS" model, which on the basis of cobalt Mössbauer results suggested that the MoS_2 crystallites are decorated with cobalt atoms at substitutional sites at the crystal edges (42).

8.3.4. Adsorption of Bromine and Krypton on Grafoil

Stern and co-workers have made an extensive study of the adsorption of bromine and krypton on graphite (43-45). By using grafoil as a substrate, detailed information could be obtained on the orientation of these adsorbates with respect to the carbon hexagons of the graphitic substrate. Grafoil is an exfoliated conglomerate of small graphite crystallites in a common orientation. About 30% of the total surface area of 22 m^2 g^{-1} is due to the random oriented crystallites, while the remaining area is on crystallites with basal planes parallel to the macroscopic sheet surface with an rms deviation of 15°. By stacking the grafoil sheets in a cell with their normals horizontal, the angle of the polarization vector with respect to the sheet surface can be varied by rotation of the cell about the vertical axis. When the polarization vector is normal to the planes the contribution of scattering atoms along the normal is emphasized, while with polarization parallel to the plane the scattering of atoms in the plane is enhanced. Thus, by using grafoil as a substrate the relative position of absorbing and scattering atoms can be determined without having to use single crystals.

In the studies of bromine on grafoil (43,44) the factorization of the EXAFS in Br—Br and Br—C contributions was made easy by the fact that the former contribution is important for $k > 6$ Å^{-1}, while the latter is only important for $k < 6$ Å^{-1}. Furthermore, the Br—Br contribution is more important in the parallel spectrum, while the Br—C contribution is more important in the perpendicular orientation (Fig. 8.8). Bromine cannot only be adsorbed at the graphite surface, but it can also be intercalated between graphite layers. The EXAFS spectra of adsorbed bromine at coverages of 0.6 and 0.9 monolayer show a Br—Br scattering that is strongly reduced when the polarization is out of plane, proving that bromine is molecularly adsorbed parallel to the basal planes. In the Fourier transforms only one peak is observed at $R = 2.31$ Å for adsorbed

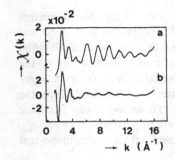

Figure 8.8. EXAFS of bromine adsorbed on graphitic Grafoil sheets with x-ray polarization (*a*) parallel and (*b*) perpendicular to the sheets (44).

bromine. For intercalated bromine $R_{Br-Br} = 2.44$ Å and at 100 K there is an indication for a second bromine atom at 4.1 Å. The Debye–Waller factor for adsorbed bromine is higher than for bromine vapor. For $\Theta = 0.6$ it has the same temperature dependence, but for $\Theta = 0.9$ the temperature dependence is stronger. A full analysis is not possible because the amplitude function of bromine on carbon apparently is different from that of gaseous bromine because of differences in many electron effects. The anisotropy in the EXAFS, as well as in the $1s \rightarrow 4p$ edge line is strongest at low temperature, when the molecules are better aligned.

The analysis of the Br—C EXAFS gave a distance of $R_{Br-C} = 2.9$ Å and a coordination number of $N = 6 \pm 1.5$ at 100 K (with decreasing R with increasing temperature) for the 0.6 and 0.9 monolayer coverages, while $R_{Br-C} = 2.6$ Å and $N = 3$ for the intercalated bromine. The sharp peak in the Fourier transform of the Br—C EXAFS measured at low temperature indicates that each bromine atom in Br_2 is predominantly situated on a specific site. Otherwise, a whole range of Br—C distances would occur and the Br—C EXAFS would be smeared out. Together with the available LEED information Heald and Stern came to the conclusion that the bromine atoms are almost above the centers of adjacent basal plane hexagons. The coordination number of $N = 6$ confirms this. The observed Br—Br distance of 2.31 Å in the adsorption state indicates a stretching of 0.03 Å of the molecule relative to the gas phase, while in the intercalated bromine the interaction with the substrate is so large that the stretching has gone so far (2.44 Å) to almost perfectly match with the distance between the centers of two adjacent hexagons (2.456 Å).

While the orientation of bromine at $\Theta = 0.6$ and 0.9 is nearly identical, at the low coverage of 0.2 and at room temperature the molecule is oriented differently. One atom is bound to a basal plane hexagon with a Br—C distance of 2.4 Å, while the other end of the molecule is free to flop around. The Br—Br distance is the same or slightly less than in the gas phase.

EXAFS measurements of krypton on grafoil were carried out by Bouldin and

Stern in a special aluminium cell with integral windows (45). Measurements were performed at coverages of 0.1, 0.2, and 0.35 monolayer and at 10 and 100 K. In the analysis it was necessary to include the third moment of the radial distribution function of each shell of atoms, because of anharmonicity of the vibrations. For krypton on grafoil μ_0 could be measured directly from the krypton vapor, so that no Fourier filtering or spline fitting was required. Only the $\Theta = 0.35$ sample contained Kr—Kr oscillations. To extract these from the mixture of Kr—Kr and Kr—C oscillations a difference spectrum of $\chi_\parallel$ and $\chi_\perp$ was made. Since $\chi_\parallel$ contained most of the Kr—Kr oscillations it was found that $\chi_\parallel - 0.35 \chi_\perp$ was almost free of Kr—C oscillations.

The $\Theta = 0.1$ sample showed more disorder at 10 K and a smaller anisotropy than at higher coverage. This indicates a tighter binding to the substrate for the low coverage sample. The angle between Kr—C and the surface normal is somewhat larger than predicted for the hole or bridge position of krypton above the center of the hexagon or between two carbon atoms. This also suggests that the dominant site for adsorption at higher coverage is the hexagonal hole, but that a fraction of the krypton atoms is adsorbed on tighter bonding defect sites, such as steps and edges.

Distances of $R_{Kr-Kr} = 4.26$ Å with $N = 1.8 \pm 0.5$ and $R_{Kr-C} = 3.6$ Å with $N = 7 \pm 2$ were found. The Kr—Kr distance of 4.26 Å is the expected $(\sqrt{3} \times \sqrt{3})30°$ distance, since this is the smallest distance between centers of carbon hexagons that is not excluded by the hard-sphere radii of the krypton atoms. The coordination number of 1.8 is higher than expected on the basis of random distribution and points to ordering.

The bromine and krypton adsorption studies fully demonstrate the power of EXAFS. EXAFS is able to directly characterize the adsorption site and to distinguish changes in adsorption as a function of coverage.

8.4. EXAFS OF METAL CATALYSTS

Most of the EXAFS studies of catalysts have dealt with metal catalysts, in conformity with the fact that metal catalysts are the most widely used catalysts in the industry. Metals of the $5d$ series (osmium, iridium, platinum, gold) and $4d$ series (ruthenium, rhodium, palladium) have been studied most. While the K- and L_3-EXAFS of $3d$ and $5d$ metals, respectively, can be studied at most synchrotron storageing facilities, the K-EXAFS of $4d$ metals can only be studied when the stored energy of the ring is high (Stanford, Hamburg, Cornell) or when a wiggler magnet is available to shift the critical energy upwards (Stanford, Daresbury, Frascati). In the following the EXAFS on metal catalysts will be discussed under three headings, EXAFS of monometallic catalysts, EXAFS of bimetallic catalysts, and edge structures. The edge structures of metal catalysts is discussed here instead of in Chapter 11 in order to keep the discussion of metal catalysts together.

8.4.1. Monometallic Catalysts

8.4.1.1. Structural Properties

Via, Sinfelt, and Lytle (46) have studied several Group VIII noble metal catalysts. Their results on monometallic catalysts will be discussed in this section, while the results on bimetallic catalysts and on white lines are presented in Sections 8.4.2 and 8.4.3, respectively. All experiments were performed on catalysts prepared via impregnation of SiO_2 or Al_2O_3 with aqueous solutions of chlorometallic acids, followed by drying and reduction by dihydrogen at 773 K (46,47). After storage in air the catalysts were rereduced in the EXAFS cell with H_2 at 698 K. This cell was described in Section 8.2.2. The analysis of the data followed the standard method. Phases and backscattering amplitudes were determined (46) from the bulk reference compounds Os, Ru, Cu, Ir, Pt, and $Ir_4(CO)_{12}$. The results obtained for the reference compounds demonstrated that the first-shell coordination number N and Debye–Waller factor $\Delta\sigma^2$ can be determined satisfactorily, even though these parameters have a high degree of correlation. The uncertainties in these parameters were estimated to be about 20 and 10%, respectively, on an absolute scale. The $\Delta\sigma^2$ data obtained for bulk platinum at elevated temperatures were in excellent agreement with calculations of the mean squares displacements of the platinum atoms about their equilibrium positions by means of a Debye model.

All catalysts studied had a very high degree of metal dispersion ($>70\%$), as determined by chemisorption studies. Nevertheless, the EXAFS spectra of the catalysts were very similar to those of the corresponding bulk metals (Fig. 8.9). Of course the oscillations were smaller because of lower N and higher σ^2 values in the catalysts. A quantitative analysis showed that for these catalysts $7 < N < 10$, while the σ values of the catalysts were about twice as large as those of the bulk metals. The interatomic distances differed by less than 0.01 Å from the corresponding bulk values, except for the Pt/Al_2O_3 catalyst whose R was lowered by 0.017 Å. Since this catalyst had also the lowest N value, the contraction in the interatomic distance may be due to the large fraction of surface atoms in this catalyst.

Measurements over a large temperature range were carried out by Marques et al. for a Pt/SiO_2 catalyst (48). They also improved the data analysis by correcting the EXAFS function of the first coordination shell for the dependence of the phase factor by comparison with platinum foil. In this way the side lobe in the radial distribution function obtained by Fourier transformation of the EXAFS function was removed and the least squares fitting procedure was made

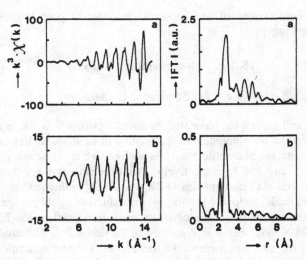

Figure 8.9. EXAFS ($k^3\chi(k)$ and magnitude of (k^3) Fourier transform ($|FT|$) of (a) bulk platinum and (b) a 1 wt.% Pt/Al$_2$O$_3$ catalyst (46).

to converge much easier. The results of the analysis demonstrated large differences between catalyst and platinum foil. R_{Pt-Pt} decreased strongly above 400 K for the catalyst and only slowly for the foil, while the Debye–Waller factor was larger for the catalyst and its temperature dependence was steeper too. The apparent contraction of the interatomic distance in the catalyst with temperature seems to be caused by an inadequate representation of thermal disorder in the simple EXAFS theory.

The EXAFS spectra of the monometallic catalysts (46) were further analyzed by Greegor and Lytle to obtain the coordination numbers (N_2 and N_3) of the second and third shell of neighboring atoms (49). Since N_2 and N_3 are much more sensitive to the size and shape of the metal particles than the coordination number of the first shell (N_1) [for instance $N_2 = 0$ for a one atom thick (111) layer of atoms], the morphology of the metal particles might in principle be determined by comparing measured values with theoretical predictions for different geometrical shape models (spheres, cubes, disks). The method should be most sensitive for small particles below 20 Å and especially for such small particles the normal methods of electron microscopy and x-ray diffraction are not applicable. Furthermore, small particles are also very interesting from a fundamental point of view because they may have a shape and structure different from that of bulk metal particles. On the other hand, the measurement of N_2 and N_3 by EXAFS is difficult because the contribution of the second and third shells to the EXAFS signal becomes very small for small particles. In addition, especially for small particles, one might expect a nonnegligible spread in R_2

and R_3 because of differences in interatomic distances between bulk and surface atoms. As a result the corresponding peaks in the radial distribution function are broad and may even be lost completely when they become too broad, because of the truncation of the data below $k = 3$ Å^{-1} (38, 39). These problems indeed prevented Greegor and Lytle of reaching unambiguous conclusions. They concluded that the osmium particles in Os/SiO$_2$ were most likely disk shaped, and that the ruthenium, iridium, and copper particles in M/SiO$_2$ catalysts were most likely spherelike. For the platinum particles, however, all shapes appeared equally possible. The study of Greegor and Lytle (49) shows that the method of determining the morphology of small metal particles by EXAFS, although promising in theory, is very difficult in practice. High quality data measured to high-k values should be available before attempting a quantitative analysis of the second and third shells.

The average coordination number of metal atoms in the small metal particles on the supports of the catalysts is often low, as demonstrated in the studies discussed previously. Therefore, a contraction of the interatomic distances from the bulk metal value is anticipated. Nevertheless, in most cases the contraction was smaller than 0.05 Å and only two studies reported more dramatic contractions. Renouprez and co-workers reported values of 0.13 for Pt/SiO$_2$ and 0.2 Å for Pt/Y–zeolite (50–52). Their results are based on primary data with low S/N ratios and are in disagreement with the much smaller values observed by others for platinum catalysts. Unambiguous contractions have been reported by Apai et al. for copper and nickel particles made by metal evaporation on amorphous carbon substrates (53). The interatomic metal–metal distance was found to decrease with decreasing coverage of the substrate, as did the threshold energy of the K-edge [similar to shifts observed in x-ray photoemission (54)]. Maximum contractions observed were 0.22 Å for copper and 0.25 Å for nickel. At the lowest metal coverages the interatomic distances approached those of diatomic particles. In agreement with this the EXAFS amplitude decreased with decreasing coverage because of the fact that at low coverage only a small fraction of the metal atoms were present in the form of particles and most atoms were present as isolated atoms.

8.4.1.2.   Catalyst Preparation

The EXAFS technique has only been used in a few studies of catalyst preparation up to now and it seems as if one is not fully aware yet of the potential of the technique in this area. Very successful EXAFS studies of the preparation of Ni/SiO$_2$ and Co/TiO$_2$ catalysts via hydrolysis of a mixed solution of a metal alkoxide and ethyl orthosilicate or n-propyl orthotitanate were published by Tohji et al. (23, 24). These studies were performed with the aid of a laboratory EXAFS apparatus with a rotating anode x-ray generator. The nickel on SiO$_2$

(23) and cobalt on TiO_2 (24) catalysts were made by dissolving the appropriate metal nitrates in ethylene glycol, followed by addition of ethyl orthosilicate, filtering, and washing of the resulting gel, followed by drying, calcination, and reduction by hydrogen. Nickel or cobalt K-EXAFS spectra were taken after each preparation step. The Fourier transform of the nickel K-EXAFS spectrum of the gel formed by addition of ethyl orthosilicate to the solution of nickel nitrate in ethylene glycol showed one peak at $R = 2.05$ Å belonging to the $Ni^{2+}-O^{2-}$ distance (23). Although the ^1H NMR spectrum indicated that the $Ni[OSi(OEt)_3]_2$ species had been formed, the silicon atoms in the second shell could not be seen with EXAFS. After drying at 383 K a second peak at $R = 3.30$ Å showed up, however. Inverse Fourier transformation and comparison with a calculated transform showed good agreement when silicon was taken as scatterer, but no agreement when nickel was taken as scatterer atom. The reason that the second silicon shell is not seen in the gel is, according to the authors, due to the uncertainty in the position of the silicon atom. In the gel rotation of the O—Si group around the Ni—O bond is possible, while in the dried gel the position of the silicon atoms relative to the nickel atoms is fixed. It is hard to visualize, however, how a rotation without a change in bond length can make the second shell disappear. Another explanation might therefore be that in the gel there are oxygen atoms of water molecules as well as silicon atoms at a distance of about 3.3 Å from the Ni^{2+} ion. Since the phases of oxygen and silicon differ by nearly π radians their contributions to the EXAFS signal will cancel each other. Drying removes the water molecules and makes the Ni—Si contribution observable.

Calcination of the dried sample led to a spectrum resembling that of NiO, with Ni—O and Ni—Ni peaks in the Fourier transform. Differences in the relative magnitudes of these peaks compared to NiO were attributed to a small contribution of Ni—O—Si species. This and additional IR and electron microscopy results led to the conclusion that the NiO particles formed in the alkoxide method were smaller than 20 Å and had silicate groups at their surface. After reduction the Fourier transform of the EXAFS spectrum indicated that metallic nickel particles had been formed. The analysis of the EXAFS spectra of nickel catalysts prepared via the classic impregnation route demonstrated that much larger NiO and nickel particles were formed in this method than in the alkoxide method.

In the analogous preparation of a cobalt on TiO_2 catalyst no Co—O—Ti bond formation was observed after addition of $Ti(OC_3H_7)_4$ to a solution of cobalt nitrate in ethylene glycol and drying of the resulting gel at 383 K (24). Apparently the Co—O—Ti structure is not as rigid as the M—O—Si structure. After calcination at 723 K the Fourier transform showed four peaks at the same positions as those of Co_3O_4, but with magnitudes decreasing rapidly with distance. Thus after calcination at 723 K small Co_3O_4 particles were formed, of about 8-Å diameter. Calcination at 973 K, however, led to the formation of $CoTiO_3$ (24).

Lagarde et al. have published EXAFS data on Pt/Al_2O_3 catalysts during different stages of preparation (55, 56). They observed that after impregnation of the support with H_2PtCl_6 or H_2PtBr_6, followed by drying, the EXAFS was almost equal to that of chloro- or bromoplatinic acid in solution. After calcination at 803 K the EXAFS had changed and analysis showed that most of the chlorine and probably all of the bromine had been removed from the first coordination shell of platinum. Apparently, the PtX_6^{2-} complex had been transformed in PtO_2. On the other hand, calcination at 973 K led to the appearance of the EXAFS of metallic platinum. This is in agreement with the known thermodynamic instability of platinum oxides above 873 K. After reduction with H_2 at 753 K a Pt—Pt peak was observed in the Fourier transform with $N = 6$. The Debye–Waller factor quoted was, however, hardly different from that of platinum foil, which is very surprising for small metal particles. Lagarde et al. also compared second- and third-neighbor shells with shape models for metal particles. However, because of very poor signal to noise, their conclusions based on these higher-shell peaks seem highly premature.

Further platinum L_3-EXAFS studies were performed on bimetallic Pt—Re and Pt—Rh on Al_2O_3 catalysts (57, 58). Because of the presence of the rhenium L_2-edge at about 400 eV above the platinum L_3-edge, the platinum EXAFS data for the Pt—Re/Al_2O_3 catalyst were restricted to $k < 9$ Å^{-1}. The addition of a second metal salt (NH_4ReO_4 or $RhCl_3$) had a strong influence on the state of the Pt^{4+} ions in the catalyst. Whereas in the monometallic catalyst the $PtCl_6^{2-}$ complex stayed intact during drying at 383 K, the first neighbor shell of Pt^{4+} in the dried bimetallic catalyst consisted of both chlorine and oxygen ions. For the dried Pt—Re catalyst $N_{Pt-Cl} \approx 2$ and for the dried Pt—Rh catalyst $N_{Pt-Cl} \approx$ 4. After calcination in air at 803 K the bimetallic catalysts, as well as the monometallic catalyst, showed about 2 Cl ions in the first shell. Surprisingly, in all cases (dried and calcined) the total number of atoms in the first shell around the Pt^{4+} ions was found to be about eight. All these results are intriguing, but require many more experiments before a definite conclusion can be drawn. For instance, the changes in the first coordination sphere of Pt^{4+} by the addition of the second metal salt may be simply due to changes in the Pt^{4+} and Cl^- concentration or to changes in the pH. It is obvious that further studies along the lines described by Lagarde et al. (57, 58) might provide very valuable information on the state of the metal ions on the catalyst support after various catalyst treatment steps.

8.4.1.3. Gas Adsorption

Admission of oxygen to metal catalysts has been studied by several groups. Joyner reported only one peak in the Fourier transform of the EXAFS function of Pt/SiO_2 and Pt/Al_2O_3 catalysts oxidized at 773 K and assigned it to a Pt—O distance (59). The fact that no second- or third-shell distances were observed was explained as being due to the amorphous character of the platinum oxide particles.

Nandi et al. exposed Pt/SiO_2 (60) and Pd/SiO_2 (61) catalysts to air and demonstrated that for catalysts with a low dispersion the Fourier transform of the platinum L_3, respectively, palladium K-EXAFS consists of a metal–metal peak, while the highly dispersed catalysts show a metal–oxygen peak. When the dispersion was about 60%, exposure to air resulted in crystalline Pt_3O_4 or PdO particles with small metallic cores, while for 80% dispersion the particles were fully oxidized. These results are in good agreement with temperature programmed oxidation and reduction studies of well-dispersed metal catalysts, which showed that oxidation of metal particles is diffusion limited (62). As a consequence small metal particles may be totally oxidized already at room temperature, while large particles can only be completely oxidized at elevated temperatures. Exposure of large particles to air room temperature only leads to oxidation of the outer layers.

Fukushima et al. have published results (63) on the adsorption of dioxygen on two reduced Pt/Al_2O_3 catalysts with H/M chemisorption values of 0.47 and 1.14 (mean particle size 26 and < 10 Å, respectively). EXAFS spectra measured at 90 K in an *in situ* cell before and after gas admission showed dramatic influences of catalyst dispersion and of the way in which dioxygen was admitted. Thus admission of 10-Torr O_2 at 77 K to the reduced Pt/Al_2O_3 catalyst with H/M = 1.14 changed the Fourier transform significantly. The metallic Pt—Pt peak decreased in intensity by about 30% and shifted 0.25 Å to a shorter distance and a new Pt—O peak appeared at 2.0 Å. Slow warming of the sample to room temperature changed the Fourier transform further. The Pt—Pt peak disappeared and only Pt—O and Pt—Pt peaks of platinum oxide could be observed. This demonstrated that while O_2 admission at 77 K leads to chemisorption and only to a slight oxidation and disruption of the platinum particles, warming up to 300 K under dioxygen causes complete oxidation of the small (< 10 Å) particles. The same treatment (O_2 admission at 77 K and warming up to 300 K during 45 min) did not disrupt the 26-Å platinum particles of the catalyst with H/M = 0.47. The EXAFS spectrum was almost the same as that after reduction and the Pt—Pt peak in the Fourier transform underwent < 10% reduction in intensity. On the other hand, direct room temperature admission of O_2 to this catalyst with H/M = 0.47 induced substantial oxidation and disruption. The metallic Pt—Pt peak was reduced by 50% and shifted by −0.2 Å and new Pt—O and Pt—Pt peaks of PtO_2 appeared (Fig. 8.10). These results of Fukushima et al. are in perfect accordance with the results obtained by Nandi et al. and demonstrate that oxygen chemisorption is not restricted to the outermost layer of metal atoms, but leads to oxidation of a skin of several atom layers.

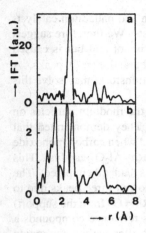

Figure 8.10. Magnitude of (k^3) Fourier transform of (a) a reduced Pt/Al$_2$O$_3$ catalyst (H/Pt = 0.47, d = 26 Å) and (b) Pt/Al$_2$O$_3$ after admission of 10 Torr O$_2$ at 298 K (63).

Oxygen adsorption studies on ruthenium catalysts by Lytle et al. (64) confirm the results for platinum and palladium. Thus after dioxygen admission at 298 K to a Ru/SiO$_2$ catalyst the Fourier transform of the ruthenium EXAFS showed a peak at the metal Ru—Ru distance, as well as a Ru—O peak, demonstrating that at 298 K chemisorption had taken place. After exposure of the catalyst to O$_2$ at 673 K the metal Ru—Ru peak had disappeared and two new peaks due to Ru—O and Ru—Ru distances like in RuO$_2$ appeared. Thus at 673 K full oxidation took place. Similar results have been published by Cox (65).

Fukushima et al. also admitted carbon monoxide to their Pt/Al$_2$O$_3$ samples and observed new peaks in the Fourier transform around 1.5 and 3.6 Å for the best dispersed catalyst [due to Pt—C and Pt—(C)—O distances, probably] and a shift of −0.12 Å for the metallic Pt—Pt peak (63). The fact that the EXAFS spectrum of the less well dispersed catalyst did not change on admission of carbon monoxide—similar to its invariableness to dioxygen admission—was ascribed to much larger metal particle size. For platinum particles that have 50% of their atoms at the surface one calculates that on average N(Pt—Pt) = (12 + 9)/2 = 10.5, while N(Pt—O) = 3/2 = 1.5 for chemisorption of oxygen atoms in hollow sites on the metal surface. Together with the much lower backscattering amplitude for oxygen this leads to very low ratios of Pt—O and Pt—Pt peaks in the Fourier transform.

The substantial decrease in Pt—Pt distance when absorbing dioxygen or carbon monoxide was claimed to be real. Renouprez has observed a similar effect upon dioxygen adsorption on Pt/Y-zeolite and Pt/SiO$_2$ catalysts (50–52). As noted by Burwell [cf. discussion at the end of Fukushima's article (63)] this shortening is in disagreement with x-ray diffraction results, which showed that the lattice constants of passivated platinum and palladium catalysts differed by <0.01 Å from that of well-reduced catalysts. We therefore suggest that the shift of the Pt—Pt peak upon gas adsorption is not real but is caused by an interference between (nearly) coinciding Pt—Pt and Pt—O peaks. A thorough analysis of the imaginary part of the Fourier transform may solve this problem.

The influence of adsorbed gases on the topology of the rhodium particles on Al_2O_3 has been studied by van't Blik et al. (66–68). They demonstrated that the Rh-Rh oscillation in the rhodium K-EXAFS vanished when carbon monoxide was adsorbed on a highly dispersed and well-reduced Rh/Al_2O_3 catalyst. This proved that disruption of the small rhodium particles had taken place. The Fourier transform of the resulting EXAFS spectrum showed three peaks due to Rh—C, Rh—O (oxygen from the carbonyl group) and Rh—O (from the support) distances. With the aid of $[Rh(CO)_2Cl]_2$ and Rh_2O_3 as reference compounds a full analysis of the data could be made showing that after carbon monoxide adsorption each rhodium atom was surrounded by two carbon monoxide molecules and three oxygen ions from the support (Fig. 8.11). From complementary ESR, XPS, and IR studies it was concluded that the oxidation state of rhodium was $1+$. Thus $Rh(CO)_2^+$ species had been formed that were attached to three oxygen ions of the support. This EXAFS study proved that a former suggestion put forward on the basis of IR experiments on carbon monoxide adsorbed on Rh/Al_2O_3 catalysts was true and settled the question whether fully reduced rhodium particles existed at all before carbon monoxide adsorption. The $Rh(CO)_2^+$ species could be transformed into rhodium particles by desorption at 573 K (68). The EXAFS spectrum showed the reappearance of slightly sintered rhodium particles. Heating a Rh/Al_2O_3 catalyst to 523 K under flowing carbon monoxide resulted in a very complicated EXAFS spectrum, but nevertheless it could be established that during the Boudouard reaction (2 CO → CO_2 + C) the Rh—Rh bonds were disrupted. On the other hand, heating of a Rh/Al_2O_3 catalyst under flowing synthesis gas (CO + H_2) kept the catalyst in the metallic state, although some sintering occurred as judged from the value of N.

The disruption of the rhodium particles by carbon monoxide proved to be dependent on the metal particle size. Whereas an extremely well dispersed 0.6 wt.% Rh/Al_2O_3 catalyst with $N_{Rh-Rh} = 3.7$ showed no Rh—Rh EXAFS contribution after adsorption of carbon monoxide, the EXAFS of a 1 wt.% Rh/Al_2O_3 catalyst ($N_{Rh-Rh} = 5.9$) and of two Rh/TiO_2 catalysts with rhodium loadings of 0.5 and 1 wt.% ($N_{Rh-Rh} = 4.4$ and 5.9, respectively) not only showed the presence of $Rh(CO)_2^+$ species after carbon monoxide adsorption, but also still the presence of some Rh—Rh bonds (68–70). Apparently the larger particles present in these catalysts cannot be broken up by carbon monoxide adsorption, because the heat of adsorption of carbon monoxide is only sufficient to break a limited number of Rh—Rh bonds (71). The support does not influence

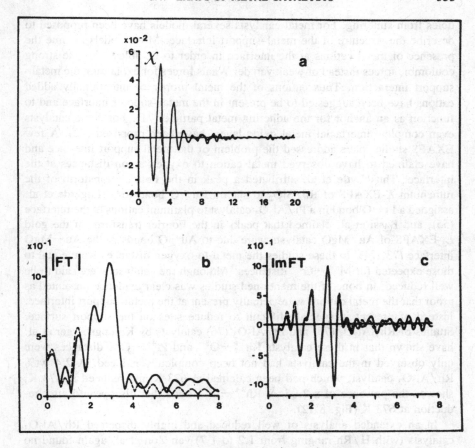

Figure 8.11. (*a*) Rhodium *K*-EXAFS of a Rh(0.57 wt. %)/γ-Al₂O₃ catalyst after CO admission at room temperature, (*b*) and (*c*) magnitude ($|FT|$) and imaginary part, respectively, of the (k^3) Fourier transform (corrected for the Rh—O phase shift). The solid lines represent the data and the dotted lines represent the calculated Rh—(C)—O + R—O EXAFS (67).

the structure of the Rh(CO)₂⁺ species formed during carbon monoxide adsorption. The distance between the rhodium cation and the O²⁻ anions of the support is the same for γ-Al₂O₃ and TiO₂ supported systems (2.12 Å), and also the number of Rh⁺—O²⁻ bonds is equal to three for both supports (67, 69, 70).

8.4.1.4. Metal–Support Interaction

A very important issue in catalysis is the interaction between the catalytically active phase and support, since it is this interaction that prevents catalyst particles from sintering. For metal catalysts several models have been proposed to describe the structure of the metal–support interface. Most models assume the presence of metal cations at the interface in order to be able to invoke strong coulombic forces instead of weak van der Waals forces for explaining the metal–support interaction. Thus cations of the metal proper or intentionally added cations have been suggested to be present in the metal–support interface and to function as an anchor for the adhering metal particle (71). For some catalysts even complete interfacial metal oxide layers have been proposed (72). A few EXAFS studies have addressed the problem of the metal-support interface and have claimed to have observed metal cation to oxygen anion distances at the interface. Thus Lytle et al. attributed a peak in the Fourier transform of the ruthenium K-EXAFS of Ru/SiO_2 to such a $Ru-O$ bond (64). Lagarde et al. assigned a $Pt-O$ bond in a Pt/Al_2O_3 catalyst to platinum cations at the interface (55), and Bassi et al. claimed that peaks in the Fourier transform of the gold L_3-EXAFS of Au/MgO catalysts were due to $Au-O$ bonds in the $Au-MgO$ interface (73, 74). In these studies the metal-to-oxygen distances were equal to those expected for $M^{n+}-O^{2-}$ distances. Although the catalysts were said to be well reduced, in none of the mentioned studies was clear evidence presented as proof that the metal cations were actually present at the metal–support interface, instead of present in special, difficult to reduce sites on the support surface. Studies of Rh/Al_2O_3 (75) and Pt/Al_2O_3 (76) catalysts by Koningsberger et al. have shown that in these catalysts $Rh^{3+}-O^{2-}$ and $Pt^{4+}-O^{2-}$ distances were only observed in the catalysts had not been completely reduced. A 2.4 wt% Rh/Al_2O_3 catalyst, which had been calcined at 623 K and reduced at 473 K, showed the presence of a 2.05 Å $Rh^{3+}-O^{2-}$ bond that disappeared after reduction at 673 K (Fig. 8.12).

In an extended analysis of well reduced and highly dispersed Rh/Al_2O_3 catalysts (with H/Rh ranging from 1.2 to 1.7) van Zon et al. again found no evidence for $Rh^{3+}-O^{2-}$ distances, but they did find a peak at 2.7 Å in the Fourier transform of the rhodium K-EXAFS belonging to a rhodium atom to oxygen anion (Rh^0-O^{2-}) distance (77, 78). To be able to distinguish this Rh^0-O^{2-} peak from the much stronger $Rh-Rh$ peak with side lobe a modified analysis method had to be used, with a Fourier transformation corrected for phase shift and backscattering amplitude. In this way the side lobe at the low-R side of the $Rh-Rh$ peak in the radial structure function, which is caused by the nonlinear k dependence of the phase shift and the low-frequency variation in the backscattering amplitude, was removed and the Rh^0-O^{2-} peak became detectable. By subtracting the calculated $Rh-Rh$ contribution to the EXAFS spectrum a difference EXAFS spectrum was obtained, which on forward and backward Fourier transformation proved to be due to a $Rh-O$ distance of 2.7 Å. This distance is much larger than that of $Rh^{3+}-O^{2-}$ in Rh_2O_3 (2.05 Å) and must be due to a Rh^0-O^{2-} distance and not to a $Rh^{3+}-O^{2-}$ distance.

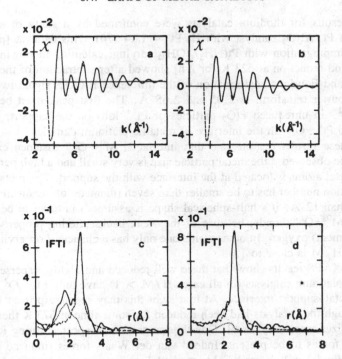

Figure 8.12. EXAFS obtained by back transformation of the Fourier transform of the experimental data (solid line) and the calculated Rh—Rh + Rh—O EXAFS: (a) T_{red} = 473 K and (b) T_{red} = 673 K. The (k^1) Fourier transform (corrected for the Rh—Rh phase shift and backscattering amplitude) of the experimental data is represented by a solid line. The corresponding Fourier transform of the calculated Rh—Rh EXAFS is given by a dashed curve; the Fourier transform of the calculated Rh—O contribution is indicated by the dotted curve: (c) T_{red} = 473 K and (d) T_{red} = 673 K. (75).

This proves that indeed structural information is obtained on interfacial rhodium atoms and that the Rh—Al$_2$O$_3$ interface apparently consists of rhodium atoms "resting" on O^{2-} anions of the support.

Also the analysis of the rhodium K-EXAFS of two Rh/TiO$_2$ catalysts with H/Rh values of 1.5 and 1.8 showed the presence of long (2.7 Å) Rh^0—O^{2-} distances, besides the Rh—Rh distance (69). Just as in the series of Rh/Al$_2$O$_3$ catalysts (77, 78), the measured number of oxygen neighbors for this Rh^0—O^{2-} bond increases with decreasing coordination number of the Rh—Rh bond. For three-dimensional particles the fraction of metal atoms present in the metal–support interface increases with decreasing particle size. This means that both on Al$_2$O$_3$ and TiO$_2$ the rhodium metal particles are three-dimensional rather than two-dimensional one atom thick rafts, because in that case the support oxygen coordination number should have been independent of particle size.

The results for rhodium catalysts were confirmed by a study of a highly dispersed Pt/Al_2O_3 catalyst with $H/Pt = 1.14$ (76). This catalyst [prepared via wet impregnation with $Pt(NH_3)_4(OH)_2$, drying, calcining in flowing O_2 at 623 K, and reduction at 623 K for 2 h] showed after subtraction of the $Pt-Pt$ EXAFS and Fourier transformation of the difference EXAFS signal two peaks in the Fourier transform at 2.05 and 2.65 Å. The first peak must be due to $Pt^{4+}-O^{2-}$ in unreduced PtO_2 particles or Pt^{4+} ions on the support, and the second to Pt^0-O^{2-} in the interface of metallic platinum particles.

All these results demonstrate that interfacial M^0-O^{2-} distance can only reliably be observed if the metal particle size is very small and a high percentage of the metal atoms is located at the interface with the support. The metal–metal coordination number has to be smaller than seven (diameter of the metal particle smaller than 12 Å, if a half-spherical shape is assumed) in order to be able to observe M^0-O^{2-} bonds, because of the weak backscattering property of the (light element) oxygen. In other words one only has a chance of observing these bonds if H/M is close to 1.0.

The EXAFS results show that these well reduced and highly dispersed rhodium and platinum catalysts (in all cases $H/M > 1$) have only M^0-O^{2-} bonds in the metal–support interface. At first sight this may seem surprising because even though the catalysts had been reduced for some time at 673 K their metal particle sizes were very small. One thus might have expected strong ionic interaction forces to be present. Indeed van der Waals forces (induced dipole-induced dipole) will never be able to explain this resistance against sintering. But ion-induced dipole interactions between interfacial O^{2-} anions and rhodium atoms are one or two orders of magnitude stronger and are of the right order of magnitude to explain the adherence of small metal particles to the support.

A supposedly special type of metal–support interaction has attracted quite some attention. This so-called strong metal–support interaction (SMSI) has been invoked to explain the fact that dihydrogen and carbon monoxide can no longer be adsorbed on a noble metal catalyst supported on carriers like TiO_2, when the catalyst is reduced at elevated temperatures ($T > 673$ K). Although at present the SMSI effect is explained by covering of the metal particles by reduced support species than by a special metal–support interaction, the name and acronym still stand (79–81). EXAFS of well-dispersed Pt/TiO_2 catalysts has been measured by Short et al. (82). Their analysis indeed showed small platinum particles to be present with $N_{Pt-Pt} \approx 6$, $R_{Pt-Pt} = 2.70$ Å (somewhat smaller than bulk platinum) and rather large Debye–Waller factors. No $Pt-Ti$ distance could be detected. On the other hand even when such distances would have been present they would have been difficult to discern using the standard method of analysis. New measurements and application of the new data analysis procedure, including the difference file technique and the phase and backscattering amplitude corrected Fourier transforms, might show interfacial Pt—Ti or Pt—O distances.

8.4.2. Bimetallic Catalysts

Ever since their first introduction in the catalytic reforming process for the manufacture of high octane gasoline, bimetallic catalysts have enjoyed large attention. Present day industrial catalysts therefore usually consist of platinum and a second metal component (iridium, rhenium, tin, or germanium) supported on Al_2O_3. In spite of many investigations the oxidation state of the second component and the structure of the bimetallic catalysts is still under debate. For the Pt—Ir system the opinions diverge least and there is more or less general agreement that both platinum and iridium are fully reduced and intimately mixed in the form of very small alloy particles. Sinfelt calls such particles bimetallic clusters to allow for the possibility of nonideal homogeneity, such as surface enrichment or phase segregation (83). In this chapter we shall adopt the word bimetallic, but we prefer the word particle instead of cluster to denote an aggregate of metal atoms, because the word cluster is already in use in modern inorganic chemistry to denote an aggregate consisting of one or several metal atoms surrounded by organic and or inorganic ligands. For the widely used Pt—Re, Pt—Sn, and Pt—Ge catalysts opinions range, however, from real bimetallic systems— in which both components are in the metallic state and intimately mixed—to systems in which only platinum is metallic and the other component is in the oxidic state (rhenium, tin, or germanium) and not in contact with the platinum particles. Because of these unsolved problems and because of the obvious industrial interest much attention is paid to the characterization of bimetallic catalysts.

Short et al. studied Pt—Re/Al_2O_3 (84). Although this study was hampered by noise and many glitches the rhenium L_3-white line intensity was found to be about a factor of two larger than that of the rhenium metal reference powder. Therefore, it was concluded that the majority of the rhenium was present in an oxidic form, probably as Re^{4+}. In agreement with this conclusion the rhenium L_3-EXAFS spectrum showed a peak at $R = 1.95$ Å (corrected for a phase shift of 0.35 Å, close to the value of 0.34 observed in PtO_2). This distance is too short to be explained by metallic Re—Re or Re—Pt distances (about 2.6 Å) and is close to the value of 1.80 Å for the Re—O distance in ReO_2. The conclusion of this study, that rhenium in a Pt—Re/Al_2O_3 catalyst is not completely reduced to the metallic state, agrees with that of some catalytic studies, but is in contradication with several other studies (85). This is not too disturbing, however, since it is known, that the reducibility of supported rhenium oxide is delicately dependent on variables such as water vapor pressure and chlorine content of the catalyst.

Sinfelt, the inventor of the industrial Pt—Ir catalyst that has been put into practice by Exxon, has published the majority of EXAFS studies on bimetallic systems together with Via and Lytle. So far, they have published detailed EXAFS studies on Ru—Cu, Os—Cu, Rh—Cu, Pt—Ir, and Rh—Ir, most of them supported on SiO_2 as well as on Al_2O_3 (86–90). In these articles they not only very carefully addressed the much more abundant problems in the analysis of bimetallic versus monometallic catalysts, but at the same time they obtained a wealth of quantitative information that greatly enlarges our knowledge of such complicated systems. In the following we shall successively discuss their method of analysis and their results.

The analysis of EXAFS data of multicomponent systems is much more complicated than that of one-component systems because the contributions of several EXAFS functions with several different scatterer atoms must be unraveled. A short look at a two-component system (such as a bimetallic catalyst, without taking into consideration the contribution of support atoms) demonstrates the rapidly increasing complexity of the analysis. If we consider the EXAFS function associated with absorber atom X and first-shell neighboring scatterer atoms X and Y, the following equation is obtained:

$$[\chi(k)]_x = [\chi^1(k)]_{xx} + [\chi^1(k)]_{xy} \tag{1}$$

with

$$[\chi^1(k)]_{xy} = [A(k)]_{xy} \sin [\Delta(k)]_{xy} \tag{2}$$

and $A(k)$ and $\Delta(k) = 2kR + \delta(k)$ as defined in Chapter 6, Section 6.3.3.1. An analogous formula can be written down for $[\chi(k)]_y$. Instead of having to determine five parameters from one experimental $\chi(k)$ function in a range of k values for a one-component system, for a two-component system one must determine 18 parameters from two experimental EXAFS functions measured at the x-ray absorption edges of components X and Y. Since the backscattering amplitude is independent of the absorber atom and because of the additivity of the absorber and scatterer contributions to the phase factor (91, 92) the number of parameters is decreased by three. Like in the case of a one-component system, it is almost impossible to disentangle all 15 parameters at once and one therefore uses information on $[F(k)]_x \exp(-2k^2\sigma_{xx}^2)$, $[F(k)]_y \exp(-2k^2\sigma_{yy}^2)$, $[\delta(k)]_{xx}$, $[\delta(k)]_{yy}$, and $[\delta(k)]_{xy}$ from reference materials to extract the following 10 parameters from the two experimental EXAFS functions: N_{xx}, N_{xy}, N_{yx}, N_{yy}, R_{xx}, R_{xy}, R_{yy}, $\Delta\sigma_{xx}^2$, $\Delta\sigma_{xy}^2$, and $\Delta\sigma_{yy}^2$

It is clear that $[F(k)]_x$ and $[F(k)]_y$, in combination with, respectively σ_{xx}^2 and σ_{yy}^2, as well as δ_{xx} and δ_{yy} can be obtained from pure X and Y materials or from supported monometallic X and Y catalysts. Sinfelt et al. have given two answers to the problem of the determination of δ_{xy}. In their study of RuCu catalysts (86) they have made use of the additivity and transferability of absorber and scatterer contributions to the phase function. By measuring the EXAFS functions of the RuO_2 and Cu_2O metal oxides and of ruthenium and copper metal the phase functions for RuCu and CuRu can be determined:

$$\delta_{RuCu} = \delta_{RuO} + \delta_{CuCu} - \delta_{CuO} \qquad (3)$$

$$\delta_{CuRu} = \delta_{CuO} + \delta_{RuRu} - \delta_{RuO} \qquad (4)$$

In their paper on Os—Cu Sinfelt et al. (87) introduced another method of determining δ_{xy}. They start with the calculation of a trial function δ_{xy} by using results of theoretical calculations by Teo and Lee (91) of the contributions of absorber and scatterer atoms to the phase shift. Then the phase shift δ_{xy} is altered by varying the threshold absorption energy E_0. In this way a series of trial functions $[\delta(k)]_{xy}$ is obtained, each of which corresponds to a different E_0. By making use of the additivity of the absorber and scatterer contributions to the phase factor the corresponding series of trial functions for $[\delta(k)]_{yx}$ can be calculated. For each set of trial phase functions for xy and yx an iterative least squares analysis is carried out for the two EXAFS functions $[\chi(k)]_x$ and $[\chi(k)]_y$. Thus for each value of E_0 12 N, R, and δ parameters are obtained, including R_{xy} and R_{yx}. In general, R_{xy} will turn out to be inequal to R_{yx}. Therefore, the additional physical constraint that R_{xy} should be equal to R_{yx} is introduced. For Os—Cu and Ir—Rh this led to a single solution for E_0, while for Rh—Cu values of E_0 between -10 and -4 eV were found acceptable. A somewhat unpleasant side effect was that the minimum in the least squares derivation plot was shallow and did not occur at the same value of E_0, which corresponds to the most reasonable physical situation at which $R_{xy} = R_{yx}$.

An alternative method would be to make use of the fact that the parameters $[\Delta(k)]_{xy}$ and $[\Delta(k)]_{yx}$ can both be obtained from the experimental $[\chi(k)]_x$ and $[\chi(k)]_y$ if the $[\chi(k)]_{xx}$ and $[\chi(k)]_{yy}$ contributions could be separated. Since $R_{xy} = R_{yx}$, subtraction of Δ_{yx} from Δ_{xy} gives

$$[\Delta(k)]_{xy} - [\Delta(k)]_{yx} = [\delta(k)]_{xy} - [\delta(k)]_{yx} \qquad (5)$$

Because

$$[\delta(k)]_{xy} + [\delta(k)]_{yx} = [\delta(k)]_{xx} + [\delta(k)]_{yy} \qquad (6)$$

both sum and difference of $[\delta(k)]_{xy}$ and $[\delta(k)]_{yx}$ are known. Consequently, both phase functions can be obtained and in turn R_{xy} as well. The advantage of this method is that, in contrast to the other methods, it does not require additional experimental or theoretical information and makes full use of all the information inherent in the EXAFS functions χ_x and χ_y.

The analysis of the Pt—Ir system proved to be even more complicated than the general type of analysis described previously (88). Besides the problem of separating the contributions of platinum and iridium scatterer atoms, which is practically impossible because of the small difference in amplitude and phase functions, there is the additional complication that the iridium and platinum L_3-edges are only 348.5-eV apart, which causes an overlap of the iridium EXAFS function on the platinum EXAFS function. Sinfelt et al. presented a solution to this problem and the related difference in zero point for k. Furthermore, because of the similarity in amplitude and phase functions of platinum and iridium, they combined the two phase terms in χ_{Pt}, as well as in χ_{Ir} by one term only. Such a combination is, however, only allowed if the additional assumption is made that the difference between the PtPt and PtIr interatomic distances, as well as that between those of IrIr and IrPt, is negligible. Although $\Delta_{PtIr} - \Delta_{PtPt} = 2k\Delta R$ is certainly very small compared to Δ_{PtPt} ($R_{PtPt} = 2.775$ Å and $R_{IrIr} = 2.714$ Å), $\Delta_{PtIr} - \Delta_{PtPt}$ is not small compared to one at all, and at high-k values (e.g., above 10 Å^{-1}) an interference effect between the sin (Δ_{PtPt}) and sin (Δ_{PtIr}) terms should make itself noticeable. The consequence of using only one phase term in χ_{Pt} and χ_{Ir} is that $R_{PtIr} = R_{PtPt}$ and that $R_{IrPt} = R_{IrIr}$. The least squares procedure then leads to a minimum fitting error for certain values of R_{PtPt} and R_{IrIr}, which in general will be unequal. Indeed Sinfelt et al. found in all three PtIr catalysts differences in R_{PtPt} and R_{IrIr}, which were well beyond the estimated uncertainties. This is, however, physically impossible, since R_{PtIr} has to be equal to R_{IrPt} and thus in their method of analysis all R_{PtPt} and R_{IrIr} distances should have been equal. Only for the PtIr alloy the difference between R_{PtPt} and R_{IrIr} was found to be smaller than the uncertainty of the method. The authors stated that equal distances are to be expected for a completely homogeneous alloy. But this is a (microscopic) misinterpretation of the (macroscopic) rule of Vegard. Although Vegard's rule states that there is a linear relationship between interatomic distance and alloy composition, this does not mean that for every alloy composition all interatomic distances are equal. On the contrary, in principle the interatomic distances R_{xx}, R_{yy}, and $R_{xy} = R_{yx}$ will be different.

Convincing evidence for the deviation from the virtual crystal approximation has been presented by Boyce and Mikkelsen (93, 94) for the ternary systems $Ga_{1-x}In_xAs$, $ZnSe_{1-x}Te_x$, $K_{1-x}Rb_xBr$, and $RbBr_{1-x}I_x$. They demonstrated that first neighbor distances remain closer to the corresponding distances found in the pure binary compounds than to the average (virtual) crystal distance determined from the lattice constant of the ternary phase. Second neighbor distances on the other hand follow Vegard's rule rather well. In view of the fact that the compressibility of Group VIII metals is even smaller than those of the covalent and ionic compounds studied by Boyce we certainly expect deviations from the virtual crystal approximation for alloys of Group VIII metals.

In view of all this we conclude that some conclusions of the work of Sinfelt et al. will have to be reinvestigated. Thus the statement (87) that a difference in R_{PtPt} and R_{IrIr} indicates that the average composition of the first coordination shell of atoms around platinum is different from that around iridium and the conclusion that in supported Pt—Ir particles there are platinum-rich and iridium-rich regions are premature, at least when based on the results of this EXAFS analysis.

Having discussed the method of analysis of bimetallic particles let us now turn to the results obtained by Sinfelt et al. (86–90). These are tabulated in Table 8.2. The results for Ru—Cu, Os—Cu, and Rh—Cu substantiate the qualitative picture that previous chemisorption and catalytic studies had produced. In these systems, in which the components are immiscible or only slightly miscible (Rh—Cu) in bulk, the bimetallic particles on the support consist of a

Table 8.2. Structural Parameters of Bimetallic Particles Supported on SiO₂ Derived from the EXAFS Data on Both X-Ray Absorption Edges

Metals	Coordination Numbers		Composition First Coordination Shell		Nearest-Neighbor Distances (Å)[b]	
RuCu	Ru	11	RuRu	0.92	RuRu	2.65
1:1	Cu	9	CuCu	0.50	CuCu	2.58
					RuCu	2.60
OsCu	Os	12.5	OsOs	0.83	OsOs	2.68
1:1	Cu	9.5	CuCu	0.49	CuCu	2.55
					OsCu	2.68
RhCu			RhRh	0.79	RhRh	2.68
1:1			CuCu	0.50	CuCu	2.62
					RhCu	2.64
RhCu			RhRh	0.92	RhRh	2.68
2:1			CuCu	0.44	CuCu	2.63
					RhCu	2.64
IrRh	Ir	11	IrIr	0.73	IrIr	2.72
1:1	Rh	10	RhRh	0.40	RhRh	2.72
					IrRh	2.72
IrRh[a]	Ir	7	IrIr	0.57	IrIr	2.75
1:1	Rh	5	RhRh	0.60	RhRh	2.71
					IrRh	2.71

[a] Carrier was Al₂O₃.
[b] Nearest-neighbor metal–metal distances in reference monometallic catalysts and metal foils were (Å): Ru/SiO₂, 2.66; Ru, 2.675; Cu/SiO₂, 2.56; Cu, 2.556; Os/SiO₂, 2.70; Os, 2.705; Rh/SiO₂, 2.690; Ir/SiO₂, 2.71; Ir/Al₂O₃, 2.70; Ir, 2.714.

core of ruthenium, osmium or rhodium covered by a layer of copper. Thus the average coordination number of copper is lower than that of the other metal component and the composition of the first coordination shell demonstrates that regions rich in ruthenium, osmium or rhodium exist (Table 8.2). In agreement with the higher miscibility for Rh—Cu it is found that rhodium exhibits a greater tendency than ruthenium or osmium to be coordinated by copper atoms. The Rh—Cu particles presumably contain some copper atoms in the interior of the particles. The fact that the CuCu interatomic distance in Rh—Cu is substantially larger than in metallic Cu—and also larger than in the Ru—Cu and Os—Cu catalyst particles—might also be taken as evidence for the tendency of copper to be present in the interior of the Rh—Cu particles.

In agreement with these results chemisorption of oxygen on a Ru—Cu/SiO$_2$ catalysts changed the copper K-EXAFS, but did not change the ruthenium K-EXAFS. As can clearly be seen in Fig. 8.13, the magnitude of the ruthenium K-EXAFS is decreased upon exposure of dioxygen to a Ru/SiO$_2$ catalyst (note the difference in left-hand and right-hand scales), but this was not the case for the Cu—Ru/SiO$_2$ catalyst. On the other hand, the magnitude of the copper K-EXAFS of Cu/SiO$_2$ as well as that of the Cu—Ru/SiO$_2$ decreased upon exposure to oxygen and the copper K-EXAFS spectrum of Cu—Ru/SiO$_2$ changed dramatically. Thus, the presence of copper in the exterior of bimetallic particles in the Cu—Ru/SiO$_2$ catalyst shields the ruthenium from the oxygen, and only copper is oxidized upon exposure of the Cu—Ru/SiO$_2$ catalyst to oxygen.

Also the Ir—Rh and Pt—Ir systems demonstrated a tendency for segregation and surface enrichment, although the metal components are completely miscible in the bulk. Nevertheless, quite a substantial enrichment of the core of these particles with iridium and of the surface with rhodium and platinum, respectively, was observed. This must be due to the much lower surface tension of rhodium and platinum than of iridium.

A full analysis of interatomic distances and Debye–Waller factors for the bimetallic particles must await further information. The only clear trend is a decrease of the R_{xx} distance and an increase of the R_{yy} distance, when the R_{xx} distance is larger than the R_{yy} distance in the pure metals or monometallic catalysts. Within the uncertainty of the measurements no more information can be obtained for the R_{xy} values than that they are about equal to $\frac{1}{2}(R_{xx} + R_{yy})$. All Debye–Waller factors are much larger than those of the bulk metals and are about as large as those of the monometallic catalyst systems. The uncertainties in σ_1^2 are large, however.

8.4.3.　White Lines

Correlations between chemical and electronic properties have always been very popular among chemists and in catalysis especially the relationship between

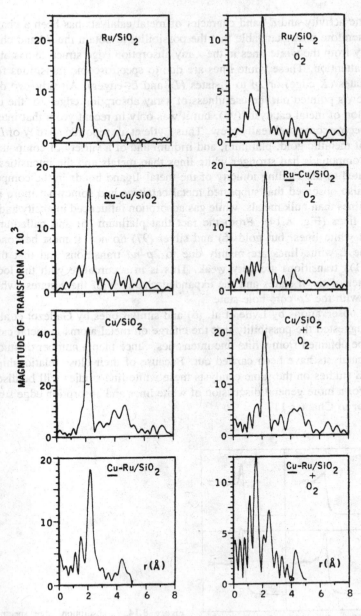

Figure 8.13. Fourier transform (magnitude) of Ru/SiO₂, Cu/SiO₂, and Cu—Ru/SiO₂ catalysts before (left-hand side) and after (right-hand side) exposure to oxygen. The metal atom for which the *K*-EXAFS was measured in the bimetallic catalysts is underlined. (86).

catalytic activity and d-band character of metal catalysts has been a challenge. It is therefore understandable that the possibility to obtain the d-band character directly from the white lines in the x-ray absorption edge structure has attracted much attention. These white lines are due to spectroscopic transitions from $1s$ to p states (K-edge) or $2p$ to d states (L_2 and L_3-edges). Already two decades ago Lewis pointed out the usefulness of x-ray absorption edges for the characterization of metal catalysts (95), but it was only in recent years that interest in this spectroscopic tool really grew. Thus Lytle et al. reported a study of L white lines of metallic gold, platinum, and iridium and of a variety of compounds (6, 96). Compounds had stronger white lines than metals and the intensities could be related to the Pauling ionicity of the metal–ligand bonds in the compounds. They also observed that supported metal catalysts had somewhat more intense white lines than bulk metals, while gas adsorption influenced intensity and shape of the lines (Fig. 8.14). From the fact that platinum (6) and palladium (97) exhibit white lines, but gold (6) and silver (97) do not, it must be concluded that the L white lines are mainly due to $2p$–nd transitions and that the $2p$–$(n + 1)s$ transitions are very weak. This is in accordance with the localized character of the d states and the expanded character of the s states, which do not fit with the $2p$ core hole state.

The observations by Lytle et al. (6) and similar ones by Gallezot et al. (98–100) suggested the possibility that the charge on metal atoms in metal catalysts could be obtained from white line intensities. Since then a number of studies of metal catalysts have been carried out. Because of their close relationship with EXAFS studies on the same catalysts these white line studies will be discussed here. For a more general discussion of white lines and absorption edge structure we refer to Chapter 11.

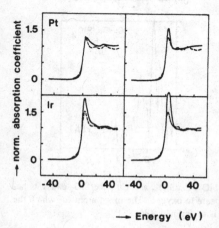

Figure 8.14. L_3-absorption edge spectra of 1 wt. % Pt/Al$_2$O$_3$ and 1 wt. % Ir/Al$_2$O$_3$ before (left) and after (right) exposure to oxygen, compared with the pure metals (dotted lines) (6).

In all cases studied, for example, platinum on SiO_2 (101), Al_2O_3 (6,101), TiO_2 (82), and Y–zeolite (98), the intensities of the L white lines were larger than those of bulk metal. Sintering, gas adsorption, reduction at low and high temperature (82,101) (because of the strong metal support interaction effect in TiO_2), and of alloying (Os–Cu) had noticeable influences on the intensities. These results have in all cases been explained as being due to an increase or a decrease in the number of holes in the $5d$ metal bands and interpreted as demonstrating increases or decreases in the electron deficiencies of supported metal particles relative to bulk metals. Thus it was concluded that small metal particles on supports are slightly electron deficient, that this deficiency is less for platinum particles on TiO_2 than on SiO_2 (82) and that osmium in Os–Cu/SiO_2 is less electron deficient than in Os/SiO_2 (87).

This interpretation of d vacancies being equal to electron vacancies has important consequences. It does not, however, take into account the possibility that an electron deficiency is not present at all, but that the number of d vacancies is just about made up by an increase in the number of s electrons via s–d rehybridization. It is well known that the electronic structure of bulk metal atoms and of isolated metal atoms is different. In general metal atoms have more d character in bulk metal, for example, platinum has the configuration $5d^9 6s^1$ in atoms and $5d^{9.7} 6s^{0.3}$ in bulk metal. Because of this one might expect that metal atoms with intermediate geometry, like in crystal surfaces or in small metal particles, have an intermediate electronic structure. Metal atoms with an incomplete shell of neighbors will have smaller local d and s bandwidths and thus have electronic structures intermediate between those of isolated atoms and fully surrounded atoms in bulk metal. All this means that a change in the metal particle size is very likely to be accompanied by a change in the state of hybridization of the metal atoms involved (102, 103).

Several theoretical calculations have demonstrated that the rehybridization model is correct (104). Also photoemission experiments have confirmed this (102). It therefore seems logical to try and explain the experimental white line results of metal catalysts by rehybridization instead of electron deficiency. Indeed all results obtained so far can be explained by rehybridization due to changes in particle size. Thus the fact that the intensity of the white lines of Pt/TiO_2 is intermediate between those of bulk platinum and Pt/SiO_2 (82) agrees with the better dispersion of platinum on SiO_2. The fact that the intensity for Os–Cu/SiO_2 is smaller than that of Os/SiO_2 but larger than that of bulk osmium (87) (Fig. 8.15), demonstrates again that in Os–Cu/SiO_2 the copper is present as an outer layer on the osmium particles. As a result the osmium atoms will on the average have more neighboring metal atoms than the osmium atoms in Os/SiO_2. Furthermore, the outcome of the study of Short et al. (82) on Pt/TiO_2 might need modification, although their conclusion that the platinum atoms on TiO_2 are slightly less electron deficient than those on SiO_2 or Al_2O_3

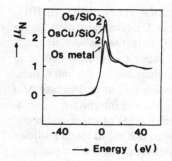

Figure 8.15. L_3-absorption edge spectra of bulk Os, Os/SiO$_2$, and Os—Cu/SiO$_2$ catalysts (87).

was already a disappointment to believers in the charge transfer explanation for the SMSI effect. The observation that the white line intensity of Pt/TiO$_2$ reduced at high temperature is smaller than that of the same catalysts reduced at low temperature, and that this intensity in turn is smaller than that of Pt/SiO$_2$, then just means that the d-electron deficiency increases in this order because of decreasing platinum particle size. Because of s–d rehybridization this does not (have to) mean that the total number of electrons on platinum also decreases in this order. Indeed the Pt/TiO$_2$ catalysts appear to have a slightly higher first-shell coordination number after high temperature reduction (82).

The question of the white line intensities is certainly not solved, but for the moment there is every reason to doubt the explanation in terms of electron deficiency. Another question that remains to be solved is the fact that not only changes in L_3 intensity have been observed, but also of L_2 intensity (82, 101, 105). This is surprising since the L_2 $2p_{1/2} - 5d_{3/2}$ transition is expected to be much weaker than the L_3 $2p_{3/2} - 5d_{3/2}$, $5d_{5/2}$ transitions, because of the fact that the large spin-orbit splitting of the platinum $5d$ band makes the holes go predominantly to the $j = \frac{5}{2}$ state (106, 107). In accordance with this the L_2-edge of bulk platinum hardly shows a white line. But if a platinum atom is situated in a position of much lower symmetry than spherical, such as at the metal surface or metal–support interface, then mixing of the d bands might occur and the L_2 band might "steal" intensity from the L_3 white line.

The formation of the core hole in the absorbing atom strongly perturbs the potential of that atom and as a result in insulators the excited states, which are localized on the perturbed atom, may be pulled down in energy below the conduction band. In that case a cluster approximation will better describe the experimental results than a delocalized band model. Therefore, Kutzler et al. (108) and Horsley (105) carried out $X\alpha$ molecular orbital cluster calculations of the absorption cross sections of K-edge transitions for some transition metal compounds. The calculations were rather successful in reproducing the experimental features. Even so, more theoretical studies are needed to increase our knowledge about near-edge structure. Especially calculations on relaxation effects in metals after the excitation of a core electron to an excited state would be of interest. Horsley performed such a calculation in the local cluster approximation (105). Interest in x-ray absorption edge spectroscopy is indeed growing in catalysis, as for instance exemplified by the study of Sham (109). By combining spin-orbit coupling with ligand field theory he was able to explain the occurrence of specific white lines in the L-edges of $Ru(NH_3)_6Cl_3$. Many more examples of edge structures are presented and discussed in Chapter 11, Sections 11.5.8, 11.5.9, and 11.7.4.

8.5. CONCLUSIONS

Even though EXAFS is a young technique already quite a few very important results have been obtained in catalysis. Especially the results on Wilkinson's catalyst (Section 8.3.1), the sulfided $Co-Mo/Al_2O_3$ catalyst (Section 8.3.3), bromine and krypton on graphite (Section 8.3.4), and the mono and bimetallic catalysts (Sections 8.4.1 and 8.4.2) are worth mentioning. In homogeneous catalysis the results have much in common with those reported on metal containing enzymes in biochemistry and *in situ* EXAFS studies allow a detailed picture to be obtained of the environment of the metal atom. In heterogeneous catalysis the best studies published so far have provided information of about the same quality as that obtained by the SEXAFS technique on adsorbates on single crystal surfaces. Therefore, it is expected that studies on highly dispersed catalysts will aid in developing a detailed picture of catalyst support interfaces and of the interaction playing a role in the bonding of catalyst particles to the support. Great difficulties will lay ahead, though, because the normal EXAFS technique is a bulk technique and thus not only metal atoms in the interface between metal particle and support will be measured but also all other metal atoms on top of those in the interface. Furthermore, the support oxygen ions have a low backscattering amplitude. For these reasons especially catalysts with very high dispersions are of interest. Such catalysts can in general only be prepared at low metal loading, which brings us to the problem of signal-to-noise ratio. Therefore, the trend in the applications of EXAFS in catalysis in the coming years will undoubtedly be to try for lower and lower catalyst loadings, which experimentally means that one ultimately has to switch from normal transmission detection to fluorescence detection.

Another area of interest for EXAFS in catalysis will be the morphology of catalyst particles with such a small size that they are not measurable with a technique like x-ray diffraction. For such small particles ($d < 3$ nm) electron microscopy also fails to be of help since if it is not used under *in situ* conditions it will only provide information of the catalyst in the passivated state. For such small particles this state is not much different from the completely oxidized state. Also, in this application of EXAFS the S/N ratio will be a problem to overcome, because the study by Greegor and Lytle (49) has already shown that unambiguous information on particle shape and size can only be inferred if good data are available on the higher coordination shells too.

The comming years will show the ultimate position of the EXAFS technique in catalysis. Like biochemistry, catalysis has been and undoubtedly will be a field in which multidisciplinary approaches are welcome and in which many techniques have found application. Of course EXAFS cannot, and has never claimed to be able to solve every problem, but in combination with other physical and chemical approaches it will certainly have a great impact on catalysis.

REFERENCES

1. S. Wanke and N. A. Dougharty, *J. Catal.*, **24**, 367 (1972).

2. R. W. Joyner, J. A. R. van Veen, and W. M. H. Sachtler, *J. Chem. Soc. Faraday Trans. I*, **78**, 1021 (1982).

3. J. Reed, P. Eisenberger, and J. Hastings, *Inorg. Chem.*, **17**, 481 (1978).

4. G. Vlaic, J. C. J. Bart, W. Cavigiolo, S. Mobilio, and G. Navarra, *Chem. Phys.*, **64**, 115 (1982).

5. R. M. Friedman, J. F. Freeman, and F. W. Lytle, *J. Catal.*, **55**, 10 (1978).

6. F. W. Lytle, P. S. P. Wei, R. B. Greegor, G. H. Via, and J. H. Sinfelt, *J. Chem. Phys.*, **70**, 4849 (1979).

7. F. W. Lytle, R. B. Greegor, E. C. Marques, D. R. Sandstrom, G. H. Via, and J. H. Sinfelt, in *Proceedings of the Advances in Catalytic Chemistry II*, Salt Lake City, May 1982.

8. E. A. Stern and S. Heald, *Rev. Sci. Instrum.*, **50**, 1579 (1979).

9. J. M. Lorntson, Ph.D. dissertation, University of Delaware, Newark, 1980.

10. B. S. Clausen, B. Lengeler, R. Candia, J. Als-Nielsen, and H. Topsøe, *Bull. Soc. Chim. Belg.*, **90**, 1249 (1981).

11. M. Boudart, R. Dalla Betta, K. Foger, and D. G. Löffler, in *EXAFS and Near Edge Structure III*, K. O. Hodgson, B. Hedman, and J. E. Penner-Hahn (Eds.), Springer-Verlag, Berlin, 1984, p. 187.

12. D. C. Koningsberger and J. W. Cook Jr., in *EXAFS and Near Edge Structure* (*Chemical Physics*, Vol. 27), A. Bianconi, L. Incoccia, and S. Stipcich (Eds.), Springer-Verlag, Berlin, 1983, p. 412.

13. J. Reed, P. Eisenberger, B. K. Teo, and B. M. Kincaid, *J. Am. Chem. Soc.*, **99**, 5217 (1977).

14. J. Reed and P. Eisenberger, *J. Chem. Soc. Chem. Commun.*, 628 (1977).

15. J. Reed, P. Eisenberger, B. K. Teo, and B. M. Kincaid, *J. Am. Chem. Soc.*, **100**, 2375 (1978).

16. R. M. Stults, R. M. Friedman, K. Koenig, W. Knowles, R. B. Greegor, and F. W. Lytle, *J. Am. Chem. Soc.*, **103**, 3235 (1981).

17. J. Goulon, E. Georges, C. Goulon-Ginet, Y. Chauvin, D. Commereuc, H. Dexpert, and E. Freund, *Chem. Phys.*, **83**, 357 (1984).

18. B. Besson, B. Moraweck, A. K. Smith, J. M. Basset, R. Psaro, A. Fusi, and R. Ugo, *J. Chem. Soc. Chem. Commun.*, 569 (1980).

19. T. Yokoyama, K. Yamazaki, N. Kosugi, H. Kuroda, M. Ichikawa, and T. Fukushima, *J. Chem. Soc. Chem Commun.*, 962 (1984).

20. F. W. Lytle, D. E. Sayers, and E. B. Moore, Jr., *Appl. Phys. Lett.*, **24**, 45 (1974).

21. R. B. Greegor, F. W. Lytle, R. L. Chin, and D. M. Hercules, *J. Phys. Chem.*, **85**, 1232 (1981).

22. R. Shulman, Y. Yafet, P. Eisenberger, and W. Blumberg, *Proc. Natl. Acad. Sci., USA*, **73**, 1384 (1976).

23. K. Tohji, Y. Udagawa, S. Tanabe, and A. Ueno, *J. Am. Chem. Soc.*, **106**, 612 (1984).

24. K. Tohji, Y. Udagawa, S. Tanabe, T. Ida, and A. Ueno, *J. Am. Chem. Soc.*, **106**, 5172 (1984).

25. R. Kozlowski, R. F. Pettifer, and J. M. Thomas, *J. Chem. Soc. Chem. Commun.*, 438 (1983); and *J. Phys. Chem.*, **87**, 5172, 5176 (1983).

26. G. Vlaic, J. C. J. Bart, W. Cavigiolo, S. Mobilio, and G. Navarra, *Z. Naturforsch.*, **36A**, 1192 (1981).

27. Y. Sato, Y. Iwasawa, and H. Kuroda, *Chem. Lett.*, 1101 (1982).

28. T. I. Morrison, A. H. Reiss, Jr., E. Gebert, L. E. Iton, G. D. Stucky, and S. L. Suib, *J. Chem. Phys.*, **72**, 6276 (1980).

29. T. I. Morrison, L. E. Iton, G. K. Shenoy, G. D. Stucky, S. L. Suib, and A. H. Reiss, Jr., *J. Chem. Phys.*, **73**, 4705 (1980).

30. T. I. Morrison, L. E. Iton, G. K. Shenoy, G. D. Stucky, and S. L. Suib, *J. Chem. Phys.*, **75**, 4086 (1981).

31. B. S. Clausen, H. Topsøe, R. Candia, J. Villadsen, B. Lengeler, J. Als-Nielsen, and F. Christensen, *J. Phys. Chem.*, **85**, 3868 (1981).

32. M. Boudart, J. Sanchez Arrieta, and R. Dalla Betta, *J. Am. Chem. Soc.*, **105**, 6501 (1983).

33. T. G. Parham and R. P. Merrill, *J. Catal.*, **85**, 295 (1984).

34. N. S. Chiu, S. H. Bauer, and M. F. L. Johnson, *J. Catal.*, **89**, 226 (1984).

35. I. Kohatsu, D. W. Blakely, and H. F. Hansberger, in *Synchrotron Radiation Research*, H. Winick and S. Doniach (Eds.), Plenum, New York, 1980, p. 417.

36. B. S. Clausen, H. Topsøe, R. Candia, and B. Lengeler, in *EXAFS and Near Edge*

Structure III, K. O. Hodgson, B. Hedman, and J. E. Penner-Hahn (Eds.), Springer-Verlag, Berlin, 1984, p. 181.

37. P. Grange, *Catal. Rev. Sci. Eng.*, **21**, 135 (1980).

38. P. Eisenberger and G. S. Brown, *Solid State Commun.*, **29**, 481 (1979).

39. G. S. Brown, in *Synchroton Radiation Research*, H. Winick and S. Doniach (Eds.), Plenum, New York, 1980, p. 387.

40. G. Bunker, in *EXAFS and Near Edge Structure III*, K. O. Hodgson, B. Hedman, and J. E. Penner-Hahn (Eds.), Springer-Verlag, Berlin, 1984, p. 268.

41. W. L. Schaich, in *EXAFS and Near Edge Structure III*, K. O. Hodgson, B. Hedman, and J. E. Penner-Hahn (Eds.), Springer-Verlag, Berlin, 1984, p. 2.

42. H. Topsøe, B. S. Clausen, R. Candia, C. Wivel, and S. Mørup, *J. Catal.*, **68**, 433 (1981).

43. E. A. Stern, D. E. Sayers, J. G. Dash, H. Shechter, ånd B. Bunker, *Phys. Rev. Lett.*, **38**, 767 (1977).

44. S. M. Heald and E. A. Stern, *Phys. Rev. B*, **17**, 4069 (1978).

45. C. Bouldin and E. A. Stern, *Phys. Rev. B*, **25**, 3462 (1982).

46. G. H. Via, J. H. Sinfelt, and F. W. Lytle, *J. Chem. Phys.*, **71**, 690 (1979).

47. J. H. Sinfelt, G. H. Via, and F. W. Lytle, *J. Chem. Phys.*, **68**, 2009 (1978).

48. E. Marques, D. R. Sandstrom, F. W. Lytle, and R. B. Greegor, *J. Chem. Phys.*, **77**, 1027 (1982).

49. R. B. Greegor and F. W. Lytle, *J. Catal.*, **63**, 476 (1980).

50. B. Moraweck, C. Clugnet, and A. J. Renouprez, *Surf. Sci.*, **81**, L 631 (1979).

51. B. Moraweck and A. J. Renouprez, *Surf. Sci.*, **106**, 35 (1981).

52. A. Renouprez, P. Fouilloux, and B. Moraweck, in *Growth and Properties of Metal Clusters*, J. Bourdon (Ed.), Elsevier, Amsterdam, 1980, p. 421.

53. G. Apai, J. F. Hamilton, J. Stöhr, and A. Thompson, *Phys. Rev. Lett.*, **43**, 165 (1979).

54. J. F. van der Veen, F. J. Himpsel, and D. E. Eastman, *Phys. Rev. B*, **25**, 7388 (1982).

55. P. Lagarde, in *EXAFS and Near Edge Structure* (*Chemical Physics*, Vol. 27), A. Bianconi, L. Incoccia, and S. Stipcich (Eds.), Springer-Verlag, Berlin, 1983, pp. 296 and 319.

56. P. Lagarde, T. Murata, G. Vlaic, E. Freund, H. Dexpert, and J. P. Bournonville, *J. Catal.*, **84**, 333 (1983).

57. H. Dexpert, P. Lagarde, and J. P. Bournonville, *J. Mol. Catal.*, **25**, 347 (1984).

58. D. Bazin, H. Dexpert, P. Lagarde, and J. P. Bournonville, in *EXAFS and Near Edge Structure III*, K. O. Hodgson, B. Hedman, and J. E. Penner-Hahn (Eds.), Springer-Verlag, Berlin, 1984, p. 195.

59. R. W. Joyner, *J. Chem. Soc. Faraday Trans. I*, **76**, 357 (1980).

60. R. K. Nandi, F. Molinero, C. Tang, J. B. Cohen, J. B. Butt, and R. L. Burwell, Jr., *J. Catal.*, **78**, 289 (1982).

61. R. K. Nandi, P. Georgeopoulos, J. B. Cohen, J. B. Butt, R. L. Burwell, Jr., and D. H. Bilderback, *J. Catal.*, **77**, 421 (1982).

62. J. C. Vis, H. F. J. van 't Blik, T. Huizinga, J. van Grondelle, and R. Prins, *J. Mol. Catal.*, **25**, 367 (1984).

63. T. Fukushima, J. R. Katzer, D. E. Sayers, and J. W. Cook, Jr., in *Proceedings of the 7th International Congress on Catalysis*, T. Seyama and K. Tanabe (Eds.), Elsevier, Amsterdam, 1981, p. 79.

64. F. W. Lytle, G. H. Via, and J. H. Sinfelt, *J. Chem. Phys.*, **67**, 3831 (1977).

65. A. D. Cox, in *Characterization of Catalysts*, J. M. Thomas and P. Lambert (Eds.), Wiley, London, 1981, p. 254.

66. H. F. J. van 't Blik, J. B. A. D. van Zon, T. Huizinga, J. C. Vis, D. C. Koningsberger, and R. Prins, *J. Phys. Chem.*, **87**, 2264 (1983).

67. H. F. J. van 't Blik, J. B. A. D. van Zon, T. Huizinga, J. C. Vis, D. C. Koningsberger, and R. Prins, *J. Am. Chem. Soc.*, **107**, 3139 (1985).

68. H. F. J. van 't Blik, J. B. A. D. van Zon, D. C. Koningsberger, and R. Prins, *J. Mol. Catal.*, **25**, 379 (1984).

69. D. C. Koningsberger, H. F. J. van 't Blik, J. B. A. D. van Zon, and R. Prins, in *Proceedings of the 8th International Congress on Catalysis*, Verlag Chemie, Weinheim, 1984, p. V-123.

70. D. C. Koningsberger, in *EXAFS and Near Edge Structure III*, K. O. Hodgson, B. Hedman, and J. E. Penner-Hahn (Eds.), Springer-Verlag, Berlin, p. 212.

71. Y. I. Yermakov and B. N. Kuznetsov, *J. Mol. Catal.*, **9**, 13 (1980).

72. E. G. Derouane, A. J. Simoens, and J. C. Vedrine, *Chem. Phys. Lett.*, **52**, 549 (1977).

73. I. W. Bassi, F. Garbassi, G. Vlaic, A. Marzi, G. R. Tauszik, G. Cocco, L. Galvagno, and G. Parravano, *J. Catal.*, **64**, 405 (1980).

74. I. W. Bassi, F. W. Lytle, and G. Parravano, *J. Catal.*, **42**, 139 (1976).

75. D. C. Koningsberger, J. B. A. D. van Zon, H. F. J. van 't Blik, G. J. Visser, R. Prins, A. N. Mansour, D. E. Sayers, D. R. Short, and J. R. Katzer, *J. Phys. Chem.*, **89**, 4075 (1985).

76. D. C. Koningsberger and D. E. Sayers, *Solid State Ionics*, **16**, 23 (1985).

77. J. B. A. D. van Zon, D. C. Koningsberger, H. F. J. van 't Blik, R. Prins, and D. E. Sayers, *J. Chem. Phys.*, **80**, 3914 (1984).

78. J. B. A. D. van Zon, D. C. Koningsberger, H. F. J. van 't Blik, and D. E. Sayers, *J. Chem. Phys.*, **82**, 5742 (1985).

79. S. J. Tauster, S. C. Fung, and R. L. Garten, *J. Am. Chem. Soc.*, **100**, 170 (1978).

80. J. Santos, J. Phillips, and J. A. Dumesic, *J. Catal.*, **81**, 147 (1983).

81. H. F. J. van 't Blik, P. H. A. Vriens, and R. Prins, *Strong Metal-Support Interactions*, in R. T. K. Baker, and S. J. Tauster, J. A. Dumesic (Eds.), ACS Symposium Series 298, American Chemical Society, Washington DC, 1985, p. 60.

82. D. R. Short, A. N. Mansour, J. W. Cook, Jr., D. E. Sayers, and J. R. Katzer, *J. Catal.*, **82**, 299 (1983).

83. J. H. Sinfelt, *Acc. Chem. Res.*, **10**, 15 (1977).

84. D. R. Short, S. M. Khalid, J. R. Katzer, and M. J. Kelley, *J. Catal.*, **72**, 288 (1981).

85. N. Wagstaff and R. Prins, *J. Catal.*, **59**, 434 (1979).

86. J. H. Sinfelt, G. H. Via, and F. W. Lytle, *J. Chem. Phys.*, **72**, 4832 (1980).

87. J. H. Sinfelt, G. H. Via, F. W. Lytle, and R. B. Greegor, *J. Chem. Phys.*, **75**, 5527 (1981).

88. J. H. Sinfelt, G. H. Via, and F. W. Lytle, *J. Chem. Phys.*, **76**, 2779 (1982).

89. G. Meitzner, G. H. Via, F. W. Lytle, and J. H. Sinfelt, *J. Chem. Phys.*, **78**, 882 (1983).

90. G. Meitzner, G. H. Via, F. W. Lytle, and J. H. Sinfelt, *J. Chem. Phys.*, **78**, 2533 (1983).

91. B. K. Teo and P. A. Lee, *J. Am. Chem. Soc.*, **101**, 2815 (1979).

92. P. H. Citrin, P. Eisenberger, and B. M. Kincaid, *Phys. Rev. Lett.*, **36**, 1346 (1976).

93. J. C. Mikkelsen, Jr. and J. B. Boyce, *Phys. Rev. B*, **28**, 7130 (1983).

94. J. B. Boyce and J. C. Mikkelsen, Jr., in *EXAFS and Near Edge Structure III*, K. O. Hodgson, B. Hedman, and J. E. Penner-Hahn (Eds.), Springer-Verlag, Berlin, 1984, p. 426.

95. P. H. Lewis, *J. Phys. Chem.*, **64**, 1103 (1960); **66**, 105 (1962); **67**, 2151 (1963); and *J. Catal.*, **43**, 376 (1968).

96. F. W. Lytle, *J. Catal.*, **43**, 376 (1976).

97. T. K. Sham, *Phys. Rev. B*, **31**, 1888 (1985).

98. P. Gallezot, R. Weber, R. A. Dalla Betta, and M. Boudart, *Z. Naturforsch.*, **34A**, 40 (1979).

99. P. H. Lewis, *J. Catal.*, **69**, 511 (1981).

100. R. A. Dalla Betta, M. Boudart, P. Gallezot, and R. S. Weber, *J. Catal.*, **69**, 514 (1981).

101. A. N. Mansour, J. W. Cook, Jr., D. E. Sayers, R. J. Emrich, and J. R. Katzer, *J. Catal.*, **89**, 462 (1984).

102. M. G. Mason, *Phys. Rev. B*, **27**, 748 (1983).

103. P. H. Citrin and G. K. Wertheim, *Phys. Rev. B*, **27**, 3160 (1983).

104. C. S. Wang and A. J. Freeman, *Phys. Rev. B*, **19**, 793 (1979).

105. J. A. Horsley, *J. Chem. Phys.*, **76**, 1451 (1982).

106. M. Brown, R. E. Peierls, and E. A. Stern, *Phys. Rev. B*, **15**, 738 (1977).

107. L. F. Mattheiss and R. E. Dietz, *Phys. Rev. B*, **22**, 1663 (1980).

108. F. W. Kutzler, C. R. Natoli, D. K. Misemer, S. Doniach, and K. O. Hodgson, *J. Chem. Phys.*, **73**, 3274 (1980).

109. T. K. Sham, *J. Am. Chem. Soc.*, **105**, 2269 (1983).

CHAPTER

9

AMORPHOUS AND LIQUID SYSTEMS

E. D. CROZIER

Department of Physics
Simon Fraser University
Burnaby, British Columbia, Canada

J. J. REHR and R. INGALLS

Department of Physics
University of Washington
Seattle, Washington

373

9.1. INTRODUCTION

One of the major applications of EXAFS spectroscopy is the determination of the structure of disordered materials. It cannot provide long-range information, but its ability to specify the local structure about x-ray absorbing atoms, which can be present in parts per million, has furthered the basic understanding of liquids, glasses, and amorphous films.

Disordered materials have been discussed in recent general reviews on EXAFS spectroscopy (1–3). Results for specific systems have been presented in reviews of metallic glasses (4–6), oxide glasses (7–9), liquid and amorphous metals and semiconductors (10), and aqueous solutions (11–13).

We begin this chapter with the theory of the EXAFS Debye–Waller factor. The Debye–Waller factor, which, in the small disorder limit or the harmonic approximation, is given simply by $e^{-2k^2\sigma_j^2}$, attenuates the amplitude of the EXAFS interference function $\chi(k)$. In this approximation, $\chi(k)$ for an unoriented sample is given by

$$\chi(k) = \sum_j \frac{N_j}{kR_j^2} S_0^2(k) F_j(k)\, e^{-2k^2\sigma_j^2}\, e^{-2(R_j - \Delta)/\lambda} \sin\left[2kR_j + \delta_j(k)\right] \quad (1)$$

where corrections for many-body effects have been included [Chapter 1, Eq. (61)]. Knowledge of σ_j^2 is important both for testing our fundamental understanding of EXAFS and for precise determination of coordination numbers N_j.

The mean square fluctuation in interatomic distance between the x-ray absorbing atom and the jth atom, σ_j^2, includes a dynamical term due to the thermal motion of the atoms and a static term due to structural disorder. A lattice dynamical theory of σ_j^2 is developed for crystalline materials in Section 9.2. Methods of extending the calculation to noncrystalline materials by evaluating the projected density of normal modes of vibration are indicated. The simpler Einstein and Debye approximations are discussed.

Equation (1) is a valid approximation for some amorphous films, glasses, and liquids in which a high degree of local order is preserved by covalent bonding or a strong ion–ion interaction. Frequently, the degree of disorder is larger and $\chi(k)$ must be represented by the more general equation:

$$\chi(k) = \sum_j S_0^2(k) F_j(k) \int P(r_j)\, \frac{e^{-2r_j/\lambda}}{kr_j^2} \sin\left[2kr_j + \delta_j(k)\right] dr_j \quad (2)$$

where $P(r_j)\, dr_j$ is the probability of finding the jth species in the range r_j to $r_j + dr_j$. The distribution function may be asymmetrical in systems in which the static arrangement of atoms is dominated by excluded volume effects such as in amorphous solids, metallic glasses (14, 15), superionic conductors (1, 16), and liquid metals (10, 17). In all systems with increasing temperature, anharmonic contributions to the vibrational displacements of the atoms can no longer be neglected and consequently the distribution function becomes asymmetrical; in any case the effective distribution $P(r_j)$ ($e^{-2r_j/\lambda}/r^2$) is always asymmetrical. The cumulant expansion introduced in Section 9.2.3 provides a framework for the discussion and analysis of these effects.

Section 9.3 extensively treats methods of analysis when Eq. (2) must be used. Model-dependent methods are discussed. The extent to which a model-independent $P(r)$ can be determined is examined. It is shown that, by combining the experimental $\chi(k)$ with its correct extrapolation to $k = 0$ (the "splice method" of Section 9.3.3), $\chi(k)$ can be inverted to obtain $P(r)$ even for systems with moderate disorder. It is also emphasized that the analysis of the EXAFS of second shells with thermal or structural disorder is complicated by multiple scattering, the neglect of which can lead to significant errors.

Section 9.4 is concerned primarily with experimental considerations for amorphous and liquid systems, although Section 9.4.3 on data acquisition is of more general interest.

A comparison of the complementary techniques of diffraction and EXAFS and their strengths and weaknesses for the examination of disordered systems are given in Section 9.5. The use of EXAFS in obtaining the effective two-body interaction potential is included in Section 9.5.3.

Although a detailed review is not given of the numerous EXAFS papers that have been published in the fields of amorphous films, glasses, and liquids, experimental results or model calculations are used to illustrate the topics covered.

9.2. DEBYE–WALLER FACTOR

9.2.1. Harmonic Approximation

The Debye–Waller factor can be viewed as a result of averaging the single-scattering EXAFS formula over many near-neighbor pairs with a given pair distribution function $P(r)$. If factors other than the sine function in $\chi(k)$ are assumed to have negligible variation with r_j, the main effect is given by

$$\text{Im} \langle e^{i2kr_j} \rangle = \text{Im} \int P(r_j) e^{i2kr_j} dr_j \tag{3}$$

where Im is the imaginary part. For later reference we may replace $P(r)$ by the effective distribution function $P(r_j, \gamma)$, which incorporates the smoothly varying

EXAFS amplitude prefactors:

$$P(r_j, \gamma) = \frac{P(r_j)\, e^{-2\gamma r_j}}{r_j^2} \tag{4}$$

where $P(r_j)$ is the pair distribution and γ is the inverse of the mean free path. The general behavior of Eq. (3) is conveniently discussed in terms of cumulant averages [(18, 19), Section 9.2.3, and Chapter 1, Section 1.3.2]. However, if the disorder is small or Gaussian in character, one can use the approximation

$$\langle e^{i2kr_j} \rangle = e^{i2k\langle r_j \rangle - 2k^2\sigma_j^2} \tag{5}$$

where

$$\sigma_j^2 = \langle (r_j - \langle r_j \rangle)^2 \rangle \tag{6}$$

is the mean square variation in bond length. The quantity σ_j^2 has two sources: (a) $\sigma_j^2(T)$ arising from thermal vibration and (b) $\sigma_j^2(s)$ from structural disorder, which is essentially temperature independent.

In this section we consider $\sigma_j^2(T)$. This quantity is obtained from a thermal average of Eq. (6) with a given lattice Hamiltonian H_{vib}. By the Born–Oppenheimer approximation, ion motion is negligible during the photoemission process, so that the thermal average should be carried out in the ground state of the system, *prior* to absorption. In performing this calculation we consider only the case of small vibrations, for which a harmonic Hamiltonian is adequate. Anharmonic effects and related corrections are discussed in Section 9.2.3. For definiteness let us consider an absorbing atom at the origin and a neighbor at site $\mathbf{R}_j$. Defining $\mathbf{u}_j$ as the instantaneous displacement of the atom at site $\mathbf{R}_j$, the change in radial distance to first order is

$$\delta R_j \cong (\mathbf{u}_j - \mathbf{u}_0) \cdot \hat{\mathbf{R}}_j \tag{7}$$

Accordingly, the mean square vibrational amplitude is given by

$$\begin{aligned}
\sigma_j^2 &= \langle [(\mathbf{u}_j - \mathbf{u}_0) \cdot \hat{\mathbf{R}}_j]^2 \rangle \\
&= \langle (\mathbf{u}_j \cdot \hat{\mathbf{R}}_j)^2 \rangle + \langle (\mathbf{u}_0 \cdot \hat{\mathbf{R}}_j)^2 \rangle - 2\langle (\mathbf{u}_j \cdot \hat{\mathbf{R}}_j)(\mathbf{u}_0 \cdot \hat{\mathbf{R}}_j) \rangle
\end{aligned} \tag{8}$$

This shows that the Debye–Waller factor in EXAFS differs in several respects from that encountered, say, in x-ray scattering or in the Mössbauer effect, where the Debye–Waller factor is given by $e^{-q^2 u^2/2}$ and $u^2 = \langle (\mathbf{u}_0 \cdot \mathbf{q})\rangle^2$, q being the momentum transfer. In EXAFS σ_j^2 is shell dependent, being sensitive to the displacement–displacement correlation function $\langle (\mathbf{u}_j \cdot \hat{\mathbf{R}}_j)(\mathbf{u}_0 \cdot \hat{\mathbf{R}}_j) \rangle$, which decays slowly with distance. This correlation is such that only the modes contributing to radial motion are important for σ_j^2; contributions from the long-wavelength acoustic modes, for example, are suppressed. As a result σ_j^2 is always less than the sum of separate contributions from sites $\mathbf{0}$ and $\mathbf{R}_j$.

It is useful to express σ_j^2 in terms of normal vibrational modes. The lattice displacements are given by

$$\mathbf{u}_i = (M_i)^{1/2} \sum_\lambda q_\lambda\, \varepsilon_i(\lambda) \tag{9}$$

where M_i is the mass of the atom at site $\mathbf{R}_i$, q_λ are the normal coordinates, and $\varepsilon_i(\lambda)$ are the normalized eigenvectors of the "Dynamical matrix," $\mathbf{D} = (\partial^2\Phi/\partial\mathbf{u}_i\partial\mathbf{u}_j)/(M_iM_j)^{1/2}$, where Φ is the interatomic potential:

$$\sum_j \mathbf{D}_{ij}\, \varepsilon_j(\lambda) = \omega_\lambda^2\, \varepsilon_i(\lambda) \tag{10}$$

Substituting Eq. (9) into Eq. (8) we obtain

$$\sigma_j^2 = \frac{\hbar}{2\mu_j} \sum_\lambda \left[\left(\frac{\mu_j}{M_j}\right)^{1/2} \varepsilon_j(\lambda) - \left(\frac{\mu_j}{M_0}\right)^{1/2} \varepsilon_0(\lambda) \right] \frac{\coth \beta\hbar\omega_\lambda}{\omega_\lambda} \tag{11}$$

where $\mu_j = 1/(1/M_0 + 1/M_j)$ is the reduced mass for bond pair $(\mathbf{0}, \mathbf{R}_j)$. Equation (11) has the following interpretation. The terms in brackets, denoted by $p_j(\lambda)$, are the normalized probabilities that an initial "displacement state" is in mode λ; such a state is defined by the displacement field

$$\mathbf{u}_i = \begin{cases} \hat{R}_j\,(\mu_j/M_j)^{1/2} & i = j \\ -\,\hat{R}_j(\mu_j/M_0)^{1/2} & i = 0 \\ 0 & \text{otherwise} \end{cases} \tag{12}$$

$$\dot{\mathbf{u}}_i = 0 \quad \text{all } i$$

Accordingly, one can define a *projected density of modes*

$$\rho_j(\omega) = \sum_\lambda p_j(\lambda)\, \delta(\omega - \omega_\lambda) \tag{13}$$

which properly weights the contribution of each mode to relative vibrational motion (Fig. 9.1). The weights are proportional to the mean square compression of the bond $(\mathbf{0}, \mathbf{R}_j)$ in a given mode.

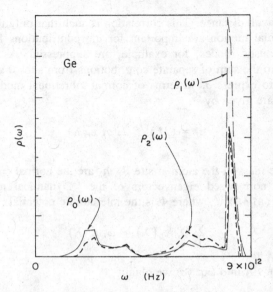

Figure 9.1. Typical projected densities of modes $\rho_j(\omega)$ contributing to the EXAFS Debye–Waller factor for shell j; $\rho_0(\omega)$ is the phonon spectrum. The results shown are based on lattice dynamical calculations for c-Ge (18).

With this definition Eq. (11) gives:

$$\sigma_j^2 = \frac{\hbar}{2\mu} \int \frac{d\omega}{\omega} \, \rho_j(\omega) \coth\left(\frac{\beta\hbar\omega}{2}\right) \tag{14}$$

In addition to having a direct physical interpretation, Eq. (14) is convenient for numerical calculations. All of the lattice dynamical structure is contained in the projected density ρ_j. The remaining Bose–Einstein thermal factor is a smooth function of ω (except at $\omega = 0$) varying between $2k_BT/\hbar\omega^2$ and $1/\omega$ at high and low temperature, respectively. Thus σ_j^2 should be independent of fine details in the vibrational spectrum. Furthermore, for any physical lattice Hamiltonian, $\rho_j(\omega)$ depends only on the *local* vibrational structure in the vicinity of the bond $\mathbf{0} - \mathbf{R}_j$.

Calculations of σ_j^2 based on accurate lattice dynamical models are possible only for certain crystals or small molecules, where the force constants can be known with precision. A number of such calculations have been carried out (18, 20) (see Fig. 9.2), they are useful as tests of techniques for measuring or calculating σ_j^2.

Since one of the main applications of EXAFS is to probe noncrystalline

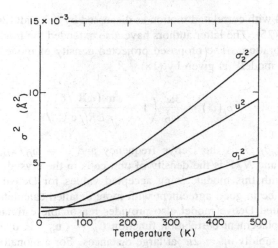

Figure 9.2. Mean square radial vibrations σ_j^2 for the first and second shells of c-Ge versus temperature T; u^2 is the mean square vibration amplitude of a single atom.

materials, it is also desirable to have methods that are tailored to permit a local determination of σ_j^2. Two such methods have been suggested: One of these (18) is based on the *recursion method* of Haydock et al. (21). In this approach $\rho_j(\omega)$ is calculated in successive approximations that encompass more and more of the crystal lattice. This method leads to δ-function representations of the projected density, that is, $\rho_j(\omega) = \Sigma_n w_n \delta(\omega - \omega_n)$, and yields an N-point Gaussian quadrature formula for σ_j^2:

$$\sigma_j^2 = \frac{\hbar}{2\mu} \sum_r \frac{w_n}{\omega_n} \coth\left(\frac{\beta\hbar\omega_n}{2}\right) \tag{15}$$

The second of these (18, 22, 23) is the *equation of motion* method in which the equations of lattice motion are integrated numerically, using for initial conditions the state in Eq. (11) for a large cluster of atoms in the vicinity of the bond $(0 - \mathbf{R}_j)$. It can be shown that $\rho_j(\omega)$ is simply the cosine Fourier transform of the relative displacement for $t \geq 0$, $[(\mathbf{u}_j(t) - \mathbf{u}_0(t)] \cdot \hat{\mathbf{R}}_j$, given a unit displacement at $t = 0$. This latter method is especially useful for disordered materials, since the normal modes need not be explicitly calculated.

9.2.2. Einstein and Debye Approximations

In the absence of knowledge of the microscopic force constants, the Einstein or Debye approximation may be used to estimate σ_j^2. This model must be modified slightly to deal with correlated motion as discussed by Schmidt (24) and by Beni and Platzman (25). The latter authors have also extended the model to deal with anisotropic vibrations. The proposed projected density of modes for the "correlated Debye model" is given by (18)

$$\rho_j(\omega) = \frac{3\omega^2}{\omega_D^3} \left(1 - \frac{\sin(\omega R_j/c)}{\omega R_j/c} \right) \tag{16}$$

Here $\omega_D = k_B \theta_D / \hbar$ is the Debye frequency and $c = \omega_D/k_D$, where $k_D = (6\pi^2 N/V)^{1/3}$ and N/V is the density of unit cells in the crystal. The values of σ_j^2 obtained with this model, using accepted values for Debye temperatures, were found to be in good agreement with more detailed calculations (20, 26).

The correlated Debye model also provides reasonable estimates of the displacement–displacement correlation function $C_R = \langle \mathbf{u}_0 \cdot \hat{\mathbf{R}} \, \mathbf{u}_j \cdot \hat{\mathbf{R}} \rangle$ (25). C_R indicates how rapidly $\sigma_j^2 \to 2u^2$ at large distances. For a monatomic crystal

$$C_R = 1 - \frac{\sigma_R^2}{2u^2} = \begin{cases} \dfrac{\sin(\omega_D R/2c)}{\omega_D R/2c} & T \to 0 \\[2ex] \dfrac{S_i(\omega_D R/c)}{\omega_D R/c} & T \to \infty \end{cases} \tag{17}$$

The displacement–displacement correlation function is also of importance to directional phenomena in crystals such as channeling, blocking, and focuson propagation. In this connection the temperature dependence of C_R using experimental phonon distributions has been calculated for molybdenum and niobium (27).

One of the simplest ways of fitting thermodynamic data is to use an *Einstein model* for the spectrum, in which the spectral weight is concentrated at a single frequency,

$$\rho_j(\omega) = \delta(\omega - \omega_E^j) \tag{18}$$

This model can provide a good approximation for σ_j^2 over a wide temperature range. It is particularly appropriate for crystals such as germanium, for which $\rho_j(\omega)$ is strongly peaked at the optical end of the spectrum (Fig. 9.1). If ω_E^j is regarded as a fitting parameter, its value lies between $(m_{-1})^{-1}$ and $(m_{-2})^{-2}$ where m_n is the nth power moment of $\rho_j(\omega)$. For "rule of thumb" numbers (18), one can take the average of these values obtained from the correlated Debye model. This yields $\omega_E = \frac{3}{4}\omega_D$ for the first shell of the body-centered

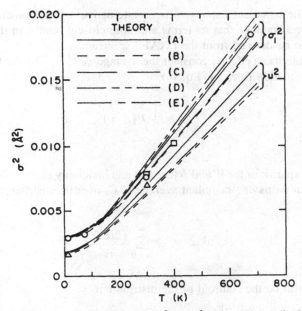

Figure 9.3. Mean square vibrational amplitudes u^2 and σ_1^2 for the first shell of copper versus temperature as calculated from various force constant models [A (28), B (29), and C (single, central force constant model)], from the correlated Debye model (D), and from the Einstein model (E) and as determined from experiment for σ^2 ○ (30), □ (31), and for u^2 △ (29).

cubic (bcc) and face-centered cubic (fcc) lattices, ω_D for the diamond lattice; and $\frac{2}{3}\omega_D$ for subsequent shells. For u^2 the rule of thumb is $\omega_E = \frac{3}{5}\omega_D$.

A comparison of experimental results with theoretical calculations is shown in Fig. 9.3 (20) for polycrystalline copper. The sensitivity to different force-constant models is apparent. For amorphous semiconductors such as a-Ge, different force constant models yield σ_1^2 that differ by approximately 10%. To the extent that the experimental σ^2 can be determined to a better accuracy, σ^2 can be used to distinguish between different force constant models.

9.2.3. Non-Gaussian Corrections

Our discussion in Section 9.2.1 was limited to the case of small vibrations or small disorder in which the EXAFS Debye–Waller factor is determined solely by the mean square fluctuations in near neighbor distances. Anharmonic terms or non-Gaussian disorder not only change the Debye–Waller factor, they can affect phase shifts and hence distance determinations as well. These effects have been treated using a cumulant expansion of the Debye–Waller factor in powers of k (18). The cumulant expansion is an alternative to the moment expansion (32) with the advantage that its terms are more closely related to the structural parameters to be derived from the EXAFS spectrum.

To illustrate these effects, consider the average $\langle e^{i2k(r-\bar{R})} \rangle$ over the effective distribution function $P(r, \gamma)$, Eq. (4)

$$\langle e^{i2k(r-\bar{R})} \rangle = \int P(r, \gamma) \, e^{i2kr} \, dr$$
$$= e^{-W + i\phi} \tag{19}$$

Systematic expansions for W and ϕ (the real and imaginary parts of the exponent) are defined in terms of "cumulant averages" C_n over the distribution (33), that is,

$$-W + i\phi = \sum_{n=0}^{\infty} \frac{(2ik)^n}{n!} C_n \tag{20}$$

If $\bar{R}$ is taken to be the centroid of the distribution

$$\bar{R} = \frac{\int P(r, \gamma) \, r \, dr}{\int P(r, \gamma) \, dr}$$

then the first few cumulants expressed in terms of the power moments of the distribution are as listed in Chapter 1, Eq. (33). A more extensive listing plus the recursion formulas to generate higher cumulants in terms of the power moments are given in ref. (19). Cumulants for several simple normalized distributions are tabulated in Table 9.1.

It follows from Eqs. (19) and (20) that the amplitude and phase of the EXAFS interference function for a single-shell containing N atoms of the same type may be written as

$$\ln \left[\frac{k\chi(k)}{NS_0^2(k)\, F(k)} \right] = C_0 - C_2 \frac{(2k)^2}{2!} + C_4 \frac{(2k)^4}{4!} + \cdots \tag{21a}$$

$$\phi - \delta(k) = 2k\bar{R} - \frac{(2k)^3}{3!} C_3 + \frac{(2k)^5}{5!} C_5 + \cdots \tag{21b}$$

Table 9.1. Cumulants and Transforms of Simple Effective (Normalized) Distributions [a]

Distribution	Cumulants

Gaussian (symmetric)

$$P_G(r) = (2\pi\sigma^2)^{-1/2} \exp\left[-(r - R_0)^2/2\sigma^2\right]$$

Centroid $= R_0$
$C_2 = \sigma^2$
$C_n(n > 2) = 0$

Delta function

$$P_D(r) = \frac{1}{N} \sum_{n=1}^{N} \delta(r - R_n)$$

$$\equiv \frac{1}{N} \sum_{n=1}^{N} \delta(r - \Delta_n - \overline{R})$$

Centroid $= \dfrac{1}{N} \sum_n R_n \equiv \overline{R}$

$$C_2 = \frac{1}{N} \sum_n \Delta_n^2$$

$$C_4 = \frac{1}{N} \sum_n \Delta_n^4 - \frac{3}{n^2} \sum_n \sum_{n'} \Delta_n^2 \Delta_{n'}^2$$

Weighted exponential

$$P_x^m(r) = \frac{B^{m+1}}{m!} (r - R_0)^m \exp\left[-B(r - R_0)\right]$$

$$\quad \text{for } r \geq R_0$$
$$= 0 \quad \text{for } r < R_0$$
$$\quad \text{for } m = 0, 1, 2$$

$$k\chi'(k) = \int P_x^m(r) \sin 2kr\, dr$$
$$= (1 + z^2)^{-(m+1)/2} \sin\left[2k\overline{R} + (m+1)\right.$$
$$\left.(-z + \tan^{-1} z)\right]$$
$$\text{where } z = 2k/B$$

Centroid $= R_0 + \dfrac{(m + 1)}{B}$

$$C_n(n > 1) = \frac{(m + 1)(n - 1)!}{B^n}$$

[a] This table was taken in part from G. Bunker (19).

where $\overline{R}$ is the centroid of the distribution. With this choice $C_1 = 0$. If the distribution is normalized then $C_0 = 0$ also.

The utility of the cumulant expansion is now apparent. Corrections to the Debye–Waller factor involve only the even cumulants of the effective distribution. The first such correction term is

$$W = +2\sigma^2 k^2 - \tfrac{2}{3} C_4 k^4 \tag{22}$$

where $C_2 = \sigma^2$. Similarly the odd cumulants contribute only to the EXAFS phase. This separation into even and odd power series in k permits the determination of the cumulants by a straightforward extension of the log-ratio method. Specific examples are given in Section 9.3.

The cumulants are defined for the effective distribution $P(r, \gamma)$. In order to recover the real distribution $P(r)$, additional corrections must be made for the distance dependence of the amplitude prefactor $e^{-2\gamma r}/r^2$ Eq. (4). One obtains an additional phase shift corresponding to the change in the peak position, that is,

$$\phi \rightarrow \phi - \frac{4kC_2}{\overline{R}} (1 + \gamma \overline{R}) \qquad (23)$$

This effect is linear in k and is important when the disorder is large (17, 19, 32).

In cases where structural disorder is present, the distribution function is a convolution of thermal and radial distribution functions. To the extent that the thermal average is independent of position, the cumulants $C_n(T)$ for the thermal and $C_n(s)$ configurational averages are, by the convolution theorem, additive;

$$C_n = C_n(T) + C_n(s) \qquad (24)$$

This property allows these terms to be separated from one another. It also permits one to model a range of asymmetric distribution functions by suitable convolution of the simple distribution functions listed in Table 9.1.

Another effect of anharmonicity is to alter the temperature dependence of the thermal vibrations. For a harmonic Hamiltonian, σ_j^2 grows linearly with T at high temperature; that is, $\sigma_j^2 = (kT/\mu_j)m_{-2}$ is the inverse second moment of $\rho_j(\omega)$. Because of thermal expansion, however, the force constants and hence the phonon frequencies will shift slightly. Also cubic, quartic, and higher-order terms in the lattice Hamiltonian must be taken into account. Both of these effects are comparable in order of magnitude but partially cancel each other. Also both give corrections to σ_j^2 quadratic in T at high temperature; that is,

$$\sigma_j^2 \rightarrow \sigma_j^2 (1 + aT + \cdots) \qquad (25)$$

For copper the magnitude of the constant a is estimated to be $0.05/\theta_D$, which is consistent with experimental data (20).

9.3. THE EFFECTS OF DISORDER

The structural information of interest in an EXAFS experiment appears in the interference function $\chi(k)$ through a Fourier transform of the effective distribution function $P(r, \gamma)$. The finite range of the data, particularly the necessity in the present state of development of EXAFS to omit the low-k data means that $P(r, \gamma)$ cannot be recovered by simply taking the inverse Fourier transform of $\chi(k)$. The extent to which this is a problem depends upon the shape of the distribution measured in terms of its cumulants. When cumulants higher than the second are negligible, that is, in the small disorder limit, Eq. (2) for $\chi(k)$ reduces to the standard EXAFS expression, Eq. (1). A Gaussian distribution yields the same result provided $\sigma^2/\overline{R}^2 \ll 1$. In this case the structural parameters R_i, N_i, σ_i^2, and identity of the ith species are obtained by well-established methods discussed in Chapter 6. These methods have provided accurate results for some amorphous materials, some liquids, and some aqueous solutions where a high degree of local order has been preserved about the x-ray absorbing atoms (1–5, 7, 10, 12, 34–39). But in general amorphous and liquid systems are expected to possess moderate to large disorder and the application of Eq. (1) in their analysis can lead to significant errors in R_1, N_1, σ_1^2 and even in the identities of nearest neighbors.

In this section disorder is intended in the general sense, including both static structural disorder and dynamical effects due to the thermal motion of the atoms. For simplicity the discussion will be restricted to a single shell of N identical atoms, with the exception of Section 9.3.3. Compositional disorder, that is the presence of more than one atomic species, will be included when referring to specific systems. It will be assumed that the mean free path is k independent. For nearest neighbors this is justified. For most disordered systems of interest in this chapter the structural information is limited to nearest neighbors. However, in some amorphous films, glasses, and concentrated electrolytes there is sufficient local order that second-shell structure can be resolved and the k dependence of the mean free path must be considered.

Methods of detecting asymmetrical distribution functions are presented in Section 9.3.1. The application of the cumulant expansion to systems with low to moderate disorder and its limitations with increasing disorder are discussed in Section 9.3.2. A model-independent method of reconstructing $P(r)$ is given in Section 9.3.3 and model-dependent methods in Section 9.3.4. Section 9.3.5 illustrates the difficulties in obtaining information beyond the first shell in disordered materials. The need to consider explicitly static and thermal structural disorder in multiple-scattering calculations for the second shell is discussed in Section 9.3.6. Finally, Section 9.3.7 is concerned with the determination of the inner potential E_0 when asymmetry is present. Throughout Section 9.3 the topics covered are illustrated with detailed model calculations as well as with experimental data.

9.3.1. Dectection of Asymmetrical Distributions

This section is concerned with asymmetrical distribution functions, that is, functions for which the odd cumulant C_3, and possibly higher ones, cannot be neglected. In the case of systems with moderate to large disorder the errors due to the neglect of asymmetry are usually evident by a comparison of the deduced structural parameters with those expected on the basis of chemical or physical intuition of the system. However, the effects of asymmetry can be subtle. This section begins with an example in which neglect of asymmetry produces significant errors even though an apparently reasonable fit to the data is obtained with the standard Gaussian approximation. Then two methods of detecting asymmetry are described and illustrated by reference to experimental data.

For simplicity let $P(r)$ be the effective distribution function shown in Fig. 9.4. It is given by the convolution $P_G \otimes P_x^0$ where P_G is a normalized Gaussian of mean square width $\sigma_i^2 = 0.005$ Å^2 centered about $R = 0$ and $P_x^0 = Be^{-B(r-R_0)}$ is a normalized exponential, nonzero for $r \geq R_0 = 2.19$ Å with width parameter $B = 10$ Å^{-1}. A related functional form has been used in modeling metallic glasses (4, 36, 37, 39). P_x^0 can be regarded as an approximation for the first peak of the radial distribution function for dense packing of hard spheres of diameter R_0. The convolution with P_G simulates a softening of the repulsive potential and allows a broadening of the peak due to thermal motion

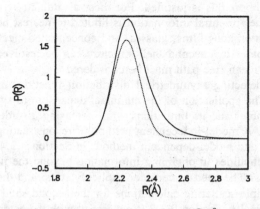

Figure 9.4. The asymmetric effective distribution function $P_G \otimes P_x^0$ and the best fit obtained by incorrectly assuming the Gaussian approximation ($\cdots$).

and static structural disorder. The values of R_0, B, and σ^2 correspond to those used to describe the distribution of cobalt atoms about phosphorus in amorphous CoP (37).

For convenience we write the reduced interference function

$$k\chi'(k) = \int P(r) \sin 2kr \, dr \qquad (26)$$

obtained by eliminating the phase shift $\delta(k)$, the amplitude $NS_0^2(k) \, F(k)$ and the mean free path dependence from the EXAFS interference function $\chi(k)$. By the convolution theorem and the results given in Table 9.1, Eq. (26) becomes

$$k\chi'(k) = \frac{e^{-2k^2\sigma_1^2}}{\left[1 + (2k/B)^2\right]^{1/2}} \sin\left[2k\overline{R} - \frac{2k}{B} + \tan^{-1}\left(\frac{2k}{B}\right)\right] \cdots \qquad (27)$$

Figure 9.5 shows $\chi'(k)k^3$ and the best fit obtained by incorrectly assuming the Gaussian low-disorder limit

$$k\chi'(k) = Ne^{-2C_2k^2} \sin 2k\overline{R} \cdots \qquad (28)$$

The distribution function obtained by fitting over the range $3.8 \text{ Å}^{-1} < k < 11.8 \text{ Å}^{-1}$ is compared with the actual distribution in Fig. 9.4. The fit coordination number and second cumulant C_2 are two small by 20 and 48%, respectively, and the mean bond length $\overline{R}$ is too short by 0.04 Å. In k space, the fit

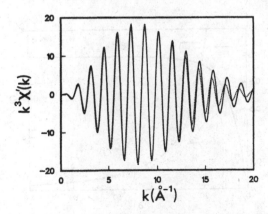

Figure 9.5. $k^3\chi'(k)$ for the $P_G \otimes P_x^0$ of Fig. 9.4 and the best fit obtained by incorrectly assuming the Gaussian approximation, Eq. (28) ($\cdots$).

is in error in amplitude at low k, but this is disguised by the k^3 weighting of $\chi'(k)$. The effect of asymmetry on the phase is not significant at low k, contributing only $\sim 1\%$ to the total phase at $k = 6$ Å^{-1}. At high k the characteristic signature of an asymmetric distribution function is apparent; the period of the asymmetrical distribution increases relative to that of the pure sinusoid. The amplitude is also clearly incorrect at high k. However, poor signal/noise, monochromator crystal glitches (see Chapter 3, Section 3.4.4), the presence of next-nearest neighbors, the proximity of other x-ray absorption edges, and transform artifacts frequently cause the EXAFS analyst to deemphasize the high-k region. If one considers only the data in the range over which the fit was made, the agreement in the phase is excellent and the agreement in the amplitude is better than frequently shown in the literature, not only for amorphous and liquid systems, but also for crystalline systems that possess low disorder. Thus, this example shows that errors can easily occur in the structural parameters of disordered systems unless the analysis procedures are designed specifically to include cumulants higher than the second.

An indication of the errors expected as a function of the asymmetry parameter B is given in Fig. 9.6. The effective values of N and C_2 were obtained by incorrectly using Eq. (28) to represent Eq. (27) and determining the intercept and slope of the best straight line in a plot of the logarithm of the amplitude versus k^2. The values of N and C_2 are always smaller than the model values 1.0 and $\sigma_1^2 + 1/B^2$ (Table 9.1). Similarly the interatomic distance, obtained by assuming that the phase of Eq. (27) is given by $2k\overline{R}$, is always smaller than $\overline{R}$; when B equals 30, 15, 10 or 5 Å^{-1} the distance is smaller by 0.005, 0.024, 0.050, or 0.141 Å. For zinc, Fig. 9.7, it has been shown that errors as large

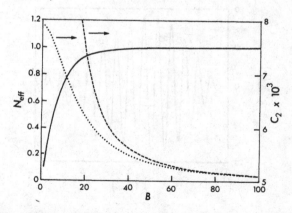

Figure 9.6. The effective coordination number (—) and second cumulants ($\cdots$) obtained by fitting a Gaussian model, Eq. (28), to the $k\chi'(k)$ of the asymmetric distribution function $P_G \otimes P_x^0$. The exact value of $C_2 = \sigma^2 + B^{-2}$ is also indicated (---).

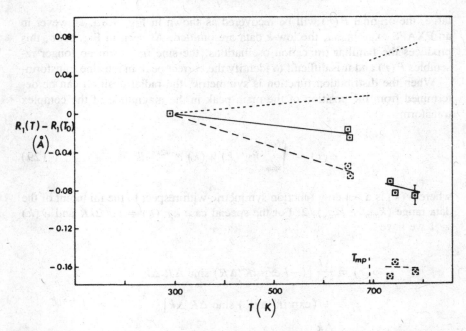

Figure 9.7. The temperature dependence of the nearest-neighbor distance in polycrystalline and liquid zinc. $R_1(T)$ estimated from thermal expansion data is indicated by (---), from the magnitude of the complex transform ▣, and from the sine transform ⊡ defined in Eq. (31). The solid (—) and dashed lines (---) are drawn only as a visual aid. [Taken from Fig. 4, ref. (17).]

as 0.16 Å can be made in the nearest-neighbor distance if the assumption of a Gaussian distribution is incorrect (17, 40).

In testing an EXAFS data set for the presence of asymmetry it is preferable to begin with an analysis of the phase. If an asymmetrical distribution is analyzed in the Gaussian approximation the percentage error in the coordination number will normally be larger than that in the interatomic distance. However, analysis of the phase can more reliably detect asymmetry because the phase is more chemically transferable than the amplitude. Also varying the inner potential E_0 will partially compensate for lack of phase transferability but will not improve the amplitude transferability. The test for asymmetry can be made in r space by a Fourier transform method or in k space by curve fitting the phase to the cumulant expansion, Eq. (21b).

The Fourier transform method will be illustrated using the effective distribution function $P(r) = P_G \otimes P_x^0$. As indicated by Eq. (26), $k\chi'(k)$ is the sine transform of $P(r)$. If the inverse sine transform of Eq. (26) is taken then, provided the low-k cutoff limit k_{min} equals zero and the upper cutoff k_{max} is large, the original $P(r)$ will be recovered as shown in Fig. 9.8a. However in an EXAFS experiment, the low-k data are omitted. As seen in Fig. 9.8b, this produces the familiar truncation oscillations, the sine transform no longer resembles $P(r)$ and it is difficult to identify the correct peak in the sine transform.

When the distribution function is symmetric, the radial position can be determined from the position of the main peak in the magnitude of the complex transform

$$\phi(r) = \int_{k_{min}}^{k_{max}} k\chi'(k)\, w(k)\, e^{-i2kr}\, dk \tag{29}$$

where $w(k)$ is a window function symmetric with respect to the midpoint of the data range $(k_{min} + k_{max})/2$. For the special case $k\chi'(k) = \sin 2k\overline{R}$ and $w(k) = 1$ we have

$$\phi(r) = \frac{\Delta k}{2} \left[(-i \exp iK'\Delta R) \operatorname{sinc} \Delta R\, \Delta k \right.$$

$$\left. + (\exp iK'\Delta R') \operatorname{sinc} \Delta R' \Delta k \right]$$

$$\simeq \frac{\Delta k}{2} \operatorname{sinc} \Delta R\, \Delta k(\sin K'\Delta R - i \cos K'\Delta R) \tag{30}$$

where $\Delta k = k_{max} - k_{min}$, $K' = k_{max} + k_{min}$, $\Delta R = \overline{R} - r$, $\Delta'R = \overline{R} + r$. The sinc functions $(\sin x/x)$ centered about $r = -\overline{R}$ contribute less than 1%

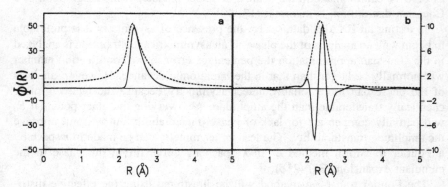

Figure 9.8. The Fourier transform of the reduced interference function $\chi'(k)$. The sine transform (—) and the magnitude (---) of $\int_{k_{min}}^{k_{max}} k\chi'(k)\, w(k)\, e^{i2kr}\, dk$, where $k\chi'(k)$ is given by Eq. (27) with $B = 4$, $R_0 = 2.30$ Å, and $\sigma_i^2 = 0.005$ Å^2. (a) $w(k) = 1.0$, $k_{min} = 0.0$ Å^{-1}, and $k_{max} = 20$ Å^{-1}. (b) $w(k)$ is a 10% Gaussian window, $k_{min} = 3.8$ Å^{-1} and $k_{max} = 16$ Å^{-1}.

for $\overline{R} \geq 2$ Å and have been neglected in Eq. (30). When $r = \overline{R}$, then Re ϕ = 0, where Re is the real part, Im $\phi = -|\phi|$ and the peaks coincide. For symmetric distribution functions the peak coincidence provides a check on the correct location of the inner potential E_0 (41).

For an asymmetric distribution, because of the inclusion of the cosine transform, the peak in $|\phi|$ will not coincide with the peak in $-$Im ϕ, even for the unattainable case of $k_{min} = 0$ [Fig. 9.8(a)] (provided E_0 has been correctly assigned) (17). Thus the coincidence of the peaks provides a sensitive test for the existence of asymmetry in the distribution function. In analyzing experimental data the complex transform taken is

$$\phi(r) = \int_{k_{min}}^{k_{max}} k\chi(k) \, e^{i2kr} \frac{e^{-i\delta(k)}}{B_i(k)} \, w(k) \, dk \tag{31}$$

where $B_i(k) = F_i(k) \, e^{-2r_i/\lambda} \, S_{0i}^2(k)$. Both $\delta_i(k)$ and $B_i(k)$ should be obtained by fitting to a suitable reference, although it may be adequate in some cases to neglect $S_{0i}^2(k)$ and to approximate $F_i(k) \, e^{-2r_i/\lambda}$ by theoretical values such as those of Teo and Lee (42). This procedure was applied in a study of polycrystalline and liquid zinc (17). The separation of the peak positions in Im ϕ and $|\phi|$ was readily detected as indicated in Fig. 9.7.

Crystalline germanium (Fig. 9.9) provides an example where the distribution function does not become asymmetric until high temperatures. At 1063 K anharmonicity in the effective two-body potential produces the displacement of the peak in Im ϕ from that in $|\phi|$ and causes the unequal strength of the side lobes in Im ϕ. $|\phi|$ indicates an apparent contraction of 0.044 Å and Im ϕ an apparent contraction of 0.026 Å relative to the thermally expanded crystal at the same temperature. In comparison the symmetry of the plots for amorphous germanium at 83 K indicates that the Gaussian or low-disorder limit is applicable (35).

In the numerical evaluation of Eq. (31) normally a 10% Gaussian window function $w(k)$ is used; that is, $w(k)$ equals 1.0 at the midpoint of the data range decreasing to 0.1 at k_{min} and k_{max}. A comment on removing the phase with the fast Fourier transform (FFT) may be appropriate. Since the FFT algorithm implicitly assumes that the first point k_{min} in the array of numbers being transformed corresponds to $k = 0$, the correct phase of $\delta(k)$ is maintained by "padding" the beginning of the array with N' zeros such that the $N' + 1$ point corresponds to k_{min}. Also, to obtain the smooth curves shown in Fig. 9.9 interpolation was performed with the FFT by the usual procedure of "padding" with zeros above k_{max} to permit a 2048 point transform. Since the mean interatomic separation $\overline{R}$ will not in general coincide with a bin frequency of the FFT, our algorithm determines the peak positions from the vertices of parabolas fitted to Im ϕ and $|\phi|$ in the vicinity of the peak. As a cross check a linear fit is made to Re ϕ and $\overline{R}$ is determined from the zero crossing.

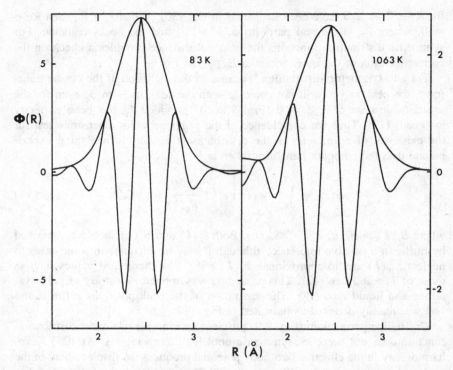

Figure 9.9. The magnitude and the sine transform of the complex transform, Eq. (31), for a-Ge at 83 K and c-Ge at 1063 K. [Taken from Fig. 1, ref. (35).]

Analyzing data via Eq. (31) has the advantage that the resolution is improved. In r space the amplitude function $B_i(k)$ is convolved with the distribution function $P(r)$, broadening it and frequently complicating the separation of neighboring peaks. It is particularly helpful with backscattering atoms of high atomic number where nonlinear k dependences in $F(k)$ and $\delta(k)$ produce side lobes about the main peak. The case of platinum is shown in Fig. 9.10 (43). In catalytic studies the large side lobe on the small r side of the main peak in Fig. 9.10a obscures any interaction of platinum with oxygen in the substrate. The narrower peak of Fig. 9.10b was obtained by removal of the Pt–Pt $\delta(k)$. Removal of the Rh $F(k)$ and Rh–Rh $\delta(k)$ permitted a reliable identification of the support oxygen in a study of dispersed rhodium on alumina catalyst (44).

Asymmetry in the distribution function can be determined directly in k space by fitting the phase to the cumulant expansion Eq. (21b). However, if a suitable reference system exists for which the asymmetry contribution to its phase is insignificant, the most direct method is to examine $\Phi_x - \Phi_{\text{ref}}$, the difference

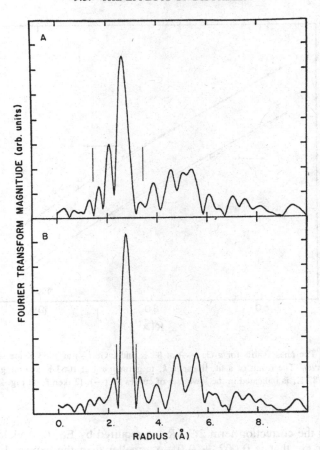

Figure 9.10. Fourier transform magnitude versus radial distance for a 100 K platinum metal foil: (a) Unmodified transform and (b) theoretical Pt–Pt phase shift (41) removed prior to transforming. [Taken from Fig. 1, ref. (43).]

between the total phase of the unknown and the reference. As an example, consider again the case of the nearest neighbors in amorphous and crystalline germanium. The phase terms $\Phi - 2k\overline{R}_a$, where $\overline{R}_a$ is the nearest-neighbor distance in a-Ge, are plotted in Fig. 9.11. Since the Gaussian approximation is applicable to a-Ge, the curvature apparent in $\Phi_c - 2k\overline{R}_a$ for c-Ge indicates the contributions of anharmonicity to the phase at high temperature. Fitting $\Phi_c - \Phi_a$ to the functional form

$$\Phi_c - \Phi_a = 2k(R_c - \overline{R}_a) - \tfrac{4}{3}C_3 k^3 \tag{32}$$

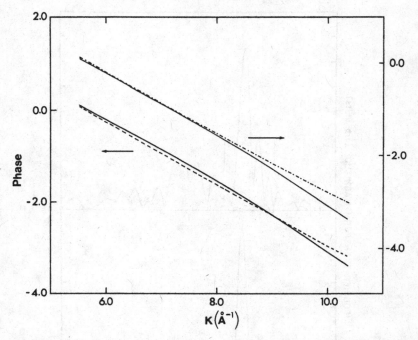

Figure 9.11. The phase shifts for a-Ge (-··-) at 83 K and c-Ge (—) at 1063 K are shown in the two upper curves. The result of a fit, linear in k, to germanium at 1063 K assuming a reference phase shift at 83 K, is indicated in the lower set of curves by (---). [Taken from Fig. 2, ref. (35).]

and adding the correction term $2C_2/\overline{R}_a$ as required by Eq. (23) yields a value of $\overline{R}_c$ for c-Ge that is 0.002 ± 0.01-Å smaller than the expected thermally expanded value (35). If the asymmetry in the distribution function for c-Ge is neglected and $\Phi_c - \Phi_a$ is fitted to the lower set of curves in Fig. 9.11, then the nearest-neighbor distance is found to be incorrectly contracted by 0.026 ± 0.01 Å.

In conclusion it has been emphasized in this section that incorrect structural parameters will be obtained if an EXAFS spectrum of an asymmetric distribution function is analyzed using the standard Gaussian approximation. Thus it is recommended that *all* thermally and structurally disordered systems be tested for asymmetry. An r-space test using a Fourier transform analysis provides a useful complement to the k-space determination of cumulants higher than the second, particularly when the third and higher cumulants are small and the quality of data may obscure the asymmetry.

9.3.2. Cumulant Expansion

The logarithm of the Fourier transform of any probability distribution $P(r, \gamma)$ may be expanded in powers of k to give the cumulant expansion (see Section 9.2.3 and Chapter 1, Section 1.3.2). It provides the correct functional form for analytically extending EXAFS data from k_{min} to $k = 0$. For systems with low to moderate disorder it permits a model-independent method of EXAFS analysis.

The cumulants can be converted into the corresponding moments of the distribution function (19). Once the cumulants have been determined they can then be used to distinguish competing models proposed for the system under investigation. This has the obvious advantage that the differentiation between models proceeds *after* the EXAFS analysis has been completed. The analysis is not hampered by the precondition that suitable models must exist prior to the analysis.

The odd and even cumulants are decoupled, contributing separately to the phase and the logarithm of the amplitude of $k\chi'(k)$, respectively, Eq. (21). They are determined by a simple extension of the usual method for systems specified by a Gaussian distribution; the phase difference and the logarithm of the ratio of the amplitude between the unknown and a reference are fitted to power series in k. The separation of the cumulants into odd and even power series has the advantage that it reduces the correlation between parameters—for example, between the k and k^3 terms in the expansion. "The correlation is much easier to deal with because of the minimal number of parameters, the smooth nature of the functions involved, and the linear dependence of the fit function on the cumulants. This linearity permits a global analysis of the correlations in parameter space and makes it easy to identify and describe the entire volume in the space that matches the data within a specified error bar. This is a strong advantage over nonlinear least squares fitting, for which only a local analysis is feasible" [page 441, ref. (19)].

The cumulant method has limitations. First, it is necessary to obtain the amplitude and phase by Fourier filtering methods. This introduces transform artifacts, but these can be reduced substantially by the usual procedure of transforming the unknown and reference data sets under identical conditions. However, in the case of two shells separated by small radial distances, interference in r space between the frequency contributions poses a severe problem. Treating the contributions of disorder effects by the cumulant expansion and fitting two shells simultaneously can introduce more adjustable parameters than the number of degrees of freedom allowed by the data

$$N_{\text{free}} \simeq \frac{2 \, \Delta R \, \Delta K}{\pi}$$

where ΔK and ΔR are the window widths in k space and r space. However, see Section 9.3.3 where it is suggested that two shells can be analyzed under a restricted set of conditions. The second limitation, which will be discussed in the remainder of this section, is the breakdown of the cumulant expansions with increasing k and increasing disorder.

The use of cumulants will be examined using the model effective distribution function $P(r) = P_G \otimes P_x^0$ used in Section 9.3.1. Other choices for the weighted exponential P_x^m are given in Table 9.1. However, for moderate disorder the differences in the convolved functions are small.

The results of fitting the logarithm of the amplitude of the model function to a power series in order to estimate the cumulants are given in Table 9.2 for different k-space fitting ranges, different values of the disorder parameter B and a fixed value $\sigma^2 = 0.005$ Å^2. The coefficients were fitted to two cumulants C_2 and C_4 under the assumption that the coordination number was known and, to four cumulants $C_0 = \ln N$, C_2, C_4, and C_6, assuming that the coordination number was unknown. By reference to Fig. 9.6 and Table 9.2 it is seen that the choice $B = 25$ Å^{-1} approaches the low disorder limit where only small errors in N and C_2 result from assuming the Gaussian model. $B = 16$ Å^{-1} represents moderate disorder and $B = 10$ Å^{-1} approaches the region of large disorder where the experimental data is significantly attenuated at large k and multiple scans are required to acquire quality data, even using synchrotron radiation.

The cumulant expansion equations, which were derived in Section 9.2.3 for a general distribution, can be obtained directly for the model function P_x^0 by expanding

$$\tan^{-1} z = z - \frac{1}{3} z^3 + \frac{1}{5} z^5 + \cdots \qquad z^2 < 1$$

$$\ln(1 + z^2) = z^2 - \frac{z^4}{2} + \frac{z^6}{3} + \cdots \qquad z^2 < 1 \qquad (33)$$

Thus with $z = 2k/B$ the cumulant expansion diverges for $k > B/2$. With $B = 10$ and a fitting range $3 \le k \le 12$ Å^{-1}, the sixth cumulant C_6 is 13.3 times the value of the exact function evaluated at 12 Å^{-1}. The correspondingly large deviation, 98%, of the fit coefficient from the exact value of C_6 is not unexpected. However, the following problem exists; polynomials in k provide deceptively good fits to the data, but the coefficients obtained depend upon the range of the fit and only approach the correct cumulants in the limit of low k. Moreover, goodness of fit criteria can be misleading. For example, we calculated the value of C_n for which χ^2, the sum of the square of the residuals, was twice its best fit value when the other C_n were fixed at their best fit values. This value is listed in Table 9.2 as a percentage deviation from the best fit coefficient and is always small even in the case of $B = 10$ Å^{-1} and $k_{max} = 12$ Å^{-1}.

The problem is illustrated further in Fig. 9.12. The unknown $k\chi'(k)$ was constructed from the model $P_G \otimes P_x^0$, Eq. (27), with $B = 10$ Å^{-1}, $R_0 = 2.19$ Å^{-1}, $\sigma_{1x}^2 = 0.005$ Å^2. In order to simulate the analysis of real data a Gaussian reference $k\chi'(k)$ was created. Both the unknown and the reference were Fourier transformed and then inverse transformed using k-space and r-space windows of 17 Å^{-1} and 2 Å. The difference in phase and the log ratio of the amplitude for the unknown and the reference filtered data sets were fitted to obtain the cumulants C_2, C_3, and C_4. The $k\chi'(k)$ reconstructed from the fit coefficients according to

$$k\chi'(k) = \exp(-2C_2 k^2 + \tfrac{2}{3} C_4 k^4) \sin (2kR - \tfrac{4}{3} C_3 k^3) \qquad (34)$$

is overlaid in Fig. 9.12a. The good agreement at low k occurs because the analysis proceeded as if the coordination number were known. The divergence of the fit $k\chi'(k)$ at high k emphasizes that the cumulant expansion is a power series expansion about $k = 0$, which will break down at sufficiently large k. Over the range of the fit $3.4 \leq k \leq 11.8$ Å^{-1}, good agreement exists. Yet as seen from Table 9.2, the deviation of the fitted C_4 from the exact value is 87%. Thus problems would exist if the magnitude of the cumulants were used as the basis for accepting or rejecting physical models for the unknown system.

It is relevant to examine the effect that known errors in the cumulants have upon the overall shape of the distribution function. $P(r)$ can be reconstructed from the cumulant coefficients by taking the inverse sine transform of $k\chi'(k)$ with $k_{min} = 0$ and $w(k) = 1.0$. Approximately the correct functional form is obtained with $k_{max} = 20$ Å^{-1}, Fig. 9.12b. However, the nonconvergence in the cumulant fit at high k produces unsatisfactory side lobes in $P(r)$. A better reconstruction is obtained by restricting k_{max} to the minimum in the amplitude of the fit $k\chi'(k)$, which occurs near 14 Å^{-1}. Figure 9.12 was deliberately constructed to exaggerate the errors that can occur in fitting data with large disorder to a cumulant expansion. In fact, with $B = 10$ Å^{-1} only data for $k < 5$ Å^{-1} should have been used in determining the cumulants. Nevertheless the basic shape of $P(r)$ is apparent. With decreasing amounts of disorder the cumulant method should converge to the correct effective $P(r)$.

An extensive study of anharmonicity in polycrystalline CuBr provides an example of the application of the cumulant method (45). It was selected for discussion because a comparison is possible with an independent EXAFS study in which the analysis proceeded in r space by fitting to a theoretically motivated distribution function (46–49). The difference in phase and the log ratio of the amplitudes between the unknown and a reference CuBr at 72 K were fitted to

Table 9.2. Comparison of Cumulants

Exact Cumulants[a]	Fractional Weight[b]	Fit Deviation (%)[c]	χ² % Deviation[d]	Fractional Weight[b]	Fit Deviation (%)[c]	χ² % Deviation[d]
$B = 25$ $k_{max} = 12$ Å^{-1}				$k_{max} = 5$ Å^{-1}		
C_2 0.006	1.08	−1.2	0.13	1.02	−0.08	0.01
C_4 1.54×10^{-5}	−0.12	−4.6	2.8	−0.02	−13.4	0.4
N 1	−	−0.06	0.07	−	−0.001	0.001
C_2 0.0066	1.08	−0.5	0.01	1.02	−0.02	0.00
C_4 1.54×10^{-5}	−0.12	−21	0.1	−0.02	−2.6	0.00
C_6 4.92×10^{-7}	0.07	−67	0.6	0.002	−26.0	0.02
$B = 16$ $k_{max} = 8$ Å^{-1}				$k_{max} = 5$ Å^{-1}		
C_2 0.0089	1.15	−2.5	0.2	1.07	−0.7	0.04
C_4 9.16×10^{-5}	−0.25	−48	2.7	−0.09	−28	0.8

C_n	2	C_n^{fit}	% dev	χ^2	C_n^{fit}	% dev	χ^2
N	1		− 0.15	0.2		− 0.02	0.02
C_2	0.0089	1.15	− 1.5	0.01	1.07	− 0.4	0.00
C_4	9.16×10^{-5}	−0.25	−26	0.1	−0.04	−11.0	0.01
C_6	7.15×10^{-6}	0.17	−72	0.4	0.02	−50.0	0.05

C_n	2	C_n^{fit}	% dev	χ^2	C_n^{fit}	% dev	χ^2
C_2	0.015	1.80	−22	1.7	1.25	− 4.6	0.2
C_4	0.0006	−3.46	−87	7.6	−0.42	−51.0	1.6
	$B = 10 \; k_{max} = 12$ Å^{-1}				$k_{max} = 5$ Å^{-1}		
N	1		− 4.5	5.0		− 0.4	0.4
C_2	0.015	1.80	−20	0.2	1.25	− 3.9	0.00
C_4	0.0006	−3.46	−78	0.5	−0.42	−34.0	0.02
C_6	0.00012	13.30	−98	1.4	0.28	−78.0	0.09

[a]Cumulants calculated exactly from the model $P_x^0 \otimes P_G$ are given in column 2. The fit coefficients C_n^{fit} were obtained by fitting the exact equation $y(k) = -2\sigma^2 k^2 - \frac{1}{2}\ln(1 + 2k/B)$ over the range $k_{min} = 3$ Å^{-1} to the k_{max} indicated.

[b]The fractional weight of a given term is evaluated at $k = k_{max}$. It is given by $C_n(2k)^n/n!y(k)$.

[c]The fit deviation is given by $100 \times (C_n^{fit} - C_n^{exact})/C_n^{exact}$.

[d]χ^2 is the value of C_n for which the sum of squares of the residuals is twice its minimum value when the other C_n^{fit} are fixed at their best fit values. The percentage deviation of the χ^2 value of C_n from the best fit C_n is listed in columns 5 and 8.

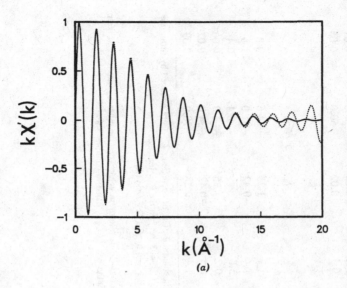

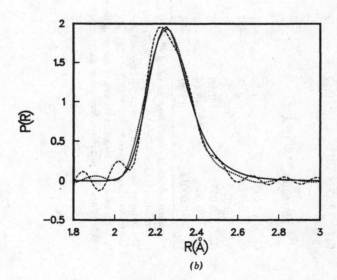

Figure 9.12. (a) $k\chi(k)$ for the asymmetric distribution function $P_G \otimes P_x^0$ and the result (---) of fitting the model with a fourth-order cumulant expression over the range 3.4–12 Å^{-1}. (b) A comparison of the reconstructed $P(r)$ with the actual $P(r)$ (—). The curves ($\cdots$) and (---) are obtained by inverting the cumulant fit $k\chi'(k)$ of Fig. 9.12(a) with $k_{\max} = 12$ and 20 Å^{-1}, respectively.

obtain the average cumulants $C_2^{fit} = 0.011$ Å$^2 \pm 15\%$, $C_3^{fit} = 0.37 \times 10^{-3}$ Å$^3 \pm 14\%$, $C_4^{fit} = 0.72 \times 10^{-4}$ Å$^4 \pm 17\%$ for the copper and bromine EXAFS spectra at 295 K. The phase differences and the cumulant fits are shown in Fig. 9.13. To place CuBr in the context of the present article, solving the equations relating the disorder parameter B to C_3 and C_4 gives the average values, $B \approx 17$ and 24 Å^{-1} for the models $P_G \otimes P_x^0$ and $P_G \otimes P_x^2$. Thus CuBr is a system of moderate disorder for which the cumulant expansion breaks down at a k_{max} of 8–12 Å^{-1}. The cumulants were obtained by fitting over the range 4.5 Å $< k <$ 13.4 Å^{-1} and consequently the deviation of the C^{fit} from the correct values, at least on the basis of Table 9.2, may exceed the amount implied by the error bars assigned in the least-squares fitting of the data. Neglecting this possible limitation on the magnitudes of the cumulants, two models for the cumulants were considered (45). It was concluded that the data were consistent with an effective anharmonic single-particle potential for the copper ions. A

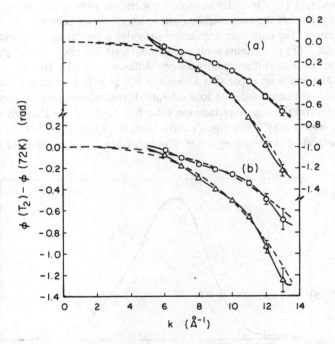

Figure 9.13. Difference between single-shell phases at T_2 and 72 K for CuBr: (*a*) copper-edge data, $T_2 = 216$ K (circles), 295 K (triangles); (*b*) bromine-edge data, $T_2 = 206$ K (circles), 295 K (triangles). The dashed lines correspond to least-squares fits to the data including only a linear and cubic term in the fit equation. Error bars used in the fitting are shown where they are larger than the symbol size. [Taken from Fig. 4, ref. (45).]

second model, in which copper atoms were permitted to occupy off-center sites was rejected because the model cumulants did not agree with experimental values and more importantly, because reasonable values of the model cumulants resulted in a phase and amplitude that could not be reconciled with the experimental phase and amplitude.

The experimental cumulants were used to reconstruct the distribution function shown in Fig. 9.14 by inverting Eq. (34). The inverse transform was taken using for k_{max} the location of the minimum in the function $\exp(-2k^2 C_2^{fit} + \frac{2}{3}k^4 C_4^{fit})$. This exceeds slightly the maximum value of k, 13.4 Å^{-1}, to which the data were fitted but it has the advantage of producing a minor reduction in the magnitude of the side lobes produced by the truncation of the $kx'(k)$ at high k and evident in the "tails" of $P(r)$ in Fig. 9.14.

An alternative method of determining the $P(r)$ function has been described by Boyce et al. (46). In essence their method is to fit the r-space transform of the EXAFS interference function to a model distribution (see Section 9.3.4). Their resulting $P(r)$ for CuBr at room temperature (48) is also shown in Fig. 9.14. It is seen that the two methods are in excellent agreement.

In summary, the cumulant expansion provides a model-independent specification of the $P(r)$ of systems with low to moderate disorder. The fit cumulants C_n^{fit} can be used to differentiate between different models. However, the cumulant expansion is an expansion about $k = 0$ that will diverge at high k. Thus for systems with semimoderate to moderate disorder low-order polynomial fits to the phase and the log amplitude over the full range of the EXAFS data will yield C_n^{fit}, which may differ significantly from the exact C_n. There is no satisfactory method of determining when this breakdown occurs unless a model is

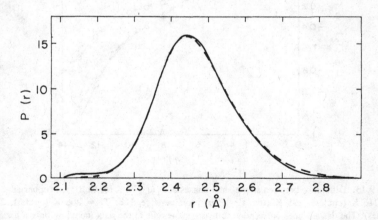

Figure 9.14. The reconstructed $P(r)$ for CuBr at 295 K. The cumulant $P(r)$ was obtained from the C_n^{fit} by taking the inverse sine transform of Eq. (34) with $k_{min} = 0$ Å^{-1} and $k_{max} = 15.6$ Å^{-1}. The dashed curve was obtained by fitting the r-space transform of $\chi(k)$ to a model $P(r)$ (48).

invoked. Thus the magnitudes of the C_n^{fit} should be used with caution in differentiating between models. A preferable method is to compare the experimental amplitude and phase directly with those calculated from the model $P(r)$. Indeed, if it is desired to parameterize the amplitude and phase for future comparisons, then as demonstrated by Alberding (50), a better fit with fewer parameters can be achieved using polynomials in even powers (for the log of the amplitude) and odd powers of k (for the phase) expanded about the midpoint of the data range rather than about $k = 0$. Nevertheless the C_n^{fit} can still be used to reconstruct the $P(r)$ by the simple expedient of inverting equations like Eq. (34). This provides satisfactory $P(r)$ for systems with low to moderate disorder. The next section illustrates an alternative method of applying the cumulant expansion that may be satisfactory even for systems with a large degree of disorder.

9.3.3. Model-Independent Determination of $P(r)$

If the reduced EXAFS interference function $\chi'(k)$ is extrapolated to $k = 0$ via the cumulant expansion, then the experimental $\chi'(k)$ plus the extrapolated portion can be inverted to reconstruct the distribution function. This "splice method" has the advantage that the actual experimental data is used for the EXAFS range rather than a fitted function, which inevitably will introduce errors. Since the cumulant expansion is used only in the low-k region, it will be shown that the method can be applied to systems with greater disorder than was possible using the methods of Section 9.3.2. Moreover, since the cumulant expansion is valid for any distribution function, it will also be shown that the splice method has the potential of providing a model-independent reconstruction of the effective $P(r)$ of two closely separated shells (of the same atomic composition) that produce visible "beats" in $\chi'(k)$.

Model functions only were used for calculations in this section. To simulate the analysis of real data, the unknown and a Gaussian reference data set were transformed to r space and then inverse transformed using a r-space window $\Delta R = 2.0$ Å^{-1}. The cumulants were then obtained by fitting appropriate polynomials to the phase difference and the logarithm of the ratio of the amplitudes between the unknown and the reference over the range $k_{min}^{fit} \leq k \leq k_{max}^{fit}$. It was assumed in all cases that the coordination number was unknown. We obtained the even cumulants C_0, C_2, C_4, and C_6 and the odd cumulants C_3 and C_5. The correction term for the difference between the mean distance $\bar{R}$ of the unknown and reference was also included in the fits. The cumulants were used to construct a $k\chi'(k)$ for $0 \leq k \leq k_s$, which was then joined to the filtered model $k\chi'(k)$ at the splice point k_s. Finally the $P(r)$ was reconstructed by taking the inverse sine transform of the spliced $k\chi'(k)$ with the lower limit $k_{min} = 0.0$ Å^{-1} and the upper limit k_{max}.

The asymmetric model distribution function shown in Fig. 9.15 has the same parameters $B = 10$ Å^{-1}, $R_0 = 2.19$ Å, and $\sigma^2 = 0.005$ Å^2 as used in Figs.

9.4 and 9.12. The cumulants were obtained by fitting over the range $k_{min}^{fit} = 3.4$ Å^{-1} to (a) $k_{max}^{fit} = 12$ Å^{-1} and (b) $k_{max}^{fit} = 10$ Å^{-1}. The splice occurred at 3.4 Å^{-1} and the upper limit of the inverse sine transform was $k_{max} = 16$ Å^{-1}. The cumulants for curve (a) were obtained by fitting over the same range as used in Fig. 9.12b. The marked suppression of the ripple in Fig. 9.15, relative to that in Fig. 9.12b, occurs because the actual (filtered) $k\chi'(k)$ is used at high k and the upper limit of integration k_{max} is extended to a region where $k\chi'(k)$ is small. The further improvement in the $P(r)$ of curve (b) is obtained by reducing k_{max}^{fit}, which reduces the deviation between the C_n^{fit} and exact C_n in accord with the trends shown in Table 9.2.

The next two examples involve the superposition of two Gaussians. In this case both the amplitude and phase depend upon N_1, R_1, σ_1^2, N_2, R_2, σ_2^2, and k. The amplitude of $k\chi'(k)$ can be written (51)

$$A^2(k) = a_1^2 e^{-4k^2\sigma_1^2} + a_2 e^{-4k^2\sigma_2^2} + a_1 a_2 e^{-2k^2(\sigma_1^2 + \sigma_2^2)} \cos(2k\,\Delta R) \quad (35)$$

where $a_i = N_i/R_i^2$ and $\Delta R = R_1 - R_2$. The first three even cumulants obtained from a Taylor's expansion about $k = 0$ are (50):

$$C_0 = \ln(a_1 + a_2)$$

$$C_2 = \frac{(a_1 + a_2)(a_1\sigma_1^2 + a_2\sigma_2^2) + a_1 a_2(\Delta R)^2}{(a_1 + a_2)^2}$$

$$C_4 = \frac{a_1 a_2}{16(a_1 + a_2)^4} [48(a_1 + a_2)^2(\sigma_1^4 - 2\sigma_1^2\sigma_2^2 + \sigma_2^4)$$
$$+ 24(\Delta R)^2(a_1^2 - a_2^2)(\sigma_2^2 - \sigma_1^2) + (\Delta R)^4(a_2^2 - 4a_1a_2 + a_1^2)]$$

$$(36)$$

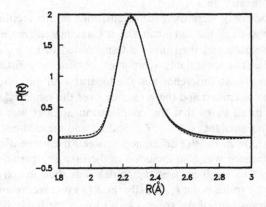

Figure 9.15. The reconstructed $P(r)$ for a single-shell asymmetric distribution function using the splice method. The original $P(r)$ (—) is compared with the $P(r)$ reconstructed when the cumulant fit range extends to (a) 12 Å^{-1} (---), (b) 10 Å^{-1} ($\cdots$) and splicing occurs at $k_s = 3.4$ Å^{-1} in each.

The first example consists of two Gaussians, $N_1 = 0.5$, $R_1 = 2.24$ Å, $\sigma_1^2 = 0.005$ Å^2; $N_2 = 0.5$, $R_2 = 2.34$ Å, $\sigma_2^2 = 0.020$ Å^2, selected to approximate the asymmetric distribution $P_G \otimes P_x^0$. The splice method was applied with $k_s = 3.4$ Å^{-1}, $k_{min}^{fit} = 3.4$ Å^{-1}, and $k_{max}^{fit} = 10$ Å^{-1}. Negligible difference existed between the original and reconstructed $P(r)$. The reconstructed $P(r)$ is compared in Fig. 9.16 with the reconstructed asymmetric $P(r)$, curve (b), of Fig. 9.15. These simulations suggest that the splice method has the potential of permitting a model-independent differentiation between models that may differ only slightly.

The two Gaussians of the final example are specified by $N_1 = 0.5$, $R_1 = 2.24$ Å, $\sigma_1^2 = 0.005$ Å^2 and $N_2 = 0.5$, $R_2 = 2.44$ Å, $\sigma_2^2 = 0.010$ Å^2. Again there is good agreement between the original $P(r)$ and the reconstructed $P(r)$ as shown in Fig. 9.17. For this model, beating between the two frequencies is readily apparent in $k\chi'(k)$ with a pronounced minimum occurring in the amplitude near $k = \pi/2 (R_1 - R_2) = 7.9$ Å^{-1}. In order to obtain reasonable cumulant fit coefficients C_n^{fit} it is necessary to avoid such minima. The cumulants were obtained with $k_{min}^{fit} = 3.4$ Å^{-1} and $k_{max}^{fit} = 6.4$ Å^{-1}. The splice point was taken at $k_s = 3.8$ Å^{-1}.

When real data are involved there are limitations to the splice method. The reconstruction depends upon the accuracy with which the reduced interference function $\chi'(k)$ can be extracted from $\chi(k)$. Thus a well-characterized reference system is necessary to provide accurate EXAFS phase shifts $\delta_i(k)$ and amplitude reduction terms $B_i(k) = S_{0i}^2(k) F(k) e^{-2r/\lambda(k)}$. If a reference system is not available, limited success may be possible using theoretical estimates for δ_i and $B_i(k)$. In this case it will also be necessary to include corrections for the finite energy resolution of the EXAFS monochromator that introduces an additional

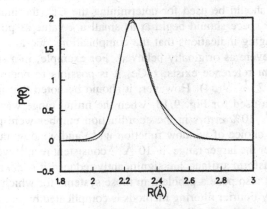

Figure 9.16. The reconstructed $P(r)$ for two distributions using the splice method. The reconstructed asymmetric $P(r)$, curve (b) of Fig. 9.15 ($\cdots$), is compared with the reconstructed $P(r)$ resulting from the superposition of two Gaussians with parameters given in the text.

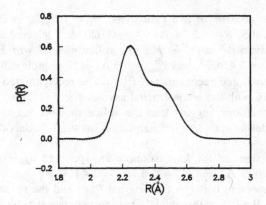

Figure 9.17. The distribution resulting from the superposition of two Gaussians compared with the reconstructed $P(r)$ obtained via the splice method ($\cdots$).

exponential damping factor $M(k)$ (52). Moreover, when the log-ratio method is used to determine the even cumulants for the construction of the extrapolated portion of $k\chi'(k)$, the coefficient of k^2 is the difference $C_2 - \sigma_{ref}^2$. Thus an accurate value of C_2 requires the absolute value of the EXAFS Debye–Waller factor of the reference system. This is most accurately obtained by temperature-dependent measurements as, for example, in c-Ge and a-Ge (35, 53). The linear term in the phase must also be corrected, as per Eq. (23), but the correction is small. These problems are not unique to the splice method; knowledge of $B_i(k)$ and $M(k)$ is required before curve fitting to k-space or r-space model functions (see Chapter 6).

Transform artifacts also limit the accuracy of the splice method. Since only the low k region should be used for determining the C_n^{fit}, the initial transform from k space to r space should begin at as small a k value as possible. There are some encouraging indications that the complications due to many-body effects are not as severe as originally believed. For example, in a study of a-Ge where an excellent reference exists, c-Ge, it is possible to begin the EXAFS analysis at $k_{min} \approx 2$ Å^{-1} (54). However, it should be noted that in the case of the simulated data used for Fig. 9.15, when the fitting range was restricted to $3 \leq k \leq 6$ Å^{-1}, 10% errors in the coordination number were produced depending upon the choice of window function $w(k)$ and window ranges Δk and ΔR. By fitting over the larger range, 3–10 Å^{-1} consistent results were achieved. The effects of transform artifacts are significantly reduced if N is known. Transform artifacts will also pose a problem in those systems for which the isolation of the first shell by Fourier filtering methods is complicated by interference with second and higher shells. Finally, in reconstructing $P(r)$ by Fourier inversion of $k\chi'(k)$, it is necessary that the upper integration limit be as large as possible in order to suppress ripple in $P(r)$. Thus the splice method is severely limited by interrupting absorption edges.

In principle many of the limitations indicated previously can be handled by careful experimental and analysis procedures, in which case the splice method has an intrinsic advantage over model-dependent methods. In those cases where an accurate reconstruction of $P(r)$ is not possible, the splice method may still provide the insight necessary for model building. The splice method will not work for systems with very large disorder because of the breakdown of the cumulant expansion at even small values of k.

9.3.4. Model-Dependent Determination of $P(r)$

When large disorder exists, methods based on the cumulant expansion are inappropriate and recourse must be made to fitting to model functions. Modeling procedures must also be used when a single shell contains more than one atomic species (compositional disorder) and when neighboring shells, containing different atomic species, cannot be isolated without transform-interference effects occurring. The fit functions chosen are multiparameter functions that either are of sufficient generality that they hopefully permit a good approximation to the actual distribution or are more specific functions anticipated from prior physical and chemical knowledge of the local structure. Since there is usually significant correlation between the fit parameters it is essential that a thorough search be made in multidimensional fitting-parameter space to test the uniqueness of the best fit. A general discussion of fitting procedures has been presented in Chapter 6. Fitting may occur in either k space or r space. In this section one application of each will be discussed briefly.

Boyce and co-workers (1, 46) have presented a general formulation for the determination of the distribution function by an r-space fitting procedure. They express ϕ, the Fourier transform of $k\chi(k)$, as the sum over each backscattering species i of a convolution of the real distribution function $P_i(r)$ and a peak function ξ_i. ξ_i is the Fourier transform of the product of the window function $w(k)$ and a term including the phase shift and amplitude reduction factor $B_i(k)$. It is assumed to be independent of structure and temperature. It is obtained from a reference system at low temperature where $P_i(r)$ is Gaussian and well characterized.

Models of $P_i(r)$ with adjustable parameters are then fit to the r-space data at other temperatures. Their "best" fit criterion is that there be a minimum in

$$R^2 = \frac{1}{2N} \sum_N \left[\frac{\left[\mathrm{Re}(\phi - \phi_m) \right]^2}{\left(\mathrm{Re}\ \phi \right)^2 + \left[\mathrm{Re}\ (d\phi/dr) \right]^2} + \frac{\left[\mathrm{Im}(\phi - \phi_m) \right]^2}{\left(\mathrm{Im}\ \phi \right)^2 + \left[\mathrm{Im}\ (d\phi/dr) \right]^2} \right]$$

$$(37)$$

where the sum extends over all N points in the range of the peak(s) being fit, ϕ is the transform of the data, and ϕ_m is the transform of $P_i \otimes \xi_i$. This is a weighted fractional difference least squares where the $d\phi/dr$ term is added to enhance the sensitivity of R^2 to variations in the shape of $P_i(r)$ at the expense of some of its sensitivity to position (46). This criterion is superior to minimizing only the sum of residuals $\Sigma \left[\text{Re} \, (\phi - \phi_m) \right]^2 + \left[\text{Im}(\phi - \phi_m) \right]^2$. In general E_0 must also be adjusted. This is inefficient in r space because the calculated $k\chi(k)$ corresponding to each new E_0 must be Fourier transformed before comparison of ϕ_m with the unknown ϕ. It would appear that a k-space procedure [Section 9.3.5, Eq. (43)] may offer numerical advantages. Finally the fitting procedure is repeated for different models. Most of the data of Boyce et al. on liquids, glasses, and amorphous materials have been analyzed in this fashion.

They have reported extensive work on superionic conductors that exhibit pronounced asymmetry in $P_i(r)$ due to the mobility of the cations at high temperatures. This is evident in Fig. 9.18 for AgI where the narrow Gaussian at 77 K becomes broad and asymmetric at 198°C. The best fit to $P(r)$ was obtained from the excluded volume model. In this model the cations are excluded from spherical regions around each anion. The hard-sphere exclusion is somewhat softened by a Gaussian convolution function. The asymmetry that develops is analogous to that indicated in Section 9.3.1 for the model $P_G \otimes P_x^0$.

When fitting to models in k space it is efficient numerically if the Fourier transform of the model $P(r)$ can be represented by a simple analytical function. This is the case for the functions listed in Table 3.1; the Gaussian P_G, the weighted exponential P_x^m, and their convolution. The function P_x^2 has been

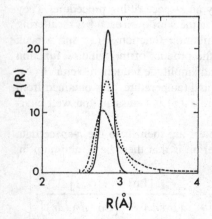

Figure 9.18. First-neighbor pair correlation functions for AgI deduced from the silver K-edge EXAFS spectra at 77 K (—), 22°C ($\cdots$), and 198°C (---). The increase in asymmetry at elevated temperatures is described by the excluded volume model. [Taken from Fig. 8, ref. (47).]

applied to liquid zinc (17) and metallic glasses (3, 55). Metallic glasses have also been analyzed with Gaussians and the convolution $P_G \otimes P_x^0$ (4, 11, 36, 39). Of course it is always possible to construct a model reduced interference function

$$k\chi'(k)_m = \sum_j P_j \sin 2kr_j \qquad (38)$$

where $P_j = P(r)\Delta r$ is the weight assigned to the jth interval of a continuous distribution $P(r)$. This was done, for example, in an analysis of the contribution of multiple-scattering effects arising from dynamic fluctuations in bond angles due to thermal motion. It was assumed that $P(r) = z^{-1} \exp[-y^2/\sigma^2]$ where $y = (r - r_p) \exp[a(r - r_p)]$ with r_p the peak of the distribution and z a normalization constant. This distribution is well behaved for small values of a. For $a = 0$ it is Gaussian, $a > 0$ skews the distribution above its peak, and $a < 0$ skews it below the peak (56).

Throughout Section 9.3 it has been emphasized that it is essential to test all EXAFS data sets for the presence of asymmetrical distributions and that failure to include the effects of asymmetry can result in significant error in EXAFS-deduced structural parameters. In the field of metallic glasses there are papers, published since the identification of the asymmetry problem, in which tests for asymmetry have evidently not been made although the structural disorder is large. In conflict with this, other authors have found that it is essential to assume asymmetrical distributions, using functions such as $P_G \otimes P_x^0$.

The studies of $Ni_{66}Y_{33}$ and $Cu_{60}Zr_{40}$ amorphous alloys are relevant (39). The magnitudes of the Fourier transforms of $k^3\chi(k)$ for the yttrium K-edge are shown in Fig. 9.19a for the amorphous and crystalline Ni—Y alloys. The decrease in the amplitude and the contracted position of the main peak of the amorphous alloy relative to the crystalline are strong indicators of asymmetry effects. The analysis used empirical phase shifts $\delta_i(k)$ and amplitude reduction terms $B_i(k)$ obtained from elemental foils and their chemical transferability was confirmed by the quality of fit to crystalline Ni_2Y alloy. The inner potential E_0 was specified to ± 2 eV. The asymmetrical model $P_G \otimes P_x^0$ was tested and found to be un-satisfactory. The "best" fits to the nickel and yttrium Fourier filtered $k^3\chi(k)$ were obtained with models in which the nearest-neighbor distributions were composed of two Gaussian subshells. The asymmetry of the composite distri-butions is apparent in Fig. 9.19b.

The ability of the EXAFS technique to distinguish between asymmetrical models such as $P_G \otimes P_x^0$ and a superposition of two Gaussian subshells was illustrated earlier with the splice method, Section 9.3.3, Fig. 9.16. The splice method could be applied to the nickel K-edge of a-$Ni_{66}Y_{33}$ since Fourier filtering would permit isolation of a composite shell of the same atomic species. But it would be unsatisfactory for the yttrium K-edge since the main peak, Fig. 9.16a

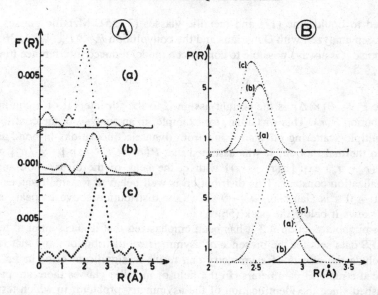

Figure 9.19. (A) The magnitude of the Fourier transform of $k^3\chi(k)$ in amorphous (a)-(b) and crystalline (c) Ni_2Y at the nickel K-edge (a) and the ytterbium K-edge (b)-(c). The arrow indicates the 3.40 Å Y—Y distance found in Ni_2Y by diffraction techniques. (B) Nickel pair distribution function around nickel atoms (upper curves) (a) 2 Ni at 2.40 Å with σ = 0.08 Å. (b) 4 Ni at 2.55 Å with σ = 0.12 Å. (c) total asymmetrical distribution function (the maximum is at 2.45 Å). Nickel pair distribution function around ytterbium atoms (lower curves). (a) 9 Ni at 2.71 Å with σ = 0.15 Å. (b) 3 Ni at 3.05 Å (same σ). (c) Total asymmetrical distribution function (12 Ni at an average distance of 2.80 Å). [Taken from Figs. 4 and 9 of ref. (39).]

curve b contains yttrium atoms at ~3.40 Å in addition to the nickel atoms at shorter distances (i.e., compositional disorder). The Y—Y distribution was included in a three-shell Gaussian model, but because the number of free parameters is limited by the k-space and r-space windows of the transforms, the Y—Y inclusion required prior structural information from complementary diffraction studies. It should be noted that the diffraction studies were in agreement with the EXAFS results although they did not resolve the two subshell structure shown in Fig. 9.16b because of a small maximum momentum transfer q (see Section 9.5.1). For example, EXAFS finds 9 nickel atoms at 2.71 Å and 3 nickel atoms at 3.05 Å in agreement with the diffraction results of 12 nickel atoms at an average distance of 2.80 Å. Sadoc et al. (39) have argued that their subshell model is sterically reasonable; in a compact structure the number of atoms that surround a given central atom at a distance R, must be smaller if R is shortened and the remaining atoms must be pushed to a second subshell at a larger distance.

Finally, the ability to distinguish between different models depends upon the available k-space range of the data and its quality. In a study of a-CoP it was found that the function $P_G \otimes P_x^0$ and a model of two Gaussian subshells provided distribution functions for cobalt atoms about the x-ray absorbing phosphorus atoms that were essentially equivalent (37). The shape of the overall distribution is important and not the functional form that represents it. Nevertheless, from the viewpoint of structural hypotheses the subshell model may be more appealing.

9.3.5. Next-Nearest Neighbors

Although the next-nearest neighbors of crystalline materials can be quantitatively examined using the EXAFS technique, this is frequently prevented by the additional broadening of the distribution that accompanies the transition to the amorphous or liquid states. This is observable, for example, in the magnitude of the Fourier transforms of the yttrium K-edge in amorphous and crystalline $Ni_{66}Y_{33}$ (Fig. 9.19a). This is also the usual situation in amorphous and liquid semiconductors such as Ge, Se, and As_2Se_3 (10), simple liquid metals (10), molten salts (57, 58), and aqueous solutions (11, 12). Amorphous germanium is illustrated later. Examples where second-nearest neighbor information has been obtained are given in Sections 9.3.6 and 9.6.

The EXAFS data from a-Ge are consistent with a continuous random network (CRN) model. The histogram shown in Fig. 9.20 was obtained from a 519-atom, fully relaxed, tetrahedrally coordinated, CRN model (59). The model did not include any dynamical contributions due to the thermal motion of the atoms. The mean square width of the first peak in the histogram, 0.003 $Å^2$, is in agreement with the static structural disorder contribution σ_{ss}^2 extracted from one temperature-dependent study of a-Ge and c-Ge (60) but 66% larger than found in a different sample (35). The model was used to construct

$$\chi(k) = \frac{F(k)}{k} \sum_j \frac{e^{-2R_j/\lambda}}{R_j^2} \sin\left[2kR_j + \delta(k)\right] \tag{39}$$

using a linear phase shift approximation for $\delta(k)$, a parameterized Lorentzian for $F(k)$ (61), and a constant mean free path $\lambda = 4.6$ Å (62). Figure 9.20(b) shows the full $k\chi(k)$ obtained by summing Eq. (39) for all pairs of atoms in the 519-atom cluster. Also shown is the summation over all atoms except nearest neighbors. It is apparent that the second- and higher-nearest neighbors make a measurable contribution to the full $k\chi(k)$ only in the low k region. They are responsible for the modulation of the full $k\chi(k)$ at $k \sim 2$ $Å^{-1}$, although the exact shape of the envelope function will depend upon a more realistic approximation for $F(k)$ and $\delta(k)$ at low k.

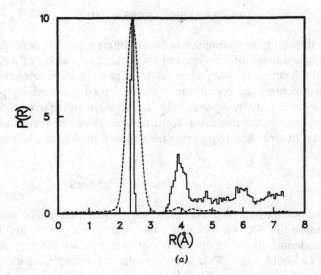

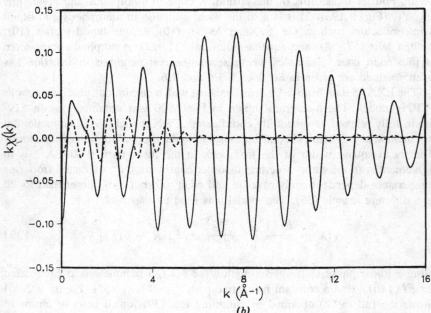

Figure 9.20. (*a*) A continuous random network model for *a*-Ge. The histogram was constructed from a 519 atom, fully relaxed, tetrahedrally coordinated, CRN model (59). The dashed line is the magnitude of the Fourier transform of the full $k\chi(k)$ shown in (*b*) obtained with integration limits $k_{min} = 3.9$ Å^{-1}, $k_{max} = 16.4$ Å^{-1} and a 10% Gaussian window. The abscissa of the transform has been shifted such that the first peaks of the transform and histogram coincide. (*b*) The interference function for *a*-Ge constructed from Eq. (39) including all pairs of atoms (—) and second and higher neighbors only (---).

Omitting the low-k data severely limits the ability of EXAFS to characterize second and higher shells. This is illustrated in Fig. 9.20a where the magnitude of the $k\chi(k)$ transform, obtained by integrating over a typical EXAFS range, is compared with the histogram. The second shell is visible, but in real data for a-Ge, even at 83 K, when a comparable k range was used, the disorder was too large to permit determination of structural parameters. In c-Ge at temperatures above 780 K the large thermal disorder similarly prevented accurate second-shell structural determinations (35). In the analysis of such systems where the second-shell contribution is so weak, mixing of the Fourier components of the first and second shells poses a severe problem. This can be minimized by fitting a model to the Fourier filtered first-shell $k\chi_1(k)$, which can then be subtracted from the full $k\chi(k)$. However, it is obvious from the histogram of Fig. 9.20a that separation of the second and third shells is not possible.

For systems where the disorder of the second shell is large, it is essential that the lowest possible value of k_{min} be used. This means that it is even more important to select a correct reference system; theoretical calculations for the required EXAFS parameters are presently inadequate. In the case of germanium, a suitable reference for a-Ge is the crystalline phase. There are indications that k_{min} can be reduced to approximately 2.0 Å^{-1} (54). However, at least for the CRN model of Fig. 9.20, this lower value of k_{min} produces only a marginal improvement in the analysis of the second shell; the peak height in the transform is increased by only about 50%. The second shell must be less disordered if EXAFS is to be useful.

9.3.6. Multiple Scattering

Some glasses and concentrated solutions possess sufficient local order that EXAFS can provide second-shell information. Beyond the first shell there are multiple-scattering pathways between a backscattering and absorbing atom. Consequently the single-scattering formulas assumed so far in this chapter may become inadequate, particularly when an intervening first-shell atom, x-ray absorbing atom, and the photoelectron backscatterer of interest approach a colinear arrangement (63). The geometry is indicated in Fig. 9.21 where A is the absorber and C is the backscatterer whose distribution relative to A is required. Because the forward scattering amplitude from atom B, $F_B(\beta, k)$, can be large, the interference function for the AC pair contains contributions from the paths A—B—C—A and A—B—C—B—A that cannot be neglected. Moreover, as the positions of the three atoms change due to static disorder or thermal motion, the multiple-scattering corrections can vary significantly. This is an important effect, which was not included in the earlier discussion of multiple scattering, Chapter 1, Section 1.5.2 and is the subject of this section.

Including multiple-scattering pathways, the contribution to $\chi(k)$ for the AC pair with the instantaneous configuration of Fig. 9.21 is

$$\chi_{AC}(k, \mathbf{r}) = \frac{1}{kr_{AC}^2} F_C(\pi, k) \sin\left[2kr_{AC} + \delta^{(1)}(k)\right]$$

$$+ 2\frac{\hat{\mathbf{r}}_{AB} \cdot \hat{\mathbf{r}}_{AC}}{kr_{AB}r_{BC}r_{AC}} F_B(\beta, k) F_C(\gamma, k)$$

$$\cdot \sin\left[k(r_{AB} + r_{BC} + r_{AC}) + \delta^{(2)}(\beta, \gamma, k)\right]$$

$$+ \frac{F_B^2(\beta, k)}{kr_{AB}^2 r_{BC}^2} F_C(\pi, k) \sin\left[2k(r_{AB} + r_{BC}) + \delta^{(3)}(\beta, k)\right] \quad (40)$$

where $\delta^{(1)}(k) = 2\delta_A(k) + \phi_C(\pi, k)$, $\delta^{(2)}(\beta, \gamma, k) = 2\delta_A(k) + \phi_B(\beta, k) + \phi_C(\gamma, k)$, $\delta^{(3)}(\beta, k) = 2\delta_A(k) + 2\phi_B(\beta, k) + \phi_C(\pi, k)$, $F_C(\gamma, k)$ is the magnitude of the scattering amplitude from atom C through the angle γ and $\phi_C(\gamma, k)$ is the phase shift produced by that scattering (64, 65). In writing Eq. (40) the mean free path and other amplitude reduction terms have been omitted for brevity.

Teo has shown that if the scattering angle γ approaches π and the polarization dependence of the scattering can be neglected (64, 66), Eq. (40) can be written as

$$\Omega_B(\beta, k) \frac{F_C(\pi, k)}{kr_{AC}^2} \sin\left[2kr_{AC} + \delta^{(1)}(k) + \omega_B(\beta, k)\right] \quad (41)$$

where the multiple-scattering contributions are contained in the amplitude and phase terms $\Omega_B(\beta, k)$ and $\omega_B(\beta, k)$. Both these terms depend strongly on the

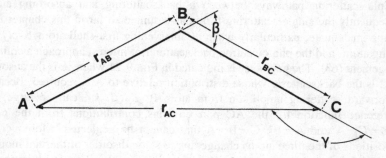

Figure 9.21. Arrangement of atoms for which multiple-scattering affects the EXAFS determination of the distance r_{AC}.

angle β and are less sensitive to r_{AB}, r_{BC}, and r_{AC}. Teo has used this dependence on β to provide a basis for quantitative determination of bond angles.

The treatment of disorder is complicated by multiple scattering. One complication arises because the thermal motion of the three atoms follow their normal modes. Boland and Baldeschwieler (67) have shown how a normal mode analysis leads to distinct Debye–Waller factors in each of the three terms of Eq. (40). The Debye–Waller factor for each term corresponds to the mean variation of the related multiple-scattering path. In some cases it may be justified to approximate the three σ^2's by a single value throughout. This may be justified if the factors for each path are nearly equal or if the backscattering from one path dominates. In such cases Eq. (41) may be used and simply multiplied by $\exp(-2\sigma^2 k^2)$ to account for disorder.

The strong variation of $\Omega_B(\beta, k)$ or, equivalently, $F_B(\beta, k)$, introduces another complication. The scattering amplitude $F_B(\beta, k)$ is strongly peaked for β near 0. Thus when the motion or disorder is large and the bond angle is nearly linear, this factor is very nonlinear over the range of possible molecular positions. In this case the distribution must be explicitly integrated:

$$\chi(k) = \int P(\mathbf{r}_A, \mathbf{r}_B, \mathbf{r}_C) \, \chi_{AC}(k, \mathbf{r}) \, d\mathbf{r}_A \, d\mathbf{r}_B \, d\mathbf{r}_C \qquad (42)$$

When disorder is small and $F_B(\beta, k)$ [or equivalently $\Omega_B(\beta, k)$] varies linearly over the range of the distribution, then β may be replaced by its mean value $\bar{\beta}$ and Eq. (42) reduces to the standard single-scattering EXAFS equation with the difference that the amplitude and phase are modified by the multiple-scattering correction terms $\Omega_B(\bar{\beta}, k)$ and $\omega(\bar{\beta}, k)$. When these conditions are not met, then the results depend on the details of the system and numerical integration of Eq. (42) may be the best solution, though in some cases a closed-form replacement for the Debye–Waller factor may be derived (68).

The distribution of angles may be large due to a static disorder. For example, x-ray scattering measurement and model calculations of SiO_2 glass have found an average bond angle ($180° - \bar{\beta}$) for Si—O—Si of 144° with a full width at half-maximum (FWHM) of 37° (69). In cases like this EXAFS analysis may proceed by numerical integration of Eq. (41). This has been done using an integrated multiple-scattering (IMS) approach (56). The IMS approach is applicable to systems where the disorder is static or arises from dynamic fluctuations in β due to thermal motion. IMS calculations indicate that when the bond angle exceeds $\sim 140°$, significant errors will occur if β is fixed at its mean value. On the other hand, in GeO_2 glass the bond angle is 130° ± 6.5°, the contributions of multiple scattering are small and EXAFS analysis using the single-scattering formula is possible (70).

The effect that thermal motion has on the multiple-scattering contributions to the $\chi(k)$ for the $A-C$ pair has also been considered using detailed normal mode analyses (67, 68, 71). Boland and Baldeschwieler (71) find that the change in the scattering angle β induced by thermal vibrations produces a large change in the scattering amplitude, particularly as β approaches zero. Yet they conclude that it is sufficient to use the multiple-scattering analysis method of Teo described previously but modified to treat correctly the system's normal modes. This is not surprising since in their analysis they neglect the strong nonlinearity in $F_B(\beta, k)$ retaining only the first-order term in a Taylor series expansion about the mean scattering angle $\bar{\beta}$ [Eq. (10) of ref. (71)].

On the other hand, Alberding and Crozier (68) in a normal mode analysis retain the nonlinearity in $F_B(\beta, k)$ and find that $\chi(k)$ is sensitive to thermal motion. An example is given in Fig. 9.22 for a hypothetical bridged $Fe-O-Fe$ complex where the bending and stretching force constants are 1 and 30 mdyne Å^{-1}. Because the photoelectron path distances are nonlinear in the bending coordinate, Gaussian distributions of the harmonic normal coordinates become asymmetric distributions of path lengths. It is obvious from the simulation that neglecting this asymmetry and the nonlinearity in $F_B(\beta, k)$ can result in significant errors in the $A-C$ structural parameters. Moreover, this work also showed that the multiple-scattering effects depend sensitively on the ratio of the bending and stretching force constants. Depending on the ratio, the amplitude of $\chi(k)$ may actually show an increase with temperature instead of the more usual decrease.

Multiple scattering imposes a further restriction upon the choice of a reference compound for determination of $\delta(k)$, $F(k)$, $S_0^2(k)$, and $\lambda(k)$. As indicated previously, when the bond angle $(180° - \beta)$ exceeds approximately $140°$, the total amplitude factor and phase shift depend sensitively on the relative positions of the atoms as well as their chemical nature. Therefore, unless the geometry of the model system exactly matches that of an unknown system, a direct transfer of phase and amplitude functions is incorrect. A possible solution is to use theoretical multiple-scattering functions to correct empirical functions for changes in geometry of the unknown system and for differences in dynamic and static disorder (56).

In conclusion, the analysis of the EXAFS of second shells with thermal or static structural disorder is complicated by multiple-scattering contributions. Significant errors in the specification of the $A-C$ distribution function can occur if multiple scattering is neglected.

9.3.7. Determination of E_0

Before a quantitative analysis of $\chi(k)$ can proceed it is necessary to determine the inner potential E_0. In the ideal case, E_0 would be transferred directly from a reference system that is chemically identical to the unknown and for which

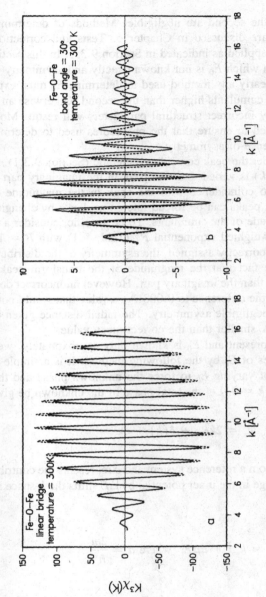

Figure 9.22. Simulated iron-edge EXAFS spectra where integration of all multiple-scattering effects is included (solid curve) compared to neglecting variation of forward scattering amplitude and asymmetry of path lengths over the range of thermal motion. Simulations are for 300 K for (*a*) the linearly bridged case $\beta = 0°$ and (*b*) the bent case $\beta = 30°$. [Taken from Fig. 2 of ref. (68).]

cumulants higher than the second are negligible. Methods of determining E_0 under such conditions are discussed in Chapter 6. Tests and corrections for asymmetry can then be applied as indicated in Section 9.3.1. In this section we will consider the case in which E_0 is not known exactly and asymmetry effects cannot be neglected. Clearly any method used to determine E_0 must explicitly include contributions of cumulants higher than the second. Otherwise an incorrect E_0 and subsequently incorrect structural parameters will result. Moreover caution must be exercised to ensure that the procedures used to determine E_0 do not obscure the presence of asymmetry.

As an example consider the peak coincidence method (Section 9.3.1). In this method the phase of $\chi(k)$ is varied until the peak in the imaginary part of the transform is rotated into coincidence with the peak in the magnitude of the transform Eq. (31). The peaks can be forced into coincidence by changing E_0, regardless of the magnitude of the cumulants. To be specific, consider a model $\chi(k)$ created from the weighted exponential P_x^2 (Table 9.1) with $B = 10$ and $R_0 = 2.09$. When E_0 is correctly assigned, the asymmetry of the distribution is clearly indicated by the fact that the magnitude of the transform peaks at a position 0.09 Å smaller than the imaginary part. However an incorrect decrease in E_0 of -6 eV reduces the difference to 0.008 Å, a value that would normally be assumed to indicate negligible asymmetry. The radial distance given by the peak positions is 0.15 Å smaller than the correct mean value.

When asymmetry is present and E_0 is known only approximately, we have obtained improved values of E_0 by the following method. It is a simple modification of the method of varying E_0 to force the unknown phase and the reference phase to agree at $k = 0$. Let the total phase of the unknown be given by

$$\Phi_x = 2k\overline{R}_x + \delta(k) + ck^3 \tag{43}$$

where $\delta(k)$ is obtained from a reference system and c represents the contribution from asymmetry. A change in the inner potential by ΔE shifts the k-space scales according to

$$k^2 = k'^2 + a\Delta E \quad \text{where} \quad a = \frac{2m}{\hbar^2} \tag{44}$$

By the binomial expansion

$$k \simeq k' + \frac{a\Delta E}{2k'} \tag{45}$$

Define the phase difference $y(k) = \Phi_x - \delta(k)$. Then

$$y(k') = \left(2\overline{R} + \frac{3}{2} ca\Delta E\right)k' + ck'3 + \frac{\overline{R}a\Delta e}{k'} \tag{46a}$$

The phase difference can then be fitted to the functional form

$$y(k') = a_1 k' + a_3 k'^3 + \frac{a_{-1}}{k'} \tag{46b}$$

By equating the coefficients in Eqs. (46a) and (46b) the first iterate to ΔE is given by

$$\Delta E \approx - \frac{2a_{-1}}{aa_1(1 - 3a_{-1}a_3/a_1^2)}$$

$$\approx - \frac{2a_1}{aa_1} \tag{47}$$

This value is used in Eq. (44) and the calculations repeated until successive values of ΔE differ by less than 0.01 eV. The fit coefficient a_1 given by the final iteration provides the estimate $R = a_1/2$ to which must then be added the correction $(2C_2/\overline{R})(1 + \gamma\overline{R})$ required by Eq. (23). This method is easily extended to include higher cumulants. Tests with model functions $P_x^0 \otimes P_G$ show a rapid convergence to the correct E_0.

We have also applied the method to real data. In a preliminary examination of the hydration of silver in aqueous solutions of $AgClO_4$ the first estimate of E_0 was obtained from the reference system, AgO powder at 77 K. The asymmetry in the solution was large and it was necessary to include a fifth-order term in Eq. (46b) (72).

It should be noted that for the case of low to moderate disorder a error in E_0 of 5 eV causes only a small perturbation in the values of the cumulants. The more serious problem is the error in the cumulants resulting from the inadequacy of the cumulant expansion at high k. For example, in the model $P_x^0 \otimes P_G$ when $B = 16$ and the amplitude is fitted over the range $3 < k < 8$ Å^{-1} for the determination of the cumulants C_0, C_2, C_4, and C_6, the percentage errors in C_2 and C_4 are -1.5 and 26% when the correct E_0 is assigned as in Table 3.2. When E_0 is in error by $+5$ eV the errors become -0.8 and 25%.

When the asymmetry is large the cumulant expansion approach is inappropriate. In this case, one may use the splice method of Section 9.3.3 or fit $\chi(k)$ using a model function. The effect that an incorrect value of E_0 has on model parameters depends sensitively on the fitting procedures used. The effect is minimized if the phase and the logarithm of the amplitude are fitted separately.

As an example, consider the model function P_x^2 for $B = 10$, $R_0 = 2.09$, $N = 1$. Assume a reference compound specified by a Gaussian distribution. First consider the phase. Fitting the phase difference

$$\Phi_x - \Phi_{ref} = 2k(R_0 - \overline{R}_{ref}) + 3 \tan^{-1}\left(\frac{2k}{B}\right) \tag{48}$$

over the range $3 < k < 16$ Å^{-1} for $\overline{R}_{ref} = 2.6$ Å with an incorrect E_0 of $+5$ eV(-5 eV) gives $R_0 = 2.091$ Å (2.089 Å) and $B = 10.73$ Å^{-1}(9.331 Å^{-1}). The corresponding error in $\overline{R}_x = R_0 + 3/B$ is -0.02 Å($+0.01$ Å). The nonlinear least squares fitting to B depends upon the value of $\overline{R}_{ref}$; if $\overline{R}_{ref} = 0$ then the fit values are $R_0 = 2.08$ Å(2.10 Å) and $B = 6.1$ Å^{-1}(16.0 Å^{-1}) for $\Delta E_0 = +5$ eV(-5 eV). Now consider the amplitude. Fitting the logarithm of the amplitude to the form

$$y(k) = \ln N - \frac{3}{2} \ln\left[1 + \left(\frac{2k}{B}\right)^2\right] \tag{49}$$

over the same range gives $N = 0.93(1.08)$ and $B = 10.3$ Å^{-1}(9.7 Å^{-1}) for $\Delta E_0 = +5$ eV(-5 eV). For a disordered system $P_x^2 \otimes P_G$ is less restrictive than simply P_x^2. Fitting the logarithm of the amplitude for this model to the form

$$y(k) = \ln N - 2k^2\sigma^2 - \frac{3}{2} \ln\left[1 + \left(\frac{2k}{B}\right)^2\right] \tag{50}$$

when $\sigma^2 = 0.010$ Å^2 gives $N = 0.90(1.13)$, $B = 10.3$ Å^{-1}(9.7 Å^{-1}), and $\sigma^2 = 010$ Å^2(0.010 Å^2) for $\Delta E_0 = +5$ eV(-5 eV). Of course, the errors are reduced by using the correct value of E_0. E_0 can be determined via Eq. (47) or by including it as an adjustable parameter in Eq. (48).

On the other hand if the phase and amplitude are fitted simultaneously to the equation

$$k\chi'(k) = N\left[1 + \left(\frac{2k}{B}\right)^2\right]^{-3/2} \sin\left[2kR_0 + 3 \tan^{-1}\left(\frac{2k}{B}\right)\right] \tag{51}$$

then $N = 1.5(0.76)$, $B = 6.2$ Å^{-1}(14.3 Å^{-1}), $R_0 = 2.09$ Å(2.09 Å) for $\Delta E_0 = +5$ eV(-5 eV). These latter values were obtained with a nonlinear least-squares routine that minimized the sum of the squares of the errors. The sum is more heavily weighted by the maxima in the sine function than the zero crossings and poorer overall fits occur. Clearly the preferable procedure is to decouple the amplitude and phase before determining the fit coefficients.

9.4. EXPERIMENTAL CONSIDERATIONS

The basic requirements of an EXAFS experiment have been discussed in detail in Chapters 4, 5, and 6. This section is concerned primarily with experimental considerations for amorphous and liquid systems. Reference compounds, samples, and sample cells are discussed. The section on data acquisition is of more general application. It includes a brief discussion of crystal "glitches" and energy scale calibration.

9.4.1. Reference Compounds

An important consideration in designing the experiment is the choice of compounds that will be used as standards for determining the phase shift, backscattering amplitude, and amplitude reduction factors. Greater accuracy is achieved if these functions are derived empirically rather than calculated ab initio. In favorable cases for systems with small disorder, the nearest-neighbor coordination number and Debye–Waller factor may be determined to an accuracy of a few percentage and changes in bond lengths to within ± 0.005 Å. For systems with larger disorder, the accuracy with which the distribution function can be reconstructed using the cumulant method (Section 9.3.2), splice method (Section 9.3.3), or curve-fitting methods (Section 9.3.4), depends directly upon the transferability of the amplitude and phase factors from the reference to the unknown and the additional factors discussed in these sections. The question of chemical transferability of amplitude and phase factors has been reviewed elsewhere (1, 2, 53, 73, 74).

In some cases, the reference compound may have an asymmetrical distribution function due to either anharmonicity in the effective two-body interaction potential or an asymmetrical static distribution of neighbors about the x-ray absorbing atom. In the former case the reference compound should be studied at least at liquid nitrogen temperatures where the anharmonic effects will be reduced. Reference compounds of the latter type should be avoided since the amplitude and phase factors due to interference effects will be characteristic of the structure rather than the individual atoms and hence cannot be transferred to the disordered system.

It should be emphasized that the reference compound should be as chemically and structurally similar to the unknown as is possible. The presence of other chemically dissimilar atoms in systems with the same nearest-neighbor AB pair can adversely affect the transferability of amplitude even for nearest neighbors. For example, when the manganese K-edge EXAFS of MnO was used as a model for the unknown $KMnO_4$, the error in the number of nearest neighbors coordinated to manganese was -30%, but when the unknown was MnO_2, the error was only 1% (74). The transferability of amplitude is particularly sensitive to the x-ray absorbing atom. For example, when the $F(k)$ for O in NiO was obtained from CoO, which has the same crystal structure, the error in the number of oxygens coordinated to nickel was 65% (74). The reference and the unknown should have the same valence and bonding states. This can be checked by examination of the near-edge structure. Ideally the first two features in the XANES of the standard and the unknown should show equal separations, but this is not always the case when a glass and its crystalline state are compared.

In extending EXAFS to nonnearest neighbors one must consider the additional problems associated with inelastic and multiple-scattering effects. Inelastic effects are normally included phenomenologically through a k dependent mean free path $\lambda(k)$ which, however, is not a universal function of k, evidently depending upon details of the atomic structure. Variations in λ as large as 100% have been observed between Ge and CuBr (53, 73). The further restrictions imposed by multiple scattering upon the choice of a reference compound have been indicated in Section 9.3.6.

A good reference compound is not always available. Then one should use a compound containing the same x-ray absorbing atom and a backscattering atom that is adjacent in the periodic table. Theory can then be used to correct for the differences. Although the absolute theoretical values for a given element may be in error, theory can provide a reasonable estimate of the difference between adjacent elements in the same row in the periodic table.

9.4.2. Samples and Cells

In transmission EXAFS measurements sample inhomogeneities such as pinholes and variation in thickness should be avoided (75–77). Although statistical considerations support an optimum signal thickness of $\mu_T x = 2.6$, where μ_T is the total linear x-ray absorption coefficient, thickness effects are less serious if $\mu_T x < 1.5$. For powdered samples with high concentrations of the x-ray absorbing atom, the particle diameter D should be such that $\mu_T D \approx 0.1$. The cross-sectional area of the x-ray beam should be less than that of the sample and should be defined by appropriate masks preceding the I_0 chamber.

Solid samples with the required thickness can usually be obtained as evaporated or sputtered films, as foils or as finely divided powders.

Aqueous solutions are readily contained in polyethylene bags positioned between two metal plates. An opening is milled in each plate and covered with a low-Z material such as aluminum or Kapton to provide a flat x-ray window. The separation of the plates can be varied to give the correct sample thickness by simply using a drive consisting of a finely threaded nut and cylindrical tube with the x-rays traveling along the tube axis. More elaborate mechanisms using micrometer drives have been described (78).

Concentrated liquids with x-ray edges of interest occurring at energies less than 10 keV can be injected into evacuated beryllium or aluminum cells provided chemical reaction with the cell does not occur. For example, molten salts of MnBr complexes were studied in cells consisting of aluminum disks bolted together with a knife edge on one face making a vacuum-tight compression seal to a Teflon gasket. The height of the knife edge was machined to provide the correct absorption path length at the x-ray edge of interest. Kapton glued to the mating faces prevented reaction of the salt with the aluminum and at the same time acted as x-ray windows (58).

When the x-ray edge of interest is at higher energies, liquids can be injected into fused silica optical cells, commercially available with fixed paths of 10 and 100 μm. X-ray windows are then formed by removing silica by grinding and/ or etching with hydrofluoric acid. Cells made of Teflon have also been used.

For high-temperature studies, graphite and boron nitride are frequently used as substrates and cell materials (34, 79). Liquid Se—Te alloys have been contained in boron nitride cells where a metal ring is compressed into a grafoil gasket to obtain the optimum cell thickness (80). If the porosity of the cell wall is a problem, then providing chemical reaction with the sample does not occur, the porosity can be reduced by first coating the walls with an evaporated layer of SiO or SiO_2.

For corrosive materials or hot liquids where containment is a problem, obtaining samples with large cross-sectional areas can pose problems. Thus it should be noted that by careful masking, smaller samples can be used. This is done in our high-pressure experiments (81, 82) where diameter of the samples is now routinely less than 1 mm.

9.4.3. Data Acquisition

Choice of the EXAFS detection technique is determined by the nature of the sample and signal/noise considerations. These questions have been discussed in detail in Chapter 4 and elsewhere (1, 2, 77, 83). For disordered systems it is important to obtain good statistics at high k. This is achieved at a synchrotron radiation source by a combination of multiple scans and different integration times for different regions of k space. The requisite number of counts at a given point can be estimated in advance by detailed signal/noise calculations. However, with the availability of on-line data analysis packages (84), it is preferable to start a multiple scan, examine the EXAFS $\chi(k)$ and the Fourier transform of the first scan while taking the second scan and then decide upon the number of additional scans. Also, since the raw data will eventually be Fourier transformed, the data acquisition time, particularly in the high k range, can be effectively decreased by taking data on a uniform grid in k space rather than at uniform increments over different ranges in x-ray energies or Bragg angles.

Crystal "glitches" pose a severe data analysis problem if they appear in the $\chi(k)$ of disordered systems at high k where the amplitude of $\chi(k)$ is weak. Experimental precautions can be taken to reduce the magnitude of the glitches (Chapter 4 and refs. 77, 85–87) but nevertheless some may still remain in the data. For example, in transmission EXAFS high-pressure experiments, the large nonlinearity introduced by the absorption of the anvils favors glitches. Also in the fluorescence EXAFS mode, glitches are troublesome, particularly when the normalizing signal is obtained from a transmission ion chamber with a different response function than the fluorescence detector. Procedures for "deglitching" data have been discussed in Chapter 6, Section 6.2.3.

When a material undergoes a change from the crystalline to the liquid or amorphous states, changes in E_0 may occur. To distinguish real chemical shifts from changes in the monochromator energy calibration, a reference energy marker should be taken with every EXAFS scan. This can be done simply by inserting a reference compound before a third ionization chamber positioned after the sample ionization chamber in transmission EXAFS experiments (88). Alternatively, a diffraction peak can be produced by positioning a silicon single crystal after the sample ionization chamber (89). This has the advantage that an absolute energy scale can be estimated for each scan. One can also use the crystal "glitches" of the EXAFS monochromator itself as recorded by I_0, the presample ionization chamber (86, 87).

9.5. DIFFRACTION AND EXAFS

Both diffraction and EXAFS have their strengths and limitations. For example, fluorescence EXAFS may be the only method applicable if the sample is at low concentration (~ 10 ppm). For more concentrated samples, the concerted use of both methods complement each other to give more complete and certain information. In the following sections a brief comparison of the methods is made. Features common to all diffraction techniques are discussed in Section 9.5.1 for monatomic systems and illustrated by results obtained by the application of both techniques. Structural information concerning second nearest neighbors and the amorphous-to-crystalline transition is summarized in Section 9.5.1.3. Methods of obtaining the partial structure factors of polyatomic systems by diffraction measurements are given in Section 9.5.2. Results obtained with diffraction and EXAFS for electrolyte aqueous solutions are compared. The usefulness of EXAFS in providing information concerning the effective two-body interaction potential in disordered systems is indicated in Section 9.5.3. Finally, some strengths and weaknesses of the EXAFS and diffraction techniques are summarized in Section 9.5.4.

9.5.1. Monatomic Systems

For monatomic systems the pair distribution function $g(r)$ of a diffraction experiment is related to the measured intensity via the structure factor $S(q)$ (90):

$$S(q) - 1 = 4\pi\rho_0 \int_0^\infty [g(r) - 1] \frac{\sin qr}{qr} r^2 \, dr \qquad (52)$$

$$g(r) - 1 = \frac{1}{2\pi^2\rho_0} \int_0^\infty [S(q) - 1] \frac{\sin qr}{qr} q^2 \, dq \qquad (53)$$

$g(r)$ is defined such that $4\pi\rho_0 g(r) r^2 \, dr$ is the number of atoms in a spherical shell of radius r where ρ_0 is the number density. The momentum transfer $\hbar q$ is calculated from $q = 2k \sin \theta$ where 2θ is the angle through which the incident photon, neutron, or electron of momentum $\hbar k$ is elastically scattered and is twice the photoelectron momentum $\hbar k$ used in EXAFS.

Due to the finite range of the experimental $S(q)$, $g(r)$ is incompletely specified, affecting the accuracy with which radial distances, coordination numbers, and peak shapes can be determined. The uncertainty in the shape of the first peak in $g(r)$ and the errors associated with the determination of coordination numbers are indicated in the following two subsections.

9.5.1.1. Resolution

In EXAFS and diffraction studies the resolution of structural features is limited by the finite range of the data. In most studies of disordered systems using conventional diffraction techniques, $S(q)$ is measured for the range $q_{min} = 0.5$ Å^{-1} to $q_{max} = 12$ Å^{-1}. The lower limit is determined, in part, by the difficulty of approaching the forward scattering direction and the decrease in the coherently scattered amplitude. The upper limit depends on signal-to-noise considerations and whether x-rays, neutrons, or electrons are being scattered. In an EXAFS study, data analysis typically begins at $k_{min} \approx 3.5$ Å^{-1}, although as indicated in Section 9.3.3, if a suitable reference exists, the lower limit may be reduced to $k_{min} \approx 2$ Å^{-1}. The upper limit k_{max} depends on the degree of disorder and the magnitude of the backscattering amplitude at high k. In systems with moderate disorder, $k_{max} \approx 14$ Å^{-1} and the resolution of EXAFS, given $q = 2k$, is approximately twice that of a typical x-ray diffraction experiment. For systems with larger disorder, the effective resolution of diffraction and EXAFS measurements become comparable. For example, in liquid mercury at room temperature, the amplitude of the oscillations both in $S(q)$ and $\chi(k)$ has decayed to zero by $q_{max} \approx 12$ Å^{-1} (91) and $k_{max} \approx 5.6$ Å^{-1} (10), respectively.

In calculating $g(r)$ via Eq. (53), the truncation of the experimental data obtained with conventional diffraction methods at $q_{max} \approx 12$ Å^{-1} poses a more serious analysis problem than the loss of low-q-data. $S(q)$ can be simply extrapolated smoothly from q_{min} to $q = 0$ or in the case of a monatomic liquid, can be extended theoretically to zero by using the thermodynamic relation between the isothermal compressibility and $S(o)$ (90).

Larger values of q_{max} can be attained with two relatively new diffraction techniques. With the energy dispersive x-ray diffraction (EDXD) method, which utilizes white radiation, a fixed diffraction angle, and an energy sensitive photon detector, $q_{max} \approx 30$ Å^{-1} has been achieved for liquid and amorphous samples (92–94). In the time-of-flight (TOF) neutron diffraction method, which is based on the principle of varying the wavelength at fixed scattering angle, measurements to $q \approx 40$ Å^{-1} are possible. The latter method was used by Suzuki et al. (95) to obtain the $S(q)$ of liquid gallium (Fig. 9.23). They used the larger range of data available to examine the effects on $g(r)$ of truncating the data at smaller q_{max}. It was found that the position of the first peak in $g(r)$ did not

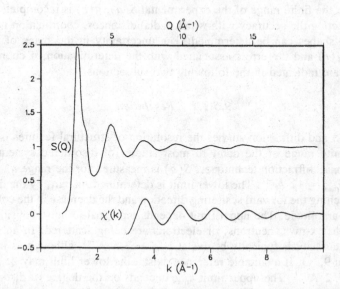

Figure 9.23. The structure factor $S(q)$ and a reduced EXAFS interference function $\chi'(k)$ for liquid gallium at 40°C. $S(q)$ was obtained via TOF neutron diffraction. [Taken from Fig. 5 of ref. (95).] $\chi'(k)$ is $k\chi(k)$ with the EXAFS backscattering amplitude removed but not the EXAFS phase shift.

change provided $q_{max} > 8$ Å^{-1}, but that the shape and height of the first peak was not established until $q_{max} > 12$ Å^{-1} and that spurious side lobes in $g(r)$ only became negligibly small when $q_{max} = 17$ Å^{-1}. Thus, for systems in which the first peak in $g(r)$ is asymmetric, such as liquid gallium, errors in structural parameters occur if $q_{max} \leq 12$ Å^{-1}.

Liquid gallium provides an example in which high resolution is necessary. The oscillations in its $S(q)$ persist to higher q than for other liquid metals such as mercury or rubidium. This may be due to a more repulsive ion–ion interaction potential in gallium or the presence of short-lifetime molecularlike associations of diatomic gallium (95, 96). Suzuki et al. found that the oscillations for $q > 10$ Å^{-1} can be fitted with a diatomic model with a bond length of 2.69 Å. A comparable distance of 2.72 Å is found from the EXAFS of liquid gallium, Fig. 9.23 when asymmetry corrections are included (72). On the other hand, in those x-ray and neutron diffraction experiments in which $q_{max} < 12$ Å^{-1}, the resolution is restricted and the first peak in $g(r)$ is specified by just a single distance of approximately 2.82 Å (97–99).

It is of interest to note that the EXAFS spectrum of liquid gallium, Fig. 9.23 can be analyzed to $k \approx 14$ Å^{-1} ($q \approx 28$ Å^{-1}) before signal-to-noise becomes a problem. In the case of $S(q)$, although data were acquired to 33 Å^{-1}, signal-to-noise placed an upper limit of $q \approx 20$ Å^{-1} on the analysis range.

9.5.1.2. Coordination Numbers

In diffraction studies of monatomic systems the coordination number is calculated from

$$N = 4\pi\rho_0 \int_{R_{min}}^{R_{max}} r^2 g(r)\, dr \qquad (54)$$

The value of N_1 is dependent upon q_{max}, the assumptions used in deconvolving the first- and second-shell contributions to the first peak in $g(r)$ and, in specifying R_{min} and R_{max}. The deconvolution is normally done by one of four different methods (100). As an indication of the convergence of these methods, in a conventional neutron diffraction study of liquid gallium at 50°C with $q_{max} = 10.3$ Å^{-1} N_1 varied from 7.2 to 9.6 with values in the range 9–9.6 preferred (98).

Analogous problems exist in EXAFS analysis as indicated in Section 9.3 and Chapter 6. Although it is frequently neglected in the literature, the effect of the choice of R_{min} and R_{max} should be included in the errors assigned to EXAFS deduced coordination numbers, particularly in cases when asymmetry in the distribution function cannot be neglected. Moreover, it should be recognized that within the range R_{min} to R_{max} there may be second nearest neighbors with a sufficiently large static disorder that they will go undetected unless the low-k data are analyzed (see Fig. 9.20b).

A high resolution TOF neutron diffraction study of a-germanium with q_{max} = 23 Å^{-1} (101) permits comparison with EXAFS studies of comparable resolution. Both techniques confirm that the nearest-neighbor distribution is Gaussian and, although differences can be expected due to variation in preparation methods, the widths of the distributions are comparable. The values found for R_1 differ only slightly; various EXAFS studies [35, Table 1 of ref. (101)] quote $R_1 = 2.45 \pm 0.01$ Å, the same as in c-germanium, whereas the TOF diffraction result 2.463 Å is consistent with conventional diffraction results that typically find that R_1 is 0.01 Å larger than in c-germanium. A more significant discrepancy exists in the value of N_1. EXAFS gives the tetrahedral coordination 4.0 with an uncertainty of $\pm 4\%$ (35) and 1.5% (102). The TOF result, $N_1 = 3.68$, while the smallest reported, agrees with the trend of conventional diffraction studies that find $N < 4.0$. The diffraction results are normally interpreted in terms of the existence of dangling bonds at the boundaries of the voids within the amorphous material. On the other hand, one of the difficulties in diffraction studies of thin films involves the determination of the average number density ρ_0, which in turn determines the coordination number. In the TOF study, ρ_0 was found by adjusting ρ_0 until the experimental correlation function $t'(r)$ oscillated evenly about zero below and immediately above the first peak (Fig. 9.24). $t'(r)$ is the convolution of the true correlation function

$$t(r) = 4\pi r \rho_0 g(r) \tag{55}$$

with a peak function that includes a window function to reduce transform artifacts, that is, side lobes. The optimum value of ρ_0 was 10% smaller than that

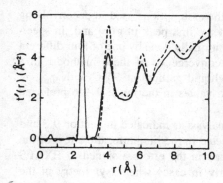

Figure 9.24. A comparison of the experimental correlation function of a-Ge obtained from TOF neutron diffraction with that calculated from a Polk 519 atom model (---). The correlation function $t'(r)$ is the convolution of the true correlation function with a peak function included to reduce transform artefacts. [Taken from Fig. 14 of ref. (101).]

of c-germanium, the same amount by which the coordination number was smaller than 4.0. In the EXAFS of germanium, and other simple amorphous semiconductors, this difficulty is avoided since a direct comparison with the corresponding crystalline phase is possible.

9.5.1.3. Next-Nearest Neighbors

It is evident from Fig. 9.24 that inclusion of the low-q diffraction data provides second- and higher-shell structural features that are not apparent in EXAFS (Fig. 9.20a). Nevertheless, the second neighbor peak at 4.0 Å is only partially resolved and contains contributions from the third nearest neighbors, which in c-germanium are at 4.67 Å. To obtain deeper insight it is necessary to adopt a model-dependent analysis. The Polk 519 atom model was found to provide the best fit. The sharpness of the second peak is attributed to insufficient strain in the model resulting in a distribution of bond angles that is too narrow (101).

The small peak at 4.8 Å in the model $t'(r)$ is noteworthy. A similar peak has been found in x-ray diffraction studies of a-Ge, a-GaAs and, a-GaSb and a structural significance has been associated with it. In some papers the prediction of this peak has been used as a criterion for preferring one structural model over another. However, Etherington et al. (101) conclude that it is a spurious peak due to truncation side lobes because a smooth window function was not applied before Fourier transforming the experimental interference functions.

It would thus appear that diffraction measurements are more suited for an examination of the amorphous-to-crystalline transition where knowledge of the second and higher shells is necessary to permit a differentiation between competing models. However, this is not the case, at least for germanium films.

For germanium films prepared by thermal evaporation (103, 104) or sputtering (105), there exists a range of substrate temperatures, above an intermediate temperature $T_i \approx 240°C$ and below the crystallization temperature T_c, for which germanium films are heterogeneous, consisting of crystallites embedded in an amorphous matrix. The crystallites were detected by Bragg peaks superimposed upon the broad background characteristics of x-ray diffraction of amorphous materials (104, 105). The crystallites were also identified by a characteristic optical mode in the Raman spectra (104). The size and number of crystallites increased as the substrate temperature T_s was increased above T_i. At T_i the size of the crystallites as determined by the width of the Bragg peaks was dependent upon preparation method, with dimensions varying from 40 Å (105) to 200 Å (104). For $T_s < T_i$, crystallites were not detected by either x-ray diffraction (104, 105) or Raman spectroscopy (104) and the structure of the amorphous films was found to be independent of T_s. However, for the same films and

$T_s < T_i$, the EXAFS did show structural changes. This is indicated in Fig. 9.25 by the evolution of the second peak in the magnitude of the transform of thermally evaporated films at substrate temperatures less than $T_i = 240°C$. The sudden appearance of the third-shell peak at $240°C$ coincides with the detection of crystallites by x-ray or Raman scattering. From a quantitative analysis of the EXAFS of the second shell of a sputtered film for which T_s was slightly below T_i it was concluded that the film was heterogeneous, 20% of which consisted of microcrystallites approximately 10 Å in size embedded in an amorphous matrix (105). The authors argued that the EXAFS detection of the microcrystallites was strong evidence for the continuous random network model for a-Ge.

These EXAFS studies are significant for two reasons. They demonstrate the sensitivity of EXAFS to structural features in disordered systems with a medium range of order, larger than the tetrahedral units but less than about 20 Å.

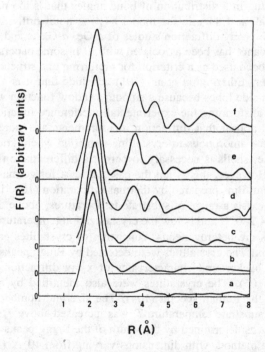

Figure 9.25. Magnitude of the Fourier transforms of the germanium K-edge of thermally evaporated films deposited on substrates with different temperatures. The substrate temperatures are (a) $T_s = 130°C$, (b) $T_s = 160°C$, (c) $T_s = 200°C$, (d) $T_s = 240°C$, (e) $T_s = 270°C$, and (f) $T_s = 320°C$. For these films, crystallites are embedded in an amorphous matrix for $240°C \leq T_s \leq 285°C$. The abscissa is not corrected for the EXAFS phase shift. [Taken from Fig. 7 of ref. (104).]

Medium range order is not readily studied by conventional diffraction measurements. Secondly, the studies emphasize that EXAFS and diffraction complement each other. Both should be applied to the same sample if possible.

9.5.2. Polyatomic Systems

In a diffraction experiment on a liquid or amorphous material containing N different atomic species, the measured intensity is a weighted sum of contributions from the $N(N + 1)/2$ different pair correlation functions. It may be expressed, on a per atom basis in the static approximation, as

$$I_a(q) = \sum_i c_i f_i^2 + \sum_i \sum_j c_i c_j f_i f_j [S_{ij}(q) - 1] \tag{56}$$

where f_i is the coherent scattering amplitude of the ith element of atomic concentration c_i (106). The partial structure factor $S_{ij}(q)$ is related to the partial pair distribution functions $g_{ij}(r)$ via the Fourier transform pairs:

$$S_{ij}(q) = 1 + 4\pi\rho_0 \int_0^\infty [g_{ij}(r) - 1] \frac{\sin qr}{qr} r^2 \, dr \tag{57}$$

$$g_{ij}(r) = 1 + \frac{1}{2\pi^2\rho_0} \int_0^\infty [S_{ij}(q) - 1] \frac{\sin qr}{qr} q^2 \, dq \tag{58}$$

$g_{ij}(r)$ is defined such that $4\pi r^2 c_j \rho_0 g_{ij}(r) \, dr$ is the number of atoms of the jth type in a spherical shell of radius r with an ith type atom at the center.

The determination of all the partial structure factors requires that the scattering amplitudes be systematically varied and $N(N + 1)/2$ independent diffraction experiments be made. This is possible in the case of binary systems for which there are only three partial structure factors. In neutron diffraction studies, the isotope dependence of f_i is utilized and measurements are made on three separate samples of the same overall atomic composition but different isotopic composition (107). This method is limited by the availability of suitable isotopes. Also at low concentrations the changes in the diffracted intensity with change in isotopic composition cannot be measured reliably.

In x-ray scattering experiments use has been made of the anomalous dispersion in the scattering amplitudes near an x-ray absorption edge (97, 108). This approach is limited to elements with atomic number greater than about 30. For lighter elements the maximum value of q is restricted by the x-ray energy; $q_{max} = 12 \text{ Å}^{-1}$ occurs at the arsenic K-edge. Also the variation of the f_i with x-ray energy is small and measurements appear to be restricted to concentrated samples. Both EXAFS and the anomalous x-ray scattering method have been applied to sputtered amorphous $Mo_{42}Ge_{58}$ films. The scattering data were analyzed over the range $0.05 < q < 18.5$ Å^{-1} at the molybdenum-edge and the range $0.5 < q < 10.5$ Å^{-1} at the germanium-edge. A broad nearest-neighbor distribution was indicated. The EXAFS was analyzed over the range 7 Å$^{-1} < 2k < 30$ Å^{-1} with a single-shell Gaussian model yielding a shorter near-neighbor distance and narrower distribution relative to the scattering results. In order to detect the additional atoms belonging to the first shell, but at larger radial distances, the EXAFS data would have to be analyzed to smaller k values. The question of extension to low k is addressed in Section 9.3.

The difference in the f_i of x rays, electrons, and neutrons, (the three radiation method), has been used to determine the S_{ij} of $Pd_{80}Si_{20}$ glass (106). The method of isomorphous substitution has also been applied. In this method the difference in the f_i is achieved by substituting an isomorphic element with the assumption that this does not change the structure of the system (106).

In all these methods the accuracy with which the $S_{ij}(q)$ can be determined depends sensitivity upon the differences in the scattering amplitudes f_i (97, 106). The accuracy with which the $g_{ij}(r)$ can be obtained is further limited by the finite range of the data as indicated in Section 9.5.1 for monatomic systems.

When a system contains more than two atomic species, the number of independent diffraction experiments that must be done increases rapidly. Glass and liquid systems containing four to six atomic species have been investigated, but attention was restricted to the determination of a subset of the pair distribution functions. A case in point is the application of the neutron diffraction, isotopic substitution method to electrolyte solutions. For a salt MX dissolved in heavy water D_2O there are 10 $g_{ij}(r)$. Intensity measurements on two solutions differing only in the isotopic state of M permit one to obtain an averaged distribution function $\overline{G}_M(r)$ that is a linear combination of the four metal partial distribution functions (107)

$$\overline{G}_M(r) = A\, g_{MO} + B\, g_{MD} + C\, g_{MX} + D\, g_{MM} + E \qquad (59)$$

The coefficients A to E depend on the concentrations, f_O, f_D, f_X and f_M-$f_{M'}$, the difference in the scattering amplitudes of the two isotopes of M. For a wide range of concentration, A and B are much larger than C and D so that the hydration of metal ions can be studied. A major advantage of this method is that the MD distribution function contributes to $\overline{G}_M(r)$ so that the orientation of the heavy water molecule hydrated to the metal ion can be determined. An EXAFS experiment cannot detect hydrogen (or deuterium) atoms in a disordered system [although the location of interstitial hydrogen can be obtained in crystals by exploiting the focused multiscattering effect (109)]. The EXAFS of the metal edge contains contributions from the M—O, M—X, and M—M distributions and, in principle, the distinctive k dependences of the EXAFS $f_i(k)$ and $\delta_i(k)$ permit identification of the coordinating species. However, in a diffraction experiment, if the radial separation of the different species is small, the partial distribution functions cannot be deconvoluted using only Eq. (59). EXAFS also has the advantage that it is applicable when the concentration of x-ray absorbing species approaches 10 ppm.

When three isotopes of M (or X) are available, three scattering experiments are sufficient to obtain $S_{MM}(q)$ [or $S_{XX}(q)$]. Measurements have been made to determine the ion–ion correlations in aqueous solutions of $NiCl_2$. However, the experiments were at the limit of the existing technique and self-consistent $g_{NiNi}(r)$ or $g_{ClCl}(r)$ could not be obtained (107).

It is clear from Eq. (56) that a single diffraction experiment cannot provide a unique determination of all the partial structure factors. The Fourier transform of $I_a(q)$ gives a total $G(r)$ that is dominated by the most concentrated species and, in the case of x-rays, because of the dependence of the atomic form factors on atomic number, is preferentially weighted towards the heavier elements. Consequently sophisticated modeling techniques have been developed to determine the structure.

A single x-ray diffraction experiment is particularly useful in those systems such as aqueous electrolyte solutions for which the total $G(r)$ usually has a single well-defined peak. A comparison of EXAFS and x-ray diffraction results for metal halide aqueous solutions, taken from the review of Licheri and Pinna (11), is shown in Fig. 9.26. Considering that the EXAFS data were analyzed only within the Gaussian or small disorder approximation, the agreement of the two techniques for the metal–oxygen (water) radial distance and coordination number is reasonable.

A comparison of results obtained with EXAFS, neutron diffraction isotopic substitution (NDIS), and single x-ray diffraction experiments is possible for 4 M $NiCl_2$ aqueous solutions. There is general agreement on the gross structure of the first coordination shell of Ni^{2+}. The number of water molecules hydrated to, and their radial distances from the nickel ion are by EXAFS 7.2 at 2.06 Å (11); 6.2 at 2.07 Å (110); by NDIS 5.8 at 2.07 Å (111); by single x-ray diffraction 6 at 2.07 Å (112).

There is also some evidence that the first coordination shell may contain Cl^- ions. A second x-ray diffraction study (113) indicated that a 3 M $NiCl_2$ solution contained equal amounts of $Ni(H_2O)_5Cl^+$ and $Ni(H_2O)_6^{++}$. This represents an average coordination of 0.5 Cl ions per Ni. In the NDIS experiment, g_{NiCl} due to the small value of the coefficient C makes only a small contribution to $G_{Ni}(r)$, Eq. (59), and evidently 0.5 Cl ions is below the detection sensitivity. The EXAFS experiments on the nickel K-edge were also insensitive to first shell Cl^- coordination. Recently, Sandstrom (114) has measured the EXAFS on the chlorine K-edge in $NiCl_2$ solution and, although quantitative analysis has not

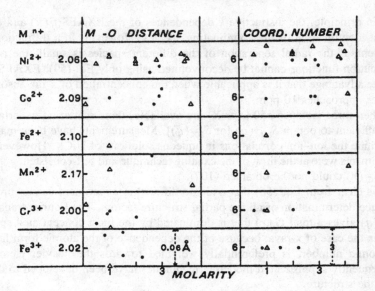

Figure 9.26. First hydration shell in metal halide MX_n aqueous solutions with $X = Cl^-$, NO_3^-, ClO_4^- (dots: x-ray diffraction; triangles: EXAFS). [Taken from Fig. 1 of ref. (11).]

been completed, claims unequivocal evidence for direct Ni—Cl bonding at the same level as indicated by the x-ray diffraction studies of Magini (113).

There is disagreement on the second coordination shell of nickel in aqueous $NiCl_2$. In the EXAFS experiment of Licheri et al. (11, 112) there was no reliable second-shell information, whereas the EXAFS work of Sandstrom (110) suggested the presence of a broad distribution of three chloride ions at an average distance of 3.1 Å. Asymmetry effects were not included in the analysis. The NDIS data indicated a second coordination of 15 M_2O molecules in the range 3.7–5.3 Å (111). Structural models fitted to a single x-ray diffraction experiment (11) are consistent with this latter result.

Aqueous solutions of $Ag(ClO_4)_2$ provide a second example in which the method of EXAFS, a single x-ray diffraction measurement and neutron diffraction from solutions with different isotopic compositions have been applied (115). The Ag—O distances (and number of coordinating oxygens) obtained by x-ray and neutron diffraction were 2.38 Å (4.0) and 2.36 Å (3.8), respectively. An initial analysis of the EXAFS data using a single-shell Gaussian model gave $R(Ag—O) = 2.31$ Å and $N(O) = 2.9$. Both these values are smaller than the diffraction results and suggest asymmetry of the pair distribution function. This has been confirmed by a cumulant analysis by Crozier and Seary (72) in which terms up to the fifth cumulant were included. In the analysis it was also necessary to include the effects of asymmetry in the determination of E_0 (Section 9.3.7). The quality of the data, both for the reference and the unknown, restricted analysis to the range $3 < k < 10$ Å^{-1}. Preliminary results are $R(\text{Ag}-\text{O}) = 2.38$ Å and $N(\text{O}) = 3.9$. Recently Yamaguchi and Boyce (116) have approximated the asymmetric distribution by a two-shell Gaussian model [$R_1(\text{Ag}-\text{O}) = 2.25$ Å, $N_1(\text{O}) = 1$, $\Delta\sigma_1^2 = 0.006$ Å^2; $R_2(\text{Ag}-\text{O}) = 2.43$ Å, $N_2 = 4$, $\Delta\sigma_2^2 = 0.02$ Å^2 where $\Delta\sigma^2$ is the difference in σ^2 between the solution and a reference powder of AgO at 77 K] with an improved residual sum of errors relative to their earlier single-shell fit. Their mean radial distance 2.34 Å is more in agreement with the diffraction studies.

9.5.3. Pair Potentials

The effective two-body interaction potential, or the pair potential $U(R)$, is the basic quantity from which the pair distribution function and other properties of liquids and disordered systems originate. Pair distribution functions, obtained from diffraction studies, have been used to check the validity of statistical mechanical theories of the liquid state, the accuracy of molecular dynamic calculations, and the accuracy of assumed forms for $U(R)$. Since EXAFS can now provide accurate pair distribution functions, theoretical calculations can be tested against EXAFS data. The usefulness of this approach is illustrated for the case of superionic conductors and liquid metals.

The EXAFS distribution function $P_{AB}(r)$ for an absorber-back scatterer AB pair is related to the radial distribution function of diffraction studies by

$$P_{AB}(r) = 4\pi r^2 \rho_B g_{AB}(r) \tag{60}$$

where ρ_B is the number density of B atoms and $\int P_{AB}(r)\,dr$ equals the total number of B atoms in the sample. Hayes and Boyce (117) have compared EXAFS data for the copper k-edge of the superionic conductor α-CuI at 743 K with the $g_{\text{CuI}}(r)$ calculated via molecular dynamics (MD) at 698 K. The initial MD result for $g_{\text{CuI}}(r)$, indicated in Fig. 9.27, increases from zero at an r value that is too short by 0.12 Å and yields a distribution that is too broad relative to the EXAFS result. A similar discrepancy was found between the EXAFS and MD calculation for AgI (1, 117). In a subsequent MD calculation, using a modified potential with a stronger core–core repulsive term, improved agreement with the EXAFS data was achieved (1).

The structure of a liquid metal is determined to a first approximation by the random packing of hard spheres. In a refinement of this, Jacobs and Anderson (118) have developed a Blip-function theory in which the repulsive portion of the ion–ion interaction potential is treated as a perturbation of a Verlet and Weiss hard-sphere fluid. They have argued that the repulsive potential can be

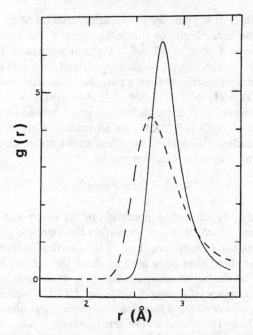

Figure 9.27. The partial distribution function $g_{CuI}(r)$ obtained from the copper K-edge EXAFS of the superionic conductor α-CuI and from molecular dynamics calculations (---). [Taken from ref. (117).]

specified by forcing the $S(q)$ deduced from their model to fit the experimental $S(q)$ at large q. Since this is the region in which the EXAFS technique is more accurate, the Blip-function theory was applied to an EXAFS study of liquid zinc. It was concluded that the model repulsive potential obtained by fitting the Blip-function theory to x-ray diffraction data was not soft enough in comparison with that expected from the EXAFS data (10, 17).

9.5.4. Summary

We conclude with a list of some strengths and weaknesses of the EXAFS and diffraction techniques:

1. EXAFS can provide the local structure about an x-ray absorbing atom that is present at a concentration level 10 ppm. The minimum concentration is two or three orders of magnitude larger for diffraction studies.

2. By using either K-edges or L-edges, EXAFS can be applied to most elements heavier than oxygen. In some cases the range in k space is

restricted by the presence of interfering edges from the same element or different elements in the material under study. The neutron diffraction substitution method is limited to a comparatively small number of suitable isotopes. It is, however, very sensitive to deuterium. The anomalous scattering of x-rays is restricted to elements with atomic number greater than about 30.

3. The r-space resolution of EXAFS is better than that achieved with most conventional diffraction studies. The lack of high-q data in the latter affects the accuracy with which radial distances, coordination numbers, and peak shapes can be obtained. However, with TOF neutron diffraction studies and energy dispersive x-ray diffraction studies, $q_{max} \approx 30$ Å^{-1} or larger can be achieved. The extent to which EXAFS is limited at low k is dependent on the existence of a suitable reference material and the accuracy of the cumulant extrapolation to k equal to zero.

4. EXAFS has been more successful in studying medium-range order than diffraction.

5. In the case of compositional disorder, when the nearest-neighbor shell contains more than one type of atom and the atomic numbers differ by more than five, the chemical signature of the EXAFS backscattering amplitude and phase can be used to deconvolve the contributions. In diffraction studies the isolation of the partial pair distribution function $g_{ij}(r)$ requires separate diffraction experiments with suitably different coherent scattering amplitudes. This is not always possible.

6. The processing of EXAFS data involves numerous steps as indicated in Chapter 6 and Section 9.3. Errors may occur in EXAFS-deduced structural parameters of disordered systems unless the analysis procedures are designed specifically to include cumulants higher than the second. But numerous corrections must also be made before a partial structure factor $S_{ij}(q)$ is extracted from the measured scattered intensity.

It is evident that the information provided by EXAFS and diffraction techniques is complementary and by a combination of both, a true structural picture of a disordered system may emerge.

ACKNOWLEDGMENTS

It is a pleasure to acknowledge helpful discussions with N. Alberding and A. J. Seary. Their preparation of many of the figures is appreciated. We wish to thank them and K. R. Bauchspiess for carefully reading the manuscript. We have benefited from discussions with J. B. Boyce, R. B. Greegor, F. W. Lytle,

J. C. Mikkelsen, and M. Plischke. We thank J. B. Boyce, T. M. Hayes, and J. C. Mikkelsen for providing their CuBr data, J. B. Boyce and T. Yamaguchi for their Ag(ClO$_4$) data, and D. C. Koningsberger for his suggestions throughout this chapter. Finally, we thank D. Young for her efforts in typing the manuscript.

REFERENCES

1. T. M. Hayes and J. B. Boyce, *Solid State Phys.*, **37**, 173 (1982).

2. P. A. Lee, P. H. Citrin, P. Eisenberger, and B. M. Kincaid, *Rev. Mod. Phys.*, **53**, 769 (1981).

3. P. Rabe and R. Haensel, "EXAFS and its Applicability for Structural Analysis," in *Festkörperprobleme xx (Advances in Solid State Physics)*, J. Treusch (Ed.), Vieweg, Braunschweig, 1980, p. 43.

4. R. Haensel, P. Rabe, G. Tolkiehn, and A. Werner, in Proceedings of NATO Advanced Study Institute, *Liquid and Amorphous Metals*, E. Lüscher and H. Coufal (Eds.), Sijthoff and Noordhoff, The Netherlands, 1980, p. 459.

5. D. Raoux, A. Flank, and A. Sadoc, in *EXAFS and Near Edge Structures*, A. Bianconi, L. Incoccia, and S. Stipcich (Eds). Springer Series in Chemical Physics, vol. 27, Springer-Verlag, Berlin, 1983, p. 232.

6. J. Wong, "EXAFS Studies of Metallic Glasses," in *Glassy Metals I*, H. J. Güntherodt and H. Beck (Eds.), Topics in Applied Physics, vol. 46, Springer-Verlag, Berlin, 1981, p. 45.

7. G. Calas and J. Petiau, *Bull. Mineral*, **106**, 33 (1983).

8. G. N. Greaves, A Fontaine, P. Lagarde, D. Raoux, and S. J. Gurman, *Nature London*, **293**, 611 (1981).

9. G. N. Greaves, in *EXAFS and Near Edge Structures*, A. Bianconi, L. Incoccia, and S. Stipcich (Eds.), Springer Series in Chemical Physics, vol. 27, Springer-Verlag, Berlin, 1983, p. 248.

10. E. D. Crozier, "Disorder Effects in the EXAFS of Metals and Semiconductors in the Solid and Liquid States," in *EXAFS Spectroscopy, Techniques and Applications*, B. K. Teo and D. C. Joy (Eds.), Plenum New York, 1981, p. 89.

11. G. Licheri and G. Pinna, in *EXAFS and Near Edge Structures*, A. Bianconi, L. Incoccia, and S. Stipcich (Eds.), Springer Series in Chemical Physics, vol. 27, Springer-Verlag, Berlin, 1983, p. 240.

12. D. R. Sandstrom and F. W. Lytle, *Ann. Rev. Phys. Chem.*, **30**, 215 (1979).

13. D. R. Sandstrom, B. R. Stults, and R. B. Greegor, "Structural Evidence for Solutions from EXAFS Measurements," in *EXAFS Spectroscopy, Techniques and Applications*, B. K. Teo and D. C. Joy, (Eds.), Plenum, New York, 1981, p. 139.

14. P. H. Gaskell, *J. Phys. C: Solid State Phys.*, **12**, 4337 (1979).

15. G. S. Cargill, *J. Non-Cryst. Solids*, **61–62**, 261 (1984).

16. T. M. Hayes, J. B. Boyce, and J. L. Beeby, *J. Phys. C: Solid State Phys.*, **11**, 2931 (1978).

17. E. D. Crozier and A. J. Seary, *Can. J. Phys.*, **58**, 1388 (1980).

18. J. J. Rehr (to be published).

19. G. Bunker, *Nucl. Instrum. Methods*, **207**, 437 (1983).

20. E. Sevillano, H. Meuth, and J. J. Rehr, *Phys. Rev. B*, **20**, 4908 (1979).

21. R. Haydock, V. Heine, and M. J. Kelly, *J. Phys. C: Solid State Phys.*, **8**, 2591 (1975).

22. P. P. Lottici and J. J. Rehr, *Solid State Commun.*, **35**, 565 (1980).

23. B. Bunker, S.-H. Chan, and J. J. Rehr (to be published).

24. V. V. Schmidt, *Bull. Acad. Sci., USSR, Ser. Phys.*, **25**, 998 (1961); **27**, 392 (1961).

25. G. Beni and P. M. Platzman, *Phys. Rev. B*, **14**, 9514 (1976).

26. G. Martens, P. Rabe, N. Schwentner, and A. Werner, *J. Phys. C: Solid State Phys.*, **11**, 3125 (1978); *Phys. Rev. B*, **17**, 1481 (1978).

27. D. P. Jackson, B. M. Powell, and G. Dolling, *Phys. Lett.*, **51A**, 87 (1975).

28. E. C. Svensson, B. N. Brockhouse, and J. M. Rowe, *Phys. Rev. B*, **155**, 619 (1967).

29. R. M. Nicklow, G. Gilat, H. G. Smith, L. J. Raubenheimer, and M. K. Wilkinson, *Phys. Rev.*, **164**, 922 (1967).

30. R. B. Greegor and F. W. Lytle, *Phys. Rev. B*, **20**, 4902 (1979).

31. W. Böhmer and P. Rabe, *J. Phys. C: Solid State Phys.*, **12**, 2465 (1979).

32. P. Eisenberger and G. S. Brown, *Solid State Commun.*, **29**, 481 (1979).

33. R. Kubo, *J. Phys. Soc. Jpn.*, **17**, 1100 (1962).

34. E. D. Crozier, F. W. Lytle, D. E. Sayers, and E. A. Stern, *Can. J. Chem.*, **55**, 1968 (1977).

35. E. D. Crozier and A. J. Seary, *Can. J. Phys.*, **59**, 876 (1981).

36. M. DeCrescenzi, A. Balzarotti, F. Comin, L. Incoccia, S. Mobilio, and N. Motta, *Solid State Commun.*, **37**, 921 (1981).

37. P. Lagarde, J. Rivory, and G. Vlaic, *J. Non-Cryst. Solids*, **57**, 275 (1983).

38. A. Sadoc, A. Fontaine, P. Lagarde, and D. Raoux, *J. Am. Chem. Soc.*, **103**, 6287 (1981).

39. A. Sadoc, D. Raoux, P. Lagarde, and A. Fontaine, *J. Non-Cryst. Solids*, **50**, 331 (1982).

40. R. Frahm, R. Haensel, and P. Rabe, in *EXAFS and Near Edge Structures*, A. Bianconi, L. Incoccia, and S. Stipcich (Eds.), Springer Series in Chemical Physics, vol. 27, Springer-Verlag, Berlin, 1983, p. 107.

41. P. A. Lee and G. Beni, *Phys. Rev. B*, **15**, 2862 (1977).

42. B. K. Teo and P. A. Lee, *J. Am. Chem. Soc.*, **101**, 2815 (1979).

43. E. C. Marques, D. R. Sandstrom, F. W. Lytle, and R. B. Gregor, *J. Chem. Phys.*, **77**, 1027 (1982).

44. J. B. A. D. van Zon, D. C. Koningsbergér, H. F. J. van't Blik, and R. Prins, *J. Chem. Phys.*, **80**, 3914 (1984).

45. J. M. Tranquada and R. Ingalls, *Phys. Rev. B*, **28**, 3520 (1983).

46. J. B. Boyce, T. M. Hayes, and J. C. Mikkelsen, Jr., *Solid State Commun.*, **35**, 237 (1980).

47. J. B. Boyce, T. M. Hayes, and J. C. Mikkelsen, Jr., *Phys. Rev. B*, **23**, 2876 (1981).

48. J. B. Boyce, private communication.

49. J. B. Boyce and T. M. Hayes, "Structural Studies of Superionic Conduction," in *EXAFS Spectroscopy Techniques and Applications*, B. K. Teo and D. C. Joy (Eds.), Plenum, New York, 1981, p. 84.

50. N. Alberding, private communication.

51. G. Martens, P. Rabe, N. Schwenter, and A. Werner, *Phys. Rev. Lett.*, **39**, 1411 (1977).

52. B. Lengeler and P. Eisenberger, *Phys. Rev. B*, **21**, 4507 (1980).

53. E. A. Stern, B. A. Bunker, and S. M. Heald, *Phys. Rev. B*, **21**, 5521 (1980).

54. C. E. Bouldin and E. A. Stern, in *EXAFS and Near Edge Structures III*, K. O. Hodgson, B. Hedman, and J. E. Penner-Hahn (Eds.), Springer Proceedings in Physics, vol 2, Springer-Verlag, Berlin, 1984, p. 273.

55. D. V. Baxter, A. Williams, and W. L. Johnson, *J. Non-Cryst.* Solids, **61–62**, 409 (1984).

56. N. Alberding and E. D. Crozier, *Phys. Rev. B*, **27**, 3374 (1983).

57. J. Wong and F. W. Lytle, *J. Non-Cryst. Solids*, **37**, 273 (1980).

58. E. D. Crozier, N. Alberding, and B. R. Sundheim, *J. Chem. Phys.*, **79**, 939 (1983).

59. D. J. Kay, M.Sc. thesis, Simon Fraser University, Burnaby, British Columbia, Canada, 1978.

60. P. Rabe, G. Tolkiehn, and A. Werner, *J. Phys. C: Solid State Phys.*, **12**, L545 (1979).

61. B. K. Teo, P. A. Lee, A. L. Simons, P. Eisenberger, and B. M. Kincaid, *J. Am. Chem. Soc.*, **99**, 3854 (1977).

62. E. A. Stern, D. E. Sayers, and F. W. Lytle, *Phys. Rev. B*, **11**, 4836 (1975).

63. P. A. Lee and J. P. Pendry, *Phys. Rev. B*, **11**, 2975 (1975).

64. B. K. Teo, *J. Am. Chem. Soc.*, **103**, 3990 (1981).

65. J. J. Boland, S. E. Crane, and J. D. Baldeschwieler, *J. Chem. Phys.*, **77**, 142 (1982).

66. B. K. Teo, "EXAFS Spectroscopy," in *EXAFS Spectroscopy, Techniques and Applications*, B. K. Teo and D. C. Joy (Eds.), Plenum, New York, 1981, p. 13.

67. J. J. Boland and J. D. Baldeschwieler, *J. Chem. Phys.*, **80**, 3005 (1984).

68. N. Alberding and E. D. Crozier, in *EXAFS and Near Edge Structures III*, K. O. Hodgson, B. Hedman, and J. E. Penner-Hahn (Eds.), Springer Proceedings in Physics, vol. 2, Springer-Verlag, Berlin, 1984, p. 30.

69. R. L. Mozzi and B. E. Warren, *J. Appl. Cryst.*, **2**, 164 (1969).

70. D. E. Sayers, E. A. Stern, and F. W. Lytle, *Phys. Rev. Lett.*, **35**, 584 (1975).

71. J. J. Boland and J. D. Baldeschwieler, *J. Chem. Phys.*, **81**, 1145 (1984).

72. E. D. Crozier and A. J. Seary, private communication.

73. E. A. Stern, B. Bunker, and S. M. Heald, "Understanding the Causes of Non-Transferability of EXAFS Amplitude," in *EXAFS Spectroscopy, Techniques and Applications*, B. K. Teo and D. C. Joy, (Eds.), Plenum, New York, 1981, p. 59.

74. P. Eisenberger and B. Lengeler, *Phys. Rev. B*, **22**, 3551 (1980).

75. E. A. Stern and K. Kim, *Phys. Rev. B*, **23**, 3781 (1981).

76. E. A. Stern, "Limitations of EXAFS: Real and Imagined," in *EXAFS for Inorganic Systems*, Science and Engineering Research Council, C. D. Garner and S. S. Hasnain (Eds.), Daresbury Laboratory, England, 1981, p. 40.

77. J. Goulon, C. Goulon-Ginet, R. Cortes, and J. M. Dubois, *J. Phys.*, **43**, 539 (1982).

78. S. C. Moss, H. Metzger, M. Eisner, H. W. Huang, and S. H. Hunter, *Rev. Sci. Instrum.*, **49**, 1559 (1978).

79. J. C. Mikkelsen, Jr., J. B. Boyce, and R. Allen, *Rev. Sci. Instrum.*, **51**, 388 (1980).

80. F. Bell, M. Cutler, and E. D. Crozier (to be published).

81. R. Ingalls, E. D. Crozier, J. E. Whitmore, A. J. Seary, and J. M. Tranquada, *J. Appl. Phys.*, **51**, 3158 (1980).

82. R. Ingalls, J. M. Tranquada, J. E. Whitmore, E. D. Crozier, and A. J. Seary, "EXAFS Studies at High Pressures," in *EXAFS Spectroscopy, Techniques and Applications*, B. K. Teo and D. C. Joy (Eds.), Plenum, New York, 1981, p. 127.

83. J. B. Hastings, "EXAFS of Dilute Systems: Fluorescence Detection," in *EXAFS Spectroscopy, Techniques and Applications*, B. K. Teo and D. C. Joy (Eds.), Plenum, New York, 1981, 1971.

84. A. J. Seary, N. Alberding, and E. D. Crozier, Stanford Synchrotron Radiation Laboratory, Report No. 83/03, 80 (1983).

85. E. A. Stern and K. Lu, *Nucl. Instrum. Methods*, **195**, 415 (1982).

86. K. R. Bauchspiess and E. D. Crozier, in *EXAFS and Near Edge Structures III*, K. O. Hodgson, B. Hedman, and J.E. Penner-Hahn (Eds.), Springer Proceedings in Physics, vol. 2, Springer-Verlag, Berlin, 1984, p. 514.

87. Z. U. Rek, G. S. Brown, and T. Troxel, in *EXAFS and Near Edge Structures III*, K. O. Hodgson, B. Hedman, and J. E. Penner-Hahn (Eds.), Springer Proceedings in Physics, vol. 2, Springer-Verlag, Berlin, 1984, p. 511.

88. F. W. Lytle, private communication.

89. J. V. Acrivos, K. Hathaway, J. Reynolds, J. Code, S. Parkin, M. P. Klein, A. Thompson, and D. Goodin, *Rev. Sci. Instrum.*, **53**, 575 (1982).

90. S. A. Rice and P. Gray, *The Statistical Mechanics of Simple Liquids*, Wiley, New York, 1965.

91. P. J. Black and J. A. Cundall, *Acta Crystallogr*, **19**, 807 (1965).

92. T. Egami, *J. Mater. Sci.*, **13**, 2587 (1978).

93. J. M. Prober and J. M. Schultz, *J. Appl. Cryst.*, **8**, 405 (1975).

94. J. C. Malaurent and J. Dixmier, *Thin Solid Films*, **32**, 370 (1976).

95. K. Suzuki, M. Misawa, and Y. Fukushima, *Trans. J. Inst. Met.*, **16**, 297 (1975).

96. N. H. March, M. Parrinello, and M. P. Tosi, *Phys. Chem. Liq.*, 39 (1976).

97. Y. Waseda, *The Structure of Non-Crystalline Materials, Liquids and Amorphous Solids*, McGraw-Hill, New York, 1980.

98. P. Ascarreli, *Phys. Rev.*, **143**, 36 (1966).

99. S. E. Rodriquez and C. J. Pings, *J. Chem. Phys.*, **42**, 2435 (1965).

100. C. J. Pings, "Structure of Simple Liquids by X-ray Diffraction," in *Physics of Simple Liquids*, H. N. V. Temperley, J. S. Rowlinson, and G. S. Rushbrooke (Eds.), Wiley, New York, 1968, p. 387.

101. G. Etherington, A. C. Wright, J. T. Wenzel, J. C. Dore, J. H. Clarke, and R. N. Sinclair, *J. Non-Cryst. Solids*, **48**, 265 (1982).

102. C. E. Bouldin and E. A. Stern, in *EXAFS and Near Edge Structures III*, K. O. Hodgson, B. Hedman, and J. E. Penner-Hahn (Eds.), Springer Proceedings in Physics, vol. 2, Springer-Verlag, 1984, p. 278.

103. F. Comin, L. Incoccia, and S. Mobilio, *Solid State Commun.*, **37**, 413 (1981).

104. F. Evangelisti, M. Garozza, and G. Conte, *J. Appl. Phys.*, **53**, 7390 (1982).

105. E. A. Stern, C. E. Bouldin, B. von Roedern, and J. Azouloy, *Phys. Rev. B*, **27**, 6557 (1983).

106. C. N. J. Wagner, *J. Non-Cryst. Solids*, **42**, 3 (1980).

107. J. E. Enderby and G. W. Neilson, *Rep. Prog. Phys.*, **44**, 594 (1981).

108. J. Kortright, W. Warburton, and A. Bienenstock, in *EXAFS and Near Edge Structures*, A. Bianconi, L. Incoccia, and S. Stipcich (Eds.), Springer Series in Chemical Physics, vol. 27, Springer-Verlag, Berlin, 1983, p. 362.

109. B. Lengeler, *Phys. Rev. Lett.*, **53**, 74 (1984).

110. D. R. Sandstrom, *J. Chem. Phys.*, **71**, 2381 (1979).

111. J. E. Enderby, *Philos. Trans. R. Soc. London Ser. B.*, **290**, 553 (1980).

112. G. Licheri, G. Paschina, G. Piccaluga, G. Pinna, and G. Vlaic, *Chem. Phys. Lett.*, **83**, 384 (1981).

113. M. Magini, *J. Chem. Phys.*, **74**, 2523 (1981).

114. D. R. Sandstrom, in *EXAFS and Near Edge Structure III*, K. O. Hodgson, B. Hedman, and J. E. Penner-Hahn (Eds.), Springer Proceedings in Physics, vol. 2, Springer-Verlag, Berlin, 1984, p. 409.

115. T. Yamaguchi, O. Lindqvist, J. B. Boyce, and T. Claeson, in *EXAFS and Near Edge Structures III*, K. O. Hodgson, B. Hedman, and J. E. Penner-Hahn (Eds.), Springer Proceedings in Physics, vol. 2, Springer-Verlag, Berlin, 1984, p. 417.

116. T. Yamaguchi and J. B. Boyce, private communication.

117. T. M. Hayes and J. B. Boyce, *J. Phys. C: Solid State Phys.*, **13**, L731 (1980).

118. R. E. Jacobs and H. C. Anderson, *Chem. Phys.*, **10**, 73 (1975).

CHAPTER

10

SEXAFS: EVERYTHING YOU ALWAYS WANTED TO KNOW ABOUT SEXAFS BUT WERE AFRAID TO ASK

JOACHIM STÖHR

IBM Almaden Research Center
San Jose, California

10.1. INTRODUCTION

Over the last 20 years surface science has developed into one of the major fields in chemistry and physics. This development was driven mostly by the need for a more detailed understanding of gas–solid, solid–solid, and to a lesser extent liquid–solid interfaces, which play the key role in catalysis, electronic device fabrication, and electrolysis. While early work in these latter areas typically employed a largely empirical approach to determine systems with a high technological efficiency, it was hoped that new surface or interface specific techniques could provide an insight into the detailed mechanisms and the general rules governing the interactions between two or more media. However, progress has been impeded not only by the complexity of the surface phenomena but by the lack of understanding of the information content of the data provided by the newly developed experimental techniques. Here, one important factor is the long development time for surface techniques (1–4).

Because of the complexity of the *phenomena* to be studied and of the *techniques* employed in these studies, experimental and theoretical surface scientists have resorted to the study of *model systems* such as single-crystal surfaces and well-characterized adsorbate overlayers (5–8).

At present we are at a state where, although many techniques have been developed, it is still difficult and in some cases impossible to obtain reliable information on model systems. Even for model systems the elucidation of the surface reaction mechanisms is sufficiently complex that only combination of different experimental and theoretical results and a check of their mutual consistency will provide a reliable picture. The importance of interplay between the various surface probes cannot be overemphasized.

The information needed for surfaces and interfaces is threefold: *composition*, *crystallographic structure*, and *electronic structure*. First, all the atomic species involved and their abundance need to be determined. Second, information is needed about their structural arrangement. Third, their electronic interactions (nature of bonding) need to be known. Of these three problems the first one can be solved most easily using techniques such as Auger electron spectroscopy (AES) (9), photoemission spectroscopy (PES) (1-3), high-resolution electron energy loss (EELS) spectroscopy (6), thermal desorption spectroscopy (TDS) (10), secondary ion mass spectroscopy (SIMS) (11), or Rutherford ion scattering spectroscopy (RISS) (12-14). The other two questions are more difficult to answer. In many ways the crystallographic and electronic structures are linked since one is the consequence of the other. From both experimental and theoretical points of view more work has been done on electronic than crystallograpic structure investigations. This is largely due to the success of PES spectroscopy, in particular polarization dependent angle-resolved UV photoemission by means of synchrotron radiation (1-3, 15-18). For many systems photoemission spectra of the valence bands or levels have provided detailed information on the bonding orbitals and have stimulated theoretical electronic structure calculations (1, 4). Other investigations probed the local bonding by means of high-resolution EELS (6, 19-21), which has proved particularly fruitful for chemisorbed molecules.

Over the last few years great emphasis has been placed on the development of new tools for surface crystallography. Such efforts are a response to the lack of basic structural information for all but the simplest systems. Low-energy electron diffraction (LEED) (8, 22-25) was the first technique developed for the determination of surface structures. However, unlike x-ray scattering, which has been successfully employed for bulk systems, the scattering of low-energy electrons cannot be described by a kinematic theory. Rather, the strong coupling between the incident electrons and the valence electrons of the surface complex leads to multiple-scattering complications. It is the problems associated with the multiple-scattering analysis of the measured LEED intensities and the ultimately limited applicability of the technique to other than periodic surface structures that triggered the search for alternate tools.

Like LEED some techniques use the *elastic scattering* of incident particles (electrons, photons, ions, or atoms) (26). Here x-ray Bragg reflection diffraction (BRD) (27) and RISS (12-14) appear especially promising because the scattering can be analyzed by a kinematic theory. *Spectroscopic* structural tools measure a signal in response to a local electronic excitation. Electrons are detected in angle-resolved photoemission extended fine structure (ARPEFS) (28), extended appearance potential fine structure (EAPFS) (29), and extended electron loss fine structure (EXELFS) (30) measurements, while ions are monitored in low-energy ion scattering (31), electron (ESD) (10, 32, 33) and photon (PSD) (34, 35) stimulated ion desorption studies, and SIMS (11). One spectroscopic technique that can use either electron (36, 37) or ion (38) detection is the surface version of the extended x-ray absorption fine structure (EXAFS) method, or SEXAFS. It is this technique that is the subject of this chapter. It will become apparent that SEXAFS has made the transition from a conceptual to a powerful applied surface structural tool. It is one of the most generally applicable and precise techniques and therefore will have a major impact on unsolved problems in surface crystallography.

A reliable surface structure technique needs to fulfill several criteria. At first it needs to be *surface sensitive*, that is, a sizable and characteristic part of the measured signal has to originate from the outermost layer(s) of a solid or liquid. Second, the technique should be *nondestructive* in the sense that the measured signal cannot contain a sizable contribution from surface sites that were modified during the course of the measurement. Besides these two fundamental criteria one may synthesize a wish list of criteria that an ideal surface structural technique should satisfy. The two basic questions in surface crystallography are *What atoms are there and how are they arranged?* Thus a technique should ideally be able to differentiate between different atoms and yield their structural arrangement in terms of bond lengths, bond angles, and coordination numbers. Quantitatively, one would like to differentiate in atomic number Z to $\Delta Z < 1$, in bond lengths R to $\Delta R/R < 1\%$, in bond angles α to $\Delta\alpha \lesssim 1°$, and in coordination number N to $\Delta N/N < 10\%$. At present no surface structural technique can fulfill all these demands but we shall see that SEXAFS satisfies most of them.

The historical development of the SEXAFS technique has its roots in three key factors: (a) the development of the bulk analogue (EXAFS) technique during the early 1970s (39), (b) the recent availability of high-brightness (flux/unit area) synchrotron radiation (40–42), (c) the recognition that the EXAFS technique can be made surface sensitive by detection of the electron yield from the sample (43, 44).

Bulk EXAFS measurements have been carried out now for over 10 years as discussed in the other chapters in this book and previous reviews (45–50). Bulk studies typically employed 4–20-keV x-ray radiation that traverses thin windows and thus conveniently allows the measurements to be performed at atmospheric pressures.

Surface EXAFS measurements require an ultrahigh-vacuum (UHV) environment ($\sim 1 \times 10^{-10}$ Torr) and a completely different technology (51–53). Such studies are significantly more demanding than the conventional bulk measurements with respect to both the x-ray source and the measurement technique. The extremely small number of surface atoms (typically 10^{15} atoms cm^{-2}) necessitates higher x-ray intensities to obtain comparable signal-to-noise ratios as for bulk EXAFS studies (typically 10^{19} atoms cm^{-2}). Furthermore, the small sample sizes in many surface studies call for a collimated or focused x-ray beam. This requirement of high brightness is ideally met by the characteristics of radiation emitted by an electron storage ring (40–42). The availability of monochromatized synchrotron radiation for the past 10 years with photon flux levels of the order 10^{10} photons s^{-1} eV^{-1} mm^{-2} or more was an important prerequisite for the development of SEXAFS (and EXAFS).

SEXAFS studies also require a different detection technique than bulk EX-AFS measurements. Because the x-ray mean free path in solids is typically several thousand angstroms (54, 55), the x-ray signal from the surface layer will always be swamped by that from the bulk. Thus new ways had to be found to measure a signal that is proportional to the absorption coefficient and originates preferentially from the outermost layer(s) of the sample.

The first experiments that indicated how to obtain the absorption coefficient of atoms located near the surface were carried out by Lukirskii and co-workers in 1964 (56, 57). Figure 10.1 is taken from their work and compares the absorption coefficient near the oxygen K-edge recorded with monochromatized bremsstrahlung in transmission through a thin BeO foil and as the electron yield from a BeO photocathode. Comparison of the absorption and electron yield *near-edge* spectrum for this case and other materials led to the conclusion that the total electron yield signal is proportional to the absorption coefficient. This proportionality was investigated more quantitatively by Gudat and Kunz in 1972 (58, 59) by use of monochromatized synchrotron radiation in the 50–150-eV spectral range. In 1975 electron yield detection was employed to investigate EXAFS features above the aluminum $L_{2,3}$-edge of solid and liquid aluminum metal by Petersen and Kunz (60). Their results shown in Fig. 10.2 clearly indicate the broad applicability of electron yield measurements. In 1978 Martens et al. (61) directly compared and analyzed K-edge EXAFS spectra of copper metal taken in transmission and by electron yield detection. The first application of electron yield EXAFS to the study of surfaces was conducted in 1978 by Stöhr, Denley, and Perfetti (37) and Stöhr (62) who investigated a 30-Å thick oxide layer on bulk aluminum metal and a thin oxide layer on nickel metal, respectively. A similar approach was used in 1979 by Bianconi and Bachrach to investigate the structure of clean aluminum surfaces (63).

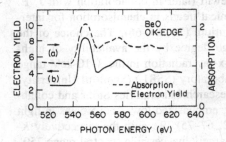

Figure 10.1. Oxygen K-edge x-ray absorption spectrum recorded in transmission through a thin foil and as the electron yield from a BeO photocathode by Lukirskii et al. (56).

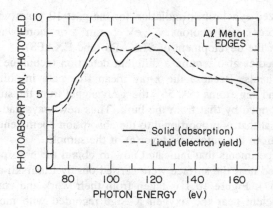

Figure 10.2. Aluminum $L_{2,3}$-edge absorption for a thin aluminum film in transmission and liquid aluminum by electron yield detection. The liquid curve was measured at about 1000 K. From Petersen and Kunz (60).

Besides the previously mentioned largely empirical development of electron yield SEXAFS, another approach was initiated by the theoretical work of Lee (43) and Landman and Adams (44) in 1976. These authors suggested the use of Auger electron detection to obtain surface EXAFS information. Since Auger electron emission is the nonradiative decay analogue of fluorescent x-ray emission, the Auger electron intensity should be directly proportional to the absorption coefficient. This fact was first used experimentally by Citrin, Eisenberger, and Hewitt (36) who in 1978 published the SEXAFS spectrum and structure determination of one-third monolayer of iodine on the (111) plane of a silver single crystal. Since 1978 total electron and Auger electron yield SEXAFS measurements have been performed on many systems and by now this technique is well established.

The development of SEXAFS from a conceptual to a powerful structural technique was largely due to the efforts of two independent groups. At Bell Laboratories Citrin, Eisenberger, and Hewitt (later in collaboration with J. E. Rowe) concentrated on establishing chemical trends in chemisorption for intermediate-Z atoms such as iodine and tellurium (36, 64–66). The choice of the adsorbate atoms was governed to a large degree by the availability of high-intensity, focussed, and monochromatic x-ray radiation in the 3–10-keV range at the Stanford Synchrotron Radiation Laboratory (SSRL) (Stanford University, USA). At Stanford and later at Exxon Research Laboratories Stöhr and collaborators, in particular Brennan and Jaeger, emphasized low-Z adsorbates in light of their technological importance (37, 62, 67–75). Because the respectively K-absorption edges fall into the previously largely inaccessible spectral range 250–

3000 eV, SEXAFS had to be developed simultaneously with bulk EXAFS (51) [note that the SEXAFS measurement on oxidized aluminum in 1978 (37) was in fact the first EXAFS measurement of a low-Z atom] and by improving existing (76) and developing new soft x-ray capabilities (77–79).

Other approaches to obtain the absorption coefficient of surface complexes have been suggested such as detection of the ion yield signal following photon stimulated desorption (PSD) from surfaces (34, 35, 80). This ion yield SEXAFS technique was first employed by Jaeger et al. (38) in 1980 to study the chemisorption of oxygen on the (100) plane of a molybdenum single crystal. The total reflection of x-rays and the associated short penetration depth (~ 30 Å) into the sample at grazing incidence angles ($<1°$) has been suggested (81–85) to be useful to determine the absorption coefficient of surface atoms. These ideas have been tested for thin film adsorbate layers (81, 85) and bulk samples (83, 84), but no structure determination of a surface complex has been performed to date. It has been suggested by Greegor, Lytle, and Sandstrom (86) that the detection of *optical* photons, which are created after a core hole deexcitation, may lead to an enhanced surface sensitivity because of the shorter sampling (escape) depth of low-energy (<10 eV) photons as compared to x-rays. Although the first EXAFS studies by means of fluorescence detection (87) were performed in 1977 the technique was not applied to surfaces until 1984 when Heald et al. (88) studied 1-2 monolyers of Au on glass. The first determination of a surface structure, namely 0.08 monolayer of sulfur on a (100) nickel surface, was carried out by Stöhr et al. in 1985 (89). In particular, Stöhr et al. (89) demonstrated that fluorescence detection can be particularly powerful and advantageous for the case of low-Z atoms on high-Z substrates, rather than for the more conventional case of a high-Z adsorbate on a low-Z substrate studied by Heald et al. (88).

This chapter discusses the basic principles underlying SEXAFS measurements carried out by the various experimental approaches in Section 10.2. It gives experimental details and points out experimental problems and pitfalls in Section 10.3. Section 10.4 outlines data analysis procedures with emphasis on differences between typical SEXAFS and EXAFS data. Applications of electron yield SEXAFS to various problems in surface crystallography are discussed in Section 10.5. Finally, Section 10.6 tries to assess the possibilities and limitations of the technique as a surface structural tool and give a perspective of future research directions. A summary of the various surfaces investigated by SEXAFS is given in tabular form in the Appendix.

10.2. PRINCIPLES OF SEXAFS

For surface crystallographic studies by means of EXAFS, the signal originating from surface atoms needs to be sizable and distinguishable from that due to the bulk. Conventional absorption experiments require thin film samples. Even if samples appropriate for surface crystallography (e.g., single crystals) could be made as ultrathin (~ 500 Å) free standing films, the surface-to-bulk signal ratio would be insufficient ($< 1\%$). Therefore, a detection technique is needed that suppresses the unwanted bulk signal relative to that from the surface or near-surface region.

10.2.1. Fluorescence Versus Auger Yield

From Fig. 10.3 it is apparent that the probability of exciting an electron from level A, that is, the absorption coefficient, is equal to the probability of creating a core hole in shell A. Therefore, any process whose statistical average is proportional to the annihilation of the core hole is also proportional to the absorption coefficient and can be used as a measure of EXAFS (43, 44, 87). The core hole in shell A can be annihilated by a radiative or nonradiative electronic transition from a lower binding energy level B. Radiative transitions result in fluorescent x-ray emission, while Auger electrons are produced in the nonradiative process. For K-edges ($1s$ initial state) the nonradiative Auger channel is strongest for low-Z atoms and decreases at the expense of the radiative x-ray fluorescence channel with increasing Z (Fig. 10.4 and Table 3.1 in Chapter 3) (90). Both channels have equal probability for $Z \approx 30$ (Zn). For L_3 ($2p_{3/2}$ initial state), L_2 ($2p_{1/2}$ initial state), and L_1 ($2s$ initial state) edges the Auger process dominates up to $Z = 90$ as shown in Fig. 10.4 (91).

In comparing various detection schemes it is useful to establish a criterion of merit. In the following we shall assume that the "surface" signal arises from a thin adsorbate layer and the "background" signal originates from the underlying bulk substrate consisting of different atoms. Let I be the incident-flux-normalized (see Section 10.3) total count rate and I_B be the background count

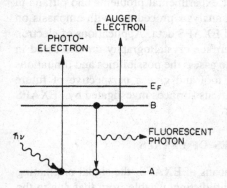

Figure 10.3. Schematic diagram of a photon absorption process by a core electron and the annihilation processes of the created core hole.

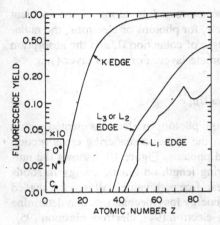

Figure 10.4. Fluorescence yield for K- and L_1-, L_2-, and L_3-edge excitations as a function of atomic number. From Refs. 90 and 91.

rate at a photon energy *above* the absorbate edge. I_B is typically obtained by extrapolation of the pre-edge signal. The strength of the surface signal I_s is then characterized by the *edge jump* $J = I_s = I - I_B$. The *edge jump ratio* $J_R = I_s/I_B$ is a measure of the surface-to-bulk or *signal-to-background ratio*. The quantity to be maximized is the *signal-to-noise ratio* $S_N = I_s/\sqrt{I}$. With these definitions

$$S_N = \left(\frac{I_s}{1 + I_B/I_s}\right)^{1/2} = \left(\frac{J}{1 + 1/J_R}\right)^{1/2} \tag{1a}$$

For integrating detection schemes the noise is best estimated from the experimental data. If we denote the noise-to-signal ratio above the edge by N_R one should optimize

$$S_N = \frac{J_R}{N_R} \tag{1b}$$

Equation (1) is the signal-to-noise ratio of the *total* surface signal. For EXAFS oscillations of amplitude ΔI_s the corresponding signal-to-noise ratio would be $S_N = \Delta I_s/I_s$. However, since for a given system the EXAFS amplitude is a fixed fraction of the jump (i.e., of I_s) the ratio of $\Delta I_s/I_s$ is a constant and in the following discussion we shall therefore use Eq. (1) as the criterion of merit. Thus the optimum detection technique is determined by a large surface signal I_s and a small background signal I_B.

The *surface* signal strength I_s can be expressed in terms of the incident x-ray intensity I_0, the detection efficiency ε_s for photons or electrons, the radiative or nonradiative yield ω_s, the solid angle of collection Ω, and the absorption coefficient μ_s and thickness x of the adsorbate layer. For a thin layer ($\mu_s x \ll 1$) we obtain (92)

$$I_s = \varepsilon_s I_0 \omega_s \Omega \mu_s x \qquad (2)$$

The *background* signal I_B originates from photons or electrons created within a depth L of the substrate. L depends on the inelastic scattering cross section and differs considerably for electrons and photons. Figure 10.5 shows the universal dependence of the electron scattering length on kinetic energy in solids (93). For most materials the measured mean free path values fall into the shaded area in Fig. 10.5. The dominant electron energy loss mechanisms that determine the shape of Fig. 10.5 consist of electron–electron (94), electron–plasmon (95), and electron–phonon (96) scattering. In contrast, x-ray scattering lengths are much larger and in the photon energy range of interest for SEXAFS (>250 eV) they are determined by the photoelectron excitation cross sections (54, 55).

In comparing values for L for electron versus photon detection schemes, one has to bear in mind that for electrons L is not simply determined by Fig. 10.5. Instead, the background is usually produced by *inelastically scattered electrons*. The effective electron escape depth L is then a complicated sum of electron scattering lengths depending on the details of the cascading process due to inelastic scattering of the high-energy primary Auger and photoelectrons inside the sample. Electrons that are excited at a distance larger than L below the surface will have thermalized their energy before reaching the surface and will

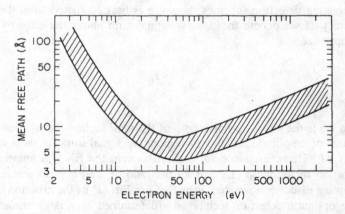

Figure 10.5. Electron mean free path (scattering length) in solids as a function of the kinetic energy of the electron.

therefore be unable to overcome the work function and escape into vacuum. L has been estimated by Gudat (59) in the 50–150-eV spectral range to be less than 50 Å for metals and semiconductors. In insulators, larger values of L are possible because electron–electron scattering is less likely at low kinetic energies. Recently, Jones and Woodruff (85) investigated the sampling depth of total electron yield measurements at the aluminum K-edge ($\sim$ 1550 eV) by preparing oxide overlayers of various thicknesses on bulk aluminum metal. Their results, shown in Fig. 10.6, give a value of $L = 65$ Å for the effective sampling depth in aluminum metal and $L = 130$ Å for aluminum oxide. At larger excitation energies, L increases because of the higher kinetic energy of the primary photoelectrons and Auger electrons. At the K-edge of copper ($\sim$ 8980 eV), L was estimated to be 500–2000 Å for copper metal by Martens et al. (61, 97).

For electron detection at x-ray incidence angles larger than the critical angle of total reflection the background signal, I_B can be written (58, 59)

$$I_B^e = \varepsilon_B^e I_0 g\Omega \left(\frac{\mu_B L}{1 + \mu_B L} \right) \tag{3}$$

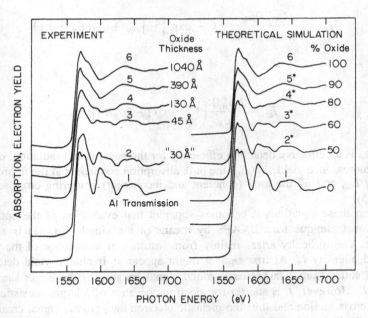

Figure 10.6. Left-hand side shows experimental EXAFS spectra taken in transmission from a 1-μm aluminum foil (1) and in total yield from the same foil (2) and from foils on which various thicknesses of anodic oxide have been deposited. On the right-hand side simulations of spectra (2*– 5*) are shown, which are derived from incoherent mixtures of spectra 1 and 6 with the percentages of oxide spectrum (6) shown. From Jones and Woodruff (85).

where ε_B^e is the electron detection efficiency, g is a smoothly varying function with photon energy describing the inelastic scattering process, and μ_B is the bulk x-ray absorption coefficient. Equation (3) arises from the fact that x-rays are absorbed by matter primarily by electron excitation and thus the number of created electrons must be proportional to the number of absorbed photons. At photon energies used for EXAFS (>200 eV) the x-ray penetration (scattering) length $1/\mu_B$ will always be much larger than the electron escape depth L. With $\mu_B L \ll 1$, Eq. (3) simplifies to

$$I_B^e = \varepsilon_B^e I_0 g \Omega L \mu_B \tag{4}$$

For photon detection the background signal arises from *fluorescent* (fl) and *scattered* (sc) photons within the bulk substrate (87, 98) of thickness d.

$$I_B^p = \varepsilon_B^p (I_{fl} + I_{sc}) \tag{5}$$

where

$$I_{fl} = \frac{I_0 \omega_B \Omega \mu_B}{\mu_B + \mu_B(E_{fl})} \left(1 - \exp\left\{-\left[\mu_B + \mu_B(E_{fl})\right] d\right\}\right) \tag{6}$$

and

$$I_{sc} = \frac{I_0 \mu_B^* \Omega}{2\mu_B} \left[1 - \exp\left(-2\mu_B d\right)\right] \tag{7}$$

Here ε_B is an effective detection efficiency for the background radiation, ω_B the bulk fluorescence yield, $\mu_B(E_{fl})$ the bulk absorption coefficient at the fluorescent energy E_{fl}, and μ_B^* the total (coherent and incoherent) scattering cross section (54, 55).

From these equations it becomes apparent that evaluation of the optimum detection technique for SEXAFS by means of the simple Eq. (1) is all but simple. The difficulty arises mainly from insufficient knowledge of the background intensity I_B. At first sight it might appear as if photon yield detection should suffer from a larger background signal because of the larger sampling depth L. However, L is shorter for electrons because of a larger inelastic scattering cross section and thus the inelastic electron background signal created in a short depth L may well be as large or larger than the background photon signal from a larger depth L. The background intensity in fluorescence detection is dependent not only on the detection scheme (98) but also on the angle of x-ray incidence (88) and the photon energy (89). If we assume for lack of better knowledge $I_B^e = I_B^p$, the optimum detection technique is determined by I_s [Eq. (2)]. Since I_0, μ_s, and x are not dependent on the detection mode, the nonradiative to radiative signal ratio is given by

$$\frac{I_s^e}{I_s^p} = \frac{\varepsilon_s^e \omega_s^e \Omega^e}{\varepsilon_s^p \omega_s^p \Omega^p} \tag{8}$$

In general, it is easier to collect a large solid angle for electrons since charged particles can be deflected (focused) without losses, but we shall assume that both detection modes can use comparable solid angles $\Omega^e \approx \Omega^p$. For K-edges the nonradiative versus radiative yield ratio ω_s^e/ω_s^p clearly favors the electron yield mode for $Z < 30$ (Fig. 10.4) (90). For L_1-, L_2-, or L_3-edges, $\omega_s^p < 0.5$ for $Z < 90$ and the electron yield mode is favored for all elements of interest (91). ε describes the detection efficiency and is determined by the detector system. For low-Z atoms the fluorescence energies and the Auger kinetic energies lie in the 200–4000-eV range. For electron multiplier detectors in this range photon detection efficiencies are typically $<20\%$, while electron detection efficiencies are 100% (99). However, photon detectors with 100% detection efficiency have recently been developed even in the soft x-ray region (100). For $Z \geq 20$ ε_s^p and ε_s^e will be comparable. In summary, the electron detection scheme will usually be superior for all cases except for K-edge SEXAFS on elements with $Z > 30$, in which case photon and electron detection are comparable. The above arguments and the convenient application of ultrahigh-vacuum compatible electron detectors has led to the preferential use of electron yield detection for SEXAFS measurements.

Finally, another advantage of the electron yield technique should be mentioned. From Eqs. (2) and (3) it is seen that the electron signal is always proportional to the absorption coefficient. Therefore, this mode can also be used for bulk samples such as model compounds. As we shall see later, it is usually important to measure both the unknown surface system and the corresponding bulk standard under the same experimental conditions. For fluorescent detection the measured signal for a thick sample ($d \rightarrow \infty$) is no longer proportional to the absorption coefficient as seen from Eq. (6). For high-Z materials this does not present a problem because one can use a thin film sample of thickness $d \ll 1/\mu_B$. However, for low-Z atoms $1/\mu_B \approx 2000$ Å and it becomes difficult to satisfy the criterion $d \ll 1/\mu_B$ experimentally.

10.2.2. Electron Yield Detection

Before we discuss the different electron yield detection schemes, some general statements can be made about the applicability of electron yield SEXAFS. Because of the size of the sampling depth L, the electron yield signal from the surface will, in general, contain only a small fraction (1–10%) from the outermost surface layer. Thus for a sample consisting of atoms B as shown in Fig. 10.7 (e.g., a clean metal surface), the effective escape depth of approximately 50 Å implies that the electron yield EXAFS above an appropriate absorption edge of atoms B will give *bulk* information. In general, *electron yield EXAFS studies are not suited for the study of clean surfaces*. From Fig. 10.5 it is seen that the surface sensitivity can be enhanced by detecting elastic Auger electrons with kinetic energies around 50 eV. However, this procedure, which was applied by Bianconi and Bachrach (63) for aluminum metal and by Comin et al. (101) for amorphized silicon, severely restricts the EXAFS range available above the absorption edge. The surface sensitivity can also be enhanced by collecting only electrons at *grazing emission angles* (102). In practice, this is most effective for the elastic Auger yield, since refraction of electrons at the surface is only well defined for elastically emitted electrons (103). For inelastically emitted electrons, especially at low kinetic energy, a grazing electron propagation direction outside the crystal cannot necessarily be related to a grazing (and therefore surface enhancing) propagation inside the crystal. At present this grazing collection procedure has not been applied for SEXAFS.

The power of SEXAFS lies in the study of adsorbate atoms A, which differ from the substrate atoms B, as illustrated in Fig. 10.7. Now we can distinguish

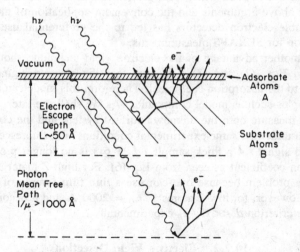

Figure 10.7. Photoabsorption and electron production in a solid consisting of substrate atoms B and an adsorbate layer A. Only electrons originating within a depth L from the surface will contribute to the total electron yield signal. Electrons originating from layer A serve as the signal for SEXAFS studies, those from atoms B give rise to unwanted background. From Ref. 53.

the EXAFS of atoms A if atoms B do not exhibit an absorption edge in the energy range of interest. The measured electron yield will have a distinguishable signal from atoms A, which is superimposed on the background due to atoms B. This is shown in Fig. 10.8a for a sample of about two monolayers of oxygen on Ni(100) (62). The small structure in the clean Ni(100) electron yield around 560 eV reflects a modulation in the monochromator transmission function (gold N_3-edge). This structure is divided out in the normalized yield spectrum in Fig. 10.8b, which now clearly shows the EXAFS above the oxygen K-edge of the oxygen atoms on the Ni(100) surface. The surface signal to bulk background ratio depends on the electron yield detection mode as discussed in detail later. The optimum detection mode maximizes the signal-to-background ratio (i.e., the edge jump ratio) and the total signal rate according to Eq. (1).

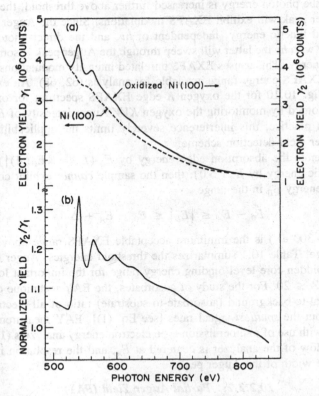

Figure 10.8. (a) Total electron yield spectra of a clean Ni(100) crystal (dashed line) and with 2–3 monolayers of NiO on the surface (solid line), (b) Ratio curve [solid/dashed line in (a)] exhibiting the EXAFS of the surface NiO layer. From Stöhr (62).

10.2.2.1. Elastic Auger Yield (EAY)

The most direct way to measure the absorption coefficient of surface atoms is by detection of elastically emitted Auger electrons (36, 43, 44). Here the intensity of a characteristic Auger transition corresponding to the core excitation of interest is monitored as a function of photon energy. Figure 10.9 illustrates schematically the energetics of electron production following excitation of a core electron. The system is assumed to consist of two core levels A and B and the valence band (VB) with binding energies (E_B) (104–106) relative to the Fermi level (E_F) as indicated. The goal is to measure the SEXAFS above the absorption edge corresponding to level A. As the photon energy is increased from below [$h\nu_1 < E_B(A)$] to just above the absorption edge [$h\nu_2 > E_B(A)$], the creation of the core hole in shell A gives rise to a characteristic Auger transition. The Auger electrons emerge from the sample with a kinetic energy E_A (9). As the photon energy is increased further above threshold, the intensity of the Auger peak will exhibit EXAFS modulations. Since the Auger electrons have a fixed kinetic energy, independent of $h\nu$, and the direct photoemission peaks move with $h\nu$ the latter will sweep through the Auger peak at some photon excitation energy. This causes EXAFS unrelated intensity modulations and thus limits the EXAFS energy range available for analysis (62, 68). An example is shown in Fig. 10.10 for the oxygen K-edge EXAFS spectrum for oxygen on Ni(100) recorded by monitoring the oxygen KVV Auger intensity of $E_A = 507$ eV (62). In practice, this interference severely limits the applicability of the elastic Auger yield detection scheme.

If we denote the absorption edge energy by E_0 ($E_0 \approx |E_B(A)|$) and the Auger kinetic energy by E_A (>0), then the sample *cannot* exhibit core levels of binding energy E_B in the range

$$E_0 - E_A \leq |E_B| \leq E_0 - E_A + \Delta \tag{9}$$

where Δ (~ 300 eV) is the minimum acceptable EXAFS range above the absorption edge. Table 10.1 summarizes the threshold energies, Auger energies, and the forbidden core-level binding energy range for the important low-Z elements $6 \leq Z \leq 20$. For the study of adsorbates, the EAY technique offers the *largest* signal-to-background (adsorbate-to-substrate) ratio of all electron yield techniques but the *smallest* signal rates [see Eq. (1)]. EAY measurements are carried out with use of photoemission-type electron energy analyzers (107). The energy window of the analyzer is centered at E_A, and the resolution is chosen to match the width of the Auger peak.

10.2.2.2. Partial Auger Yield (PAY)

Most of the elastic Auger electrons will suffer a loss of energy on their way out of the solid through inelastic scattering. Because the primary Auger kinetic

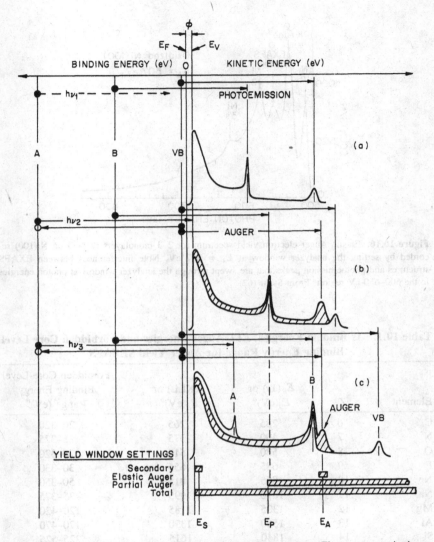

Figure 10.9. Energy level diagram and schematic photoemission spectra. The energy zero is chosen at the Fermi level (E_F) which lies below the vacuum level (E_v) by the work function ϕ. (a) $h\nu$ below the excitation threshold of core level A, (b) just above the absorption threshold but below the photoemission threshold, (c) far above threshold. The contributions to the photoemission spectra from different photoelectrons and Auger electrons are indicated. At the bottom the window settings for various electron yield detection techniques are shown. From Ref. 109.

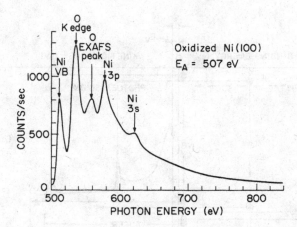

Figure 10.10. Elastic Auger electron yield spectrum for 2–3 monolayers of NiO on Ni(100) recorded by setting the analyzer window at $E_A = 507$ eV. Note interferences between EXAFS structures and photoemission peaks that are swept through the analyzer window at photon energies in the 500–650-eV region. From Stöhr (62).

Table 10.1. **1s Binding Energies, *KLL* Auger Energies and Forbidden Core Level Binding Energy Range for Auger Yield SEXAFS**

Element	Z	$E_B(1s)$ or $E_0(eV)^a$	$E(KLL)$ or $E_A(eV)^a$	Forbidden Core-Level Binding Energy Range $(eV)^b$
C	6	285	265	20–320
N	7	400	375	25–325
O	8	530	510	20–320
F	9	685	655	30–330
Ne	10	865	815	50–350
Na	11	1070	995	75–375
Mg	12	1305	1185	120–420
Al	13	1560	1390	170–470
Si	14	1840	1615	225–525
P	15	2150	1850	300–600
S	16	2470	2115	355–655
Cl	17	2825	2390	435–735
Ar	18	3205	2675	530–830
K	19	3610	2980	630–930
Ca	20	4040	3305	735–1035

^{a}Values rounded to 5-eV increments.
^{b}Assuming a minimum EXAFS energy window of 300 eV past the absorption edge.

460

energy is independent of $h\nu$, the scattering channels are also $h\nu$ independent. Therefore, the scattered Auger electron intensity (shown dashed Fig. 10.9) will exhibit the same EXAFS intensity modulations with $h\nu$ as the elastic Auger peak. This fact is utilized in the partial Auger electron yield variant (70) where only electrons of kinetic energy larger than a cutoff energy E_p are detected, as indicated in Fig. 10.9. The elastic and part of the scattered Auger electron intensity serve as the signal. The signal-to-background ratio is smaller than for the EAY detection mode because of a diminished contrast for the inelastically scattered portion of the signal (see Fig. 10.9). However, the signal rate is enhanced.

If E_p (>0) is the cutoff energy of the partial yield window, the forbidden core level binding energy range of the sample to be studied by PAY is

$$E_0 - E_p \leq |E_B| \leq E_0 - E_p + \Delta \tag{10}$$

This criterion resembles that for the EAY technique [Eq. (9)], but the important difference is that E_p can be chosen such that interference with photoemission peaks is avoided. Here photoemission peaks are not allowed to enter the partial yield window for a photon energy range Δ above the absorption edge. Usually E_p is chosen to be a few hundred electron volts such that the elastic photoemission peak associated with the absorption edge of interest (i.e., core level A in Fig. 10.9) never contributes to the measured signal. As will be discussed later, the direct photoemission intensity also exhibits intensity oscillations, which may differ from EXAFS. The PAY technique is especially suited for low-Z adsorbates since it offers good signal-to-noise ratios and avoids the interference problems associated with the EAY technique. PAY measurements are carried out with retarding electron detectors where the retarding potential (e.g., applied to a high transmission grid) is set to $-E_p$.

10.2.2.3. Total Yield (TY)

In total electron yield measurements, all electrons emitted from the sample are collected. The TY from a sample is composed of elastically emitted Auger electrons, photoelectrons, and inelastically scattered electrons. As first pointed out by Liebsch (108) and discussed in detail by Lee (43), the *elastic* photoemission intensity will, in general, not exhibit the same modulations (i.e., EXAFS) as the elastic Auger electron intensity. At first sight, this may seem surprising, since it is the scattering of the photoelectrons that gives rise to the EXAFS structure. However, opposite to the Auger electrons, the kinetic energy of the photoelectrons changes with photon energy. For nonisotropic systems like crystal surfaces, this may cause EXAFS unrelated intensity variations because of changes in the angular electron distribution seen by the detector and in the fraction of electrons lost in inelastic scattering events. Only in the limit of complete angular averaging would one expect a one-to-one correspondence between the modulation of the *elastic* photoemission intensity and those of the absorption coefficient (43).

In practice, however, only a small fraction of the photoelectrons (and Auger electrons) escapes into vacuum without suffering inelastic scattering such that the total yield (especially at the high photon energies used for EXAFS) is dominated by inelastically scattered electrons. This leads to an effectively angular-averaged (inelastic) photoelectron contribution to the TY, even for single crystal surfaces. More importantly, since the elastic photoelectrons have a smaller kinetic energy than the elastic Auger electrons (see Fig. 10.9) and the production of inelastically scattered electrons increases with the primary kinetic energy, *the total yield signal is usually dominated by inelastically scattered Auger electrons*. This is the main reason why it can be used for SEXAFS.

A detailed comparison of the EXAFS oscillations measured in transmission and by means of total electron yield detection has been made by Martens et al. (61, 97) for the K-edges of nickel (8335 eV), copper (8980 eV), and germanium (11,105 eV). While the frequency of the oscillations was found to be identical these authors reported a *reduced* amplitude (up to 30% for copper) for the TY relative to the transmission EXAFS. Recent experimental results for the aluminum K-edge (1560 eV) by Stöhr et al. (109) show that the EXAFS phase is identical for TY and EAY measurements on clean aluminum metal. The amplitude is found to be identical near the edge with an increasing smaller TY amplitude with increasing energy above the edge, corresponding to a 15% reduction at $k = 8$ Å^{-1}, relative to the EAY. Differences between transmission and electron yield EXAFS results have been proposed by theoretical studies of Noguera and Spanjaard (110). Starting from similar considerations Stöhr et al. (109) showed that the observed experimental differences can be attributed to the increasing importance of the photoelectron contributions to the TY with increasing energy above the edge. As seen in Fig. 10.9 the elastic photoemission peak A increases in kinetic energy and with it the number of inelastically scattered photoelectrons until the photoelectron and Auger electron contributions become comparable at photon energies well above the edge. It can be shown (109) that the inelastically scattered photoelectrons do not contribute to the EXAFS because the coherence between the outgoing and backscattered electron waves is lost. They do however contribute to the total electron signal from the adsorbate. Since the EXAFS is defined relative to the atomic absorption above the edge the inelastic photoelectron contribution causes a reduction of the effective EXAFS amplitude.

The TY is the most generally applicable electron yield technique. For surfaces, the EAY and TY detection modes offer many of the same advantages and disadvantages as the fluorescence technique (87) with and without photon energy discrimination, respectively (111, 112). TY offers higher count rates but a reduced signal-to-background ratio as compared to EAY detection. Experimentally, TY measurements are especially simple and consist of collecting the electrons from the sample by a biased ($+20$ to $+200$ V) electron multiplier (113).

10.2.2.4. Secondary Yield

Measurement of the low-energy inelastic portion of the total yield is another variant of electron yield SEXAFS. Such measurements were originally performed because conventional electron energy analyzers as used for photoemission could conveniently be utilized (62, 67, 68). As shown in Fig. 10.9 the energy window of the analyzer is positioned in the inelastic tail of the photoemission spectrum, at an energy E_s above the Fermi level.

Since, at the photon energies considered here, the number of elastically emitted Auger and photoelectrons is negligible as compared to the number of inelastically scattered ones, the secondary yield (SY) and total yield spectra will be identical. For both techniques, the absorption edge is observed because the Auger channel opens up at the threshold of the core transition, giving rise to elastic and, more important, inelastically scattered Auger electrons. As seen from Fig. 10.9 the core line A will be swept through the analyzer window at an energy $E_0 + E_s$, just above threshold. However, for small values of E_s (~ 10 eV), the core transition strength will still be small (114–116), and the photoemission intensity will be negligible as compared to the large intensity of the inelastic tail. Therefore, in practice, the secondary yield spectrum does not deviate from the total yield spectrum. Because of the higher obtainable count rates, the TY is generally preferable to the SY mode.

10.2.2.5. Optimization Criteria

The question arises, which electron yield detection technique should be used for a given sample. In the following we shall outline some basic selection criteria.

For a given sample the first point to be checked is whether there is an *interference problem between the various absorption edges*. This is ultimately the most crucial criterion since it determines whether a sample can be studied by SEXAFS or not. Here two cases can be distinguished. If the EXAFS energy range above the adsorbate edge of interest is limited (<300 eV) by another absorption edge the sample cannot be studied by EXAFS. An example is shown in Fig. 10.11 for the L-edges of palladium metal (117). Here the L_3 EXAFS range is limited to 160 eV by the occurrence of the L_2-edge. Figure 10.11 also represents an example of the opposite situation, namely, that an absorption edge

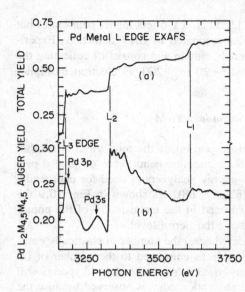

Figure 10.11. EXAFS spectra of the L-edge region of palladium metal recorded (a) in the total electron yield mode and (b) in the Auger yield mode with the window set on the $L_2M_{4,5}M_{4,5}$ palladium Auger line at 2600 eV (see Fig. 10.12). In (b) the palladium $3p$ and $3s$ photoemission peaks are swept through the Auger window just before the L_2-edge. From Stöhr and Jaeger (117).

of interest (e.g., the L_2-edge) falls just above another absorption edge of the sample (e.g., the L_3-edge). This situation can be salvaged by using the EAY technique. The trick is that the Auger line of a higher-energy edge (e.g., L_2) will usually fall at a higher kinetic energy than that of the preceding edge (e.g., L_3) as shown for palladium metal in Fig. 10.12. Thus by collecting the Auger electrons (e.g., the $L_2M_{4,5}M_{4,5}$ peak) corresponding to the higher-energy (e.g., L_2) edge the lower-energy (e.g., L_3) edge will not be seen. Figure 10.11b shows this case. The structure before the L_2-edge is mainly caused by the palladium $3p$ and $3s$ photoemission peaks that are swept through the Auger window just before the edge. Note, however, that if the L_3 EXAFS was to be measured the L_2-edge cannot be eliminated by collecting the $L_3M_{4,5}M_{4,5}$ Auger electrons because the $L_2M_{4,5}M_{4,5}$ energy is higher and thus inelastically scattered $L_2M_{4,5}M_{4,5}$ electrons will contribute (as background) to the measured $L_3M_{4,5}M_{4,5}$ signal. In fact, the L_3-edge in Fig. 10.11b is still weakly present because some L_3-edge related Auger peaks fall above the measured $L_2M_{4,5}M_{4,5}$ Auger peak. This selective Auger detection procedure does not have to be applied if the EXAFS of the preceding edge has a different frequency and weak amplitude over the EXAFS range of the higher-energy edge (118). Since the spacing between the EXAFS oscillations varies with the square root of the energy above the edge there will be significant difference in the EXAFS phases if two edges are separated by more than 100–200 eV. If the EXAFS of the preceding edge is not eliminated it will show up as a peak at low distance in the Fourier transform of the higher-edge EXAFS signal. Often this spurious

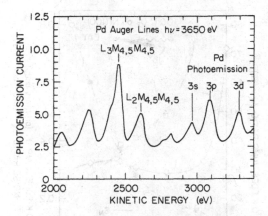

Figure 10.12. Photoemission spectrum recorded for clean palladium metal in the region of the *LMM* Auger electron peaks for $h\nu = 3650$ eV.

low-R peak (~ 1 Å) does not interfere with the real distance peaks at larger R (~ 2 Å) and can then be eliminated by a window function (see in the following section). Such considerations lead to the rules that L_3-edges can only be used for $Z > 51$ (Sb) and L_2-edges can be studied for $Z > 45$ (Rh) by EAY. L_1-edges are not very suitable for SEXAFS because of the small edge jump. Also, they cannot be enhanced by EAY detection because no strong characteristic L_1MM Auger line exists due to the dominance of the intrashell (*Coster–Kronig*) relaxation channel.

The second step in selecting the appropriate detection technique is to check for a possible *interference between photoemission peaks and the EAY or PAY window* by means of Eqs. (9) and (10). This is especially important for low-Z atoms as summarized in Table 10.1.

The next criterion of choice between all possible detection modes is the optimum *signal-to-noise ratio* S_N [Eq. (1)]. This is best determined experimentally by measurement of the absorption edge jump at the adsorbate edge. The measured edge jump J (i.e., the signal I_s) and the edge jump ratio J_R (i.e., the signal-to-background ratio) are used to calculate S_N. This is shown in Fig. 10.13 for a sample consisting of one monolayer oxygen on Ni(100). The oxygen K-edge jump ratio is lowest for the TY mode ($E_p = 0$) and highest for the EAY mode ($E_A = 508$ eV), while the K-edge jump or surface signal strength shows the opposite behavior. This trade-off between signal-to-background and the signal strength (counts per second) is also seen by comparing Figs. 10.8a and 10.10 for a thicker oxide layer on Ni(100) (62). The resultant signal-to-noise ratio is found to be highest for the TY mode ($E_p = 0$) and rather independent of the partial yield retardation voltage for $|V_R| > 100$ eV. Unfortunately, we

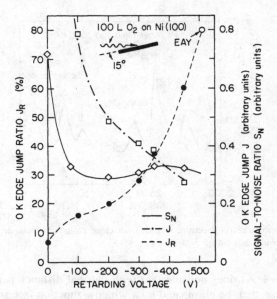

Figure 10.13. Oxygen K-edge jump, jump ratio, and signal-to-noise ratio for 100-L oxygen on Ni(100), corresponding to roughly one monolayer of oxygen, as a function of analyzer retarding potential. The retarding potential applied to the second grid of a simple electron detector (see Fig. 10.18) prevents all electrons with kinetic energy smaller than the potential to be detected. The point labeled EAY was recorded by monitoring only oxygen Auger electrons by means of a cylindrical mirror analyzer.

did not measure the EAY spectrum under the same experimental conditions and thus the difference in count rate (i.e., J) is not known. However, the EAY mode cannot be used for SEXAFS anyway because of interference between photoemission peaks and the Auger window (see Fig. 10.10).

From a signal-to-noise point of view the TY mode would clearly be favored for one monolayer oxygen on Ni(100). However, from Fig. 10.13 it is seen that the edge jump ratio or *signal-to-background* ratio is only $J_R = 7\%$ for this case. If we assume that the EXAFS amplitude ΔI_s is approximately 5% of the edge jump the SEXAFS signal-to-background ratio $S_B \approx J_R \Delta I_s / I_s$ is only 0.35%. This ratio needs to be compared to the amplitude of structures that are left over from insufficient normalization of often sizable structures in the monochromator transmission function (e.g., absorption structures due to the beamline optics, see Section 10.3). If we define a *normalization factor* $N_Q = \Delta I_N / I_N$ where I_N is the incident-flux normalized signal in an "empty cell" type experiment (e.g., the signal from a clean surface without the adsorbate) it is clear that the SEXAFS signal is only well defined for $S_B \gg N_Q$. Hence in cases of finite N_Q it is important to increase the signal-to-background ratio at the expense of the signal-to-noise ratio. As will be discussed later, this is in fact necessary for submonolayer low-Z adsorbates, such that the PAY technique with a retardation voltage of a few hundred volts (see Fig. 10.13) needs to be employed (70, 75).

Finally, it is seen from Eqs. (1), (2), and (4) that the signal-to-noise ratio is proportional to $\sqrt{I_0}$, where I_0 is the incident photon flux. Ultimately, therefore, with the future availability of significantly larger (factor 10^2–10^4) flux levels at improved storage rings (wigglers, undulators) (42) and with optimally coupled monochromators the detection mode with the largest signal-to-background ratio will be most favorable (i.e., EAY or PAY).

10.2.3. Ion Yield Detection

Ion desorption from surfaces following *electron* excitation, commonly called electron stimulated desorption (ESD) (32, 33), has been studied for many years (10). The bulk of these studies were mechanistic in nature and explained the desorption in terms of direct valence excitations by electron impact that leads to repulsive ionic final states. This Menzel–Gomer–Readhead (MGR) model (32) was challenged in 1978 by Knotek and Feibelman (KF) (119, 120) who suggested that for strongly ionic or so-called maximum-valence surface complexes, the desorption was a consequence of a *core* rather than a direct *valence* electron ionization.

The KF model, illustrated in Fig. 10.14 (53) for an arbitrary adsorbate-substrate complex, links the desorption to the creation of a core hole in an adsorbate or a surface substrate atom. The core hole may be filled by an intra- or interatomic Auger transition, which as shown in Fig. 10.14, results in holes in the valence shell. This may lead to bond breaking between the adsorbate and the substrate atoms. In the KF model a *single* core hole can produce *several* holes in the valence shell because of an Auger cascading process through intermediate levels. Thus it enables larger charge changes in the valence shell than the MGR mechanism.

It is evident that the core hole can also be produced by photon rather than electron excitation. *Photon* excitation offers the advantage that the core hole production probability is largest at threshold of the core excitation (i.e., at the absorption edge) whereas, the probability for *electron* excitation increases gradually at threshold with the maximum shifted to two to three times the threshold energy (121). This fact was utilized by Knotek, Jones, and Rehn (34) who used the increased ion desorption at a core threshold following photon excitation to unambiguously prove the existence and dominance of the KF mechanism for ionic surface complexes. The work by Knotek and co-workers gave a new perspective to the field of ion desorption. It became immediately clear that as

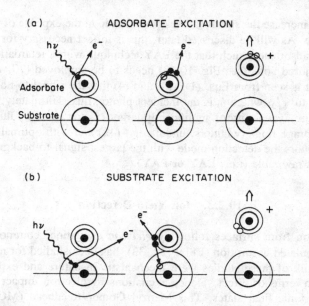

Figure 10.14. Schematic of the Knotek–Feibelman photon stimulated ion desportion process for photon absorption by (*a*) an adsorbate atom and (*b*) a surface substrate atom. Only the simplest possible case is shown. From Stöhr, Jaeger, and Brennan (53).

long as ion desorption followed the KF model, it could be directly used for surface chemistry and crystallography.

The important point is that for photon excitation the KF model directly links the number of desorbing ions to the probability that a core hole was created in the first place, that is, to the absorption coefficient of the surface atom on which the core hole was created. If the core hole is created on the *adsorbate*, the photon stimulated desorption (PSD) ion yield thus directly measures the absorption coefficient of the adsorbate atoms prior to their desorption, or if the *substrate* edge is studied, of those surface substrate atoms to which the desorbing adsorbate ions were bonded. Figure 10.15 shows an example where the O^+ ion yield for a monolayer of oxygen on Mo(100) clearly reveals the three M absorption edges of the molybdenum substrate and the K-edge of the oxygen adsorbate. The first SEXAFS measurement was performed on the same system above the molybdenum L_1-edge by Jaeger et al. (38) in 1980.

In the KF model, the ion yield γ_{ion} following the excitation of a photoelectron from a core level n is proportional to the partial absorption coefficient $\mu_n(h\nu)$ of this level according to (80)

$$\gamma_{ion} \approx (1 - f) P_n \mu_n(h\nu) \tag{11}$$

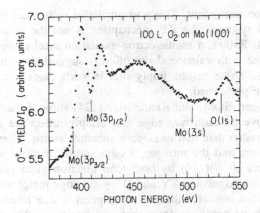

Figure 10.15. Normalized PSD O$^+$ yield in the spectral range 370–550 eV (grating: 600 lines mm^{-1}) for α-O/Mo(100). Vertical lines denote the binding energies of the 3p and 3s levels of molybdenum and the 1s level of oxygen. From Jaeger et al. (80).

Here, P_n is the probability for Auger transitions into the core hole in shell n, which result in holes in the valence region such that the surface complex is transformed into a repulsive ionic state; P_n is independent of $h\nu$ but depends on the overlap of the valence and core wave functions involved. f is the probability for reneutralization of the repelled ion either causing the recapture of the ion or its desorption as a neutral. The survival probability $(1 - f)$ is also $h\nu$ independent but depends strongly on the overlap of valence wave functions of the repelled ion and its neighbors, which leads to a high site specificity. Because of the vanishing $(1 - f)$ term for bulk atoms, the PSD signal originates exclusively from surface atoms.

From our previous discussion, it appears that ion yield SEXAFS measurements should offer the advantage over electron yield detection that the surface structure of an adsorbate complex can be investigated from both the adsorbate, as well as the substrate side by tuning to the appropriate absorption edge. Also, since the ion signal originates exclusive from the surface layer, it should, in principle, be possible to investigate the structure of clean surfaces. However, in practice, things are more complicated. Unfortunately, it has been recognized over the past years that often, especially for *covalent* surface bonds, other desorption mechanisms have comparable or even larger cross sections than the KF mechanism.

Two important mechanisms that severely limit the applicability of PSD for SEXAFS studies have been discussed in the literature. Jaeger et al. (121, 122) found that for chemisorbed diatomic molecules [CO, NO, N$_2$ on Ni(100)], the ion yield corresponding to fracture of the molecule did not follow the molecular

K-edge absorption coefficient. This was attributed to the fact that for such co-valently bonded systems, the KF mechanism is not the dominant source of desorption (123). Rather, a multielectron excitation mechanism involving core excitation coupled with valence shake off was suggested to dominate the desorption process. These results imply that in such cases ion yield SEXAFS studies cannot be performed.

Recently, Jaeger, Stöhr, and Kendelwicz (124) showed that ion yield SEX-AFS studies above the *substrate* edge may simply reflect the EXAFS of bulk substrate stoms rather than that of *surface* substrate atoms. They monitored the nickel *L*-edge jump and the nitrogen *K*-edge jump for variable-thickness am-monia multilayers on Ni(110) by both total electron yield and H^+ ion yield detection. For both edges, the TY and H^+ yield edge jumps were proportional to each other. This indicates that H^+ desorption is due to excitations by the same inelastically scattered electrons that give rise to the TY signal. Since the electrons are created either inside the nickel substrate (nickel *L*-edge) or inside the ammonia multilayer (nitrogen *K*-edge), the H^+ yield from the outermost surface layer mimics at least partially the absorption coefficient of *bulk* nickel or the *bulk* ammonia layer. Thus, photon stimulated ion desorption, in this case, is dominated by an indirect excitation channel, namely, secondary electron in-duced valence excitations in the outermost ammonia layer (e.g., by means of the MGR mechanism). This shows that at least in some cases PSD EXAFS studies above a *substrate* edge will not give the desired absorption coefficient of the surface complex, but rather simply bulk information just like the total electron yield.

Finally, another problem that at present limits the applicability of ion yield SEXAFS measurements is the usually weak signal strength. Even with the high-photon flux levels provided by synchrotron radiation (typically 10^{11} photons s^{-1}), count rates are often small for SEXAFS studies owing to the small PSD cross sections (typically 10^{-6}–10^{-9} ions per photon) (125).

10.2.4. Optical SEXAFS

The deexcitation of a core hole is accompanied by production of electrons and photons of variable energy. Besides fluorescent photons that like Auger electrons result from elastic deexcitation other low-energy photons are created. Most of these low-energy photons will originate from inelastic scattering events of the primary Auger electrons and fluorescent photons with valence electrons. Thus the energy of these photons created by electronic relaxation processes of valence electrons will typically be in the 0–30-eV range. It is exactly this spectral range in which the photon absorption coefficient of matter is largest. Thus the sampling depth for the low-energy photons may be significantly reduced relative to that for the fluorescent x-rays.

In 1979, Greegor, Lytle, and Sandstrom (86) recorded the optical EXAFS spectrum of ZnO and compared it to that in transmission. Their results demonstrated the equivalence of the EXAFS obtained by the two measurement techniques. Besides ZnO a variety of other materials were investigated. All exhibited an optical signal that followed the absorption coefficient and the signal strength was found to vary roughly with the optical escape depth.

Greegor et al. (86) saw three main advantages of the optical over the x-ray detection technique: (a) Since the signal must arise within the optical escape depth it is a near surface probe. (b) Available optics (mirrors, light-pipes) allow large solid-angle collection of optical photons. (c) In the light of the extensive use of x-ray excited optical luminescence in analytical chemistry monochromator wavelength isolation may provide signal-to-background discrimination. At present, however, optical EXAFS is not more than an interesting conceptual technique for surface crystallography.

10.2.5. Total Reflection SEXAFS

The experimental approaches to obtain surface EXAFS spectra discussed previously utilized the detection of secondary by-products. Because of the low-yield ratios ($\sim 10^{-2}$ for electrons, $\sim 10^{-7}$ for ions) per incident photon such measurements clearly require high-intensity synchrotron radiation as the excitation source. The search for a way to perform surface sensitive experiments with conventional x-ray generators was one reason for the conceptual development of total reflection EXAFS (REXAFS). The idea is simple enough. By measuring the reflectivity of a sample for incidence angles less than the critical angle for total reflection θ_c, the reflectivity is high ($> 50\%$), and the penetration depth for x-rays is low (< 30 Å). Thus, this technique combines the advantages of the high-intensity primary reflection process with the required surface sensitivity to only a few atomic layers.

REXAFS was first suggested by Denley et al. (81) who noticed that the transmission of the grazing incidence grating monochromator "grasshopper" at Stanford showed absorption structures above the carbon K-edge that closely resembled the absorption coefficient of polycrystalline graphite. Their results shown in Fig. 10.16 lead to the conclusion that the structure in the monochromator transmission function must originate from graphitelike thin film contamination layers on the totally reflecting optical elements of the beamline. Calculations indicated that it should be possible to measure the K-edge fine structure of chemisorbed monolayers of low-Z atoms.

Theoretical investigations by Fox and Gurman (82) for clean and oxidized copper films came to the same conclusion. These authors also stated that for incidence angles below the critical angle, the phase of the EXAFS oscillations measured by REXAFS is identical to that measured in transmission. Thus in-

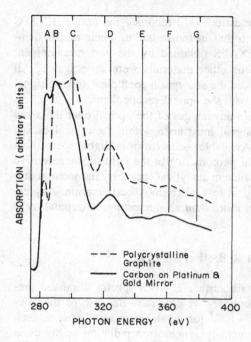

Figure 10.16. Structure around the carbon *K*-edge region in the transmission function of the grasshopper monochromator at Stanford (solid line) as compared to the absorption spectrum of a polycrystalline graphite foil (dashed line). From Denley et al. (81).

teratomic spacings should be obtainable without the need for a Kramers–Kronig analysis that normally has to be used to obtain the absorption coefficient from the reflectivity.

The first thorough measurement and analysis of REXAFS spectra were carried out by Martens and Rabe (83, 84) for the *K*-edge of copper metal. These authors pointed out [contrary to Fox and Gurman (82)] that in general REXAFS differs from the absorption EXAFS because of an angle-of-incidence dependent phase and amplitude contribution due to the *real* part (δ) of the refractive index. Because of this δ-EXAFS contribution the REXAFS fine structure differs from the absorption coefficient (which is proportional to the *imaginary* part β of the refractive index) by an additional phase shift $\tan^{-1}(f)$ and an amplitude enhancement factor $(1 + f^2)^{1/2}$. The weighting factor f determines the contribution of the δ-EXAFS relative to the β-EXAFS. It depends on the absolute values of the optical constants in the energy region above the edge, but in practice is found to be $h\nu$ (i.e., wave vector) independent. f also depends on the photon incidence angle θ. For $\theta < \theta_c$, f (and, therefore, the δ-EXAFS contribution) is small (~ 0.3) and the REXAFS and absorption EXAFS signals are similar. Figure 10.17 taken from Martens and Rabe's work (84) shows this

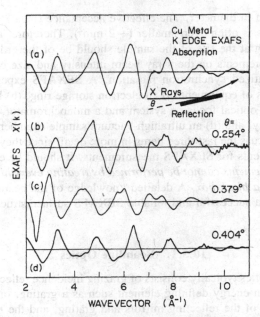

Figure 10.17. Copper K-edge EXAFS $\chi(k)$ of the absorption coefficient of copper metal versus photoelectron wave vector k and the reflection EXAFS of a copper film measured at the photon glancing angles of $\Theta = 0.254°$, $0.379°$, and $0.404°$. From Martens and Rabe (84).

similarity for $\theta = 0.254° < \theta_c = 0.33°$. However, for $\theta \gtrsim \theta_c$, f may become large and be either positive or negative leading to the phase shifts and amplitude modulations shown in Fig. 10.17 for $\theta = 0.379°$ ($f = +11$) and $\theta = 0.404°$ ($f = -1.5$). The finite contribution of the δ-EXAFS, which causes phase shift and amplitude corrections is a severe drawback of the REXAFS technique. In principle, the f weighting factor can be calculated from the optical constants. However, these are often not known well, especially in the presence of adsorbed layers.

10.3. EXPERIMENTAL DETAILS

To obtain sufficient signal-to-noise ratios in electron, fluorescence or ion yield experiments SEXAFS studies require incident flux levels of the order of 10^{10} photons s^{-1} or higher. Besides the raw monochromatic photon flux the size of the beam is important since sample sizes in surface science applications are often only 5 mm in diameter. The effective acceptance (focal) spot of electron energy analyzers is often even smaller (~ 2 mm^2). Therefore, a focused beam is needed, which at the site of the sample should be of the order of 1 mm^2 in size. Such requirements on the x-ray beam intensity and size can only be met by monochromatized synchrotron radiation. A SEXAFS experiment requires four basic pieces of equipment (a) an electron storage ring, (b) beamline optics consisting of an optical focusing system and a monochromator, (c) a reference monitor assembly, and (d) an ultrahigh-vacuum sample chamber. In the following we shall discuss the last three items in more detail since they need to satisfy certain requirements for SEXAFS measurements. It should be emphasized that *SEXAFS measurements cannot be performed by treating everything but the sample chamber as a black box*. A detailed knowledge of the beamline optics is a prerequisite for a successful and reliable SEXAFS surface structure determination.

10.3.1. Beamline Optics

The beamline optics typically consists of grazing incidence reflection optics (77, 126, 127) and an energy defining element such as a grating, or a crystal. The coating material of the reflecting mirrors and grating, and the monochromator crystals, need to be chosen carefully in order to minimize the introduction of structures into the transmitted intensity. All absorption edges and EXAFS modulations of the coating and crystal materials will be directly imposed onto the energy dependence of the reflected intensity because the absorption coefficient and reflectivity are linked through Fresnel's equations (99) (also see Section 10.2.4).

The monochromator is the heart of the beamline and is mainly responsible for the four important quantities (a) *photon flux*, (b) *spectral resolution*, (c) *spectral purity*, and (d) *spatial beam stability*. As discussed previously high-photon flux levels are needed for SEXAFS and therefore gratings and crystals should be chosen with high efficiencies or Bragg reflectivities. For SEXAFS, like EXAFS, only moderate spectral resolution (< 5 eV) is necessary. However, it should be pointed out that often the details of the absorption edge fine structure contain complementary information, as reviewed by Bianconi in Chapter 11, and thus resolution of the order of 1 eV or less is desirable.

The importance of *spectral purity* has not been sufficiently appreciated. For SEXAFS it is a crucial quantity. There are different causes for spectral impurities in the beam exiting from the monochromator. Both gratings and crystals reflect *higher orders* of the first order or principal energy. These can amount to several percent of the principal intensity. The harmonics can be suppressed by filters with an absorption edge above the first and below the appropriate higher harmonic energy or more efficiently by using a mirror reflection with a suitable cutoff energy (determined by the incidence angle and coating material) (99). For a double crystal monochromator the rocking curve (i.e., resolution) for a higher-order reflection is narrower and usually not centered on the first-order reflection. Thus, detuning of the two crystals will preferentially suppress the higher-order intensity. The problem of higher-order contamination is most difficult to deal with in the soft x-ray region (250–1500 eV) since the separation of harmonics is comparable to the length of a typical SEXAFS scan. For example, in order to carry out SEXAFS measurements above the carbon (285 eV), nitrogen (400 eV), and oxygen (530 eV) K-edges, a spectrally pure energy range from 250 to 850 eV is needed. If we choose a filter or mirror with a 850-eV cutoff energy the energy range 250–425 eV will still suffer from higher-order contamination. Thus other schemes as suggested by Hunter et al. (128) are best used.

Another source of non-monochromatic spectral contamination is *scattered light*. Again this is most problematic for the soft x-ray region where low-energy scattered photons cannot be eliminated by thick windows or filters without a significant loss of the principal monochromatic beam. For grating monochromators the grating itself is usually the main source of scattered light. For crystal monochromators we have to remember that radiation up to about 30 eV is reflected by any smooth surface. For example, in order to avoid the specular reflected and Bragg reflected radiation to be superimposed one crystal in a double crystal monochromator needs to be cut slightly asymmetrically ($\Theta_A < 1°$). The specular beam is then reflected differently by $2\Theta_A$ and can be eliminated by a collimator. Scattered light is particularly bothersome in TY measurements since low-energy electrons that may originate from low-energy photons are detected. Finally, for crystal monochromators sudden intensity changes can occur at certain energies because several Bragg reflections are satisfied simultaneously (Umwegen Regen). If these *Bragg glitches* (129) are too broad or if they are too numerous they can severely impede the data analysis. The consequences of spectral impurities will be discussed in detail in Section 10.3.4.

The *spatial stability* of the x-ray beam is very important. Because of the small acceptance spot (1–2 mm^2) of electron energy analyzers used in EAY measurements a movement of the x-ray beam caused by energy scanning of the monochromator will lead to EXAFS-like intensity changes of the measured electron yield signal. It is therefore important to carefully check the beam alignment (e.g., with a phosphor screen, see the following section) over the entire energy range of interest. For TY and PAY measurement electrons will be collected from any part of the sample that is exposed to the x-ray beam. It is therefore important that the beam be small and spatially stable such that at grazing incidence angles only the well-characterized part of the sample is exposed.

10.3.2. Reference Monitor

A typical post-monochromator experimental arrangement is shown in Fig. 10.18 (51). The monochromatic beam is first trimmed to an appropriate size by an arrangement of collimators. The collimator and a beam intensity monitor are best located in a small UHV chamber before the actual sample chamber. The beam intensity monitor is used to normalize the signal from the sample to any fluctuations, modulations, or structures of the incident x-ray beam intensity. Thus this reference monitor should (a) provide an output signal that is proportional to the number of incident photons and that is not signal-to-noise limited, (b) have a constant or at least smoothly varying quantum efficiency over the investigated energy range, (c) operate without a significant loss in transmitted intensity, and (d) not introduce any intensity modulations in the transmitted beam.

The reference monitor scheme shown in Fig. 10.18 satisfies these requirements (51). It consists of a high transmission ($\sim 80\%$) metal grid (typically 50 wires cm^{-1}), which can be coated *in situ*. The TY signal from this grid amplified by a high current channeltron electron multiplier (e.g., Galileo CEM 4716)

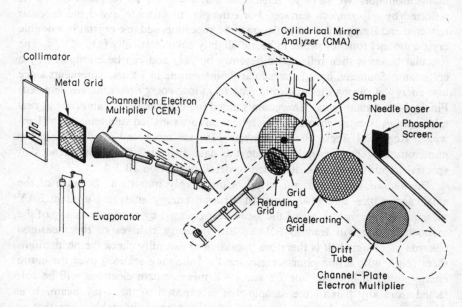

Figure 10.18. Experimental arrangement for electron and ion yield SEXAFS studies. Electrons are detected with a CMA for elastic Auger yield measurements and with a two grid retarding detector for total yield and partial Auger yield studies. Ions are detected and mass analyzed with the shown TOF detector opposite the CMA. From Stöhr, Jaeger, and Brennan (53).

Table 10.2. Coatings for I_0 Grid
Monitor

Energy Range (eV)	Material
>800	C
250–900 1,500–8,950	Cu
1,000–3,500 4,500–25,500	Ag
1,000–2,200 4,000–11,900 15,000–80,700	Au

serves as a dynamic intensity monitor. The coating material of the grid applied by *in situ* evaporation is chosen not to exhibit any absorption structures in the photon energy range of interest. Suitable materials are given in Table 10.2.

10.3.3. Sample Chamber and Detectors

The sample chamber of UHV design is equipped with the usual sample preparation and characterization tools such as sputter gun, sample cleaver, residual gas analyzer (RGA), LEED optics, gas dosing systems, evaporators, quartz oscillator (thickness monitor), and so on. A phosphor screen that can be positioned behind the sample allows one to properly focus and align the x-ray beam with respect to the focal spot of the electron energy analyzer [e.g., a cylindrical mirror analyzer (CMA)].

In our experimental arrangement (Fig. 10.18) a CMA is used for EAY measurements. The TY and PAY yield measurements are carried out with a simple retarding grid detector consisting of two hemispherical grids and a high-gain ($>10^7$) spiraltron electron multiplier (e.g., Galileo SEM 4219). For TY measurements the grids are operated at a small positive voltage ($\sim +20$ V). For PAY detection the first grid is kept at ground potential and the second grid is operated at $-E_p(\sim -300$ V). The output signal of the electron multipliers is high enough to employ current measurement techniques using a floating high voltage ($V_2 \approx +2.5$ kV) battery box (113) and a current amplifier as shown in Fig. 10.19. With a typical spiraltron gain factor of 10^7 one incident electron per second (i.e., a pulse count rate of 1 count s^{-1}) results in a spiraltron output of 10^7 electrons s^{-1} or a current of about 10^{-12} A. The instrumental noise is determined by the leakage current of the battery box and typically amounts to 1×10^{-11} A for fresh batteries. For a count rate of 10^3 counts s^{-1} the statistical

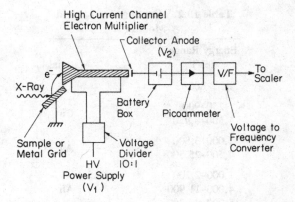

Figure 10.19. Electronics for electron yield measurements carried out in the integrating current mode. The floating battery box supplies a noise-free collector potential of 2–3 kV. From Stöhr and Denley (113).

noise is 3%. In the current mode this corresponds to about 10^{-9} A such that the instrumental noise (1%) is less than the statistical noise (3%). Thus for count rates greater than 10^3 counts s^{-1} or output currents greater than 10^{-9} A the current measurement technique can be used. The integrating current technique is advantageous not only because of its simplicity but it is not accumulation rate limited. In total yield measurements a flux of 10^{10} photons s^{-1} typically generates 10^8 electrons s^{-1} (i.e., quantum efficiency 10^{-2}) from the sample. This would correspond to 10^8 counts s^{-1}! In the current mode the spiraltron gain (i.e., the voltage V_1 in Fig. 10.19) is simply decreased until the output current is less than the manufacturer specified limit of *linear* spiraltron response ($\sim 1 \times 10^{-7}$ A for SEMs or $\sim 1 \times 10^{-6}$ A for high current CEMs). It is important to note that the statistics of the spiraltron output current is determined only by the signal from the sample and *not* the spiraltron gain (i.e., the applied HV). This is why individual SEXAFS sweeps may only be summed to obtain better statistics if the same gain (HV) is used in all sweeps.

Ion yield SEXAFS measurements are performed with a time-of-flight (TOF) detector (80) consisting of an accelerating grid, a drift tube (~ 5 cm), and a channel plate electron multiplier as shown in Fig. 10.18. Since usually *positive* ions are detected the sample is biased positively ($\sim +1.5$ kV) and the accelerating grid and drift tube are at a large negative potential (~ -1.5 kV). This insures that all ions (2π solid angle) are collected and small differences in the desorption kinetic energies can be neglected (i.e., the ion flight times are solely determined by the experimental voltages). The flatness of the channel plate furthermore ensures nearly identical ion flight path lengths such that the distribution of flight times for a given mass is not detector limited but directly reflects the storage ring pulse width. The first channel plate is operated at a larger negative potential than the drift tube to discriminate against electrons that are liberated from the drift tube wall by fluorescent and scattered x-rays.

The time-of-flight (TOF) measurements utilize the unique time structure of electron storage rings (42). Electrons travel around the ring in so-called "bunches". At the Stanford ring SPEAR it takes 780 ns for one bunch to travel around the ring. Each bunch is composed of individual "buckets" with a typical width of 500 ps (dependent on the operating conditions of the storage ring) and a separation of 2.7 ns. Thus if the ring is filled in a 3 × 3 pattern there are three equally spaced bunches (260 ns), each composed of three buckets with a total bunch width of approximately 6 ns. The TOF detector needs to be designed (80) to give a flight-time difference greater than 6 ns for two ion species that need to be discriminated. For most experiments it is important to resolve O^+ ($M = 16$) from OH^+ ($M = 17$), that is, a mass resolution requirement of $M/\Delta M = 16$ at mass 16. The ion flight times from the sample to the channel plate are measured relative to the "prompt" photon pulses of the electron storage ring. The prompt pulse in the channel plate is triggered by a scattered or fluorescent photon from the sample that reaches the detector instantly ($\ll 1$ ns). The absolute ion flight times t are then measured from the prompt pulse at $t = 0$. Furthermore, the ions can be identified self-consistently by means of the TOF relation $M_i/M_k = (t_i/t_k)^2$ where $M_{i,k}$ and $t_{i,k}$ are the masses and the flight times of the ion species i and k, respectively. A TOF spectrum of a slightly surface contaminated Mo(100) sample is shown in Fig. 10.20 (80). An ion yield EXAFS

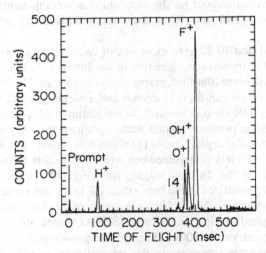

Figure 10.20. Time-of-flight spectrum at $h\nu = 550$ eV of ions desorbing from a hydrogen-contaminated Mo(100) sample after the adsorption of oxygen at 1100 K. From Jaeger et al. (80).

spectrum is recorded by monitoring the energy dependent intensity variation of a given peak in the TOF spectrum.

10.3.4. Normalization

For SEXAFS studies intensity modulations in the monochromator transmission function (68, 79) need to be suppressed such that the normalization factor N_Q is small compared to the SEXAFS signal-to-background ratio S_B (see Section 10.2.2.5). The achievable quality of normalization crucially depends on the size of the nonmonochromatic intensity component of the x-ray beam. If we measure a yield signal $Y_s = Y_s^M + Y_s^*$ from the sample and $Y_0 = Y_0^M + Y_0^*$ from the reference monitor the measured normalized yield $\gamma' = Y_s/Y_0$ will differ from the idealized normalized yield $\gamma = Y_s^M/Y_0^M$ according to (51)

$$\gamma' = \gamma + \frac{\alpha I_0^*}{I_0^M} \tag{12}$$

Here the superscripts M and $*$ stand for monochromatic and nonmonochromatic, respectively, and α depends on the difference in electron yield for the monochromatic I_0^M and nonmonochromatic I_0^* light components. The correction term $\alpha I_0^*/I_0^M$ causes problems in the normalization of structures in the incident flux ($I_0 = I_0^M + I_0^*$). The structures in I_0 arise from changes in only the monochromatic component I_0^M while the nonmonochromatic component I_0^* is essentially constant (or at least smoothly varying). Thus for a linear detector response these structures are normalized out for the monochromatic components (γ_s^M/γ_0^M) but for the measured yield ratio γ' a modulation remains because of the term $\alpha I_0^*/I_0^M$.

Figures 10.21 and 10.22 give examples of excellent and poor normalization of structures in the transmission function of the Jumbo (78, 79) and grasshopper (76) monochromators at Stanford, respectively. Figure 10.21a and 10.21b show the TY spectra of a clean Si(111) crystal and a nickel grid reference monitor. Both exhibit a $\sim 50\%$ drop in intensity in the platinum $M_{4,5}$-edge region caused by absorption in the platinum coated focusing mirror in the beamline. In addition, a monochromator crystal [InSb(111)] glitch is observed at 2125 eV. These structures are completely normalized out when the ratio is formed as shown in Fig. 10.21c and 10.21d. In fact, judging from Fig. 10.21d the platinum $M_{4,5}$-edge structure is normalized out to better than 1% in the ratio spectrum indicating a *suppression factor* $S_f = \Delta I_0/\Delta(I/I_0)$ in excess of 50. The same normalization procedure is applied in Fig. 10.22 for the SEXAFS spectrum of approximately $\frac{1}{3}$ monolayer oxygen on Ni(100) (75) using the grasshopper grating monochromator. Here there are structures in the transmission function over the whole 500–800-eV range due to the gold and platinum mirrors in the beamline (see Fig. 10.23). These structures are smaller ($\sim 5\%$) than in Fig. 10.21 but they

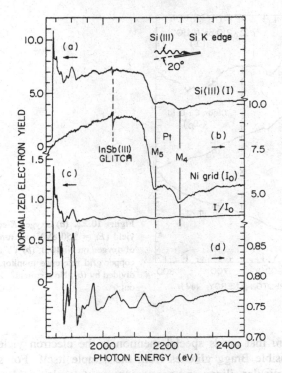

Figure 10.21. Silicon *K*-edge EXAFS recorded for a clean Si(111)7 × 7 crystal using total yield (TY) detection and InSb(111) monochromator crystals (Jumbo monochromator at Stanford). (*a*) TY from sample, (*b*) TY from nickel grid reference monitor, (*c*) ratio of (*a*) divided by (*b*), (*d*) blown up EXAFS. Note excellent cancellation of monochromator transmission structures.

remain (~1%) even after the ratio is taken. Note in particular that the broad structure X is still visible in Fig. 10.22*c*. The low suppression factor $S_f \approx 5$ indicates the presence of non-monochromatic components (~10–15%), most likely originating from imperfections of the reflection grating.

Figure 10.23 indicates how the normalization problem can be overcome. Rather than dividing the signal from the sample by the reference monitor the PAY spectrum recorded for oxygen on nickel (Fig. 10.23*a*) is divided by that for clean nickel (Fig. 10.23*b*). In both measurements the scattered light contribution is kept the same since the scattered light yield originates almost exclusively from the bulk nickel sample. Therefore, the scattered light correction in Eq. (12) cancels and the absorption structures due to platinum and gold are eliminated (Fig. 10.23*c*). Note that for $\frac{1}{3}$ monolayer of oxygen on Ni(100) the EXAFS amplitude is only about 1% thus requiring a normalization factor of the order of 0.1%, that is, a suppression factor of $S_f \approx 50$.

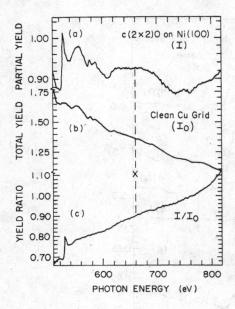

Figure 10.22. (*a*) Oxygen *K*-edge partial Auger yield (E_p = 350 eV) spectrum of about $\frac{1}{3}$ ML of oxygen on Ni(100), (*b*) TY signal from clean copper grid reference monitor, (*c*) ratio of (*a*) divided by (*b*). Note structure X does not divide out.

One problem that needs special mention is the electron yield modulation caused by possible Bragg glitches from the sample itself. For single crystal samples (in particular silicon and germanium crystals) the incident x-ray beam may satisfy the Bragg condition for high Miller index planes at a particular incidence angle and photon energy. This will cause a derivative-like glitch in the electron yield signal that can only be eliminated from the SEXAFS range under study by rotation of the sample. Such glitch problems are often a great nuisance in polarization dependent studies since they may preclude the investigation of the sample over a rather wide angular range (~20°). To minimize the problem it is advantageous to use a sample mount that allows azimuthal rotation of the sample (about the surface normal) since the condition for the Bragg reflection depends on both the polar and azimuthal sample orientation with respect to the x-ray beam.

10.4. DATA ANALYSIS

10.4.1. EXAFS Formalism

For *s* initial states (*K*- or L_1-edges) the EXAFS $\chi(k)$ for a nonisotropic system with Gaussian disorder is given by (43, 130–132):

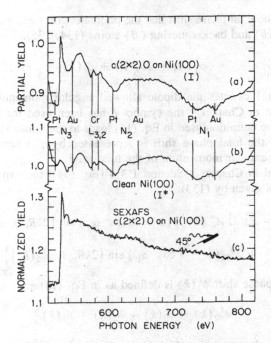

Figure 10.23. (*a*) Oxygen *K*-edge spectrum as in Fig. 10.22*a*. (*b*) spectrum recorded under same conditions as (*a*) for clean Ni(100) crystal. (*c*) ratio (*a*) divided by (*b*). Now excellent cancellation of all monochromator transmission structures exists and SEXAFS of oxygen on Ni(100) is revealed.

$$\chi(k) = -3 \sum_i \sum_j \cos^2 \alpha_{ij} A_i^*(k) \sin\left[2kR_i + \phi_1^i(k)\right] \qquad (13)$$

This equation can be obtained by combining Eqs. (28) to (34) with Eqs. (35) and (47b) of Chapter 1. The index *i* labels atoms of the same atomic number that have the same distance R_i from the absorbing atom. The index *j* refers to individual atoms in a given neighbor shell *i*. α_{ij} is the angle between the electric field vector **E** of the x-rays at the central atom site and the vector $\mathbf{r}_{ij}$ from the central atom to the *j*th atom in the *i*th shell. The amplitude parameter $A_i^*(k)$ is given by

$$A_i^*(k) = \frac{1}{kR_i^2} F_i(k) e^{-2\sigma_i^2 k^2} e^{-2R_i/\lambda_i(k)} \qquad (14)$$

$F_i(k)$ is the backscattering amplitude of the neighbor atoms and the exponential terms in Eq. (14) are the Debye–Waller-like term and the damping term due to inelastic scattering [mean free path $\lambda(k)$] of the photoelectrons. The term $\phi_1^i(k)$

in Eq. (13) is the total phase shift that the photoelectron wave experiences from the absorbing (δ) and backscattering (β) atoms (134, 135).

$$\phi_1^i(k) = 2\delta_1(k) + \beta_i(k) \tag{15}$$

Here the index 1 denotes the dipole allowed angular momentum of the final state. Note that in Chapter 1 the symbol δ has been used for the total phase shift and that the definition used in Eq. (15) leads to the minus sign in Eq. (13). In this chapter the total phase shift is represented by ϕ_ℓ where ℓ denotes the dipole allowed angular momentum of the final state.

As discussed in Chapter 1 Section 1.3.4 [Eq. (54)] the expression for L_2- and L_3-edges is given by (133):

$$\chi(k) = \sum_i \sum_j A_i^*(k)\left\{\tfrac{1}{2}(1 + 3\cos^2\alpha_{ij})\sin\left[2kR_i + \phi_2^i(k)\right]\right.$$
$$\left. + M_{02}(1 - 3\cos^2\alpha_{ij})\sin\left[2kR_i + \phi_{02}^i(k)\right]\right\} \tag{16}$$

Here the total phase shift $\phi_2^i(k)$ is defined as in Eq. (15) and

$$\phi_{02}^i(k) = \delta_0(k) + \delta_2(k) + \beta_i(k) \tag{17}$$

M_{02} is a matrix element ratio which has been calculated to be about 0.2, independent of atomic number and wavevector (134). This value has been confirmed experimentally (133).

For *isotropic* systems like cubic, polycrystalline, or disordered materials the second interference term vanishes upon angular integration and for N_i neighbors in the ith shell we obtain

$$\overline{\chi(k)} = \sum_i N_i A_i^*(k)\sin\left[2kR_i + \phi_2^i(k)\right] \tag{18}$$

Note that in this case *the EXAFS phase shift is simply given by* $\phi_2^i(k)$.

In previous L_3-edge SEXAFS studies (36, 64–66, 235) Eq. (16) has been approximated by the expression,

$$\chi(k) = \sum_i \sum_j \left[0.5 + c + (1.5 - 3c)\cos^2\alpha_{ij}\right]$$
$$\times A_i^*(k)\sin\left[2kR_i + \phi_2^i(k)\right] \tag{19}$$

where the case $c = 0$ describes the case of a negligible interference term. Alternately, $c = 0.2$ corresponds to the assumption $\phi_{02}(k) = \phi_2(k)$ and M_{02}

= 0.2 in Eq. (16). For a more general discussion see Section 10.4.2.3. By comparison with Eq. (13) for s initial states we can write a generalized EXAFS expression for *both* K- and L-edges

$$\chi(k) = (-1)^{\ell} \sum_i N_i^* A_i^* (k) \sin \left[2kR_i + \phi_{\ell}^i (k) \right] \tag{20}$$

where ℓ denotes the final-state angular momentum ($\ell = 1$ for K and L_1 and $\ell = 2$ for L_2- and L_3-edges).

The *polarization dependent* effective coordination number N_i^* is of particular importance for SEXAFS and will be discussed in detail later.

In Eq. (20) the effective coordination number N_i^* for N_i atoms in the ith neighbor shell for K- and L_1-edges is given by

$$N_i^* = 3 \sum_{j=1}^{N_i} \cos^2 \alpha_{ij} \tag{21}$$

and for L_2 and L_3 we have

$$N_i^* = (0.5 + c) N_i + (1.5 - 3c) \sum_{j=1}^{N_i} \cos^2 \alpha_{ij} \tag{22}$$

The $\cos^2 \alpha_{ij}$ term contributes to the EXAFS amplitude only if the electric field vector E has a sizable component along an internuclear axis from the absorbing to a neighbor atom. For inherently anisotropic systems like surfaces the polarization dependence of the EXAFS signal therefore provides an extremely powerful tool to sort out neighbor atoms that are located in different directions from the central atom. For K- or L_1-edges the E vector can be envisioned as a *search light* revealing all neighbors in a given direction. The angular anisotropy is weakened for L_2- and L_3-edges by the isotropic term $0.5N_i$ in Eq. (22). For isotropic systems like cubic, polycrystalline, or amorphous materials an average over all angles yields $N_i^* = N_i$. Note that we can also force the condition $N_i^* = N_i$ for certain E vector orientations (see the following pages).

For single-crystal materials with lower than cubic symmetry or in the case of oriented molecules or atoms on surfaces, Eqs. (21) and (22) need to be evaluated for an assumed model geometry. The two most often encountered surface structural units and the relevant structural parameters are shown in Fig. 10.24. The $\cos^2 \alpha_{ij}$ factor in Eqs. (21) and (22) can be expressed in terms of the angle Θ between the E vector and the surface normal and the angle β between the adsorbate–substrate internuclear axis and the surface normal. With the def-

(a) Three-Fold Hollow Site

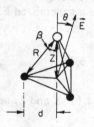

(b) Two-Fold Hollow Site

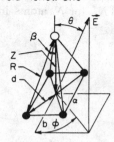

Figure 10.24. Two structural units commonly encountered for an adsorbate (white atom) on a substrate (black atom). **E** is the electric field vector, characterized by spherical coordinates Θ from the surface normal and ϕ in the surface plane. β is the bond angle with the surface normal.

inition

$$\sum_{j=1}^{N_i} \cos^2 \alpha_{ij} = f(\Theta, \beta) \qquad (23)$$

we obtain for chemisorption in an atop *site* ($N_i = 1$)

$$f(\Theta, \beta) = \cos^2 \Theta \qquad (24)$$

For the *threefold hollow site* ($N_i = 3$) pictured in Fig. 10.24(a) we obtain (52, 53)

$$f(\Theta, \beta) = 3 \cos^2 \Theta \cos^2 \beta + 1.5 \sin^2 \Theta \sin^2 \beta \qquad (25)$$

and for the *twofold hollow site* ($N_i = 4$) in Fig. 10.24b:

$$f(\Theta, \beta) = 4\left[\cos^2 \Theta \cos^2 \beta + \sin^2 \Theta \sin^2 \beta \left(\frac{a^2}{4d^2} \cos^2 \phi + \frac{b^2}{4d^2} \sin^2 \phi \right) \right]$$

$$(26)$$

Other sites of interest are special cases of Eq. (26). The *fourfold hollow site* with $N_i = 4$ is defined by $a = b$ and

$$f(\Theta, \beta) = 4 \cos^2 \Theta \cos^2 \beta + 2 \sin^2 \Theta \sin^2 \beta \qquad (27)$$

For the *twofold bridge site* characterized by $N_i = 2$ we obtain from Eq. (26) ($b = 0$, $a = 2d$)

$$f(\Theta, \beta) = 2 \cos^2 \Theta \cos^2 \beta + 2 \sin^2 \Theta \sin^2 \beta \cos^2 \phi \qquad (28)$$

Note that for all cases with higher than twofold symmetry around the surface normal we obtain $N_i^* = N_i$ for Θ or $\beta = 54.7°$, the so-called magic angle.

By inversion of Eqs. (24)–(28) we can determine the *bond angle* β (Fig. 10.24) from the experimentally derived absolute effective coordination number $N_i^*(\Theta)$ or from the ratio $P = N_i^*(\Theta_1)/N_i^*(\Theta_2)$ obtained from a polarization dependent study. In most cases the *ratio* can be determined more accurately ($<10\%$) than the *absolute value* for reasons discussed in the following section and, as an example, for the *fourfold hollow* site we obtain

$$\beta = \cos^{-1} \left(\frac{\sin^2 \Theta_1 - P \sin^2 \Theta_2}{P(3 \cos^2 \Theta_2 - 1) - 3 \cos^2 \Theta_1 + 1} \right)^{1/2} \qquad (29)$$

For a measured amplitude ratio P of 10% uncertainty β can be determined with an accuracy of about 1°.

10.4.2. Some Aspects of SEXAFS Data Analysis

10.4.2.1. Amplitude

Determination of N_i^* requires comparison of the total amplitude $N_i^* A_i^*(k)$ [Eq. (20)] with a model system, a procedure that relies on amplitude transferability. Both the Debye–Waller and inelastic loss terms in $A_i^*(k)$ [Eq. (14)] are inherently determined by the chemically sensitive valence electrons, even at larger k (>4 Å^{-1}) values. That is why even in favorable cases coordination numbers can only be determined to an accuracy of about 10% (136, 137). (See also Chapter 1, Section 1.6.) If we define a total amplitude function $A_i(k) = N_i^* A_i^*(k)$ the determination of N_i^* is usually done by plotting $\ln[A_s(k)/A_u(k)]$ as a function of k^2 (See Chapter 6, Section 6.3.5.). Here $A_s(k)$ is the amplitude function corresponding to a given neighbor shell of a standard and $A_u(k)$ is that of the system under investigation. Both are obtained from inverse Fourier transformation. From Eq. (14) we obtain

$$\ln\left[\frac{A_s(k)}{A_u(k)}\right] = \ln\left(\frac{N_s^* R_u^2}{N_u^* R_s^2}\right) + 2k^2(\sigma_u^2 - \sigma_s^2) + 2\left(\frac{R_u}{\lambda_u} - \frac{R_s}{\lambda_s}\right) \quad (30)$$

For an isotropic electron mean free path $\lambda_u \approx \lambda_s \approx \lambda$, the last term is negligible because $2|R_u - R_s| \approx 0.2$ Å $\ll \lambda \approx 5$ Å. Equation (30) also assumes that the backscattering amplitude $F_u(k) = F_s(k)$. A linear plot of $\ln[A_s(k)/A_u(k)]$ versus k^2 yields the unknown coordination number N_u^* from the ordinate intercept at $k^2 = 0$ and $\sigma_u^2 - \sigma_s^2$ from the slope.

Most problems associated with amplitude transferability are overcome by comparison of relative SEXAFS amplitudes obtained for the same sample at different angles of x-ray incidence with respect to the surface. In this case both the Debye–Waller and electron-loss terms should be similar and a plot of ln $[A(\Theta_1)/A(\Theta_2)]$ versus k^2 should give a constant for all k values. Alternatively one can simply plot the amplitude ratio versus k yielding directly the ratio of effective coordination numbers [assuming $R(\Theta_1) = R(\Theta_2)$],

$$\frac{A(\Theta_1)}{A(\Theta_2)} = \frac{N^*(\Theta_1)}{N^*(\Theta_2)} \quad (31)$$

Often, especially for K- and L_1-edges, the experimentally determined ratio of $N^*(\Theta_1)/N^*(\Theta_2)$ is sufficient to distinguish between various model geometries.

10.4.2.2. Nonseparable Neighbor Shells

In the previous discussion we have assumed that the measured distance and amplitude correspond to a well-defined neighbor shell of separation R_i and coordination number N_i. If two or more neighbor shells are separated by only a small difference in distance ($\Delta R \leq 0.3$ Å) the observed peak in the Fourier transform will contain contributions from several shells. In this case an average distance

$$R(\Theta) = \frac{\sum_i \dfrac{N_i^*(\Theta)}{R_i}}{\sum_i \dfrac{N_i^*(\Theta)}{R_i^2}} \quad (32)$$

is measured where we have assumed similar amplitude functions $A_i^*(k)$ for the contributing neighbor shells. The average effective coordination number $N^*(\Theta)$

is given by

$$N^*(\Theta) = \sum_i \left(\frac{R(\Theta)}{R_i}\right)^2 N_i^*(\Theta) \qquad (33)$$

The contribution of more than one shell to the measured EXAFS thus can result in a *polarization dependent* measured distance $R(\Theta)$ and affect the effective coordination number $N^*(\Theta)$.

10.4.2.3. Interference Term for L_2- and L_3-Edges

The EXAFS equation for L_2- and L_3-edges given by Eq. (16) contains two terms characterized by different phase shifts $\phi_2(k)$ and $\phi_{02}(k)$. In general, these phase shifts will be significantly different as discussed by Teo and Lee (134). In principle, analysis of the measured EXAFS signal could be carried out by use of calculated phase shifts $\phi_2(k)$ and $\phi_{02}(k)$. However, curve fitting procedures would have to be applied and the accuracy of calculated phase shifts is often inappropriate, especially at low k. On the other hand, the phase shift derived from a model compound corresponds to $\phi_2(k)$ [Eq. (18)] and use of this phase shift [i.e, assuming $\phi_2(k) = \phi_{02}(k)$] may lead to a derived distance that could be significantly in error ($\Delta R \geq 0.05$ Å) (138). The relative contribution of the second interference term in Eq. (16) can be assessed by comparison of the relative amplitudes. We rewrite Eq. (16) as

$$\chi(k) = \sum_i A_i^*(k) N_i \Big\{ C_2 \sin\left[2kR_i + \phi_2^i(k)\right] + C_{02} \sin\left[2kR_i + \phi_{02}^i(k)\right] \Big\}$$

$$(34)$$

where N_i is the number of atoms in the ith neighbor shell and

$$C_2 = \frac{1}{2}\left(1 + \frac{3}{N_i}\sum_{j=1}^{N_i} \cos^2 \alpha_{ij}\right) \qquad (35)$$

and

$$C_{02} = M_{02}\left(1 - \frac{3}{N_i}\sum_{j=1}^{N_i} \cos^2 \alpha_{ij}\right) \qquad (36)$$

The bothersome second term in Eq. (34) is negligible if $C_2 \gg C_{02}$. The simplest case is $C_{02} = 0$ which happens for

$$\frac{3}{N_i} \sum_{j=1}^{N_i} \cos^2 \alpha_{ij} = \frac{3}{N_i} f(\Theta, \beta) = 1 \tag{37}$$

The function $f(\Theta, \beta)$ is given by Eqs. (24)–(28) for various chemisorption geometries. For all sites with higher than twofold symmetry around the surface normal (e.g., the atop, threefold, and fourfold sites) the condition (37) is satisfied for either Θ or β being equal to the magic angle 54.7°. Thus if either the E vector is at an angle of 54.7° with respect to the surface normal or for a bond angle with respect to the surface normal $\beta = 54.7°$ the second term in Eq. (34) vanishes by symmetry. Thus, using a phase shift $\phi_2(k)$ derived from a model an unambiguous distance determination can always be carried out by choosing $\Theta \approx 55°$ for the unknown surface system. Note that for this magic angle the absolute coordination number of the adsorbate is directly determined by amplitude comparison with the model since $N^* = N$.

To obtain more general rules of when the second term in Eq. (34) needs to be considered we have plotted in Fig. 10.25 the amplitude ratio C_2/C_{02} as a function of β or Θ assuming $M_{02} = 0.2$ (138). If we fix the bond angle $\beta = 0°$ ($\beta = 90°$) the solid (dashed) curve represents the variation of the ratio C_2/C_{02} *as a function of the* E *vector orientation* Θ. For example, the solid curve is representative for the atop site and the dashed curve represents chemisorption in the surface plane. Because of the equivalence of β and Θ in the expression $f(\Theta, \beta)$ for the threefold and fourfold sites [Eqs. (25) and (27)] Fig. 10.25 can also be read to display the ratio C_2/C_{02} *as a function of the bond angle* β for $\Theta = 0°$ (grazing incidence, solid curve) and $\Theta = 90°$ (normal incidence, dashed curve). If we demand C_{02} to be < 10% of C_2 we find that the second interference term in Eq. (34) can be neglected for all Θ and β values that fall into the unshaded area of Fig. 10.26. Here we have assumed that there is higher than twofold symmetry about the surface normal. For Θ and $\beta < 42°$ C_{02} is negative and its absolute value is less than 20% of C_2. For large β (>70°) and small Θ (<30°) or vice versa the interference term can no longer be neglected since it may be 40% of that of the leading term (e.g., for $\beta = 90$ and $\Theta = 0°$). With these general considerations we can establish some guidelines for *distance* and *amplitude* corrections for the case of a nonnegligible C_{02} term.

Distance analysis with a single phase shift $\phi_2(k)$ derived from a model may lead to significant errors (~0.05 Å). This error will be largest for systems with chemisorption bonds parallel or perpendicular to the surface plane. In polari-

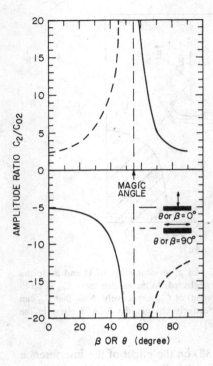

Figure 10.25. Ratio of the L_3- and L_2-edge amplitude factors C_2 and C_{02} defined by Eqs. (35) and (36) as a function of Θ and β defined in Fig. 10.26. We have assumed higher than twofold symmetry around the surface normal and $M_{02} = 0.2$. C_{02} is a measure of the interference term contribution to the $L_{2,3}$-edge EXAFS, which affects both the amplitude and phase. Note $C_{02} = 0$ for the magic angle Θ or $\beta = 54.7°$.

zation dependent studies the interference term will for such cases contribute less for **E** along the bond direction than for **E** perpendicular to it. Note that in addition C_{02} changes sign between parallel and perpendicular polarization directions that is equivalent to changing the phase shift $\phi_{02}(k)$ by π. This typically has the effect of introducing ΔR corrections of opposite sign. *Hence the derived distance will exhibit a polarization dependence.* The correct distance is always obtained if Θ is chosen to be close to the magic angle $54.7°$.

To assess the contribution of the interference term to the EXAFS *amplitude* we consider two extreme cases. First, if $\phi_2(k) = \phi_{02}(k)$ the two terms simply add and the EXAFS is given by Eq. (19) with $c = M_{02} = 0.2$. Secondly, if $\phi_2(k)$ and $\phi_{02}(k)$ are very different two peaks will result in the Fourier transform. The EXAFS corresponding to the dominant peak will then be described by Eq. (19) with $c = 0$. Hence in general the measured amplitude will be described by Eq. (19) with $0 \le c \le 0.2$. For determination of the absolute coordination number, Θ should be chosen to be close to the magic angle since all complications are eliminated and the real coordination number is directly

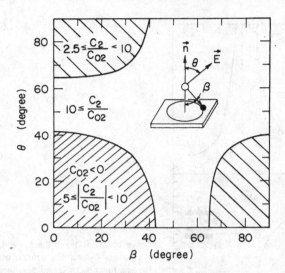

Figure 10.26. $L_{2,3}$-edge amplitude factor ratio C_2/C_{02} for all combinations of Θ and β. In the unshaded area C_{02} is less than 10% of C_2 and can be neglected. In the shaded areas C_{02} is up to 20% (lower left) and up to 40% (upper left and lower right) of C_2, respectively. Note that C_{02} can also be of opposite sign as C_2 (lower left), which effectively changes the EXAFS phase of the interference term by 180°. From Stöhr and Jaeger (138).

measured ($N^* = N$). Model calculations (138) on the effect of the interference term will be presented in Section 10.5.4.2 for the silver on Si(111) surface.

10.5. APPLICATIONS

10.5.1. Problems in Surface Crystallography

Most research in surface science has its roots in two fundamental technological problems. One of these is the understanding of reaction mechanisms at *semiconductor, metal and oxide surfaces and interfaces* and of the electronic and crystallographic structure of the products. The understanding of the interaction at solid-solid interfaces such as Schottky-barriers, heterojunctions and supported catalysts is the key issue in the successful *fabrication* of *electronic devices and efficient catalysts*. The second problem is a better understanding of the principles underlying *heterogeneous catalysts*, in particular, of the interactions and reactions between gases and solid surfaces (mostly metals and supported metals). In both areas of electronic device fabrication and heterogeneous catalysis initially largely empirical approaches were employed in the search for better and more efficient materials. The surface science approach that gradually evolved over the last 20 years tried to isolate certain key issues and study them under (idealized) experimental conditions that eliminated the interference of other simultaneous mechanisms. As mentioned in the introduction this approach served a dual purpose of not only reducing a complex real problem to a simpler idealized one but it also aided in the development of surface techniques. Clearly, in the long run, surface science research has to be able to put together the individual pieces of the puzzle and allow the whole picture to be seen. At the present time this has not been achieved and many problems are yet unanswered even for idealized systems.

In the following we shall present some areas where SEXAFS has contributed to a better understanding of existing problems. Unfortunately, and this is true for any technique, in some cases published SEXAFS work has confused rather than solved an issue. We shall discuss *clean surfaces* first and then cover *gas–solid surface reactions* and *solid–solid interfaces*.

10.5.2. Clean Surfaces, Al(111) and Al(100)

As discussed in Section 10.2.2, SEXAFS is not well suited for the determination of clean surface structures. Therefore, it is not surprising that only one such investigation has been published. Bianconi and Bachrach (BB) (63) investigated the interplanar spacing between the first and second layer of clean Al(100) and Al(111) surfaces relative to the respective spacings in bulk aluminum. Because the outermost surface layer exhibits broken bonds on one side, the plane of atoms may be expected to either move closer to the second layer (contraction) or away from it (expansion). Bianconi and Bachrach investigated the aluminum metal $L_{2,3}$-edge (72.7 eV) extended structure as shown in Fig. 10.27 and achieved a high surface sensitivity ($L \approx 3$ Å) by collecting electrons of 45-eV kinetic energy (see Fig. 10.5). By comparing the position of the EXAFS peak around $h\nu = 100$ eV (Fig. 10.27) for high ($E_k = 45$ eV) and low ($E_k = 4$ eV, and using a bulk aluminum film measured in transmission) surface sensitivity spectra, BB obtained the surface Al–Al bond length relative to that in the bulk (2.86 Å).

The spectra were recorded at grazing ($\sim 10°$) x-ray incidence, with the **E** vector nearly perpendicular to the surface. For a given aluminum atom in the surface layer, BB concluded that this geometry eliminated any EXAFS contribution from the six nearest neighbors in the surface plane itself and only the three nearest neighbors in the second layer were seen. The extracted distance from the surface-sensitive spectra was thus assumed to represent the nearest-neighbor distance between first and second layer aluminum atoms. From this distance the interplanar spacing Z is readily calculated. For Al(100) no relaxation of Z from the bulk value was found within 0.05 Å. For Al(111) a *contraction*

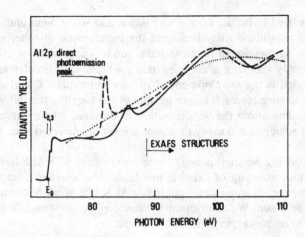

Figure 10.27. Surface soft x-ray absorption spectrum of the Al(111) surface. The solid line is obtained using the constant final-state energy $E_A = 45$ eV. The dashed line is obtained using $E_A = 4$ eV probing a much thicker layer and approximates the bulk absorption. The dotted line is the atomic like aluminum $2p$ absorption cross section used to extract the EXAFS modulations. From Bianconi and Bachrach (63).

of $\Delta Z = 0.19 \pm 0.06$ Å or 8.1% relative to the bulk ($Z = 2.338$ Å) was derived.

The absence of evidence of relaxation for Al(100) is in good accord with earlier LEED intensity measurements (139). However, two independent LEED studies (140, 141) reported a slight *expansion* of 2.5% for the Al(111) surface, in striking disagreement with the work of BB. More recently, Jona, Sondericker, and Marcus (142) performed a more refined analysis of their old LEED intensity data and confirmed the original LEED result.

The SEXAFS result of BB has been criticized in a recent review by Lee et al. (49) in that it used only a single EXAFS oscillation in an energy region close to the edge (< 50 eV). In k space this energy region corresponds to $k < 3.5$ Å and the EXAFS theory is generally not valid. However, let us for the sake of argument assume that the EXAFS formalism can be applied as assumed by BB. For $L_{2,3}$-edges the EXAFS expression is given by Eqs. (16) or (34) and for the (111) surface geometry the angular term $\Sigma_i \cos^2 \alpha_{ij}$ can be evaluated by means of Eq. (25). It is then easily seen that even for grazing x-ray incidence the six neighbor atoms in the surface plane contribute almost equally to the EXAFS signal as the three neighbor atoms in the second plane. In fact, the amplitude factors [Eqs. (35) and (36)] for the surface plane are $C_2 = 3$ and $C_{02} = 1.2$ and for the second plane $C_2 = 4.5$ and $C_{02} = -0.6$. Besides giving a large contribution to the measured SEXAFS signal the six surface plane neighbors also give rise to a sizable interference term ($C_{02} = 1.2$). This interference term is characterized by a different phase shift and will thus influence the derived neighbor distance. The data analysis of BB has neglected the important interference term contribution and their results are thus not credible.

10.5.3. Gas–Solid Reactions

This section will review SEXAFS investigations of surface complexes that are formed through reaction of gases with clean single crystal surfaces. Rather than using a chronological order of investigation we shall discuss the surface complexes in order of increasing atomic number Z of the adsorbate and for a given adsorbate in order of increasing Z of the substrate. Structural results are summarized in the Appendix.

In discussing gas–surface reactions a few terms need to be defined. The *overlayer coverage* is usually given in fractions (or multiples) of the number of atoms per square centimeter of the surface substrate layer. Thus *1 monolayer* (ML) corresponds to an equal amount of adsorbate atoms as there are in the outermost substrate layer ($\sim 1 \times 10^{15}$ atoms cm^{-2}). The most precise coverage determinations today involve Rutherford ion backscattering (12–14) measurements. The gas dosing of the clean sample is usually described in units of (exposure pressure times time) where *1 Langmuir* (1 L) corresponds to 10^{-6} Torr s.

In most cases the gases of interest consist of molecules that adsorb directly or dissociate upon reaction with the surface. Besides bromine on graphoil, which was studied in a transmission EXAFS experiment by Heald and Stern [(143), cf. Chapter 8, Section 8.3.4)], only few SEXAFS measurements have been carried out on chemisorbed molecules. Stöhr et al. investigated the chemisorption of the methoxy (CH_3O) and formate ($HCOO$) groups produced by decomposition of methanol (CH_3OH) and formic acid ($HCOOH$) on Cu(100) (144) and Puschmann et al. studied formate on Cu(110) (145). SEXAFS related work has been published involving x-ray absorption fine near-edge structure (XANES) studies of chemisorbed CO, NO, and N_2 on Ni(100) (146) NH_3 on Ni(110) (124), CO, CH_3OH, and HCOOH on Cu(100) and O_2 and various hydrocarbons on Pt(111) (147). At lower temperatures (≤ 100 K) molecular multilayers with well-defined crystal structures can be formed on top of a substrate surface but only one SEXAFS study of H_2O ice has been performed (148).

Dissociative chemisorption, in the simplest cases, results in ordered overlayers with periodicities determined by the two-dimensional surface symmetry of the substrate. The *overlayer symmetry* is usually characterized by its LEED spot pattern as discussed in detail in the LEED literature (22–25). Often the dissociative chemisorption of gases stops at monolayer coverage after depletion of the substrate bonding orbitals. In other cases the adsorbate atoms penetrate the substrate surface and form a surface compound. In certain cases this surface compound may be quite thick (~ 100 Å), a good example being oxide formation at metal and semiconductor surfaces. Since it is often the electronic and crystallographic properties of such surface compounds that are of fundamental technological importance it is clear that it is important to possess a technique that allows one to follow the progressive gas–solid reaction even in the absence of long-range structural order. This capability, as shown later, is one of the important assets of SEXAFS.

Before we review SEXAFS studies of surface complexes formed by gas–solid reactions, the importance of model compounds needs to be pointed out. Accurate structure determinations of surface complexes crucially depend on the EXAFS phaseshifts and amplitudes derived from such "standards". Because oxygen chemisorption will be extensively discussed in the following Sections we show in Fig. 10.28 EXAFS spectra of various oxides which have been used to derive some of the phaseshifts listed in Table 10.3. These phaseshifts will be the basis for all structural studies of oxygen on surfaces discussed below.

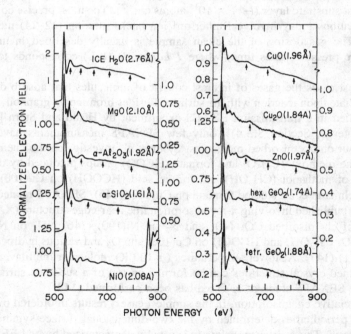

Figure 10.28. Total yield EXAFS spectra above the oxygen K-edge for a variety of model compounds. Samples were single crystals or were prepared from powders. H_2O ice was condensed at 90 K. The nearest-neighbor distances are indicated in brackets. Note different EXAFS frequencies as indicated by arrows for similar compounds with different distances (e.g., CuO and Cu_2O).

Table 10.3. Some Experimentally Derived EXAFS Phase Shifts for Low-Z Absorbers (Eq. 15)

Absorber A	Backscatterer B	Model Compound	R_{A-B}	Experimentally Derived Phase $\phi = X - Yk^a$		k-Range of Data	
				X	Y	k_{min}	k_{max}
C	C	Graphite	1.42	3.94	0.34	3.1	11.3
O	O	Ice	2.76	1.91	0.56	4.0	8.4
F	F	LiF	2.84	3.38	0.70	2.5	8.4
F	Na	NaF	2.31	4.99	1.06	2.4	7.6
O	Mg	MgO	2.10	4.85	0.94	1.7	9.4
O	Al	α-Al_2O_3	1.915	3.83	0.74	2.3	8.8
N	Si	Si_3N_4	1.72	4.48	0.64	1.6	9.6
O	Si	α-SiO_2	1.61	4.40	0.65	2.3	9.1
O	Ni	NiO	2.084	7.70	0.75	2.7	8.5
O	Cu	CuO	1.96	6.58	0.54	2.7	9.7
O	Cu	Cu_2O	1.84	6.17	0.46	3.7	10.0
O	Zn	ZnO	1.97	6.29	0.45	4.6	9.1
N	Ga	GaN	1.94	6.45	0.47	3.8	10.8
O	Ge	Tetrahedral GeO_2	1.88	8.07	0.58	3.6	8.2
O	Y	Y_2O_3	2.25	6.20^b	0.10^b	2.2	9.4
O	Er	Er_2O_3	2.24	7.48^b	0.95^b	2.4	9.4
O	Yb	Yb_2O_3	2.21	6.94^b	0.89^b	2.4	9.4

[a] $k = 0$ corresponds to inflection point of K-edge.
[b] For the more pronounced peak at larger R in the Fourier transform.

10.5.3.1. Metal Substrates

Ordered adsorbate layers have long been studied by LEED (8, 22–25) and from these studies valuable insight has been gained on preferred bonding sites and surface bond lengths. With few exceptions atomic adsorbates preferentially occupy sites where the next metal atom would have been located in a layer-by-layer extension of the substrate, that is, the hollow sites as shown in Fig. 10.29 for the (100) surface of a fcc substrate. Correlations have also been established by Mitchell (149, 150) and Madhukar (151) between observed surface bond lengths and the bond-order concept developed by Pauling (152) for bulk materials. In light of the success of LEED structure determinations for such systems the question arises why to do SEXAFS. Three arguments can be made: (a) Since no technique is free of potential errors a given system should be studied by more than one technique, (b) SEXAFS provides a better accuracy ($\Delta R \leq 0.02$ Å) for bond distances than LEED ($\Delta R \leq 0.10$ Å) and (c) in contrast to LEED, SEXAFS can study the progressive structural development past the stage of ordered overlayer formation. In the following we shall present experimental results of SEXAFS studies that will illustrate these points.

O/Al(111). One of the experimentally and theoretically most extensively studied adsorbate–substrate interactions is the progressive oxidation of Al(111) surfaces. From a theoretical point of view this system is favorable because the bonding involves only s and p orbitals. The experimental interest in this system was largely stimulated by the photoemission and LEED study of Flodström et al. (153) in 1978. This study showed the existence of an ordered 1 × 1 oxygen overlayer for the initial interaction of oxygen with Al(111) in contrast to disordered oxide formation on both Al(100) and Al(110) surfaces. Since then the

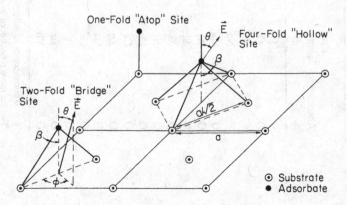

Figure 10.29. Chemisorption sites on a fcc (100) surface.

O/Al(111) system has been investigated by many experimental techniques such as work function measurements (154, 155), angle resolved photoemission (156–158), LEED intensity measurements (159–162), SEXAFS (71–73, 163, 164), quantitative Auger spectroscopy (165), and electron energy loss spectroscopy (EELS) (166, 167). On the theoretical side the different calculational schemes involved a simple adatom–jellium model (168, 169), a self-consistent field $X\alpha$ scattered wave molecular orbital approach (170), a self-consistent linear combination of Gaussian orbitals calculation (171, 172), a self-consistent linearized augmented plane wave method (173, 174), a Hartree–Fock cluster calculation (175), and a calculation of the LEED electron scattering potential by means of a discrete variation method applied to a cluster (176).

Starting with the experimental situation, all published work agrees with the basic findings of Flodström et al. (153) that a chemisorbed oxygen state with a 1×1 overlayer structure exists on Al(111) at a coverage of approximately one monolayer. There is also agreement on the existence of another oxidelike state that increases in abundance with increasing oxygen coverage. Both states may exist simultaneously over a certain coverage range, at least above monolayer coverage. This oxidelike state is characterized by a different aluminum $2p$ chemical shift (2.7 eV) and a different EELS frequency (166) (105 meV) than the chemisorbed state (1.4 eV and 80 meV, respectively). Whereas the chemisorbed oxygen state is usually associated with an *overlayer*, the oxidelike state is thought to correspond to an oxygen *underlayer*. Experiments also agree that for heavy oxygen exposure a third oxygen state exists that closely resembles *amorphous* Al_2O_3. The formation of this state is accompanied by a blurring and finally disappearance of the 1×1 LEED spot pattern.

There are differences among the experimental studies with respect to what exposure corresponds to monolayer coverage. Core-level photoemission studies associate a monolayer of oxygen with 150-L exposure (153), Auger studies with approximately 55 L (165), and a combined LEED plus Auger study with 100 L (162). There is further disagreement on the relative abundance of the chemisorbed and oxidelike states at different exposures. Core-level photoemission spectra (153), vibrating-capacitor work function measurements (155), and quantitative Auger studies (165) indicate that the chemisorbed state dominates or forms exclusively below about 50-L exposure. In contrast, EELS data indicate the presence of comparable amounts of chemisorbed and oxidelike phases for exposures as low as 2 L (166). Furthermore, UPS and EELS measurements find a irreversible transformation with time (~ 2 h) of part of the chemisorbed into the oxidelike phase. This transformation had previously been observed by photoemission at room (158, 177) and elevated (153) temperatures ($\sim 200°C$).

To complicate matters, other oxygen phases have been suggested to exist in addition to the three phases mentioned. Bachrach et al. (164) find pressure-dependent aluminum $2p$ core-level spectra. At exposure pressures of 2×10^{-7}

Torr and < 100-L oxygen exposure a phase characterized by a 0.9-eV chemical shift was observed in addition to the normal 1.4-eV shift for chemisorbed atomic oxygen produced at 1×10^{-6} Torr pressure. The 0.9-eV shifted peak was assigned to molecular oxygen. In contrast, EELS studies (166) find no evidence for molecular oxygen under the same experimental conditions used by Bachrach et al. (164). Finally, from comparison of self-deconvoluted aluminum $L_{2,3}$ VV Auger spectra for 25-L oxygen exposure with the calculated density of states for different chemisorption geometries, Soria et al. (162) postulate a low-coverage phase for 0–30-L exposure. In their model the oxygen atoms are located in the threefold hollow either in or slightly underneath ($-0.5 \text{ Å} \leq Z \leq 0 \text{ Å}$) the aluminum surface plane.

Structural investigations focused on the 1×1 LEED structure formed in the exposure range 100–150 L. Three independent LEED intensity studies published in 1979–1980 came to the conclusion that in this coverage range oxygen chemisorbs in the threefold hollow site and derived distances $Z = 1.54$ (159), 1.33 ± 0.08 (160), and 1.46 ± 0.05 Å (161), respectively, above the outermost plane of aluminum atoms. These values correspond to O—Al distances of 2.26, 2.12 ± 0.05, and 2.20 ± 0.04 Å, respectively. Two total yield SEXAFS measurements in the same exposure range (100–150 L) by Johansson and Stöhr (JS) (71) in 1979 and by Bachrach, Hansson, and Bauer (164) (BHB) in 1981 derived O—Al distances of $R = 1.79 \pm 0.05$ and 1.92 ± 0.05 Å, respectively. A more detailed account of the Letter by Johansson and Stöhr (71) was published by Stöhr et al. (72) in 1980. Both SEXAFS studies did not determine the chemisorption site but the distances would correspond to values of $Z = (0.7^{+0.10}_{-0.15})$ Å and $Z = 0.98 \pm 0.10$ Å in the threefold hollow site, respectively. In both cases the samples were characterized by aluminum $2p$ photoemission spectra and the chemisorbed species (1.4-eV shift) was found to dominate the photoemission spectra with an intensity 2–3 times that of the oxidelike species (2.7-eV shift). SEXAFS measurements using total yield detection were also performed at 50-L oxygen coverage by Norman, Brennan, Jaeger, and Stöhr (73) in 1981. Again aluminum $2p$ core-level photoemission spectrum were used to characterize the sample and showed no appreciable oxidelike component but only the chemisorbed phase (1.4-eV chemical shift). The O—Al distance was derived to be $R = 1.76 \pm 0.03$ Å, corresponding to $Z = 0.60 \pm 0.10$ Å for chemisorption in the threefold hollow site. Upon heating to 200°C for 10 min the aluminum $2p$ photoemission spectrum indicated conversion of the chemisorbed to the oxidelike phase. For this sample the O—Al distance was determined to be $R = 1.75 \pm 0.03$ Å. The original SEXAFS data for the two phases are compared to the spectrum for bulk α-Al_2O_3 (corundum) in Fig. 10.30.

In light of the gross discrepancy between the three LEED results with the early SEXAFS results of Johansson and Stöhr the LEED data were reexamined by Jona and Marcus (178) in 1980. These authors showed that the LEED in-

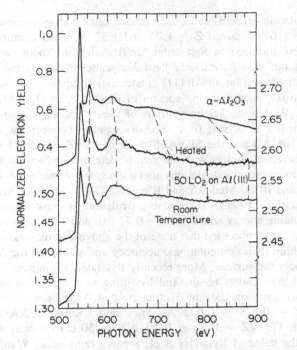

Figure 10.30. EXAFS spectra above the oxygen K-edge for α-Al_2O_3 (corundum) and 50-L oxygen on Al(111) at room temperature and after heating to 200 °C for 10 min. Note the change in energy positions of the EXAFS maxima for the oxygen on Al(111) surface as compared to the bulk Al_2O_3 spectrum. From Norman et al. (73).

tensity data could be accounted for by values of $Z = 0.7$ Å as well as $Z \approx 1.4$ Å. The reason is that for insufficiently large experimental data sets, good fits may be obtained at multiples of the interplanar spacing Z, the so-called multiple coincidence problem in LEED analyses. The disparity was resolved by new LEED work by Soria et al. in 1981 (162). For 90-L oxygen coverage a value of $Z = 0.73 \pm 0.05$ Å was derived corresponding to an O—Al distance of $R = 1.80 \pm 0.02$ Å. At 150-L coverage Soria et al. find $Z = 0.8 \pm 0.03$ Å or $R = 1.83 \pm 0.02$ Å. Although the small error bars on the spacings derived in this latest LEED study appear questionable, the Z and R values are nevertheless consistent with the SEXAFS work of JS. Reanalysis of the early LEED work by Martinson et al. (160) (who deduced $Z = 1.33$ Å) by Neve et al. 1982 (176) also favors the SEXAFS ($Z = 0.7$ Å) over the previous LEED ($Z = 1.33$ Å) value.

Before publication of the structural LEED and SEXAFS work two theoretical calculations were available that discussed chemisorption of oxygen on Al(111).

In a simple adation–jellium model calculations using energy minimization Lang and Williams (168) obtained $Z = 1.75$ and 1.32 Å when aluminum pseudopotentials were included to first order (169). Salahub, Roche, and Messmer (170) carried out a self-consistent field $X\alpha$ scattered-wave molecular orbital calculation for oxygen on an Al(111) cluster and compared the calculated valence bands to the photoemission spectra of Flodström et al. (153) in the 10–150-L range. Within the accuracy limits of theory and experiment agreement was found for $0.5 \le Z \le 1.0$ Å. The discrepancy between the early LEED work ($Z \approx 1.4$ Å) and the SEXAFS work of JS ($Z = 0.7$ Å) stimulated a variety of theoretical studies with the goal to determine which value (if any) is correct. The most recent calculations also address the disparity of the SEXAFS results of JS and BHB. Mednick and Kleinman (171) reported a self-consistent linear combination of Gaussian orbitals calculation of work function and aluminum $2p$ binding energy shifts for $Z = 0.7, 1.0,$ and 1.4 Å above a six-layer Al(111) film. They concluded that none of the above Z values gave satisfactory agreement within their computational accuracy and suggested the oxygen layer should lie under the surface. More recently Bylander, Kleinman, and Mednick (172) revised their earlier results and by fitting to the experimental vibrating capacitor work function data by Hofmann et al. (155) for <50-L exposure determined $Z = 0.578 \pm 0.032$ Å. They thus confirm SEXAFS results of Norman et al. (73) ($Z = 0.60 \pm 0.10$ Å) for 50-L exposure, although the accuracy of the value of Bylander et al. seems exaggerated. Wang, Freeman, and Krakauer (173, 174) calculated the work function, core shifts, and valence band dispersion for oxygen on Al(111) at $Z = 0.70$ and 1.33 Å by a self-consistent linearized augmented plane wave method. By comparison to experiment they concluded that the SEXAFS value of 0.7 Å by JS is the correct one. Finally, Cox and Bauschlicher (175) predicted Z by minimizing the total Hartree–Fock energy in a LCAO self-consistent field molecular orbital calculation. They find $Z = (0.70 \pm 0.10)$ Å and state that this value is consistent only with JS's SEXAFS results and irreconcilable with both the early LEED value (~ 1.4 Å) and the SEXAFS value of BHB (0.98 ± 0.10 Å) within the quoted experimental and theoretical error bars.

With the conflict between the early LEED and SEXAFS work settled by LEED reanalysis and by supporting theoretical calculations the discrepancy of the two SEXAFS results of JS ($R = 1.79 \pm 0.05$ Å) and BHB ($R = 1.92 \pm 0.05$ Å) obtained for 100–150-L exposure remains unexplained. We note, however, that of the corresponding Z values for the threefold hollow site (JS: $0.70^{+0.10}_{-0.15}$ Å, BHB: 0.98 ± 0.10 Å) only the value of JS is supported by the latest LEED results and the theoretical investigations. Furthermore, the value found by JS is supported by the careful later SEXAFS study by Norman et al. (73) at lower coverage (50 L) who find $R = 1.76 \pm 0.03$ Å or $Z = 0.60 \pm 0.10$ Å for the chemisorbed phase. The present author believes the results of

BHB to be incorrect because of poor data quality and data analysis problems. From a comparison between the Fourier transforms of the data obtained by BHB (cf. Fig. 10.31) and JS (cf. Fig. 10.32) it is apparent that the Fourier transform peaks of JS are significantly narrower. This is due to a larger data range in k space, because BHB had to cutoff their data before the iron L-edge because of a spurious iron contamination. Also the signal-to-noise quality of the JS data is better, as can be seen from the low noise in the JS transforms at distances larger than 4 Å. The peak at approximately 4.5 Å in the BHB transform is unphysical and has to be due to noise.

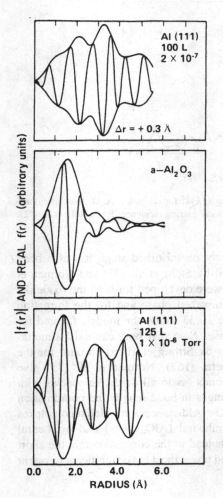

Figure 10.31. Magnitude and real part of Fourier transform of oxygen K-edge SEXAFS spectra for two different oxygen exposures of Al(111) and an amorphous aluminum oxide standard. From Bachrach, Hansson, and Bauer (164).

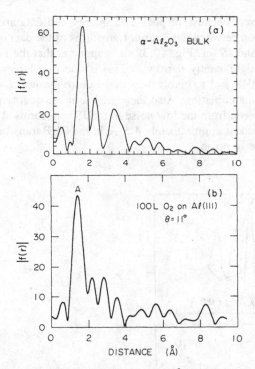

Figure 10.32. Fourier transforms of EXAFS spectra $\chi(k)k^2$ for (a) bulk α-Al_2O_3 (corundum) and (b) 100-L oxygen on Al(111) recorded at $\Theta = 11°$ x-ray grazing incidence. From Stöhr et al. (72).

The progressive oxidation past the early chemisorbed stage has also been studied by SEXAFS as discussed in detail by Stöhr et al. (72) and Norman et al. (73). In these studies detailed models were developed for both the oxidelike underlayer that may accompany the chemisorbed phase and for the formation of thicker amorphous oxide layers. Figure 10.33 shows the models favored for the chemisorbed and initial oxidelike phase. Note that both can exist simultaneously. This model was recently shown by Strong et al. to account for the frequencies observed in their EELS spectra (167). Norman et al. (73) give arguments that for predominantly ionic bonds (as in all oxidelike species with a large aluminum $2p$ shift of 2.7 eV) changes in bond length and coordination numberaredirectlyrelated.ThusthederivedO$-$Aldistancesforoxidelikecomplexes can be used to discriminate between tetrahedral (AlO_4 units) and octahedral (AlO_6 units) coordination. Such arguments lead to the conclusion that the short O$-$Al bond (179, 180) for the chemisorbed phase must have significant *covalent*

(a) Chemisorbed Phase

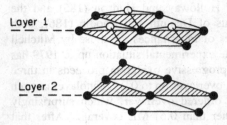

Layer 1 ____

Layer 2 ____

(b) Oxide-Like Phase

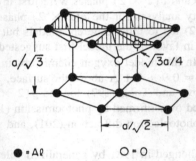

$a/\sqrt{3}$ $\sqrt{3}a/4$

$a/\sqrt{2}$

● = Al   ○ = O

Figure 10.33. Structural models for (a) chemisorbed oxygen on Al(111), (b) subsurface oxidelike phase that increases with coverage. In both cases the O—Al distance is approximately 1.75 Å. From Norman et al. (73).

character. This has indeed been confirmed by all theoretical calculations (172–175).

Finally we note that the oxidation of Al(111) has also been investigated by aluminum $L_{2,3}$ XANES measurements by Bianconi et al. (181, 182) who regarding oxide formation came to similar conclusions as Norman et al. (73) from SEXAFS measurements. Changes of the oxygen K-edge fine structure with progressive oxidation were also studied by Stöhr et al. (72).

The determination of accurate structural parameters for chemisorbed oxygen on Al(111) by JS not only served as a stimulus and testing ground for different theoretical approaches but it also represented a first-time challenge to LEED that until then had served as the most reliable surface structural tool. With the subsequent revision of the early LEED results it appears that SEXAFS has established itself as a new powerful surface structural tool.

O/Ni(100). The chemisorption of oxygen on Ni(100) and the oxide formation at higher oxygen exposures has been studied by many experimental techniques starting with the early LEED work of Farnsworth and Madden in 1961 (183) and MacRae in 1964 (184). The primary two experimental references for this system are the LEED/Auger study of Holloway and Hudson (185) and the extensive LEED intensity measurements of Demuth and Rhodin (186), both published in 1974. Other kinetic studies of importances are those by Mitchell et al. (187) and Norton et al. (188). The experimental situation up to 1979 has been reviewed by Brundle (189). The progressive oxidation proceeds in three steps. Formation of a $p(2 \times 2)$ ordered overlayer up to 0.25 monolayer oxygen coverage is followed by a $c(2 \times 2)$ ordered overlayer, which surprisingly corresponds to only 0.33 ± 0.03 (rather than 0.5) ML coverage. After that oxide formation sets in by formation of two-to-three layer thick islands that grow epitaxially on the Ni(100) mesh.

The structures of both the $p(2 \times 2)$ and $c(2 \times 2)$ phases were first investigated by LEED. Early LEED intensity analyses for the $c(2 \times 2)$ phase in 1973 by three different groups (190–192) led to three different models but the controversy was later settled (193–196) in favor of the model first suggested by Demuth, Jepsen, and Marcus in 1973. In this model oxygen chemisorbs in the fourfold hollow position at a distance $Z = 0.9 \pm 0.1$ Å above the surface. The same structure was derived by van Hove and Tong (197) for the $p(2 \times 2)$ phase. This structural model was adopted or confirmed by photoemission (198, 199), low-energy ion scattering (200), photoelectron diffraction (201), and various theoretical studies (202–204).

This structure determination was challenged in 1981 by generalized valence bond electronic structure calculations by Upton and Goddard (UG) (205). Energy minimization resulted in two states with equilibrium distances of $Z = 0.88$ and 0.26 Å for the oxygen atom above the fourfold hollow site. Upon examining the character of the wave functions and associated charge distributions the state with $Z = 0.88$ Å was denoted a low-coverage *radical state* while that with $Z = 0.26$ Å was labeled a *precursor oxide state* corresponding to higher coverage. Electron energy loss results were then used to make a 1:1 assignment between the $Z = 0.88$-Å state with the $p(2 \times 2)$ and the $Z = 0.26$-Å state with the $c(2 \times 2)$ oxygen on Ni(100) configuration. Previous EELS measurements by Anderson (206) and Ibach et al. (207) had revealed an unexpectedly large frequency shift of 14 meV between the $p(2 \times 2)$ and $c(2 \times 2)$ phases. In comparison, only a 2-meV EELS shift was observed between the same LEED structures for the sulfur on Ni(100) system (206). Using the potential energy curve produced by UG, Rahman, Black, and Mills (RBM) (208) carried out a lattice dynamics calculation and concluded they could account for the EELS shift if they used UG's conjecture and parameters of a structural change between the $p(2 \times 2)$ and $c(2 \times 2)$ phases.

Both UG and RBM pointed out that previous structural investigations had not been unambiguous and in some cases even contradictory. Reanalysis of the $c(2 \times 2)$ LEED data by Tong and Lau (209) indeed revealed an ambiguity between configurations with $Z = 0.9$ Å and $Z \approx 0$ Å due to a multiple coincidence problem. Photoelectron diffraction data (201) for the $c(2 \times 2)$ phase favored $Z = 0.9$ Å but an analysis was not carried out for $Z < 0.5$ Å because of computational difficulties. Low-energy ion scattering (200) studies also favored $Z = 0.9$ Å but the results on Z were of low accuracy and also only limited comparison with various surface structures was made. Most importantly, x-ray photoemission studies involving the measurement and scattering analysis of azimuthal photoelectron diffraction (APD) patterns by Petersson et al. (210) gave best agreement for $Z = 0.1 \pm 0.1$ Å for $c(2 \times 2)$ and $Z = 0.8 \pm 0.1$ Å for $p(2 \times 2)$ oxygen on Ni(100).

The structural problem raised by UG and RBM initiated near-edge (XANES) (211) and EXAFS measurements on the oxygen on Ni(100) system by Stöhr, Jaeger, and Kendelewicz (SJK) (75). The first indication of the local structural *equivalence* of $p(2 \times 2)$ and $c(2 \times 2)$ oxygen on Ni(100) came from the XANES spectra of the oxygen K-edge region shown in Fig. 10.34. These spectra show a pronounced polarization dependence of the oxygen K-edge fine structure for low exposures. Below 50-L exposure, that is, in the region between the $p(2 \times 2)$ (maximum around 1.5 L) and $c(2 \times 2)$ (maximum around 20–30 L)

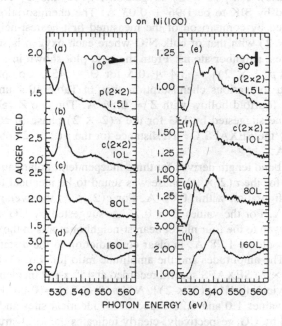

Figure 10.34. Near-edge fine structure spectra around the oxygen K-edge for increasing oxygen coverage on Ni(100) and two different x-ray incidence angles. Note that spectra show strong polarization dependence at low coverage that vanishes at high coverage due to cubic NiO formation.

phases, the spectra and polarization dependence are unchanged. Above 50 L the spectral features change due to oxide formation and the polarization dependence has almost vanished at 160-L exposure. At 280 L (not shown) the XANES spectrum has become completely isotropic and is identical to that recorded for cubic bulk NiO (211). Since the edge fine structure is more easily recorded than a complete SEXAFS spectrum Fig. 10.34 represents a nice example of how an unknown system can be characterized first by XANES before addressing certain key issues by SEXAFS. A recent calculation of the XANES spectral features and their polarization dependence by Norman et al. (212) using a LEED-like multiple-scattering approach (see Chapter 2) shows that structural information can indeed be derived from such data. The calculations favor $Z = 0.9$ Å over $Z = 0.26$ Å for $c(2 \times 2)$ oxygen on Ni(100).

Polarization dependent PAY SEXAFS measurements by SJK in 1982 (75) confirmed the local structure equivalence between the $p(2 \times 2)$ and $c(2 \times 2)$ phases implied by the XANES results. The raw PAY SEXAFS spectrum, recorded with $E_p = 350$ eV at $\Theta = 45°$ for the $c(2 \times 2)$ phase has been shown earlier in Fig. 10.23.

Reasons for choosing the PAY detection mode have also been given earlier (Section 10.2.2.5). For $p(2 \times 2)$ oxygen on Ni(100) the O—Ni bond length was determined by SJK to be 1.96 ± 0.03 Å. The chemisorption site can be directly obtained by comparison of the measured first-nearest-neighbor amplitude for $p(2 \times 2)$ with that of bulk NiO where each oxygen is surrounded by six nickel nearest-neighbor atoms. From the ratio plot shown in Fig. 10.35 we determine $N^*[p(2 \times 2)] = 3.4 \pm 0.5$ for $\Theta = 45°$. Comparison to the calculated ratio for various chemisorption sites in Table 10.4 unambiguously determines the fourfold hollow with $Z \approx 0.86$ Å. The two Z values in Table 10.4 are the one suggested by UG for the $c(2 \times 2)$ phase (0.26 Å) and the one implied by the SEXAFS O—Ni distance for the $p(2 \times 2)$ phase, namely, 0.86 ± 0.07 Å.

The O—Ni bond length derived for three independent, polarization dependent measurements for the $c(2 \times 2)$ phase was found to be identical to that for the $p(2 \times 2)$ configuration within 0.01 Å. For $c(2 \times 2)$ the average distance is 1.95 ± 0.03 Å. For the values $Z = 0.26$ Å suggested by UG and RBM the distance of oxygen to the four nickel nearest-neighbor atoms in the surface plane (Fig. 10.29) would be 1.78 Å, in clear contradiction to the distance measured by SEXAFS. The amplitudes and the amplitude ratio [cf. Eq. (31)] for the $c(2 \times 2)$ and $p(2 \times 2)$ SEXAFS spectra recorded at 45° x-ray incidence are shown in Fig. 10.36 yielding $N^*[c(2 \times 2)]/N^*[p(2 \times 2)] = 0.96 \pm 0.06$. Comparison to the values 1.0 and 1.28 expected for identical sites and the different sites suggested by UG, respectively, clearly indicates the local structural equivalence.

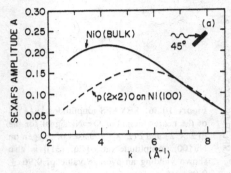

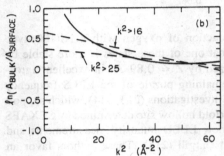

Figure 10.35. (a) SEXAFS amplitudes $|\chi(k)k|$ of the nearest-neighbor O—Ni signals for bulk NiO and $p(2 \times 2)$ oxygen on Ni(100), recorded at 45° x-ray incidence. (b) Natural logarithm of amplitude ratio of (a) according to Eq. (30). Straight line approximations for different low-k cutoffs are shown dashed. Ordinate intercept at $k = 0$ yields absolute nickel coordination number of oxygen on the surface.

Table 10.4. Experimental Versus Calculated Effective Coordination Number N^* for $p(2 \times 2)$ Oxygen on Ni(100) for $\Theta = 45°$

			Fourfold Hollow with $Z(\text{Å}) =$	
Experimental	Onefold Atop	Twofold Bridge	0.26	0.86
3.4 ± 0.5	1.5	2.4	4.6	3.6

The SEXAFS value $Z = 0.86 \pm 0.07$ Å for both phases is in good agreement with the LEED value 0.9 ± 0.1 Å and the value 0.88 Å found by UG for the low-coverage radical state. The SEXAFS result disagrees with the variations in structural behavior suggested by UG and RBM and by the APD study by Petersson et al. (210). Also, the large EELS frequency shift remains a puzzle.

Recent theoretical work, after publication of the SEXAFS results, indicates deficiencies in the calculations of UG and RBM. Bauschlicher et al. (213) also

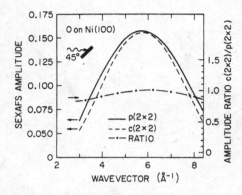

Figure 10.36. SEXAFS amplitudes $|\chi(k)k|$ of the nearest-neighbor O—Ni signal for the $p(2 \times 2)$ and $c(2 \times 2)$ phases of oxygen on Ni(100). Amplitude ratio [Eq. (31)] is also shown yielding an average value of 0.96 ± 0.06.

find two low-lying states for the interaction of oxygen with a Ni(100)-type cluster (Ni_5O). However, they argue that one of these states is more stable by ≈ 1 eV and that this state is characterized by $Z \approx 0.89$ Å, in excellent agreement with the SEXAFS value. The remaining puzzle of the EELS frequency shift has recently led to new theoretical investigations (213, 214), which support the SEXAFS result. The symmetric fourfold hollow site determined by SEXAFS (75) has recently been challenged by new LEED intensity measurements and calculations by Demuth, DiNardo, and Cargill (215). These authors favor an asymmetric fourfold hollow chemisorption geometry where oxygen is shifted laterally by 0.3 ± 0.1 Å with resulting bond lengths of 1.75 ± 0.05 Å and 2.14 ± 0.08 Å to two nickel atoms, respectively. This places the oxygen atom 0.80 ± 0.025 Å above the top nickel layer. Demuth et al. (215) claimed that this result could also be made compatible with the SEXAFS data of Stöhr et al. (75). It was argued that because of the limited k range ($k \leq 8.6$ Å^{-1}) of the data the beating in the SEXAFS amplitude arising from the two different O—Ni distances could not be observed. Furthermore, Demuth et al. pointed out that the empirical NiO phase shift used by Stöhr et al. in their analysis (Table 10.3) may be inadequate since it differs from the calculated one (134). Thus by assuming 9% larger O—Ni distances than implied by their LEED results to account for the "error" of the NiO phase shift and by adjusting Δ by 10 eV Demuth et al. calculated a SEXAFS signal that agreed with the measured data. We have recently tested the NiO phase shift by analyzing EXAFS data on Ni_2SiO_4 and found it to be extremely accurate. Thus we believe the use of this phase shift for the oxygen on Ni(100) case to be justified. Model calculations by us for the two different geometries shown in Fig. 10.37 demonstrate that SEXAFS can readily distinguish between the two proposed hollow sites and comparison with the data clearly establishes the symmetric hollow as the oxygen site in agreement with the published SEXAFS results.

The only remaining discrepancy with the SEXAFS structure determination is the experimental photoelectron diffraction work of Petersson et al. (210). This

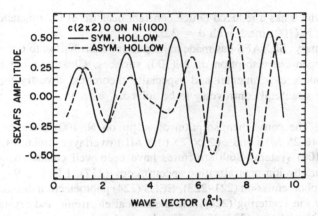

Figure 10.37. Calculated SEXAFS signals for the symmetric fourfold hollow site [ref. (75)] and asymmetric hollow site [ref. (215)] of oxygen on Ni(100). The calculations used the empirical NiO phase shift (Table 10.3) and $\Delta = 0$. An x-ray incidence angle of 45° was assumed. The two sites can clearly be distinguished and the symmetric hollow site is favored by comparison with experiment.

disparity can be lifted by two possible explanations. First, it is well established that APD when performed at high electron kinetic energies (≥ 200 eV) is determined by forward scattering and thus is much more sensitive to oxygen close to rather than well above ($Z > 0.5$ Å) the surface plane. Thus the measurements of Petersson et al. (210) may well have been dominated by a minority oxygen species. The error bars for the SEXAFS bond length would allow up to 15% of a minority species close to ($Z < 0.3$ Å) the surface plane. Second, Petersson et al. state that they observe $Z = 0.8 \pm 0.1$ Å for oxygen coverages up to 0.3 ML, while the value $Z = 0.1 \pm 0.1$ Å corresponds to about 0.5 ML. They equate 0.5 ML with 15-L oxygen exposure. This is clearly in conflict with published kinetic studies (189) and Rutherford backscattering studies (216) since 0.5 ML should correspond to 60–70-L exposure and thus should already involve oxide formation [note that the $c(2 \times 2)$ phase at saturation corresponds to only 0.33 ML]. If we assume Peterson et al. coverage values to be correct and the exposure figures to be inaccurate their results are no longer at odds with the SEXAFS studies, since for 0.5 ML coverage NiO nucleation has already begun and in this case the oxygen atoms indeed lie in the surface plane.

The oxidation past the $c(2 \times 2)$ chemisorption stage, that is, oxide nucleation has also been studied by SEXAFS (211). For 90-L exposure, corresponding to 0.85 ± 0.10 ML coverage the formation of NiO islands is observed by the second nearest-neighbor (O—O) distance. From both the O—Ni and O—O distances it is seen that the thin NiO layer formed on Ni(100) has a reduced lattice constant ($a = 3.96$ Å) relative to bulk NiO ($a = 4.168$ Å). This can be readily explained by the formation of an epitaxial NiO interface layer on

Ni(100), which has a reduced lattice constant because of the mismatch with the underlying Ni(100) mesh with $a = 3.524$ Å.

In summary, SEXAFS has made an important contribution to the understanding of oxygen chemisorption on Ni(100) surfaces. Clearly, these results will stimulate more experimental and especially theoretical investigations but we believe the structural controversy to be solved.

S/Ni(100). The early chemisorption of sulfur on Ni(100) is characterized by $p(2 \times 2)$ (0.25 ML) and $c(2 \times 2)$ (0.5 ML) overlayer structures, similar to the O/Ni(100) system. Both structures have been well characterized over the last 10 years by ion neutralization spectroscopy (217), LEED (191, 192, 197, 218–220), photoemission (221–223), EELS (224) photoelectron diffraction (225), low-energy ion scattering (226), and theoretical electronic and crystallographic structure calculations (227–229). The local chemisorption geometry for the $c(2 \times 2)$ phase was determined as early as 1973 by Demuth, Jepsen, and Marcus (192). Similar to the $c(2 \times 2)$ oxygen on the Ni(100) system there was, however, considerable discrepancy between the structures suggested from early LEED intensity analysis by two independent groups (191, 192). (192). This controversy has been settled (219, 220) in favor of the model derived by Demuth et al. (192). The structure for $p(2 \times 2)$ sulfur on Ni(100) was determined by van Hove and Tong in 1975 (197) and found to be locally equivalent to that for the $c(2 \times 2)$ phase. Sulfur is found to chemisorb in the fourfold hollow site at a distance $Z = 1.3 \pm 0.1$ Å above the surface, corresponding to a S—Ni bond length of 2.19 ± 0.06 Å. This result has been confirmed by low-energy ion scattering (226) and photoelectron diffraction (225) measurements (PD) although in the PD analysis smaller error bars were quoted, namely, $Z = 1.30 \pm 0.04$ Å corresponding to $R = 2.19 \pm 0.02$ Å. The above chemisorption geometry is also consistent with the results of angle resolved photoemission measurements (222, 223). On the theoretical side the generalized valence bond technique has been applied by Walch and Goddard (227) to determine the electronic and crystallographic structure of a Ni_4S cluster yielding a S—Ni bond length of 2.21 Å. Upton and Goddard find a S—Ni distance of 2.15 Å in the fourfold hollow site using a larger SNi_{20} cluster (228). The electronic structure of sulfur on the low index faces of nickel has also been investigated by a self-consistent local-density molecular cluster calculation (229). The picture emerging from all these investigations is that the chemisorption of sulfur on Ni(100) is well understood except for small disparities in the S—Ni distance (~ 0.05 Å).

In 1981 Brennan, Stöhr, and Jaeger (BSJ) (74) investigated the $c(2 \times 2)$ sulfur on the Ni(100) system by SEXAFS. A more extensive discussion of this study was given by Brennan (230) and by Stöhr, Jaeger, and Brennan in 1982 (53). The system was chosen for the very reason that it appeared to be one of the best understood ones in surface science. It thus allowed to test the SEXAFS

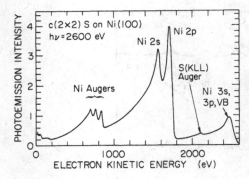

Figure 10.38. Photoemission spectrum for $c(2 \times 2)$ sulfur on Ni(100) recorded at $h\nu$ = 2600 eV with a cylindrical mirror analyzer operated in the nonretarding mode. Analyzer throughput and resolution increase with kinetic energy. For elastic Auger yield SEXAFS measurements the analyzer is set on the sulfur *KLL* Auger peak at 2100 eV.

technique and critically assess its reliability and accuracy as compared to the other surface structural techniques applied previously. The SEXAFS studies were carried out by means of EAY detection. Figure 10.38 shows the photoemission spectrum of the $c(2 \times 2)$ sulfur on the Ni(100) surface recorded at $h\nu$ = 2600 eV. The electrons were energy analyzed with a CMA operated in the nonretarding mode where the energy resolution and therefore the analyzer throughput are proportional to the electron kinetic energy. This explains the missing inelastic tail ($E_k < 50$ eV) and the broad linewidths at high kinetic energy. Figure 10.39 illustrates why the EAY rather than the TY detection mode was chosen. For the EAY the signal-to-noise ratio [Eq. (1)] is $S_N = 125$ with a signal-to-background ratio or edge jump ratio $J_R = 50\%$ while for the TY mode we obtain $S_N = 14$ and $J_R = 1.4\%$, respectively. The EAY SEXAFS spectrum is seen to be of bulklike quality due to the high photon flux provided by the monochromator ($\sim 10^{11}$ photons s^{-1}) (79).

The SEXAFS spectra recorded for three polarization directions were analyzed by comparison to the TY EXAFS spectrum of bulk NiS that served as a model compound. The Fourier transforms shown in Fig. 10.40 are dominated by the first-nearest-neighbor S—Ni peak. The $\Theta = 90°$ spectrum clearly reveals a second weaker peak around 3.8 Å. This peak is also observed for the two other polarization directions although it is close to the noise level for $\Theta = 10°$.

By detailed analysis of the phase and amplitude of the first nearest-neighbor peak BSJ accurately determined the local structure of $c(2 \times 2)$ sulfur on nickel. In agreement with the structure determination by LEED sulfur was found to chemisorb in the fourfold hollow site with a S—Ni bond length of $R = 2.23 \pm 0.02$ Å corresponding to $Z = 1.37 \pm 0.03$ Å. In addition to the analysis of the first nearest-neighbor SEXAFS amplitude (see the following pages) the chemisorption site was independently derived by the second neighbor distance for the $\Theta = 90°$ spectrum. This peak corresponds preferentially to the fourth nearest-neighbor nickel shell consisting of the 8 second nearest neighbors in the surface plane.

The SEXAFS study of BSJ is significant because it allowed one to assess the

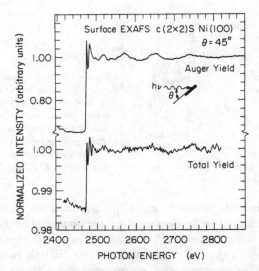

Figure 10.39. Comparison of SEXAFS spectra for $c(2 \times 2)$ sulfur on Ni(100) recorded in the elastic Auger yield (EAY) and the total electron yield (TY) modes. Note that the edge jump ratio is approximately 50% for the EAY mode but only ~1.5% for the TY mode. The noise in the TY spectrum is not completely statistical but has a component from the crystal monochromator motion [Ref. (79)].

potential and reliability of structure determinations for low-Z atoms. The high signal-to-noise ratio of the data allowed for the first time the determination of the complete chemisorption geometry from a combination of first and second neighbor distances without the necessity of amplitude analysis. With increasingly higher photon flux levels becoming available it will be possible to record SEXAFS spectra with comparable or better signal-to-noise ratios for all elements. Thus the determination of higher neighbor distances will be a valuable asset of future SEXAFS studies.

In addition, the sulfur on Ni(100) SEXAFS study revealed the reliability of chemisorption site determination from analysis of either the polarization dependent amplitude ratio [Eq. (31)] or the determination of the absolute effective coordination number N^* by amplitude transferability from a model compound [Eq. (30)]. As shown in Table 10.5 experimental and calculated amplitude ratios for the fourfold hollow agree to better than 5% with experimental error bars of 10%. The comparison between experimental and calculated absolute N^* values for the fourfold hollow agree to better than 10% with conservative experimental error bars of <25%. This larger error bar originates from limitations of amplitude transferability as discussed for bulk systems by Stern et al. (136) and Eisenberger and Lengeler (137). Three different straight line fits to the amplitude ratio between the bulk NiS and the $\Theta = 90°$ sulfur on Ni(100) spectra [Eq.

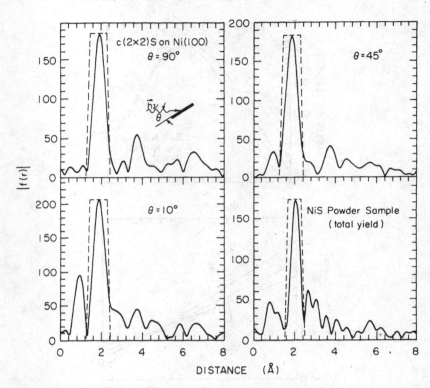

Figure 10.40. Fourier transforms of the raw SEXAFS signals $\chi(k)k^2$ for $c(2 \times 2)$ sulfur on Ni(100) recorded at different x-ray incidence angles and for bulk NiS. The peaks around 3.8 Å in the S/Ni(100) transforms correspond to the S—Ni second nearest-neighbor distance on the surface as discussed in the text. From Brennan (230).

Table 10.5. Experimental Versus Calculated Coordination Numbers and Ratios for $c(2 \times 2)$ Sulfur on Ni(100)

Incidence Angle (deg)	Experimental	Fourfold Hollow	Twofold Bridge	Onefold Atop
10/90	1.16 ± 0.10	1.20	4.31	∞
10/45	1.15 ± 0.10	1.09	1.59	1.94
10	4.42 ± 1.04^a	4.49	4.03	2.91
45	3.77 ± 0.79^a	4.13	2.53	1.50
90	3.94 ± 0.75^a	3.75	0.94	0

^{a}Relative to $N = 6$ for bulk NiS.

515

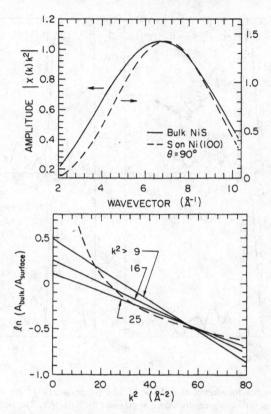

Figure 10.41. (a) SEXAFS amplitudes $|\chi(k)k^2|$ of the S—Ni first neighbor signal for bulk NiS and for sulfur on Ni(100) recorded at normal x-ray incidence. (b) Plot of the natural logarithm of the amplitude ratio of bulk NiS and $c(2 \times 2)S$ on Ni(100) at $\Theta = 90°$ versus k^2 according to Eq. (30). Three straight line fits for different low-k-cutoff values are shown. Ordinate intercept at $k = 0$ yields the absolute nickel coordination number of sulfur on the surface. From Stöhr, Jaeger, and Brennan (53).

(30)] are shown in Fig. 10.41 similar to the case of oxygen on Ni(100) discussed earlier and shown in Fig. 10.35. Again, the relatively large error bars for the derived N^* value for sulfur on Ni(100) relative to $N^* = 6$ for bulk NiS arises from the curvature of the amplitude ratio curve.

One aspect of the amplitude analysis for surfaces that has not been discussed in the literature is the direct determination of the vertical spacing Z of the adsorbate above the surface plane. With accurate polarization dependent amplitude ratios $P = N^*(\Theta_1)/N^*(\Theta_2)$ not only the chemisorption site but also Z can be determined quite accurately. For the fourfold hollow site we use Eq.

(29) to first calculate the bond angle β with respect to the surface normal. Using $P = N^*(10°)/N^*(90°) = 1.16 \pm 0.10$ for $c(2 \times 2)$ sulfur on Ni(100) we derive $\beta = 52.7 \pm 1.2°$. From knowledge of the nickel lattice constant $a = 3.524$ Å we find $d = 1.762$ Å (Fig. 10.24) and $Z = d \tan \beta = 2.31 \pm 0.10$ Å in excellent agreement with the value $Z = 2.37 \pm 0.03$ Å determined from the S—Ni first nearest-neighbor distance.

Cl/Cu(100). The geometry and electronic structure of $c(2 \times 2)$ chlorine on Cu(100) was investigated in 1982 by Citrin et al. (231) in a combined study utilizing SEXAFS, angle-resolved photoemission, spectroscopy (ARPES), and a self-consistent electronic structure calculation. This combined scheme allowed the structural parameters derived from the SEXAFS data to be taken as input parameters for a surface linear augmented plane wave calculation of a layered slab. The calculated dispersion of chlorine-induced surface states and resonances was then compared to those from previously published and newly measured ARPES spectra.

The chemisorption geometry for $c(2 \times 2)$ chlorine on Cu(100) as derived by SEXAFS consists of a simple chlorine overlayer in fourfold hollow sites with a bond length of 2.37 ± 0.02 Å corresponding to $Z = 1.53 \pm 0.03$ Å. As for sulfur on Ni(100) the SEXAFS measurements above the chlorine K-edge (2820 eV) were carried out by elastic Auger yield detection using the chlorine KLL Auger line. CuCl with $N = 4$ and $R = 2.341$ Å served as a model compound. The Fourier transform revealed the nearest-neighbor Cl—Cu distance (2.37 Å) and a weaker polarization dependent distance corresponding to a combination of four third-nearest-neighbor copper atoms in the second layer and to eight fourth-nearest neighbor copper atoms in the surface plane. This second observed distance falls at 4.31 ± 0.04 Å for $\Theta = 90°$ and 4.26 ± 0.05 Å for $\Theta = 5°$ and confirms the chemisorption site determined from the SEXAFS amplitude. With the structural parameters determined by SEXAFS, good agreement was obtained between the measured and calculated dispersion of chlorine-induced valence features. This study is a good example of how the structural parameters determined by SEXAFS serve as input for first principles theoretical calculations of surface response functions. Many of the self-consistent calculations carried out for extended periodic surface geometries depend on such input since they do not minimize total energies.

Citrin et al. used their results for $c(2 \times 2)$ chlorine on Cu(100) to estimate the bond length and geometry of $c(2 \times 2)$ chlorine on Ag(100), which had previously been studied by LEED (232). They suggest that for this system $Z = 1.50$ Å, which is shorter than the LEED value $Z = 1.67 \pm 0.10$ Å. A previous electronic structure calculation for chlorine on Ag(100) (233) similar in nature to that used for $c(2 \times 2)$ chlorine on Cu(100), had used too long a Cl—Ag bond length and lead to the suggestion of a structural model in conflict with

LEED. Citrin et al. suggest that this disparity might vanish if the calculation was repeated with the Cl—Ag distance estimated by them. Unfortunately, the $c(2 \times 2)$ chlorine on Ag(100) SEXAFS cannot be measured by the EAY technique because of interference between photoemission peaks and the chlorine *KLL* Auger peak (see Table 10.1). The $c(2 \times 2)$ chlorine on Ag(100) SEXAFS has recently been measured using TY detection by Lamble and King (234). This study determined the Cl-Ag distance to be 2.69 ± 0.03 Å, which corresponds to $Z = 1.75 \pm 0.05$ Å. The SEXAFS study confirms the LEED result but is of higher accuracy.

Te/Cu(100) and Cu(111). Tellurium overlayers on Cu(100) and Cu(111) were investigated in 1982 by Comin et al. (235) by means of SEXAFS. On Cu(100) tellurium was found to form a $p(2 \times 2)$ overlayer ($\frac{1}{4}$ ML coverage) in agreement with earlier LEED work. The long-range arrangement of tellurium on Cu(111) was found to correspond to a $(2\sqrt{3} \times \sqrt{3})R\,30°$ pattern and $\frac{1}{3}$ ML coverage. SEXAFS measurements in the TY mode were carried out above the tellurium L_3-edge (4340 eV). This allowed an energy range of 270 eV to be utilized before the L_2-edge (4610 eV) interfered.

The SEXAFS data for $\Theta = 90°$ for the two surfaces are shown in the upper panel of Fig. 10.42. The lower panel compares the SEXAFS amplitudes corresponding to the first-nearest-neighbor peak taken at different angles of x-ray incidence. Note the opposite polarization dependence. From this polarization dependence and by phase and amplitude comparison with the model compound Cu_2Te Comin et al. derived the *local* structure around the chemisorbed tellurium atoms. In their amplitude analysis they assumed $\phi_2(k) = \phi_{02}(k)$ and used Eq. (19) with $c = 0.2$.

For $p(2 \times 2)$ tellurium on Cu(100) Comin et al. find chemisorption in the usual fourfold hollow site with a first-nearest-neighbor bond length of 2.62 ± 0.02 Å, corresponding to $Z = 1.90 \pm 0.02$ Å (cf. Fig. 10.43). This value is more accurate and larger than that derived from LEED intensity analysis ($Z = 1.70 \pm 0.15$) (236). Despite the good signal-to-noise ratio of the data no reliable higher neighbor distances were observed. Comin et al. attribute this to increased vibrational amplitudes for the more weakly bound high-Z atoms as compared to the low-Z atoms sulfur and chlorine discussed previously.

The fact that the amplitude of the data for tellurium on Cu(111) at 90° is greater than that at $\Theta = 20°$ implies that the Te—Cu bond is oriented predominantly in the surface plane. From comparison of the derived first nearest-neighbor distance $R = 2.69 \pm 0.02$ Å to that between copper atoms in the (111) surface plane (2.56 Å) it is seen that the amplitude behavior cannot be satisfied without atomic rearrangement. The suggested structure for $(2\sqrt{3} \times \sqrt{3})R\,30°$ tellurium on Cu(111) is shown in Fig. 10.43 and consists of tellurium atoms replacing copper surface atoms in a nearly substitutional manner ($Z = 0.84 \pm$

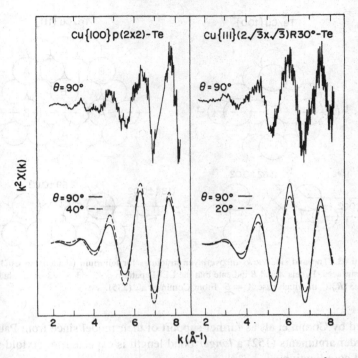

Figure 10.42. Upper: Background-subtracted raw SEXAFS data multiplied by k^2. Lower: Polarization-dependent filtered data of first nearest neighbors. Filtered data were obtained using window functions spanning 1.2–3.4 Å around the dominant Te—Cu peak in the Fourier transformed data (not shown). The intensity scales (arbitrary) are not the same for the (111) and (100) surfaces. From Comin et al. (235).

0.02 Å above the surface plane) so as to accommodate the longer Te—Cu than Cu—Cu bond. Each tellurium atom is surrounded by six first-nearest-neighbor copper atoms in the surface plane. In Fig. 10.43 we have added labels A and B to the tellurium positions pictured by Comin et al. to account for the $(2\sqrt{3} \times \sqrt{3})R\,30°$ long-range overlayer geometry consistent with $\frac{1}{3}$ ML coverage. SEXAFS only gives the short-range geometry. Comin et al. note that the SEXAFS first-nearest-neighbor bond length is completely isotropic to better than ± 0.005 Å for $\Theta = 90°$ and $\Theta = 20°$. As for $p(2 \times 2)$ tellurium on Cu(100) no higher nearest-neighbor distances were observed.

The findings of the SEXAFS study are unusual in that they deviate from the expected threefold hollow chemisorption site. As pointed out by Comin et al. (235) only a few cases that involve chemisorption of low-Z atoms are known where the adsorbate does not occupy the highest symmetry metal hollow sites (237, 238). The increased Te—Cu bond length on Cu(111) over that on Cu(100)

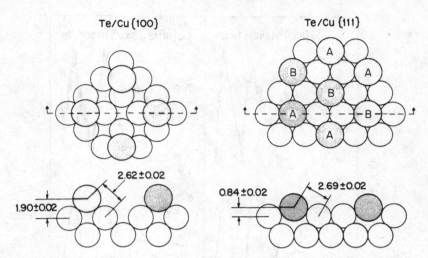

Figure 10.43. Top and side views of overlayer structures for tellurium (shaded) on Cu(100) and Cu(111) surfaces. Labels A and B indicate that the LEED pattern is $(2\sqrt{3} \times \sqrt{3})R30°$, rather than $(\sqrt{3} \times \sqrt{3})R30°$ in which case $A = B$. From Comin et al. (235).

is quoted by Comin et al. as further support of their model since from Pauling's bond order arguments (152) a *longer* bond length is expected for sixfold versus fourfold coordination.

The data quality lends strong credibility to the basic structural models proposed by Comin et al. (235). In particular, we note that the site could be independently determined from relative and from absolute coordination numbers despite the weaker L_3- than K-edge anisotropy. For the $p(2 \times 2)$ tellurium on Cu(100) structure the bond angle β (Fig. 10.24b) is calculated to be 43.5° from the derived distances. In this case the interference term is negligibly small (cf. Fig. 10.26) and amplitude analysis with $c = 0$ and $c = 0.2$ will yield nearly identical results. In addition the distance analysis can be carried out with the phase shift $\phi_2(k)$ derived from the model compound. For $(2\sqrt{3} \times \sqrt{3})R\,30°$ tellurium on Cu(111) we obtain $\beta = 72°$. For $\Theta = 90°$ the interference term is again negligible (cf. Fig. 10.26) but for $\Theta = 20°$ we calculate $C_2/C_{02} = 5.9$ or a 17% contribution from the interference term. It can be shown that this does not significantly affect the amplitude analysis and thereby the site determination. However, it is surprising that the distances determined for $\Theta = 90$ and 20° are the same within ± 0.005 Å. This result would follow if for tellurium $\phi_2(k)$ and $\phi_{02}(k)$ (note that the difference depends only on the absorber atom) were almost identical. Theoretical calculations indicate, however, that $\phi_2(k) \neq \phi_{02}(k)$ and an analysis carried out by us finds that distances derived with theoretical phase shifts ϕ_2 and ϕ_{02} can differ by as much as 0.1 Å [see the following section for Ag/Si(111)]. An alternate explanation involves cancellation of the phase shift correction by the structural distortion necessary for the $(2\sqrt{3} \times \sqrt{3})R\,30°$ LEED pattern. The $(2\sqrt{3} \times \sqrt{3})R\,30°$ LEED pattern implies that there are two inequivalent tellurium sites or that each tellurium atom is surrounded by nonequivalent copper atoms. Thus one would expect at least two different Te—Cu bond lengths with different bond angles β. This would result in a polarization dependent distance in EXAFS that might cancel the interference term correction. The detailed determination of the long-range structure will have to await future measurements by another technique such as Rutherford ion scattering, x-ray Bragg reflection diffraction, or LEED. From these considerations it appears that in future SEXAFS studies using *K-edges* a slight polarization dependence of the first nearest-neighbor distance might help to link the *short-* and *long*-range structure of surface complexes.

I/Cu(100) and Cu(111). The structure of chemisorbed iodine on Cu(100) and Cu(111) was investigated in 1980 by Citrin, Eisenberger, and Hewitt (CEH) (65) using TY SEXAFS. Besides being the first structure determination of this system the goal was to also test the reliability and accuracy of amplitude analysis using good signal-to-noise SEXAFS data. The previous study of I/Ag(111) (to be discussed) had still suffered from a large noise component and had not utilized the advantage of relative amplitude analysis by measurement of the polarization dependence. Improvements in data quality were achieved by using TY rather than EAY detection on the basis of the much larger TY collection efficiency. Citrin et al. (65) pointed out that for reliable amplitude analysis care has to be exercised in positioning the sample in the x-ray beam as discussed in Section 10.3.1. Particularly at grazing incidence angles, modifications in the SEXAFS amplitudes may be observed if part of the signal arises from the edge of the sample where strong gradients in adsorbate concentration may exist. This problem is more severe for TY than EAY measurements since in the latter case the electron analyzer usually has a restricted acceptance area defined by the electron optics.

The SEXAFS data for $\frac{1}{3}$ ML iodine on Cu(111) are shown in Fig. 10.44 in comparison with bulk γ-CuI, which served as a model compound. The surface data are of good signal-to-noise quality. For iodine on Cu(111) the LEED spot pattern corresponded to $(\sqrt{3} \times \sqrt{3})R\,30°$ while a $\frac{1}{4}$ ML $p(2 \times 2)$ pattern was investigated for the Cu(100) surface. The $(\sqrt{3} \times \sqrt{3})R\,30°$ iodine on Cu(111) overlayer was found to be characterized by a bond length of 2.66 ± 0.02 Å as compared to a distance of 2.69 ± 0.02 Å for $p(2 \times 2)$ iodine on Cu(100). In their amplitude analysis CEH originally used Eq. (19) with $c = 0$, which in an erratum was changed to $c = 0.2$. Because of the rather symmetric chemisorption geometry ($30° < \beta < 50°$) this did not affect the final result as shown in Table 10.6. Using both relative and absolute coordination numbers iodine was found

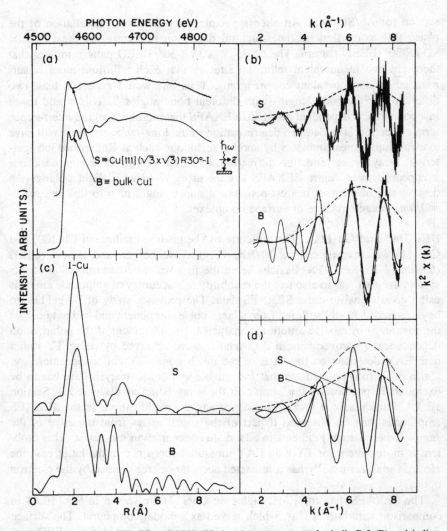

Figure 10.44. (a) Raw EXAFS and SEXAFS data in energy space for bulk CuI (B) and iodine adsorbed on Cu(111) (S). (b) Raw data in k space after background subtraction and multiplication by k^2. The amplitudes have been normalized to the edge jumps. (c) Fourier transforms of data from (b). The double peaks at $R \approx 3.3$ and 3.9 Å in CuI are due to I—I. (d) Retransformed data from (c), after filtering with a window function $0.8 \leq k \leq 3.6$ Å^{-1}. The filtered data are also shown in (b). From Citrin, Eisenberger, and Hewitt (65).

Table 10.6. Calculated Versus Experimental Values of $N_S = 2N^*/3$ for Iodine on Copper: Calculated Values are for $c = 0.2$ $(c = 0)$.

Θ (deg)	Atop	Bridge	Hollow	Experimental
Cu(111)				
90	0.47(0.33)	1.07(0.90)	1.67(1.46)	1.6 ± 0.2
20	1.00(1.22)	1.76(2.05)	2.53(2.89)	3.0 ± 0.6
20/90	2.13(3.65)	1.65(2.28)	1.52(1.98)	1.9 ± 0.4
Cu(100)				
90	0.47(0.33)	1.06(0.89)	2.42(2.24)	1.9 ± 0.2
20	1.00(1.22)	1.75(2.04)	3.09(3.38)	2.8 ± 0.6
20/90	2.13(3.65)	1.66(2.29)	1.28(1.51)	1.5 ± 0.3

to chemisorb in the hollow sites on both Cu(100) and Cu(111). CEH point out that for their L_3-edge measurements it is the combination of *both* relative and absolute coordination numbers that allow the site to be determined unambiguously. In the development of SEXAFS as a reliable structural tool the study of CEH showed for the first time that amplitudes may be transferred between bulk models and adsorbate systems with accuracies of better than 25%. This conclusion was verified for K-edges and low-Z adsorbates a year later by Brennan, Stöhr, and Jaeger for sulfur on Ni(100) (74) as discussed earlier.

I/Ag(111). The study of iodine on Ag(111) by Citrin, Eisenberger, and Hewitt (CEH) in 1978 (36) represents the first SEXAFS investigation of a well-characterized adsorbate system. Therefore, the main importance of this work is not so much the structure determination of the $\frac{1}{3}$ ML ($\sqrt{3} \times \sqrt{3}$)R 30° iodine on Ag(111) system but its impact on the field of surface crystallography by introducing a novel powerful technique.

Nevertheless, CEH did determine the structure of iodine on Ag(111) with previously unachieved precision. The data are shown in Fig. 10.45 and are arranged to facilitate comparison with Fig. 10.44, which presents comparable spectra for iodine on Cu(111) recorded more than two years later. The raw data in the upper left corner in both cases are for $\frac{1}{3}$ ML chemisorbed iodine and demonstrate the progression of the SEXAFS technique between 1978 and 1980. The double peak structure in the Fourier transform for I/Ag(111) arises from a Ramsauer–Townsend resonance in the silver backscattering amplitude. For analysis purposes the low-R shoulder can be eliminated by a window function leaving the signal due to the main peak, only. The smooth curves in Fig. 10.45d correspond to the backtransformed signal of the dominant transform peak. The I—Ag bond length was derived to be 2.87 ± 0.03 Å by using γ-AgI as a model compound. The chemisorption site was determined by amplitude comparison with the model using Eq. (30). The effective coordination number was

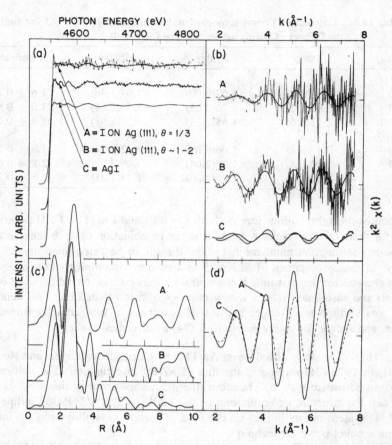

Figure 10.45. (*a*) Raw data in energy space for three different I/Ag systems. Curves *A* and *B* are EAY SEXAFS (Auger) spectra; curve *C* is an EXAFS (transmission) spectrum. Curve *C* is overlaid on curve *A*. (*b*) Raw data in momentum space after background subtraction and multiplication by k^2. Smooth curves are retransformed and filtered data [see (*d*)]. (*c*) Fourier transforms in distance space of raw data from (*b*). The peaks at 2.6 Å have been arbitrarily set equal in height. (*d*) Normalized, retransformed data from (*c*) after filtering with a window from $1.6 \leq R \leq 3.8$ Å. Note differences in phase and amplitude. From Citrin, Eisenberger, and Hewitt (36).

calculated according to Eq. (22) with $c = 0$. The results shown in Table 10.7 clearly favor chemisorption in the threefold hollow site on Ag(111).

The SEXAFS results support the chemisorption site determined previously by LEED intensity analysis (239). The LEED value for the I—Ag distance 2.80 ± 0.15 Å is also confirmed within its experimental error. However, the SEXAFS error bar is a factor of five smaller.

Table 10.7. Calculated Versus Experimental Values of $N_S = 2N^*/3$ for Iodine on Ag(111): Calculated Values are for $c = 0$

Atop	Calculated Bridge	Hollow	Experimental rel. to AgI
1.21	1.99	2.81	3.2 ± 0.8

10.5.3.2. Semiconductor Substrates

Semiconductor surfaces differ in many ways from metal surfaces. The class of semiconductors that is of particular interest in the electronics industry consists of tetrahedrally bonded units (240). Cleaving of such a sample results in a surface with broken bonds, usually referred to as dangling bonds. In contrast to metals the arrangement of atoms near the cleavage plane may be modified considerably from that in the bulk. In fact, in many cases the structure of clean semiconductor surfaces is not understood or at least still controversial, such as the 2×1 and 7×7 structures formed for Si(111) surfaces (240). From a chemisorption point of view it may be expected that the directional bonding in semiconductors may lead to simple overlayer geometries that result from saturation of the dangling surface bonds. However, such ideas fail in many cases. Often chemisorption leads to disordered surface structures as, for example, the interaction with oxygen (241, 242). In such cases the structural analysis becomes complicated and many of the conventional structural tools, which depend on long-range periodicity, like LEED, cannot be applied. Even for ordered overlayers techniques that probe long-range order may be overwhelmed by the structural complexities introduced by the underlying substrate. SEXAFS offers the advantage that it is a local probe and as such samples the structure within about 5 Å of an adsorbed atom.

In the following we shall discuss some first SEXAFS studies of chemisorption phenomena on Si(111) and GaAs(110). The local structure of the Si(111) surface is shown in Fig. 10.46. For Si(111) we also show several adsorbate bonding configurations. The surface structure of clean GaAs(110) has been thoroughly studied by LEED (243) and theoretical calculations (244) and found to involve bond angle changes. There is evidence that upon chemisorption the surface might "heal" to the bulk configuration (245).

O/Si(111) 2 × 1. The oxidation of the Si(111) surface has been the subject of numerous experimental and theoretical investigations. The most detailed experimental information has been supplied by photoemission (246–256), electron energy loss spectroscopy (241, 246, 257), and absorption edge fine structure measurements (258, 259). Photoemission clearly established the existence of a

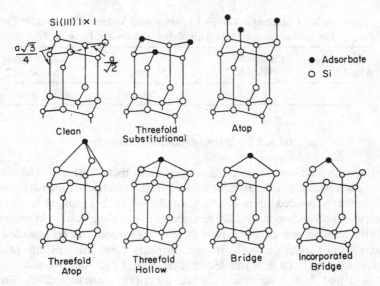

Figure 10.46. Model of the unreconstructed Si(111) 1 × 1 surface (clean) and various chemisorption geometries that may be encountered.

complicated multiple-step oxidation process of the cleaved Si(111) 2 × 1 surface. Similar to the oxygen on Al(111) case (153) several different chemical shifts of the silicon 2p core line could be distinguished with increasing oxygen dosage and dosing procedure. The presence of hot filaments, in particular, was found to lead to a more rapid oxidation that has been attributed to the production of excited oxygen, to be denoted O_2^* (241). Chemical shifts between 0.9 and 4.5 eV have been reported for the silicon 2p line and correlated with changes in the valence band (248–256). The bonding of oxygen on Si(111) has also been studied by generalized valence bond (245) and extended tight bonding (254) calculations. However, the crystallographic structure of the various oxygen phases on the Si(111) 2 × 1 surface identified by photoemission remained unsolved. LEED studies were impeded by the fact that the sharp LEED spot pattern of the clean surface rapidly disorders upon oxygen exposure (241).

The oxidation phase characterized by a 2.5 eV silicon 2p shift was studied by SEXAFS in 1979 by Stöhr et al. (68, 69). Spectra were recorded above the oxygen K-edge by collecting the SY with a CMA (E_k = 3 eV) and at a x-ray grazing incidence angle of 10° (E vector close to the surface normal). The spectra were compared to a thick thermally grown SiO_2 layer on Si(111), which served as a standard. Later measurements showed that the O—Si bond length and local oxygen coordination of such a sample closely resembles crystalline quartz α-SiO_2. Figure 10.47 shows the Fourier transforms $F(r)$ using the anal-

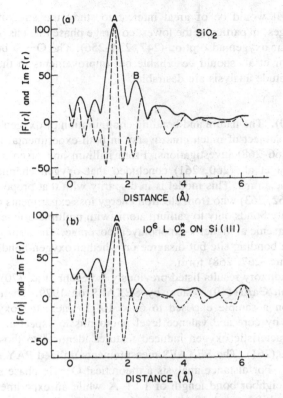

Figure 10.47. The SEXAFS data of (a) thermally grown SiO_2 and (b) of oxygen on Si (111).

ysis procedure of Lee and Beni with the O—Si phase shift derived from SiO_2. Peak B in Fig. 10.47a is due to the O—O second-nearest-neighbor distance in SiO_2. For oxygen on Si(111) corresponding to an exposure of 10^6 L O_2^* and a 2.5-eV chemical shift the O—Si bond length was derived to be 1.65 ± 0.03 Å, 0.04-Å longer than in bulk SiO_2. Amplitude analysis by comparison to SiO_2 yielded $N^* = 1.1$. Stöhr et al. (69) pointed out that this N^* value was derived from a plot according to Eq. (30), which yielded an unexpectedly (possibly nonphysical) large difference between the relative mean square displacement σ^2 on the surface and for the bulk, the bulk displacement being *larger*. If taken at face value the N^* value could eliminate all but three possible adsorbate geometries.

It appears that the study of Stöhr et al. should be repeated since improvements in many respects are now possible. It should be quite simple to improve the signal-to-noise ratio by TY detection and in addition study the polarization dependence. It would be of great interest to study the structure of different oxidation stages, in particular the lowest coverage phase that has been associated with molecular oxygen adsorption (247, 254–256). The O—Si bond length derived by Stöhr et al. should be reliable but improvements in the absolute and relative amplitude analysis are desirable.

O/GaAs(110). The nature and bonding configuration of oxygen on GaAs(110) has been the subject of much controversy in both experimental (260–265) and theoretical (266–268) investigations. From gallium and arsenic $3d$ core level shifts Pianetta et al. (260, 261) concluded that oxygen chemisorbs only on surface arsenic atoms. This model is in disparity with that proposed by Ludeke and Koma (262, 263) who from electron energy loss experiments concluded that oxygen initially bonds only to gallium atoms with multiple bridge bonds to both gallium and arsenic at higher (~ monolayer) coverage. Theoretical studies favor arsenic as the bonding site but disagree on whether oxygen bonds in molecular (266) or atomic (267, 268) form.

The contradictory results listed previously led Stöhr et al. (70) to investigate the oxygen on GaAs(110) system by SEXAFS on 1979. Their studies were carried out on a sample exposed to 6×10^9 L unexcited oxygen that was characterized by core and valence level photoemission spectra. These spectra revealed characteristic oxygen induced features identical to those reported by Pianetta et al. (261). The SEXAFS measurements utilized PAY detection with $E_p = 325$ eV. For distance analysis a theoretical O—Ge phase shift yielded a first-nearest-neighbor bond length of 1.51 Å while an experimentally derived O—Ni phase shift gave 1.70 Å. In lack of a more appropriate phase shift a precise distance could not be derived. From comparison of the measured back-scattering amplitude corresponding to the first-nearest-neighbor peak to that measured to oxygen on nickel Stöhr et al. concluded that there was no sizable O—O contribution as expected for molecular oxygen. They, therefore, concluded that oxygen chemisorbs in atomic form on GaAs(110).

The SEXAFS data suffered from a large noise component. This resulted from the fact that the spectra were recorded under unfavorable storage ring conditions (parasitic time, ~ 7-mA electron current) with a marginal photon flux of less than 10^9 photons s^{-1}. It is therefore clear that much better data could now be recorded as demonstrated for $p(2 \times 2)$ and $c(2 \times 2)$ oxygen on Ni(100). We recently reanalyzed the data using a more suitable phase shift derived from bulk GaN. We obtain a first-nearest-neighbor bond length of 1.59 Å, almost exactly in between the two values quoted earlier. Thus it appears that the oxygen first-nearest-neighbor bond length on GaAs(110) is close to 1.6 Å. A very conservative estimate would be 1.6 ± 0.1 Å. At this time it appears that the amplitude analysis performed in our 1979 paper cannot be trusted. If molecular oxygen was chemisorbed it would probably give too weak an EXAFS contribution to be detectable, at least for the low-quality published data.

A more convincing argument for atomic versus molecular oxygen chemisorption can, however, be made from the derived bond length, even with the large error bars of ± 0.1 Å. Partially stimulated by the short bond length of less than 1.7 Å predicted by the early SEXAFS study the bonding between oxygen and gallium or arsenic atoms on the surface was considered in more detail by molecular orbital considerations. Lucovsky and Bauer (LB) (269) pointed out that if oxygen was bonded to a single surface arsenic atom a bond length shorter by about 0.2 Å would be expected relative to that (~ 1.80 Å) in multicoordinated compounds such as As_2O_3. This was attributed to π as well as the normal σ bonding contribution in the situation where oxygen terminates an arsenic dangling bond. Lucovsky and Bauer predict an $O-As$ bond length on GaAs(110) of 1.62 ± 0.04 Å. An *ab initio* generalized valence bond calculation of Barton et al. (268) comes to a similar conclusion and predicts 1.63 Å for the $O-As$ bond length on the surface. Since a considerably longer bond length is expected for a chemisorbed oxygen molecule the derived SEXAFS first-nearest-neighbor distance appears to rule out molecular chemisorption. As summarized by Barton et al. (268) a consistent picture of experimental and theoretical results is now believed to be achieved. In this picture oxygen bonds in atomic form to surface arsenic atoms.

The SEXAFS study by Stöhr et al. should certainly be repeated. With better signal-to-noise data as a function of polarization it should be easy to accurately determine the chemisorption bond length and chemisorption site.

Cl/Si(111) 7 × 7 and Ge(111) 2 × 8. The chemisorption of chlorine on Si(111) 7 × 7 and Ge(111) 2 × 8 surfaces first received attention in 1976 when Schlüter et al. (270) suggested different chemisorption geometries for the two systems from polarization dependent valence band photoemission studies (271). Chlorine was proposed to chemisorb in the *atop* ionic site in Si(111) and in the *threefold* covalent site on Ge(111). Both surfaces have been investigated by Citrin, Eisenberger, and Rowe (272) by means of EAY SEXAFS. Since at the time of this writing these results have not been published, these measurements will not be discussed in detail. A detailed discussion is not needed anyway as revealed by the SEXAFS spectra in the chlorine K-edge shown in Fig. 10.48. In both cases the SEXAFS oscillations are only pronounced at grazing x-ray incidence ($\Theta = 10°$) and essentially nonexistent at normal incidence. This clearly indicates that in *both* cases chlorine chemisorbs in the onefold atop site, a finding confirmed by the strong polarization dependence of the near-edge fine structure. The shown results are a beautiful example of the strong polarization dependence of the K-edge EXAFS and fine structure for anisotropic chemisorption geometries.

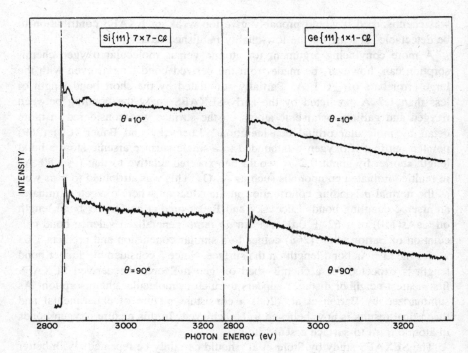

Figure 10.48. SEXAFS spectra for chlorine chemisorbed on Si(111) 7 × 7 and on Ge(111) 2 × 8 surfaces characterized by a 7 × 7 and 1 × 1 LEED pattern, respectively. Note the strong polarization dependence of the edge threshold structure and the EXAFS amplitude both revealing an on top chemisorption geometry. From Citrin, Rowe, and Eisenberger (272).

Te and I on Si(111) 7 × 7 and Ge(111) 2 × 8. The short-range structure of tellurium and iodine on Si(111) 7 × 7 and Ge(111) 2 × 8 surfaces was investigated by Citrin, Eisenberger, and Rowe (CER) in 1982 (66). The goal of this study was to test the ideas of bonding trends expected from simple chemical bond saturation arguments for monovalent iodine and divalent tellurium. Three of the studied cases were found to be consistent with such arguments while tellurium on Ge(111) occupying a threefold site was not. The TY SEXAFS study of CER made use of polarization dependent first-and second-nearest-neighbor distances and relative and absolute coordination numbers to establish the local chemisorption geometry. The data are, in general, consistent with a locally unreconstructed substrate in the presence of the adsorbates. For I/Si(111) the second-nearest-neighbor distance to the silicon atoms in the second layer is found to be greater by 0.1 Å than for an unrelaxed Si(111) 1 × 1 substrate, suggesting outward relaxation of the surface silicon atoms.

The SEXAFS data and Fourier transforms for three cases are shown in Fig.

10.49. Note the differences in first-nearest-neighbor amplitudes and phases as a function of polarization for the three cases. The calculated absolute and relative effective coordination numbers N^* (assuming $c = 0.2$) for different chemisorption geometries (see Fig. 10.46) and taking first- and second-nearest-neighbor contribution into account are compared to the measured ones in Table 10.8.

For I/Si(111) a 7 × 7 LEED pattern corresponding to ~ 1 ML was observed. Using a vapor SiI(CH$_3$)$_3$ model compound with an I—Si distance of 2.46 ± 0.02 Å yielded the I—Si first-nearest-neighbor distance on the surface as 2.44 ± 0.03 Å. Table 10.8 shows that iodine chemisorbs in the *atop site* as expected for a monovalent atom.

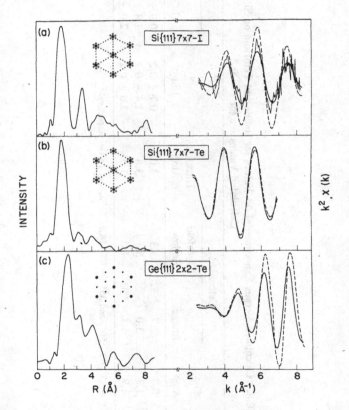

Figure 10.49. Fourier-transformed SEXAFS data (left) taken at $\Theta = 90°$ with corresponding observed LEED patterns. Background-subtracted raw data at $\Theta = 90°$ are also shown in (a). Polarization-dependent filtered data of first nearest neighbors (right) taken at $\Theta = 90°$ (solid) and $\Theta = 35, 50,$ and $35°$ (dashed) for (a), (b), and (c), respectively. Filter functions span 1.0–3.0 Å for (a) and (b) and 1.2–3.4 Å for (c). Note differences in amplitude and phase dependence with Θ between (a), (b), and (c). From Citrin, Eisenberger, and Rowe (66).

Table 10.8. Calculated Versus Experimental N^* Values for Iodine and Tellurium on Si(111) 7 × 7 and Ge(111) 2 × 8: Calculated Values are for $c = 0.2$

Incidence Angle (deg)	Calculated				Experimental			
	Atop	Bridge	Threefold Hollow	Threefold Atop	I/Si	I/Ge	Te/Si	Te/Ge
90	0.7	2.7	5.3	3.0	0.7 ± 0.2	0.9 ± 0.2	3.0 ± 0.4	2.9 ± 0.4
θ^a	1.3	3.0	5.1	3.8	1.4 ± 0.3	1.8 ± 0.4	2.9 ± 0.4	4.0 ± 0.5
$\theta/90$	1.9	1.1	1.0	1.3	2.0 ± 0.2	2.0 ± 0.3	1.0 ± 0.2	1.4 ± 0.2

[a]Corresponds to $\theta = 35°$, $40°$, $50°$, and $35°$ for I/Si, I/Ge, Te/Si, and Te/Ge, respectively.

532

The results for I/Ge(111) at approximately 1 ML coverage and a 1 × 1 LEED pattern are similar as for Si(111). Here GeI$_4$ vapor served as a model compound with a I—Ge bond length of 2.50 ± 0.03 Å. Iodine is also found to chemisorb on top of a surface germanium atom with a bond length of 2.50 ± 0.04 Å, as in the vapor model.

Figure 10.49 indicates a different chemisorption geometry for Te/Si(111) characterized by approximately 1 ML coverage and a 7 × 7 LEED pattern. Here almost no polarization dependence of the SEXAFS signal is observed indicating a bond angle β (Fig. 10.24) close to the magic angle 54.7°. Analysis using SiI(CH$_3$)$_3$ as a model, which according to CER gives only small and correctable bond length ($\Delta R \approx 0.01$ Å) and amplitude ($\leq 10\%$) errors, gives an experimental Te—Si surface bond length of 2.47 ± 0.03 Å. Amplitude analysis excludes all but the *bridge site* although the threefold atop site can only be excluded from comparison of the measured ($N^* = 2.9 \pm 0.4$) and calculated ($N^* = 3.8$) absolute N^* values for $\theta = 50°$ (cf. Table 10.8). The distinction between these sites is supported by bond length considerations. For both the bridge and threefold atop sites, the experimentally determined distance is a superposition of first- and second-nearest-neighbor distances. The measured distance of 2.47 ± 0.03 Å would imply that for the threefold atop site the tellurium atom is unphysically close ($R \approx 1.8$ Å) to the silicon atom directly underneath. Therefore, tellurium on Si(111) around monolayer coverage occupies the twofold bridge site with a first-nearest-neighbor distance of 2.44 ± 0.03 Å. Tellurium on Si(111) has also been studied by Comin, Citrin, and Rowe in 1983 (273) at lower coverage. At 0.5 ML coverage a 3 × 1 and 2 × 2 LEED pattern is observed that changes to a ($\sqrt{3} \times \sqrt{3}$)R 30° pattern at 0.25 ML coverage. For these two cases a larger Te—Si bond length of 2.51 ± 0.04 Å has been determined. Again tellurium chemisorbs in the bridge site but from second-nearest-neighbor distances the Si—Te—Si bridge is found to be tilted. The identification of the bridge site is the first of its kind and can be rationalized by the divalent nature of tellurium.

Tellurium on Ge(111) forms a 2 × 2 LEED pattern at about 0.5 ML coverage. The SEXAFS signal (Fig. 10.49) of the first-nearest-neighbor shell exhibits a larger amplitude at grazing than normal x-ray incidence and a polarization-dependent nearest-neighbor distance with a 0.08-Å larger distance at normal incidence ($R = 2.75 \pm 0.04$ Å). Using GeI$_4$ as a model, amplitude analysis (Table 10.8) clearly establishes the *threefold atop* chemisorption site. Iodine is found to "rest" on the silicon atom in the second plane at a distance of 2.45 ± 0.04 Å with three silicon atoms in the surface plane as second-nearest-neighbor at 2.78 Å. This explains the large measured average distance and the polarization dependence. This is a rather unusual chemisorption site with complex bonding that had never been observed before.

The results of CER for the first time allowed the local orbital picture of bonding at semiconductor surfaces to be tested. Previous structural work had failed because of lack of sufficient long-range order and/or the complexities of the information content of the experimental data. Of the simple chemisorption sites possible on Si(111) and Ge(111) surfaces three different ones were identified, none of which corresponds to the most common threefold hollow site observed for (111) metal surfaces. Furthermore, of the four cases studied three can be simply explained by valency and directional bonding arguments while one cannot.

For the atop chemisorption of iodine on Si(111) and Ge(111) CEH do not report any polarization dependence of the first-nearest-neighbor distance. This is very surprising according to Eqs. (34–36), which yield a 40% contribution ($C_2/C_{02} = 0.5/0.2$) of the interference term C_{02} for $\theta = 90°$ and a 13% contribution of opposite sign ($C_2/C_{02} = -1.5/0.2$) for $\theta = 35°$. Thus for $\phi_2 \neq \phi_{02}$ a polarization dependence of >0.02 Å is expected for the first-nearest-neighbor distance.

10.5.4.   Solid–Solid Interfaces and Schottky Barriers

The formation of solid–solid interfaces is one of the most interesting, complex, and technologically important phenomena in surface science. Viewed from a purely *academic point of view* such systems are in many ways more challenging than the surface complexes formed by gas–solid reactions (chemisorption) because of their immense structural diversity and complexity. If a contact is made by evaporating atoms A onto a clean substrate B a variety of structures may be formed. With increasing reactivity between A and B one may envision the following,

1. Clusters of A on B.
2. Epitaxial growth of A on B.
3. Solid solution formation (A into B or vice versa).
4. Ordered overlayer of A on B (with or without reconstruction of the surface of B).
5. Disordered overlayer of A on B (with or without reconstruction of the surface of B).
6. Compound formation.

Moreover, two or more of these structures may exist simultaneously. From an *applied point of view* the physics and chemistry of solid–solid interfaces play the key role in many of the technologically most important areas. They determine the characteristics and quality of devices in the electronics industry, the activity of supported catalysts in the oil industry, or the efficiency of energy conversion by solar cells.

Of the many important problems related with solid–solid interfaces the one that seems to have been studied the most is the formation of metal semiconductor interfaces (i.e., Schottky barriers) (240, 274, 275). Such systems have been thoroughly characterized in terms of their *electronic* structure and properties. Here the fundamental parameter is the *Schottky barrier height* (276). The energy barrier between a metal and an *n*-type semiconductor is defined as the energy necessary to promote an electron from the Fermi level to the conduction band. It has been recognized that the simple Schottky picture (277), which defines the barrier height in terms of energy states of the individual metal and semiconductor components, cannot account for the measured values. Because the position of the Fermi level at the interface determines the barrier height the problem reduces to the question: What parameters are responsible for energetically "pinning" the Fermi level when a metal contact is being made? At present this question has not been satisfactorily answered, most probably because different mechanisms are responsible for different systems. For example, for III–V compound semiconductor substrates Spicer et al. (278) have argued that defect states created during the very early stages of interface formation pin the Fermi level. For silicon, metal deposition in many cases results in compound silicide formation (274, 275) and the barrier height has been suggested to be determined by the chemical reactivity (279, 280), that is, the local bonding, of the deposited metals. Other models emphasize the stoichiometry and crystallographic structure of the interface (281, 282). Whatever the model, it appears that the initial stages of metal deposition (<2 monolayers) are most important and account for macroscopic barrier heights.

In sharp contrast to the wealth of literature from photoemission and Auger electron spectroscopy studies (240, 274, 275) (which yield information on electronic valence band states and interface composition) very few *structural determinations* have been carried out for the early stages of Schottky barrier formation. LEED has been employed to characterize ordered metal overlayers on the (111) and (100) surfaces of silicon and germanium by "spot patterns" (283, 284) but on these substrates no structure determination by dynamical intensity measurements and multiple-scattering analysis has been reported. Only the high-temperature (1 × 1) aluminum on GaAs(110) structure (285), the $\sqrt{3} \times \sqrt{3}$ silver on Si(111) (286), and the aluminum on Ge(111) (287) surfaces have been the subject of a quantitative structural analysis by LEED. Other structural investigations involved RHEED (288, 289) but the analysis of the measured intensities has not been developed to the point where microscopic atomic geometries can be determined. Low (290, 291) and medium (292) energy ion scattering have been applied for the $\sqrt{3} \times \sqrt{3}$ silver on Si(111) phase and epitaxial silicide formation of palladium on Si(111), respectively. The long-range structural complexity of the clean and metal covered surface and/or the problems associated with disorder (lack of periodicity) have been the main impediments of structural investigations. Because of its short-range sensitivity SEXAFS can overcome these problems and the following section will discuss some first results. In the future the combination of such structural data with data on electronic properties is hoped to lead to a more global understanding of Schottky barrier formation and to a better understanding of the parameters governing Schottky barrier heights.

Before we turn to the discussion of experimental SEXAFS results it is useful to examine what energy ranges are best suited for the study of metal adsorption edges. The free-electronlike *sp metals* sodium, magnesium, aluminum, and potassium are best studied above their K-edges, which fall into the range 1050–3700 eV (104). The L-edges may be employed for near-edge studies but are unsuitable for SEXAFS investigations. For the *3d metals* calcium ($Z = 20$) to zinc ($Z = 30$) the L-edges fall into the difficult energy region 350–1200 eV. The proximity of the respective L_3- and L_2-edges (~ 3 eV for calcium and ~ 23 eV for zinc) and the small separation of the L_1-edges (< 150 eV) furthermore, complicate matters (293, 294). Thus, except for the study of near-edge effects (295), the K-edges (~ 4040 eV for calcium and ~ 9960 eV for zinc) are best utilized.

The *4d metals* yttrium ($Z = 39$) to tin ($Z = 50$) can partly be investigated above the L-edges as shown in Fig. 10.50 for rhodium to tin (118). For the heavier 4d metals the L_3–L_2 spin-orbit splitting and the separation to the L_1-edge have become large enough that SEXAFS measurements become feasible above the L_2- and L_1-edges. Note that the L_1-edge is equivalent to the K-edge, a fact which has been proved experimentally by Apai and Stöhr (118). However, for the lighter 4d elements the L_1-edge is Coster–Kronig broadened and for $Z < 45$ (Rh) the K-edges should be used for SEXAFS. These fall in the energy range 17,040 eV for yttrium to 22,120 eV for ruthenium. However, at these high energies often focused x-ray beams are not available because of the limited acceptance of the focusing mirrors. All heavier metals such as the *5d's, the rare earths* and *actinides* need to be studied above the L-edges that fall in the energy range 5500 eV (La, $Z = 57$) to 20,500 eV (Th, $Z = 90$).

The first application of SEXAFS to the study of Schottky barriers was carried out in 1982 by Stöhr and Jaeger (SJ) (117) who investigated the 4d metals silver and palladium on Si(111) 7 × 7. From previous experimental data using a variety of techniques it was expected that silver and palladium would react very differently with silicon. The SEXAFS study confirmed this and showed that these two systems are textbook examples of the different surface structures that can be formed during the early stages (1–3 monolayers) of metal deposition. The observed structures range from epitaxial growth of silver on Si(111) at room

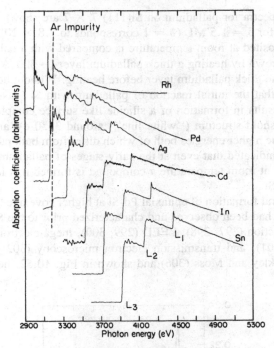

Figure 10.50. *L*-edge EXAFS spectra for the heavier *4d* transition metals recorded in transmission through thin films supported on graphite. The intensity jump around 3200 eV is due to an argon impurity in one of the ionization chambers that served as intensity monitors. Note the changes in the white line threshold structures for all three edges and the energy separations between the edges.

temperature and periodic silver-overlayer ($\sqrt{3} \times \sqrt{3}$) plus silver-cluster formation at elevated temperatures to compound (Pd$_2$Si) formation for the Pd/Si(111) system. In the following we shall review the early SEXAFS results of SJ and supplement them with more recent data by Stöhr et al. (296) for the Ag/Si(111) system. All measurements were performed using the JUMBO monochromator (78, 79) with InSb(111) crystals.

10.5.4.1. Pd/Si(111) 7 × 7

SEXAFS spectra for Pd/Si(111) by SJ (117) were recorded by EAY detection utilizing the palladium $L_2M_{4,5}M_{4,5}$ Auger line at 2600 eV as shown in Fig. 10.12. This resulted in a significant enhancement of the palladium L_2-edge jump (3331 eV) and reduction of the preceding L_3 EXAFS as compared to the TY detection mode. As shown in Fig. 10.11 the signal-to-background ratio was improved by more than a factor of 10.

SEXAFS spectra for palladium on Si(111) 7 × 7 are shown in Fig. 10.51. The spectrum for $\delta = 1.5$ ML ($\delta = 1$ corresponds to 7.8×10^{14} atoms cm^{-2}) palladium deposited at room temperature is compared with a palladium silicide (Pd$_2$Si) film grown by heating a thick palladium layer (~ 300 Å) to 500°C for 30 min, and the thick palladium layer before heating. Without detailed analysis it is apparent that the initial reaction of palladium with Si ($\delta = 1.5$) at room temperature results in formation of a silicide like surface complex. Note especially the threshold structure ("white line") around 3330 eV and the EXAFS frequency at the higher energies both of which differ from bulk palladium metal. These results indicated that even at the early stages of palladium deposition on Si(111) 7 × 7 at room temperature a compound is formed that locally strongly resembles Pd$_2$Si.

The structural formation of epitaxial Pd$_2$Si at higher coverage (> 10 ML) and heat treatment had been observed and characterized prior to the SEXAFS study by x-ray diffraction (297, 298), LEED (299, 300), mega-electron-volt ion scattering (292, 301), and transmission electron microscopy (302–305). As suggested by Buckley and Moss (306) and shown in Fig. 10.52, hexagonal Pd$_2$Si

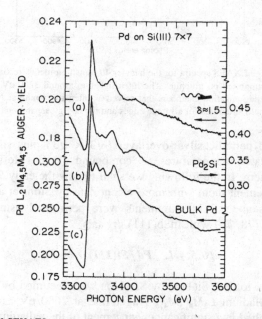

Figure 10.51. EAY SEXAFS spectra above the palladium L_2-edge for palladium on Si(111) 7 × 7; (a) 1.5 monolayers of palladium deposited at room temperature on Si(111) 7 × 7, (b) a thick layer of Pd$_2$Si grown at 500°C after deposition of 300 Å of palladium, (c) a 300 Å palladium layer deposited at room temperature. From Stöhr and Jaeger (117).

can grow on Si(111) with almost perfectly matching ($\sim 1.5\%$) lattice constants. As shown in Fig. 10.52 the 1×1 unit mesh of the Pd_2Si (0001) basal plane corresponds to a ($\sqrt{3} \times \sqrt{3}$)$R30°$ mesh of clean Si(111). This was confirmed by the nice LEED study of Okada et al. (299). The SEXAFS spectrum in Fig. 10.51b corresponds to that of such an epitaxial Pd_2Si film.

Prior to the SEXAFS study the structure of the surface complex formed at the early stages of palladium deposition at room temperature was largely unknown because of lacking sensitivity of the structural tools to the low coverage regime (<2 ML). For example, the LEED spots of the clean surface merge into a background during deposition of the first palladium layers. The LEED result is rather ambiguous and can be explained by either disordered atomic palladium chemisorption followed by metal cluster formation or by compound formation. The SEXAFS results unambiguously show the formation of a Pd_2Si-like compound around monolayer coverage. This may proceed locally by three palladium atoms replacing a surface silicon atom as in Fig. 10.52b. The replaced silicon atom (shown black in Fig. 10.52) is pushed up and rides on top of the three palladium atoms. It is coordinated by three new in-plane palladium atoms as in Fig. 10.52b. The structural results obtained from SEXAFS complement the findings on the valence band changes with palladium coverage obtained from UV photoemission (307).

The interesting "white line" phenomenon seen in Fig. 10.51 needs special discussion. At first sight, it is surprising to observe a larger white line threshold

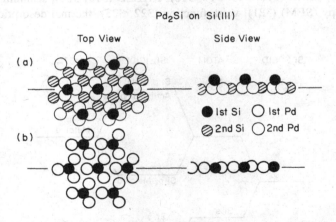

Figure 10.52. The two alternating atomic planes normal to the C_0 axis in Pd_2Si. The large clear circles are palladium atoms and the smaller darker circles represent silicon. (a) is the base plane of the hexagonal unit cell with the black silicon atoms raised half-way along the c axis, (b) is the plane of the black second layer silicon atoms. From Buckley and Moss (306).

structure for the palladium on Si(111) and Pd$_2$Si spectra than for bulk palladium metal since it appears from photoemission (307) that Pd$_2$Si has a noble metallike valence band structure. From this picture it is expected that the empty final state density above the Fermi level is larger in palladium metal and a larger white line is expected than for the silicide. The white line has been correlated to transitions from $2p$ initial to $\ell = 2$ (d-like) empty final states above the Fermi level. For $2p_{1/2}$ initial states the $j = \frac{5}{2}$ spin-orbit final-state components are dipole ($\Delta j = 0, \pm 1$) forbidden. As recently discussed by Rossi et al. (308) the observed white line difference can be explained by a larger density of unoccupied $d_{3/2}$ states in the silicide than the bulk metal. The experimental differences are in good agreement with the calculated empty $d_{3/2}$ density of states (309). As shown in Fig. 10.53 the hybridization of metal d with semiconductor sp^3 orbitals results in empty antibonding silicide states with significant d character. These are responsible for the observed pronounced white line in Fig. 10.51a and b. This observation implies that important information may be obtained from detailed coverage dependent near-edge studies of Schottky barrier systems, in particular by combining the information for all three L-edges ($2p_{3/2}$, $2p_{1/2}$, and $2s$ initial states).

10.5.4.2. Ag/Si(111) 7 × 7

The Ag/Si(111) system has been extensively studied by almost every available experimental surface technique such as photoemission (310–314), Auger (310, 315, 316), LEED (310, 315, 317, 318), RHEED (319, 320), scanning electron microscopy (SEM) (321), work function (322, 323), thermal desorption (315),

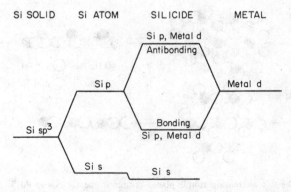

Figure 10.53. Molecular orbital bonding scheme between a silicon sp^3 orbital and the d band of a transition metal, resulting in filled bonding and empty antibonding orbitals of p and d character for the compound silicide.

electron energy loss (324, 325), electron yield (323), low-energy ion scattering (326, 327), and direct Schottky barrier height (328) measurements. In addition, a first principle $X\alpha$ cluster (329) and tight-binding (330) calculations are available. After this much experimental effort, it might be expected that the interaction of silver with Si(111) is well understood. Indeed, the growth kinetics at room and elevated temperatures appear to be well understood *above monolayer coverage*. As shown by LeLay et al. (319) the room temperature growth mode of silver proceeds in a layer-by-layer Frank–van der Merwe fashion with formation of Ag(111) planes in parallel exitaxy. Above 200°C the growth mechanism turns to a Stranski–Krastanov mode with nucleation of three-dimensional crystallites on top of an interface that displays a $(\sqrt{3} \times \sqrt{3})R30°$ LEED pattern and consists of <1 ML of silver atoms. The crystallites (clusters) grow in a fashion that even at large coverage (>20 ML) only a small fraction ($<20\%$) of the whole Si(111) surface area is covered (319). This fact is clearly seen in the beautiful SEM pictures published by Venables, Derrien, and Janssen (321). Thus even at high coverage the LEED picture is dominated by the $\sqrt{3} \times \sqrt{3}$ pattern.

The structure of Ag/Si(111) is less clear *below monolayer coverage*. The chemisorption of silver atoms at *room temperature* onto Si(111) 7 × 7 results in a gradual vanishing of the 7 × 7 into a 1 × 1 pattern with a strong background. LEED studies by Saitoh et al. (327) and Wehking et al. (310) find no evidence for any extra features up to 1.5 ML coverage when Ag(111) reflexes become detectable indicating expitaxial silver growth. In contrast, LeLay et al. (316) observe a blurred $\sqrt{3} \times \sqrt{3}$ LEED and RHEED pattern around $\frac{2}{3}$ ML silver coverage. From Auger (316) and ion scattering (327) studies a change in the silver growth process seems to occur at $\frac{2}{3}$ ML coverage that has been associated with the onset of silver metal crystallization. *The structure of silver atoms deposited at room temperature on Si(111) 7 × 7 at low coverage ($<\frac{2}{3}$ ML) has previously not been solved.* In contrast to the high temperature $\sqrt{3} \times \sqrt{3}$ structure most previous investigators did not even propose a structural model for this phase. Almost random adsorption is suggested by Gotoh and Ino from RHEED (320). From a Hubbard-like self-consistent field Hamiltonian Barone et al. (330) favored the atop over the threefold site. There is some experimental evidence for structural differences between the low coverage ($<\frac{2}{3}$ ML) room temperature and $\sqrt{3} \times \sqrt{3}$ high temperature phases (327).

Annealing a room temperature deposited Ag/Si(111) surface to above 200°C and below 600°C results in a $(\sqrt{3} \times \sqrt{3})R\,30°$ LEED pattern (319). The same pattern is also observed for silver deposition onto the heated surface and it is believed that both preparation methods give identical structures. For silver deposition onto the hot substrate the 7 × 7 LEED pattern continuously changes to $\sqrt{3} \times \sqrt{3}$ and the conversion is completed at $\frac{2}{3}$ ML coverage. Further silver deposition does not change the $\sqrt{3} \times \sqrt{3}$ pattern up to many monolayers of deposition, except for an increase in background. This has been explained as $\frac{2}{3}$ ML of silver atoms occupying periodic surface sites with a $\sqrt{3} \times \sqrt{3}$ unit mesh and the surplus of silver atoms being stored in clusters (319). Heating to $T > 600°C$ results in desorption of silver with the formation of a 3×1 LEED pattern at $\frac{1}{3}$ ML coverage (316, 320). Cooling the 3×1 structure below 200°C is found to result in a 6×1 pattern (320). Because of its ease of preparation and its stability the $\sqrt{3} \times \sqrt{3}$ LEED structure has received widespread attention and several structural models have been proposed.

The first structural models for $\sqrt{3} \times \sqrt{3}$ silver on Si(111) were suggested from coverage considerations. Spiegel in 1967 (317) and Bauer and Poppa in 1972 (315) associated the $\sqrt{3} \times \sqrt{3}$ structure with $\frac{1}{3}$ ML, Wehking et al. (1978) (310) with 1 ML and LeLay et al. (1978) (316) with $\frac{2}{3}$ ML silver coverage (1 ML = 7.8×10^{14} atoms cm^{-2}). These coverage estimates led to two distinctly different chemisorption models. Wehking et al. (310) suggested silver atoms to chemisorb on top of every surface silicon atom with *trimerization* of three adjacent silver atoms to account for an overall $\sqrt{3} \times \sqrt{3}$ symmetry. LeLay et al. (316) favored a *honeycomb model* where the silver atoms occupy threefold hollow sites and form a sixfold ring leaving an empty threefold hollow site in the center. Both the trimer and honeycomb models assumed no silicon surface reconstruction with the silver atoms riding on the surface. From analysis of low-energy ion scattering (LEIS) data Saitoh et al. in 1980 (326) proposed that the silver atoms are *embedded* in the Si(111) surface with a consequent lateral displacement of substrate silicon atoms. This model shown in Fig. 10.54 main-

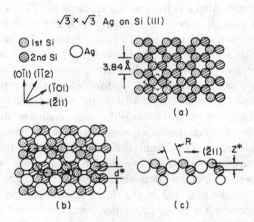

Figure 10.54. (*a*) Atomic arrangement of the clean Si(111) surface, (*b*) a model for the Si(111) $[\sqrt{3} \times \sqrt{3}]R\ 30°$ silver structure derived from low-energy ion scattering (coverage: $\frac{2}{3}$ monolayer), and (*c*) sectional view of the structural model. The silicon and silver atoms are shown with radii equal to those in their bulk state. From Saitoh et al. (326).

tains the threefold hollow honeycomb arrangement proposed by LeLay and corresponds to $\frac{2}{3}$ ML coverage. More extensive LEIS data published in 1981 by Saitoh et al. (327) in conjunction with LEED and Auger studies support this model as does a single scattering constant momentum transfer averaging (CMTA) LEED intensity analysis by Terada et al. (318) published in 1982. the CMTA LEED analysis favored the silver atoms to be 0.70-Å below the topmost silicon layer and 0.83-Å above the second silicon layer. The lateral displacement of the silicon atoms could not be determined.

The Ag/Si(111) 7 × 7 system was investigated by Stöhr and Jaeger (SJ) in 1982 (117) by means of SEXAFS. As for Pd/Si(111) the SEXAFS spectra were recorded by monitoring the $L_2M_{4,5}M_{4,5}$ Auger intensity (2720 eV). For room temperature deposition SEXAFS spectra were recorded for coverages of 2.5 ML up to the bulk metal. All spectra were found to yield identical bond lengths and amplitudes in agreement with epitaxial silver metal formation in the investigated coverage range predicted by previous experimental work (e.g., LEED). The Fourier transforms of the 2.5 ML coverage and bulk silver SEXAFS data of SJ are shown in Fig. 10.55c and d together with more recent results of Stöhr et al. (296) for lower silver coverages. Even for 1 ML coverage the transform is still dominated by the two peak structure characteristic of silver metal. Peak B represents the first-nearest-neighbor Ag—Ag distance (2.89 Å) in silver metal and its satellite at lower distance (~ 2 Å) arises from a Ramsauer–Townsend resonance in the silver backscattering amplitude (134). Using silver metal as a standard (2.89 Å) we find the Ag—Ag distance to be 2.89 ± 0.02 Å for $\delta = 2.5$ ML and 2.86 ± 0.02 Å for $\delta = 1$ ML. Closer inspection of Fig. 10.55b reveals an increased intensity of the low-distance peak around 2 Å relative to the higher-coverage data. This peak labeled A in Fig. 10.55a dominates at the lowest investigated coverage of $\delta = \frac{1}{3}$ ML. Now the characteristic silver metal peak B has disappeared in the noise. Peak A corresponds to the Ag—Si nearest-neighbor distance for the low-coverage phase of chemisorbed silver atoms. Another peak X that occurs at distances below 1 Å arises from underlying L_3 EXAFS oscillations, the background subtraction procedure and/or changes in the Auger signal strength with photon energy caused by movement of the x-ray beam with respect to the CMA focal spot. It should, therefore, be ignored.

The Ag—Si distance corresponding to peak A can be determined using an appropriate Ag—Si scattering phase shift. Unfortunately, no Ag—Si model compound exists and we have therefore used InP and Pd_2Si as models. We found the corresponding theoretical phase shifts of Teo and Lee (134) to predict the known bond lengths in the models to within 0.03 Å and therefore used the calculated Ag—Si phase shift. We note that the calculated Ag—Ag phase shift also yielded the nearest-neighbor distance in silver metal to within 0.02 Å. The calculated Ag—Si phase shifts $\phi_2(k)$ and $\phi_{02}(k)$ [Eq. (16)] are shown in Fig. 10.56. If we ignore the interference term in Eq. (16) [i.e., use the phase shift

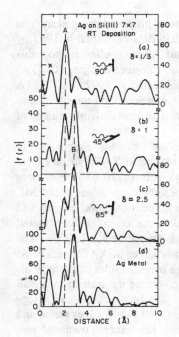

Figure 10.55. Fourier transforms of SEXAFS spectra recorded for different silver coverages ($\delta = 1$ corresponds to 7.8×10^{14} atoms cm^{-2}) on Si(111) 7 × 7. Transforms are for $\chi(k)k^2$ EXAFS signal above the silver L_2-edge and 3.1 Å$^{-1} \leq k \leq 8.3$ Å^{-1}. The satellite of the main peak B that corresponds to the Ag–Ag distance in the metal is due to a Ramsauer–Townsend resonance in the backscattering amplitude. Peak A for $\delta = \frac{1}{3}$ corresponds to the Ag–Si distance on the surface. From Stöhr et al. (296).

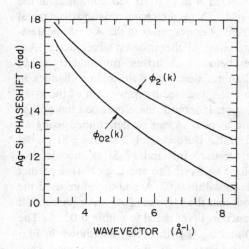

Figure 10.56. Calculated Ag–Si scattering phase shifts [ref. (134)] for a $\ell = 2$ [$\phi_2(k)$] and a mixed $\ell = 0$ and $\ell = 2$ [$\phi_{02}(k)$] final state and a $\ell = 1$ initial core state of silver.

544

$\phi_2(k)$] we derive the Ag$-$Si distance for the $\delta = \frac{1}{3}$ ML coverage case (Fig. 10.55a) measured at $\Theta = 90°$ (normal incidence) to be $R = 2.48 \pm 0.05$ Å. The same R value is obtained for $\Theta = 25°$ (grazing incidence). The SEXAFS amplitude is larger at $\Theta = 90°$ than at $25°$ with a ratio of $A(90°)/A(25°) = 1.45 \pm 0.30$. The relatively large error bars on the distance values and the amplitude ratio reflect the noise of the SEXAFS data. Note that the investigated coverage corresponds to only 2.5×10^{14} Ag atoms cm^{-2}. A discussion of the implication of these results will be given later together with the results for the $\sqrt{3} \times \sqrt{3}$ high-temperature phase.

The $\sqrt{3} \times \sqrt{3}$ Ag/Si(111) phase was also investigated in the SEXAFS study of SJ. The investigated sample consisted of 1.3 ML of silver deposited at room temperature and heated to 500°C. The Fourier transform showed both the Ag$-$Si and Ag$-$Ag distance that was attributed to a coexistence of the $\sqrt{3} \times \sqrt{3}$ phase with silver clusters. The Ag$-$Si distance was derived to be 2.45 ± 0.05 Å. The considerably larger SEXAFS amplitude at normal than for grazing incidence angles was used to determine the threefold hollow chemisorption site with the silver atoms close to or possibly underneath the surface silicon layer. SJ pointed out that such a chemisorption geometry was compatible with the observed Ag$-$Si distance only if the silicon atoms undergo a lateral displacement. The SEXAFS results for the $\sqrt{3} \times \sqrt{3}$ phase support the model in Fig. 10.54 suggested from LEIS data (326, 327). The coexisting silver clusters gave a Ag$-$Ag bond length indistinguishable from that for bulk silver. From previous EXAFS results on copper and nickel metal clusters on graphite (331) where changes in bond lengths were correlated with cluster size, SJ concluded that the silver clusters on the silicon surface must be large (>50-Å diameter). This is in good accord with other experimental results, for example, SEM (321).

In more recent SEXAFS studies of the $\sqrt{3} \times \sqrt{3}$ Ag/Si(111) surface (296) cluster formation was avoided by decreasing the silver coverage to 0.6 ML. The room temperature deposit was subsequently heated to 400°C. Using a theoretical phase shift $\phi_2(k)$ (134) the Ag$-$Si distance was determined to be 2.48 ± 0.04 Å for $\Theta = 90°$ and 2.46 ± 0.04 Å for $\Theta = 25°$. Comparison of the amplitudes of the backtransformed Ag$-$Si nearest-neighbor peaks yielded an amplitude ratio of $A(90°)/A(25°) = 1.65 \pm 0.15$. These results confirm those obtained earlier by SJ but are more accurate because of the absence of possible complications due to the presence of silver clusters in addition to the $\sqrt{3} \times \sqrt{3}$ phase.

The distance and amplitude results for the room temperature low coverage ($\delta = \frac{1}{3}$ ML) and ($\sqrt{3} \times \sqrt{3}$)R 30° heated silver phases on Si(111) can be used to derive structural models. Table 10.9 compares the observed amplitude ratio for the $\sqrt{3}$ phase with that predicted for various chemisorption geometries (cf. Fig. 10.46). As found by SJ the strong anisotropy of the SEXAFS amplitude is

Table 10.9. Experimental Versus Calculated Amplitude Ratios for Different Sites of √3 × √3 Silver on Si(111) Assuming c = 0 (c = 0.2)

Incidence Angle (deg)	Experimental Amplitude Ratio	Atop[a]	Calculated			
			Threefold Atop[a,b]	Threefold Substitution[a]	Threefold Hollow[a,c]	Threefold Hollow Embedded[d]
90/25	1.65 ± 0.15	0.29 (0.49)	0.81 (0.88)	1.29 (1.16)	1.29 (1.16)	1.71 (1.37)

[a] Assuming an unreconstructed Si(111) 1 × 1 surface and R_{Ag-Si} = 2.48 Å.

[b] The silver atoms sit 1.35 Å above the outermost silicon layer, 2.60 Å from the three surface silicon atoms and 2.13 Å above the second layer silicon atom. The calculated distance is polarization dependent: $R(25°)$ = 2.44 Å and $R(90°)$ = 2.51 Å for c = 0.2 and: $R(25°)$ = 2.42 Å and $R(90°)$ = 2.54 Å for c = 0.

[c] The silver atoms sit 0.75 Å above the outermost silicon layer, 2.34 Å from the three silicon atoms in the first layer and 2.69 Å from the three silicon atoms in the second layer. The calculated distance is polarization dependent: $R(25°)$ = 2.50 Å and $R(90°)$ = 2.48 Å for c = 0.2 and: $R(25°)$ = 2.52 Å and: $R(90°)$ = 2.48 Å for c = 0.

[d] The silver atoms sit between the first and second silicon layers 2.48 Å from the silicon atoms. The silicon layers are separated vertically by 1.25 Å.

only compatible with a bonding picture that involves Ag—Si bonds directed parallel to the surface. Clearly the atop and threefold atop sites can be excluded for both the low-coverage room temperature and $\sqrt{3}$ phases. The threefold substitutional and threefold hollow sites predict similar amplitude ratios, much closer to the experimentally observed values (Table 10.9). For the $\sqrt{3}$ structure good agreement is found between experimental and calculated amplitude ratio if the silver atom is moved between the first and second silicon layers.

In Table 10.9 we have listed calculations for both $c = 0$ and $c = 0.2$ [Eq. (19)]. The experimental amplitude ratio 1.65 ± 0.15 for the $(\sqrt{3} \times \sqrt{3})R\,30°$ phase cannot be matched for any structural geometry if $c = 0.2$ is assumed since for the limiting case of coplanar bonds the maximum amplitude ratio $A(90°)/A(25°)$ is calculated to be 1.47. This fact directly indicates that the approximation $\phi_2(k) = \phi_{02}(k)$ that leads to Eq. (19) with $c = 0.2$ is inappropriate. This is further substantiated by model calculations for the Ag/Si(111) system. Figure 10.57a shows calculated EXAFS spectra [Eq. (34)] for silver in the atop chemisorption site on Si(111) assuming a Ag—Si distance of 2.50 Å. We have used the backscattering amplitude $F(k)$ taken from Teo and Lee (134). The phase shifts $\phi_2(k)$ and $\phi_{02}(k)$ (Fig. 10.56) were taken from the same reference. Note the significant phase and k-dependent amplitude differences for grazing ($\Theta = 10°$) and normal ($\Theta = 90°$) x-ray incidence. Calculations for the coplanar threefold hollow site are almost identical to the ones shown for the atop site with a reversed polarization dependence. As shown in Figs. 10.57b and c, the corresponding amplitude ratios exhibit a strong k dependence ($>30\%$) caused by the k-dependent contribution of the interference term. This amplitude ratio predicted by Eq. (34) should be compared to that calculated from Eq. (19) with $c = 0$ and $c = 0.2$, respectively. For the atop geometry the approximation $c = 0.2$ used by Citrin et al. for iodine on Si(111) and Ge(111) [Ref. (66) and Table 10.8] predicts a ratio that is in error by about a factor of 2 around $k = 6$ Å^{-1} (middle of SEXAFS range). For the coplanar threefold geometry, $c = 0.2$ would give an amplitude ratio in error by a factor 0.65. Such errors are comparable in size to the differences in calculated amplitude ratios for $L_{2,3}$-edges (see Tables 10.6–10.9), which are used to distinguish between chemisorption sites and should be compared to typical error bars of $<10\%$ for amplitude ratios obtained from good signal-to-noise data. For the Ag/Si(111) case, the calculated amplitude ratios in Fig. 10.57 are found to be approximated quite well by the average value obtained by using Eq. (19) with $c = 0$. This validates the comparison between experimental and calculated amplitudes in Table 10.9.

The polarization dependence of the phase shown in Fig. 10.57a also leads to a *polarization dependent distance* if analysis is carried out with a single scattering phase shift $\phi_2(k)$. Note that this is usually done in practice since the phase shift derived from a model compound is used [Eq. (18)]. Analysis of the data in Fig. 10.57 (atop geometry) by use of $\phi_2(k)$ yields a distance that is

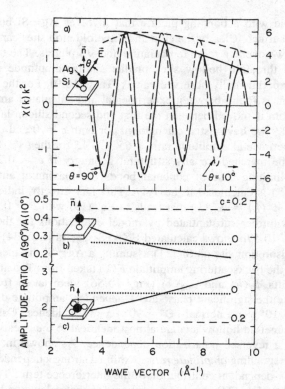

Figure 10.57. (*a*) Simulated SEXAFS signal for silver chemisorbed atop a surface silicon atom according to Eq. (34) for $\Theta = 10°$ (dashed) and $\Theta = 90°$ (solid). (*b*) Amplitude ratio of the signals in (*a*) (solid line) and the ratio expected from Eq. (19) using $c = 0$ or $c = 0.2$. (*c*) Same as (*b*) for the SEXAFS signal corresponding to silver chemisorption in a coplanar threefold hollow site. From Stöhr and Jaeger (138).

0.005 Å too *large* for $\Theta = 10°$ (note C_{02} is negative) and 0.04 Å too *small* for $\Theta = 90°$. For the coplanar threefold hollow site R is predicted correctly for $\Theta = 90°$ but is 0.035 Å too *small* for $\Theta = 10°$.

In summary, we favor the following chemisorption geometries. For the $(\sqrt{3} \times \sqrt{3})R\ 30°$ *phase* silver is embedded in the threefold hollow between the first two silicon layers with a distance of 2.48 ± 0.04 Å from a given silver to the six silicon nearest-neighbor atoms. The slightly (~ 0.02 Å) reduced distance observed for $\Theta = 25°$ arises from the fact that our analysis used a phase shift $\phi_2(k)$, which does not properly account for the interference term as discussed previously. The derived distance implies a lateral movement of surface silicon atoms, where six silicon atoms are displaced by 0.30 ± 0.10 Å toward the center of the six-ring silver honeycombs as shown in Fig. 10.54. This changes the lateral separation of the silicon atoms in the first layer from 3.84 Å to $d^* = 3.30 \pm 0.10$ Å (Fig. 10.54). We estimate the vertical spacing between the first and second silicon layers to be $Z^* = 1.36 \pm 0.30$ Å, that is, silver sits $Z = 0.68 \pm 0.15$-Å below the surface layer. This Z^* value compares to $Z^* = 0.78$ Å for the unreconstructed Si(111) 1×1 surface.

For the *low coverage* ($\delta = \frac{1}{3}$ ML) *room temperature phase* with no identifiable LEED pattern we find the same threefold hollow chemisorption site, but favor silver to sit above (but close to) the outermost silicon layer. The Ag—Si distance to the first layer atoms is shorter (~ 2.4 Å) than that to the second layer silicon atoms (~ 2.7 Å). This is consistent with the smaller amplitude anisotropy relative to the $\sqrt{3}$ phase. The average Ag—Si distance to the six first- and second-layer silicon atoms is 2.48 ± 0.05 Å. We favor a geometry where silver sits about 0.70 ± 0.25-Å above the surface with no or a small lateral displacement of the first layer silicon atoms. This model is in conflict with the atop chemisorption geometry suggested by theoretical calculations (330) and the expectation of bond formation of monovalent silver with the silicon dangling surface bond.

The SEXAFS results show a link between the room temperature and elevated temperature chemisorption sites for silver on Si(111). At room temperature, chemisorption occurs at random threefold sites or possibly already in small domains of the honeycomb structure. These domains may be too small to scatter coherently and therefore would not be seen as a spot pattern by LEED. At a certain silver coverage ($\sim \frac{2}{3}$ ML) lateral interactions between the silver atoms lead to two-dimensional growth of Ag(111) planes. This is supported by the 1 ML SEXAFS spectrum in Fig. 10.55 that is already dominated by silver metal. The silver crystallites have to be at least 2–3 layers thick to account for the small Ag—Si EXAFS contribution in Fig. 10.55*b*. Upon heating of the room temperature phase the activation barrier for incorporation of the silver atoms into the surface is overcome. It appears that this activation barrier must not be very large because of the resemblance of the room temperature and $\sqrt{3}$ chemisorption geometries. This may explain why LeLay et al. (316) observed a weak, blurred $\sqrt{3}$ structure already for room temperature deposition. The $\sqrt{3}$ structure may be regarded as a two-dimensional compound that gives rise to a sharp interface between bulk silicon and silver metal.

The structural models derived from SEXAFS provide a clear picture of the early and advanced stages of Schottky barrier formation for the Ag/Si(111) system. In particular, they reveal for the first time the arrangement of silver atoms chemisorbed at room temperature and low coverage. For the previously more intensively investigated $(\sqrt{3} \times \sqrt{3})R\,30°$ silver on Si(111) structure our model strongly supports that derived from LEIS (326, 327) and LEED and contradicts the trimer model suggested from LEED/Auger (310) and angle-resolved photoemission studies (312).

10.6. CRITIQUE OF SEXAFS AND FUTURE OUTLOOK

It appears that the main conceptual development period of electron yield SEX-AFS is over, at least as far as the use of the technique for the determination of static structures is concerned. Distances and chemisorption geometries can now be reliably established not only for surface complexes with long-range crystallographic order (> 100 Å) but also for systems where only the local (< 10 Å) structure is well defined. We have seen that distances can be determined with an accuracy of about 1% and coordination numbers to better than 20%, similar as in bulk EXAFS studies. Although the present development and capabilities of SEXAFS already constitute a significant step in the pursuit of answering surface structural problems, the other chapters in this book indicate that at best the first generation of problems has been addressed. Future refinements will allow us to obtain even more detailed and informative data and at the same time uncover inadequacies and limitations. Again the development of bulk EXAFS can serve as a guideline, and SEXAFS will be able to take advantage of the same refined analysis concepts and suffer from the same problems as bulk EX-AFS. No detailed discussion of these points is required since they are extensively covered elsewhere in this book. However, later we shall address some issues affecting the SEXAFS phase and amplitude with emphasis on surface specific applications.

The accurate determination of *distances* by use of the usual EXAFS equation assumes that the *static disorder* is small and characterized by a symmetric Gaussian pair distribution function. The distance determined by SEXAFS then corresponds to the peak of the radial distribution function for a given neighbor shell. By analogy with bulk studies of solids, static disorder is not expected to cause significant problems in SEXAFS investigations as long as it is remembered that the average distance is measured. It has been shown, however, that *thermal disorder* may well lead to erroneous distances derived by EXAFS (332). As discussed in Chapter 9 an anharmonic vibration potential will, in most cases, lead to an asymmetric pair distribution function with a sharp rise at low distance from hard core repulsion and a long tail at larger distance. The asymmetric pair distribution function corresponds to an asymmetric weighting of the EXAFS signal as a function of k (high R at low k and *vice versa*). Thus by truncation of the EXAFS at low k (~ 3 Å^{-1}) the high R contribution is lost and an apparent shrinkage in distance is observed. It may be expected that this effect is especially troublesome for surface studies because larger mean square displacements have been consistently found by LEED for clean metal surfaces relative to the bulk (8). Fortunately, anharmonic effects in SEXAFS are smaller than expected. SEXAFS studies are sensitive to *relative* mean square displacements σ_R^2 between the central and backscattering atoms (333, 334) and not like LEED to *absolute* mean square displacements σ_A^2 relative to the center of mass of the crystal. Thus, for example, if the outermost surface plane and the adsorbate vibrate in phase we obtain $\sigma_R^2 = 0$. Furthermore, in many cases adsorbate–substrate bonds at surfaces are more *covalent* and therefore stronger in nature than equivalent bulk bonds as indicated by an often shorter distance on the surface [e.g., oxygen and sulfur on Ni(100)]. In such cases the σ_R^2 values for the surface complex may actually be smaller on the surface than in the bulk! This is easily verified from the negative slopes in Figs. 10.35 and 10.41 for oxygen and sulfur on Ni(100). Even if anharmonic effects are present for a given system, this may be overcome by appropriate analysis procedures (332) and/or cooling of the sample. This has been discussed for chlorine on Ag(111) (234). In the future, these effects, which also strongly influence the amplitude in a non- Debye–Waller-like fashion, need to be studied in more detail. The disorder limitation restricts the use of SEXAFS to structures with $\sigma_R < 0.2$ Å. Because of the intrinsic anisotropy of surfaces it is to be expected that the relative mean square displacement of atoms on or near the surface can also be significantly anisotropic. For example, for a given adsorbate-substrate bond the relative vibrational amplitude parallel and perpendicular to the surface may be significantly different (335). Also, higher neighbor shells may exhibit a different magnitude and anisotropy of σ_R^2 than the first shell (336). The anisotropy is directly measured in a polarization dependent study (335, 336).

As described in Chapter 1, Section 1.5.2 *multiple-scattering* corrections of both phase and amplitude may be important for all but the first nearest-neighbor shell. Multiple scattering is particularly important as a higher neighbor atom is shadowed or partly shadowed from the absorber by an intervening atom (337). In this case the intervening atom may focus the electron wave onto the higher shell atom by forward scattering causing significant amplitude and phase corrections. This complication can in fact be turned into a benefit since it enables the determination of the forward scattering angle, for example, the bond angle between three atoms (337). This effect will be of importance for the study of molecules on surfaces. Consider, for example, the case of a diatomic molecule, CO chemisorbed via the carbon end to a surface nickel atom. SEXAFS above the carbon K-edge will yield both the C—O and C—Ni distances without multiple-scattering corrections. SEXAFS studies above the oxygen K-edge will be subject to phase and amplitude corrections for the O—Ni distance that can be analyzed to yield the Ni—C—O bond angle.

Finally, multielectron effects are adequately described by an additional k-dependent term in the EXAFS equation, $S_0^2(k)$, which reduces the amplitude by 20–30%. Stern et al. (136) find that for the $3d$-metal dichlorides S_0^2 is insensitive to changes in the chemical bonding (e.g., ionicity) and conclude that only the *atomic* S_0^2 values affect EXAFS. In SEXAFS many studies deal with *low-Z adsorbates*. Because of the large relative number of valence electrons it may be expected that in these cases S_0^2 will be sensitive to chemical effects. Furthermore, only a limited k range (≤ 300 eV) is often available for analysis, which emphasizes the contributions at low k. The range $k < 5$ Å^{-1} is the transition region between adiabatic to sudden core electron excitations (338) and $S_0^2(k)$ exhibits its largest changes. Thus we expect the amplitude effects caused by "intrinsic" multielectron excitations on the absorbing atom to be most important for SEXAFS studies on low-Z atoms. Here the effect of S_0^2 may not be simply of atomic origin and therefore cause problems in bulk to surface amplitude transferability.

The second multielectron effect that could be called "extrinsic" arises from the finite lifetime of the excited photoelectron state. The lifetime is limited by multielectron excitations of the neighboring environment of the central atom (e.g., plasmon losses, etc.) through which the photoelectron wave loses coherence with itself. Such effects are approximated by a k-dependent electron mean free path $\lambda(k)$ in analogy to that used in photoemission and Auger electron spectroscopy (see Fig. 10.5). Stern et al. (136) argue that for the first neighbor shell the effects of λ are already accounted for by the central atom term S_0^2 and the backscattering amplitude $F(k)$ [Eq. (14)]. This is very surprising because both S_0^2 and $F(k)$ are *atomic* quantities. As discussed by Eisenberger and Lengeler (137) this should not be true in general since different extrinsic losses are expected for different classes of materials where different excitation mechanisms will dominate (e.g., plasmon versus electron–hole pair excitation). For *surfaces* the loss mechanisms may be different than in the bulk (e.g., surface versus bulk plasmon excitations) and the effective mean free paths may be quite different even for carefully selected models. In addition, the mean free path may be anisotropic at surfaces (336). Again mean free path effects are emphasized by using only a restricted data range at relatively low-k (< 10 Å^{-1}) values.

It appears that multielectron effects limit the transferability of amplitudes between bulk and surface systems and more detailed studies of such effects are certainly needed. In the light of expected transferability problems the reliable determination of absolute coordination numbers for many surfaces by SEXAFS studies is remarkable. Note that in most cases the chemisorption site was independently verified by polarization-dependent amplitude comparisons for the same sample that do not suffer from multielectron complications and/or by the use of second-nearest-neighbor distances. It is believed that deviations from the linear behavior of ln $(A_{\text{Bulk}}/A_{\text{surface}})$ versus k^2 expected from Eq. (30) and observed for oxygen on Ni(100) in Fig. 10.35 and sulfur on Ni(100) in Fig. 10.41 can be attributed to multielectron effects that affect the k-dependent amplitude via the $S_0^2(k)$ or $\lambda(k)$ contributions.

Besides the previously mentioned effects that complicate the simple physical EXAFS picture usually assumed for data interpretation there are other limitations that prohibit certain systems to be studied by SEXAFS. The most severe limitation is the *interference* of adsorbate and substrate absorption edges, in particular if a substrate edge lies closely above the adsorbate edge. This case cannot be investigated by SEXAFS as discussed in Section 10.2.2.5. At present, SEXAFS studies have only been carried out for coverages of 0.1 ML or more, that is, more than about 1×10^{14} atoms cm^{-2}, with typical data collection times of a few hours. In the future it will certainly be possible to study $< 1 \times 10^{14}$ atoms cm^{-2} and reduce the time per spectrum to less than 30 min. SEXAFS measurements are possible for adsorbates with atomic numbers of $Z = 6$ or larger. The lowest Z atom, carbon, with its K-edge around 285 eV is most difficult to study by SEXAFS (339) because of the omnipresent carbon contamination structure in the monochromator transmission function (68) and the closely spaced K-edges of nitrogen (400 eV) and oxygen (532 eV), which in many cases limit the available SEXAFS range. Electron yield SEXAFS *averages* over all configurations of a given adsorbate atom on the surface unless the bond lengths are sufficiently (> 0.3 Å) different and thus may not be able to separate and determine the structures of different sites that are simultaneously present. This needs to be kept in mind especially for low-Z adsorbates where often hydrogenic molecules (e.g., OH, H_2O, CH_3, etc.) are present in addition to the species (e.g., O, C, CO, O_2) under investigation. *PSD SEXAFS* (38) can circumvent this problem by measuring the ion yield of a certain atomic or molecular species only (e.g., O^+ versus OH^+). Finally, the importance of suitable model compounds cannot be overemphasized. In order to avoid inadequacies of theoretically calculated phase shifts and amplitudes a model consisting of the same atom pair and similar coordination and bonding should be used. The near-edge structure may be a good indicator of similar bonding in the model and the unknown.

The question arises: What does SEXAFS offer with respect to other surface structural tools? The main strength of SEXAFS is its *independence from theory* based analysis. This is a definite advantage over techniques such as normal emission photoelectron diffraction (NPD) (28) and LEED, which have to rely on comparison of experimental and calculated intensity variations with excitation energy. These latter analysis procedures certainly limit the accuracy of distance determinations even with the use of reliability (R) factors (340). Other techniques that offer a similar simplicity of data analysis as SEXAFS are the ones based on kinematic scattering processes, such as Rutherford ion scattering (RIS) (12–14) and Bragg reflection diffraction (BRD) (27). These techniques are also capable of high-precision distance determinations. They differ from SEXAFS in that they are suited for the study of long-range order. The sensitivity of SEXAFS to local structural arrangements around a given atom will however be an important asset as the techniques are applied to more complex systems in the future. As discussed previously, one weakness of SEXAFS is its low sensitivity to different complexes that exhibit similar bond lengths. In this respect RIS and BRD seem to have better "resolving power." One area of surface science where SEXAFS [and XANES (146, 147)] will be able to show its full strength is the study of chemisorbed molecules. Techniques that employ electron or ion excitation or scattering are impeded by *radiation damage* problems. Photon excitation is more gentle because the damage is mainly produced by valence excitations due to secondary low-energy electrons. A photon flux of 10^{10} photons s^{-1} at $h\nu \simeq 600$ eV, that is, above the low-Z K-edges of carbon, nitrogen, and oxygen, will typically result in a secondary electron current of $< 10^{-10}$ A from a metal substrate. This would correspond to a LEED experiment with an exciting electron beam current of < 0.1 nA. For the study of molecules the *polarization dependence* of SEXAFS is another significant advantage. Even in the absence of radiation damage problems, techniques that are not governed by dipole selection rules such as the extended appearance potential fine structure (29) and possibly the extended electron loss fine structure (30) techniques will therefore be of only limited use.

This chapter would be incomplete without an outlook on future challenges. Of course, the basic challenge is to reliably determine surface structures. More important, however, than the determination of a given surface structure is the eventual understanding of *bonding concepts at surfaces*. By accurately determining the structures of model systems SEXAFS measurements help to establish reliable *theoretical concepts* that directly predict the crystallographic structures of surface complexes by energy minimization. The crystallographic input will also allow accurate calculations of surface response functions (possibly using different theoretical techniques or approximations). The match of calculated and measured response functions provides the missing link between crystallographic and electronic structure. Thus it appears that the structural results provided by SEXAFS are an important prerequisite to the general understanding of the bonding at surfaces. This discussion also points out the importance of interplay between the various experimental and theoretical techniques.

From the knowledge of bond length variations in bulk crystals, studied in detail by Pauling (152), it appears that bond lengths between a given atom pair at surfaces may typically vary by less than 0.2 Å. Therefore, to obtain meaningful information on the origin of bond length differences a technique is needed with a precision of ≤ 0.02 Å. SEXAFS provides this accuracy. Here it is interesting to mention that early LEED studies by different groups on the same system differed by as much as 0.6 Å in the adsorbate–substrate bond length, as, for example, in the case of $c(2 \times 2)$ oxygen on Ni(100) (190–192). Such discrepancies are the result of trusting incompletely developed theoretical concepts more than basic chemical intuition. Experimental results should always be checked against expectations from simple chemical concepts as found in

Pauling's (152) or Slater's (179) books.

Finally, some specific future applications of SEXAFS should be mentioned. The study of *chemisorption processes* following gas–solid reactions has only started and much more work needs to be done on atomic and molecular chemisorption systems. In my opinion, the low-Z atoms carbon, nitrogen, and oxygen and the various molecules containing these elements provide the greatest challenge. SEXAFS studies on such systems (especially carbon) are experimentally most difficult but technologically most important. Such studies in conjunction with XANES (146, 147) should be able to also monitor changes in *intramolecular* bonding relative to the gas phase as a result of the chemisorption bond. For chemisorbed molecules fluorescence detection offers exciting new possibilities (89, 341). Besides increasing the sensitivity this detection mode can be used in a non-vacuum environment. Hence the sample can be studied under reaction conditions. Hopefully, SEXAFS studies by means of ion desorption at the low-Z K-edges may be able to shed light on the structure of chemisorption complexes at minority defect sites that may well be the catalytically most important ones. Multicomponent systems such as binary or ternary substrates or coadsorption systems (e.g., coadsorption of catalytic promoter atoms) will certainly be investigated. SEXAFS studies of *solid–solid interfaces* promise significant advances in our understanding of the functioning of electronic devices and supported catalysts. Systematic studies need to be performed as a function of deposition quantity and temperature on such systems as metal–semiconductor (Schottky barriers), semiconductor–semiconductor (heterojunctions), and metal–oxide (MOS devices and supported catalysts) interfaces. Other studies may involve very weakly interacting components such as metals deposited on amorphous graphite (342). This may allow the structural study of the formation of a solid from isolated atoms to big bulk clusters. As demonstrated by Petersen and Kunz (60) electron yield SEXAFS studies are also possible in *liquids* provided the vapor pressure is not too high ($< 10^{-5}$ Torr). Thus, the chemisorption on certain liquid substrates or the structure of liquid–solid interfaces may be investigated.

Let me finish this chapter with a word of caution to future users of the SEXAFS technique. It is my belief that SEXAFS is a sufficiently powerful experimental technique that future studies may in many cases be limited not by the technique itself but rather by the lack of thoroughness of its practitioners.

NOTE ADDED IN PROOF. This chapter was originally written in the Fall of 1982. Although I have updated it as much as possible during the proofreading of the galleys, some of the developments during 1983–1987 are not adequately covered in the text. However, I have tried to include all published structural data in the Table given in the Appendix. For a concise summary of all published SEXAFS data through 1986 the reader is also referred to a recent review by Citrin (343).

ACKNOWLEDGMENTS

Synchrotron radiation experiments that are performed in a 24-hour-a-day mode (at least at Stanford) cannot be carried out without the help of capable and reliable collaborators. During the development phase of SEXAFS I have had the pleasure to work with a variety of excellent scientists, and would hereby like to thank them. In particular, two people have struggled with me almost all the way: Rolf Jaeger as a first rate collaborator and Sean Brennan as a graduate student. I would like to acknowledge them for their significant contribution. As a staff scientist and later as an outside user I always enjoyed working at SSRL because of the exceptional atmosphere created by the technical and administrative staff, and I would like to thank them for their support. In particular, I would like to thank A. Bienenstock who supported me in my struggle to push ahead as an independent scientist. Finally, I would like to thank P. Citrin, P. Eisenberger, R. Hewitt, and J. Rowe for an always open dialogue on various aspects of SEXAFS measurements.

REFERENCES

1. B. Feuerbacher, B. Fitton, and R. F. Willis (Eds.), *Photoemission and the Electronic Properties of Surfaces*, Wiley, New York, 1978.

2. M. Cardona and L. Ley (Eds.), *Photoemission in Solids I and II*, Topics in Applied Physics, Vols. 26 and 27, Springer-Verlag, Berlin, 1979.

3. I. Lindau and W. E. Spicer, in *Synchrotron Radiation Research*, H. Winick and S. Doniach (Eds.), Plenum, New York, 1980, pp. 159–221.

4. J. R. Smith (Ed.), *Theory of Chemisorption*, Topics in Current Physics, Vol. 19, Springer-Verlag, Berlin, 1980.

5. R. Gomer (Ed.), *Interactions on Metal Surfaces*, Topics in Applied Physics, Vol. 4, Springer-Verlag, Berlin, 1975.

6. H. Ibach (Ed.), *Electron Spectroscopy for Surface Analysis*, Topics in Current Physics, Vol. 4, Springer-Verlag, Berlin, 1977.

7. T. N. Rhodin and G. Ertl (Eds.), *The Nature of the Surface Chemical Bond*, North-Holland, Amsterdam, 1979.

8. G. A. Somorjai, *Chemistry in Two Dimensions: Surfaces*, Cornell University Press, Ithaca, NY, 1981.

9. L. E. Davis, N. C. MacDonald, P. W. Palmberg, G. E. Riach, and R. E. Weber (Eds.), *Handbook of Auger Electron Spectroscopy*, Physical Electronics Industries, Eden Prairie, 1978.

10. D. Menzel, "*Desorption Phenomena*," in *Interactions on Metal Surfaces*, Topics in Applied Physics, Vol. 4, Springer-Verlag, Berlin, 1975, p. 101; *Surf. Sci.*, **47**, 370 (1975).

11. A. Benninghoven, C. A. Evans, Jr., R. A. Powell, R. Shimizu, and H. A. Storms (Eds.), *Secondary Ion Mass Spectrometry SIMS-II*, Springer-Verlag, Berlin, 1979.

12. J. F. van der Veen, R. M. Tromp, R. G. Smeenk, and F. W. Saris, *Surf. Sci.*, **82**, 468 (1979).

13. L. C. Feldman, *Crit. Rev. Solid State Mater. Sci.*, **10**, 143 (1981).

14. P. R. Norton, J. A. Davies, D. K. Creber, C. W. Sitter, and T. E. Jackman, *Surf. Sci.*, **108**, 205 (1981).

15. E. W. Plummer and T. Gustafsson, *Science*, **198**, 165 (1977).

16. D. A. Shirley, J. Stöhr, P. S. Wehner, R. S. Williams, and G. Apai, *Phys. Scr.*, **16**, 398 (1977).

17. D. E. Eastman and F. J. Himpsel, *Phys. Today*, May 1981, p. 64.

18. N. V. Smith and D. P. Woodruff, *Science*, **216**, 367 (1982).

19. R. F. Willis (Ed.), *Vibrational Spectroscopy of Adsorbates*, Springer Series in Chemical Physics, Vol. 15, Springer-Verlag, Berlin, 1981.

20. H. Froitzheim, *Electron Energy Loss Spectroscopy*, Topics in Current Physics, Vol. 4, Springer-Verlag, Berlin, 1977, p. 205.

21. B. A. Sexton, *Appl. Phys. A*, **26**, 1 (1981).

22. J. B. Pendry, *Low Energy Electron Diffraction*, Academic, New York, 1974.

23. C. B. Duke, in *Surface Effects in Crystal Plasticity*, R. M. La Tanision and J. F. Fourie (Eds.), Noordhoff, Leyden, 1977.

24. F. Jona, *J. Phys. C*, **11**, 4271 (1978).

25. M. A. Van Hove and S. Y. Tong, *Surface Crystallography by LEED*, Springer Series in Chemical Physics, Vol. 2, Springer-Verlag, Berlin, 1979.

26. For a review see: P. Eisenberger and L. C. Feldman, *Science*, **214**, 300 (1981).

27. P. Eisenberger and W. C. Marra, *Phys. Rev. Lett.*, **46**, 1081 (1981).

28. S. D. Kevan, D. M. Rosenblatt, D. Denley, B. C. Lu, and D. A. Shirley, *Phys. Rev. Lett.*, **41**, 1565 (1978).

29. M. L. den Boer, T. L. Einstein, W. T. Elam, R. L. Park, L. D. Roelofs, and G. E. Laramore, *Phys. Rev. Lett.*, **44**, 496 (1980).

30. M. DeCrescenzi, L. Papagno, G. Chiarello, R. Scarmozzino, E. Colavita, R. Rosei, and S. Mobilio, *Solid State Commun.*, **40**, 613 (1981).

31. W. Heiland and E. Taglauer, *J. Vac. Sci. Technol.*, **9**, 620 (1972).

32. D. Menzel and R. Gomer, *J. Chem. Phys.*, **41**, 3311 (1964), P. A. Redhead, *Can. J. Phys.*, **42**, 886 (1964).

33. T. E. Madey and J. T. Yates, Jr., *Surf. Sci.*, **76**, 397 (1978).

34. M. L. Knotek, V. O. Jones, and V. Rehn, *Phys. Rev. Lett.*, **43**, 300 (1979).

35. R. Franchy and D. Menzel, *Phys. Rev. Lett.*, **43**, 865 (1979).

36. P. H. Citrin, P. Eisenberger, and R. C. Hewitt, *Phys. Rev. Lett.*, **41**, 309 (1978).

37. J. Stöhr, D. Denley, and P. Perfetti, *Phys. Rev. B*, **18**, 4132 (1978).

38. R. Jaeger, J. Feldhaus, J. Haase, J. Stöhr, Z. Hussain, D. Menzel, and D. Norman, *Phys. Rev. Lett.*, **45**, 1870 (1980).

39. D. E. Sayers, F. W. Lytle, and E. A. Stern, *Phys. Rev. Lett.*, **27**, 1204 (1971).

40. M. L. Perlman, E. M. Rowe, and R. E. Watson, *Phys. Today*, July 1974, p. 30.

41. C. Kunz (Ed.), *Synchrotron Radiation Techniques and Applications*, Topics in

Current Physics, Vol. 10, Springer-Verlag, Berlin, 1979.

42. H. Winick and S. Doniach, Eds., *Synchrotron Radiation Research*, Plenum, New York-London, 1980.

43. P. A. Lee, *Phys. Rev. B*, **13**, 5261 (1976).

44. U. Landman and D. L. Adams, *Proc. Natl. Acad. Sci. USA*, **73**, 2550 (1976).

45. E. A. Stern, *Contemp. Phys.*, **19**, 289 (1978).

46. P. Eisenberger and B. M. Kincaid, *Science*, **200**, 1441 (1978).

47. D. R. Sandstrom and F. W. Lytle, *Ann. Rev. Phys. Chem.*, **30**, 215 (1979).

48. P. Rabe and R. Haensel, in *Festkörperprobleme 20*, Pergamon-Vieweg, Stuttgart, 1980, p. 43.

49. P. A. Lee, P. H. Citrin, P. Eisenberger, and B. M. Kincaid, *Rev. Mod. Phys.*, **53**, 769 (1981).

50. B. K. Teo and D. C. Joy, Eds., *EXAFS Spectroscopy, Techniques and Applications*, Plenum, New York, 1981.

51. J. Stöhr, R. Jaeger, J. Feldhaus, S. Brennan, D. Norman, and G. Apai, *Appl. Opt.*, **19**, 3911 (1980).

52. J. Stöhr, in *Emission and Scattering Techniques*, P. Day (Ed.), Reidel, Dordrecht, 1981 and SSRL Report 80/07.

53. J. Stöhr, R. Jaeger, and S. Brennan, *Surf. Sci.*, **117**, 503 (1982).

54. For a listing of absorption coefficients in the soft x-ray region, see W. J. Veigele, *At. Data Tables*, **5**, 51 (1973).

55. For a listing of x-ray absorption coefficients, see W. H. McMaster, N. Kerr Del Grande, J. H. Mallet, and J. H. Hubell, *Compilation of X-ray Cross Sections*, National Technical Information Service, Springfield, MA, 1969.

56. A. P. Lukirskii and I. A. Brytov, *Sov. Phys. Solid State*, **6**, 33 (1964).

57. A. P. Lukirskii and T. M. Zimkina, *Izv. Akad. Nauk. SSSR Ser. Fiz.*, **28**, 765 (1964); A. P. Lukirskii, E. P. Savinov, I. A. Brytov, and Yu. F. Shepelev, *Bull. Acad. Sci. USSR Phys. Ser.*, **28**, 774 (1964); A. P. Lukirskii, O. A. Ershov, T. M. Zimkina, and E. P. Savinov, *Sov. Phys. Solid State*, **8**, 1422 (1966).

58. W. Gudat and C. Kunz, *Phys. Rev. Lett.*, **29**, 169 (1972).

59. W. Gudat, Ph.D. thesis (Hamburg University, 1974), Internal Report No. DESY F41-74/10 (unpublished).

60. H. Petersen and C. Kunz, *Phys. Rev. Lett.*, **35**, 863 (1975).

61. G. Martens, P. Rabe, N. Schwentner, and A. Werner, *J. Phys. C*, **11**, 3125 (1978).

62. J. Stöhr, *Jpn. J. Appl. Phys.*, **17**, Suppl. **17-2**, 217 (1978). In this paper a calculated O—Ni phase shift was used resulting in too short an O—Ni bond length. The correct O—Ni bond length is (2.04 ± 0.03) Å for the surface complex.

63. A. Bianconi and R. Z. Bachrach, *Phys. Rev. Lett.*, **42**, 104 (1979).

64. P. H. Citrin, P. Eisenberger, and R. C. Hewitt, *Surf. Sci.*, **89**, 28 (1979).

65. P. H. Citrin, P. Eisenberger, and R. C. Hewitt, *Phys. Rev. Lett.*, **45**, 1948 (1980); Erratum, **47**, 1567 (1981).

66. P. H. Citrin, P. Eisenberger, and J. E. Rowe, *Phys. Rev. Lett.*, **48**, 802 (1982).

67. J. Stöhr, *J. Vac. Sci. Technol.*, **16**, 37 (1979).

68. J. Stöhr, L. I. Johansson, I. Lindau, and P. Pianetta, *Phys. Rev. B*, **20**, 664 (1979).

69. J. Stöhr, L. I. Johansson, I. Lindau, and P. Pianetta, *J. Vac. Sci. Technol.*, **16**, 1221 (1979).

70. J. Stöhr, R. S. Bauer, J. C. McMenamin, L. I. Johansson, and S. Brennan, *J. Vac. Sci. Technol.*, **16**, 1195 (1979).

71. L. I. Johansson and J. Stöhr, *Phys. Rev. Lett.*, **43**, 1882 (1979).

72. J. Stöhr, L. I. Johansson, S. Brennan, M. Hecht, and J. N. Miller, *Phys. Rev. B*, **22**, 4052 (1980).

73. D. Norman, S. Brennan, R. Jaeger, and J. Stöhr, *Surf. Sci.*, **105**, L297 (1981).

74. S. Brennan, J. Stöhr, and R. Jaeger, *Phys. Rev. B*, **24**, 4871 (1981).

75. J. Stöhr, R. Jaeger, and T. Kendelewicz, *Phys. Rev. Lett.*, **49**, 142 (1982).

76. F. C. Brown, R. Z. Bachrach, and N. Lien, *Nucl. Instrum. Methods*, **152**, 73 (1978).

77. J. Stöhr, in Proceedings, Lithography/Microscopy Beam Line Design Workshop (Stanford, Feb. 1979), SSRL Report 79/02.

78. J. Cerino, J. Stöhr, N. Hower, and R. Z. Bachrach, *Nucl. Instrum. Methods*, **172**, 227 (1980).

79. Z. Hussain, E. Umbach, D. A. Shirley, J. Stöhr, and J. Feldhaus, *Nucl. Instrum. Methods*, **195**, 115 (1982).

80. R. Jaeger, J. Stöhr, J. Feldhaus, S. Brennan, and D. Menzel, *Phys. Rev. B*, **23**, 2102 (1981).

81. D. Denley, P. Perfetti, R. S. Williams, D. A. Shirley, and J. Stöhr, *Phys. Rev. B*, **21**, 2267 (1980).

82. R. Fox and S. J. Gurman, *J. Phys. C Solid State Phys.*, **13**, L249 (1980).

83. G. Martens and P. Rabe, *Phys. Status Solidi A*, **58**, 415 (1980).

84. G. Martens and P. Rabe, *J. Phys. C Solid State Phys.*, **14**, 1523 (1981).

85. R. G. Jones and D. P. Woodruff, *Surf. Sci.*, **114**, 38 (1982).

86. R. B. Greegor, F. W. Lytle, and D. R. Sandstrom, SSRL Report 80/01, p. VII-50 (unpublished).

87. J. Jaklevic, J. A. Kirby, M. P. Klein, A. S. Robertson, G. S. Brown, and P. Eisenberger, *Solid State Commun.*, **23**, 679 (1977).

88. S. M. Heald, E. Keller and E. A. Stern, *Phys. Lett.*, **103A**, 155 (1984); also see: B. Lairson, T. N. Rhodin and W. Ho, *Solid State Commun.*, **55**, 925 (1985).

89. J. Stöhr, E. B. Kollin, D. A. Fischer, J. B. Hastings, F. Zaera and F. Sette, *Phys. Rev. Lett.*, **55**, 1468 (1985).

90. V. O. Kostroun, M. H. Chen and B. Crasemann, *Phys. Rev. A*, **3**, 533 (1971). For low-Z see, K. Feser, *Phys. Rev. Lett.*, **28**, 1013 (1972); H. U. Freund, *X-Ray Spectrom.*, **4**, 90 (1975). M. O. Krause, *J. Phys. Chem. Ref. Data*, **8**, 307 (1979).

91. M. H. Chen, B. Crasemann, and V. O. Kostroun, *Phys. Rev. A*, **4**, 1 (1971).

92. W. Bambynek, B. Crasemann, R. W. Fink, H. U. Freund, H. Mark, C. D. Swift,

R. E. Rice, and P. Venugopala Rao, *Rev. Mod. Phys.*, **44**, 716 (1972).

93. I. Lindau and W. E. Spicer, *J. Electron Spectrosc. Relat. Phenom.*, **3**, 409 (1974).

94. R. H. Ritchie and J. C. Ashley, *J. Phys. Chem. Solids*, **26**, 1689 (1965); E. O. Kane, *Phys. Rev.*, **159**, 624 (1967).

95. J. J. Quinn, *Phys. Rev.*, **126**, 1453 (1962); L. Kleinmann, *Phys. Rev. B*, **3**, 2982 (1971).

96. J. Llacer and E. L. Garwin, *J. Appl. Phys.*, **40**, 2766 (1969); G. D. Mahan, *Phys. Status Solidi B*, **55**, 703 (1973).

97. G. Martens, P. Rabe, G. Tolkiehn, and A. Werner, *Phys. Status Solidi A*, **55**, 105 (1979).

98. J. B. Hastings, in *EXAFS Spectroscopy, Techniques and Applications*, B. K. Teo and D. C. Joy (Eds.), Plenum, New York, p. 171.

99. J. A. R. Samson, *Techniques of Vacuum Ultraviolet Spectroscopy*, Wiley, New York, 1967.

100. F. J. Zutavern, S. E. Schnatterly, E. Källne, C. P. Franck, T. Aton, and J. E. Rife, *Nucl. Instrum. Methods*, **172**, 351 (1980); D. A. Fischer, J. B. Hastings, F. Zaera, J. Stöhr and F. Sette, *Nucl. Instrum. Methods*, **A246**, 561 (1986).

101. F. Comin, L. Incoccia, P. Lagarde, G. Rossi, and P. H. Citrin, *Phys. Rev. Lett.* **54**, 122 (1985).

102. L. A. Harris, *Surf. Sci.*, **15**, 77 (1969); W. A. Fraser, J. V. Florio, W. N. Delgass, and W. D. Robertson, *Surf. Sci.*, **36**, 661 (1973); C. S. Fadley, R. J. Baird, W. Siekhaus, T. Novakov, and S. A. L. Bergström, *J. Electron Spectrosc. Relat. Phenom.*, **4**, 93 (1974).

103. G. D. Mahan, *Phys. Rev. B*, **2**, 4334 (1970); R. S. Williams, P. S. Wehner, J. Stöhr, and D. A. Shirley, *Surf. Sci.*, **75**, 215 (1978).

104. K. Siegbahn et al., *Electron Spectroscopy for Chemical Analysis, Molecular and Solid State Structure by Means of Electron Spectroscopy*, Nova Acta R. Soc. Sci. Ups. Ser. IV, **20**, 224 (1967).

105. J. C. Fuggle and N. Martensson, *J. Electron Spectrosc.*, **21**, 275 (1980).

106. B. Feuerbacher, B. Fitton, and R. F. Willis (Eds.), *Photoemission and the Electronic Properties of Surfaces*, Wiley, New York, 1978, p. 373.

107. N. V. Smith and S. D. Kevan, *Nucl. Instrum. Methods*, **195**, 309 (1982).

108. A. Liebsch, *Phys. Rev. Lett.*, **32**, 1203 (1974); A. Liebsch, *Phys. Rev. B*, **15**, 544 (1976).

109. J. Stöhr, C. Noguera, and T. Kendelewicz, *Phys. Rev. B*, **30**, 5571 (1984).

110. C. Noguera and D. Spanjaard, *Surf. Sci.*, **108**, 381 (1980).

111. J. B. Hastings, P. Eisenberger, B. Lengeler, and M. C. Perlman, *Phys. Rev. Lett.*, **43**, 1807 (1979); M. Marcus, L. S. Powers, A. R. Storm, B. M. Kincaid, and B. Chance, *Rev. Sci. Instrum.*, **51**, 1023 (1980).

112. E. A. Stern and S. M. Heald, *Rev. Sci. Instrum.*, **50**, 1579 (1979).

113. J. Stöhr and D. Denley, in *Proceedings, International Workshop on X-ray Instrumentation for Synchrotron Radiation*, H. Winick and G. S. Brown (Eds.), Stanford, April 1978, SSRL Report 78/04.

114. L. I. Johansson, I. Lindau, M. Hecht, S. M. Goldberg, and C. S. Fadley, *Phys.*

Rev. B, **20**, 4126 (1979).

115. R. Z. Bachrach and A. Bianconi, in *Proceedings of the Fifth International Conference on Vacuum Ultraviolet Radiation Physics*, M. C. Castex, M. Pouey, and N. Pouey (Eds.), Montpellier, France, 1977.

116. R. Jaeger, J. Stöhr, and T. Kendelewicz, results for nitrogen on Ni(100) and Ni(110) (unpublished).

117. J. Stöhr and R. Jaeger, *J. Vac. Sci. Technol.*, **21**, 619 (1982).

118. G. Apai and J. Stöhr, unpublished results for EXAFS studies above the *L*- and *K*-edges for the 4*d* metals rhodium to tin (unpublished) (see Fig. 10.50).

119. M. L. Knotek and P. J. Feibelman, *Phys. Rev. Lett.*, **40**, 964 (1978).

120. P. J. Feibelman and M. L. Knotek, *Phys. Rev.*, **18**, 6531 (1978).

121. R. Jaeger, R. Treichler, and J. Stöhr, *Surf. Sci.*, **117**, 533 (1982).

122. R. Jaeger, J. Stöhr, R. Treichler, and K. Baberschke, *Phys. Rev. Lett.*, **47**, 1300 (1981).

123. Results for CO/Ru(001) at lower energies (< 100 eV) were reported by T. E. Madey, R. Stockbauer, S. A. Flodström, J. F. van der Veen, F. J. Himpsel, and D. E. Eastman, *Phys. Rev. B*, **23**, 6847 (1981).

124. R. Jaeger, J. Stöhr, and T. Kendelewicz, *Phys. Rev. B*, **28**, 1145 (1983); *Surf. Sci.*, **134**, 547 (1983).

125. For a review see: Proceedings of the *First International Workshop on Desorption Induced by Electronic Transitions—DIET 1*, Springer Series in Chemical Physics, Vol. 24, Springer-Verlag, Berlin, 1983.

126. J. Stöhr, V. Rehn, I. Lindau, and R. Z. Bachrach, *Nucl. Instrum. Methods*, **152**, 43 (1978).

127. See proceedings of Conferences on Synchrotron Radiation Instrumentation: *Nucl. Instrum. Methods*, **152** (1978); **172** (1980); **195** (1982); **208** (1983); **222** (1984); **A246** (1986).

128. W. R. Hunter, R. T. Williams, J. C. Rife, J. P. Kirkland, and M. N. Kabler, *Nucl. Instrum. Methods*, **195**, 141 (1982); M. Howells, Brookhaven National Laboratory Report, BNL 31030, 1982.

129. G. Brown, in Stanford Synchrotron Radiation Laboratory X-Ray Beam Line Documentation (unpublished).

130. P. A. Lee and G. Beni, *Phys. Rev. B*, **15**, 2862 (1977).

131. G. Martens, P. Rabe, N. Schwentner, and A. Werner, *Phys. Rev. B*, **17**, 1481 (1978).

132. E. A. Stern, *Phys. Rev. B*, **10**, 3027 (1974); C. H. Ashley and S. Doniach, *Phys. Rev. B*, **11**, 1279 (1975); P. A. Lee and J. B. Pendry, *Phys. Rev. B*, **11**, 2795 (1975).

133. S. M. Heald and E. A. Stern, *Phys. Rev. B*, **16**, 5549 (1977).

134. Boon-Keng Teo and P. A. Lee, *J. Am. Chem. Soc.*, **101**, 2815 (1979).

135. P. H. Citrin, P. Eisenberger, and B. M. Kincaid, *Phys. Rev. Lett.*, **36**, 1346 (1976).

136. E. A. Stern, B. A. Bunker, and S. M. Heald, *Phys. Rev. B*, **21**, 5521 (1980).

137. P. Eisenberger and B. Lengeler, *Phys. Rev. B*, **22**, 3551 (1980).

138. J. Stöhr and R. Jaeger, *Phys. Rev. B*, **27**, 5146 (1983).

139. D. W. Jepsen, P. M. Marcus, and F. Jona, *Phys. Rev. B*, **5**, 3933 (1972).

140. P. M. Marcus, D. W. Jepsen, and F. Jona, *Surf. Sci.*, **31**, 180 (1972); D. W. Jepsen, P. M. Marcus, and F. Jona, *Phys. Rev. B*, **6**, 3684 (1972); Erratum *B*, **8**, 1786 (1973).

141. M. R. Martin and G. A. Somorjai, *Phys. Rev. B*, **7**, 3607 (1973).

142. F. Jona, D. Sondericker, and P. M. Marcus, *J. Phys. C*, **13**, L155-8 (1980).

143. S. M. Heald and E. A. Stern, *Phys. Rev. B*, **17**, 4069 (1978).

144. J. Stöhr, D. Outka, R. J. Madix, and U. Döbler, *Phys. Rev. Lett.*, **54**, 1256 (1985); D. A. Outka, R. J. Madix and J. Stöhr, *Surf. Sci.*, **164**, 235 (1985).

145. A. Puschmann, J. Haase, M. D. Crapper, C. E. Riley and D. P. Woodruff, *Phys. Rev. Lett.*, **54**, 2250 (1985).

146. J. Stöhr, K. Baberschke, R. Jaeger, R. Treichler, and S. Brennan, *Phys. Rev. Lett.*, **47**, 381 (1981); J. Stöhr and R. Jaeger, *Phys. Rev. B*, **26**, 4111 (1982).

147. For a review see: J. Stöhr, Z. *Physik*, **B61**, 439 (1985).

148. R. A. Rosenberg, P. R. LaRoe, V. Rehn, J. Stöhr, R. Jaeger, and C. C. Parks, *Phys. Rev. B*, **28**, 3026 (1983).

149. K. A. R. Mitchell, *Surf. Sci.*, **92**, 79 (1980).

150. K. A. R. Mitchell, *Surf. Sci.*, **100**, 225 (1980); *J. Chem. Phys.*, **73**, 5904 (1980).

151. A. Madhukar, *Solid State Commun.*, **16**, 461 (1975).

152. L. Pauling, *The Nature of the Chemical Bond*, Cornell University Press, Ithaca, NY, 1960.

153. S. A. Flodström, C. W. B. Martinson, R. Z. Bachrach, S. B. M. Hagström, and R. S. Bauer, *Phys. Rev. Lett.*, **40**, 907 (1978).

154. P. O. Gartland, *Surf. Sci.*, **62**, 183 (1977).

155. P. Hofmann, W. Wyrobisch, and A. M. Bradshaw, *Surf. Sci.*, **80**, 344 (1979).

156. W. Eberhardt and F. J. Himpsel, *Phys. Rev. Lett.*, **42**, 1375 (1979).

157. C. W. B. Martinson and S. A. Flodström, *Solid State Commun.*, **30**, 671 (1979).

158. P. Hoffmann, C. v. Muschwitz, K. Horn, K. Jacobi, A. Bradschaw, K. Kambe, and M. Scheffler, *Surf. Sci.*, **89**, 327 (1979).

159. R. Payling and J. A. Ramsey, *J. Phys. C Solid State Phys.*, **13**, 505 (1980).

160. C. W. B. Martinson, S. A. Flodström, J. Rundgren, and P. Westrin, *Surf. Sci.*, **89**, 102 (1979).

161. H. L. Yu, M. C. Muñoz, and F. Soria, *Surf. Sci.*, **94**, L104 (1980).

162. F. Soria, V. Martinez, M. C. Muñoz, and J. L. Sacedón, *Phys. Rev. B*, **24**, 6926 (1981).

163. L. I. Johansson, J. Stöhr, and S. Brennan, *Appl. Surf. Sci.*, **6**, 419 (1980).

164. R. Z. Bachrach, G. V. Hansson, and R. S. Bauer, *Surf. Sci.*, **109**, L560 (1981).

165. R. Michel, C. Jourdan, J. Castaldi, and J. Derrien, *Surf. Sci.*, **84**, L509 (1979).

166. J. L. Erskine and R. L. Strong, *Phys. Rev. B*, **25**, 5547 (1982).

167. R. L. Strong, B. Firey, F. W. de Wette, and J. L. Erskine, *Phys. Rev. B*, **26**, 3483 (1982).

168. N. D. Lang and A. R. Williams, *Phys. Rev. Lett.*, **34,** 531 (1975).

169. K. Y. Yu, J. N. Miller, P. Chye, W. E. Spicer, N. D. Lang, and A. R. Williams, *Phys. Rev. B*, **14,** 1446 (1976).

170. D. R. Salahub, M. Roche, and R. P. Messmer, *Phys. Rev. B*, **18,** 6495 (1978).

171. K. Mednick and L. Kleinman, *Phys. Rev. B*, **22,** 5768 (1980).

172. D. M. Bylander, L. Kleinman, and K. Mednick, *Phys. Rev. Lett.*, **48,** 1544 (1982).

173. Ding-Sheng Wang, A. J. Freeman, and H. Krakauer, *Phys. Rev. B*, **24,** 3092 (1981).

174. Ding-Sheng Wang, A. J. Freeman, and H. Krakauer, *Phys. Rev. B*, **24,** 3104 (1981).

175. B. N. Cox and C. W. Bauschlicher, Jr., *Surf. Sci.*, **115,** 15 (1982).

176. J. Neve, J. Rundgren, and P. Westrin, *J. Phys. C*, **15,** 4391 (1982).

177. P. Hoffmann, K. Horn, A. M. Bradshaw, and K. Jacobi, *Surf. Sci.*, **82,** L610 (1979).

178. F. Jona and P. M. Marcus, *J. Phys. C. (Solid State Phys.)*, **13,** L477 (1980).

179. For a convenient listing of bond lengths, see J. C. Slater, *Symmetry and Energy Bands in Crystals*, Dover, New York, 1972, pp. 308–346.

180. R. W. G. Wyckoff, *Crystal Structures*, Wiley, New York, 1964.

181. A. Bianconi, R. Z. Bachrach, S. B. M. Hagstrom, and S. A. Flodström, *Phys. Rev. B*, **19,** 2837 (1979).

182. A. Bianconi, R. Z. Bachrach, and S. A. Flodström, *Phys. Rev. B*, **19,** 3879 (1979).

183. H. E. Farnsworth and H. H. Madden, Jr., *J. Appl. Phys.*, **32,** 1933 (1961).

184. A. U. MacRae, *Surf. Sci.*, **1,** 319 (1964).

185. P. H. Holloway and J. B. Hudson, *Surf. Sci.*, **43,** 123 (1974).

186. J. E. Demuth and T. N. Rhodin, *Surf. Sci.*, **45,** 249 (1974).

187. D. F. Mitchell, P. B. Sewell, and M. Cohen, *Surf. Sci.*, **61,** 355 (1976).

188. P. R. Norton, R. L. Tapping, and J. W. Goodale, *Surf. Sci.*, **65,** 13 (1977).

189. C. R. Brundle, in *Aspects of the Kinetics and Dynamics of Surface Reactions*, AIP Conference Proceedings No. 61, U. Landman (Ed.), American Institute of Physics, New York, 1980.

190. S. Anderson, B. Kasemo, J. B. Pendry, and M. A. Van Hove, *Phys. Rev. Lett.*, **31,** 595 (1973).

191. C. B. Duke, N. O. Lipari, G. E. Laramore, and J. B. Theeten, *Solid State Commun.*, **13,** 579 (1973); *Nuovo Cimento*, **23B,** 241 (1974).

192. J. E. Demuth, D. W. Jepsen, and P. M. Marcus, *Phys. Rev. Lett.*, **31,** 540 (1973).

193. C. B. Duke, N. O. Lipari, and G. E. Laramore, *J. Vac. Sci. Technol.*, **11,** 180 (1974).

194. P. M. Marcus, J. E. Demuth, and D. W. Jepsen, *Surf. Sci.*, **53,** 501 (1975).

195. M. Scheffler, K. Kambe, F. Forstmann, and K. Jacobi, in *Proceedings of the 7th International Vacuum Congress and 3rd International Conference on Solid Sur-*

faces, Vienna, 1977, R. Dobrozemsky et al. (Eds.), Berger and Söhne, Vienna, 1977, p. 2223.

196. G. Hanke, E. Lang, K. Heinz, and K. Müller, *Surf. Sci.*, **91**, 551 (1980).

197. M. Van Hove and S. Y. Tong, *J. Vac. Sci. Technol.*, **12**, 230 (1975).

198. K. Jacobi, M. Scheffler, K. Kambe, and F. Forstmann, *Solid State Commun.*, **22**, 17 (1977).

199. C. R. Brundle and H. Hopster, *J. Vac. Sci. Technol.*, **18**, 663 (1981).

200. H. H. Brongersma and J. B. Theeten, *Surf. Sci.*, **54**, 519 (1976).

201. D. H. Rosenblatt, J. G. Tobin, M. G. Mason, R. F. Davis, S. D. Kevan, D. A. Shirley, C. H. Li, and S. Y. Tong, *Phys. Rev. B*, **23**, 3828 (1981).

202. S. P. Walch and W. A. Goddard, III, *Solid State Commun.*, **23**, 907 (1977); *Surf. Sci.*, **75**, 609 (1978).

203. C. H. Li and J. W. D. Connolly, *Surf. Sci.*, **65**, 700 (1977).

204. C. S. Wang and A. J. Freeman, *Phys. Rev. B*, **19**, 4930 (1979).

205. T. H. Upton and W. A. Goddard, III, *Phys. Rev. Lett.*, **46**, 1635 (1981); *CRC Crit. Rev. Solid State Mater. Sci.*, **10**, 261 (1981).

206. S. Anderson, *Solid State Commun.*, **20**, 229 (1976); *Surf. Sci.*, **79**, 385 (1979).

207. H. Ibach and D. Bruchmann, *Phys. Rev. Lett.*, **44**, 36 (1980); S. Lehwald and H. Ibach, in *Proceedings of the International Conference on Vibrations at Surfaces*, Namur, Belgium (1980), A. Lucas (Ed.), Pergamon Press, New York (1982).

208. T. S. Rahman, J. E. Black, and D. L. Mills, *Phys. Rev. Lett.*, **46**, 1469 (1981).

209. S. Y. Tong and K. H. Lau, *Phys. Rev. B*, **25**, 7382 (1982).

210. L. G. Petersson, S. Kono, N. F. T. Hall, S. Goldberg, J. T. Lloyd, C. S. Fadley, and J. B. Pendry, *Mater. Sci. Eng.*, **42**, 111 (1980).

211. J. Stöhr (unpublished).

212. D. Norman, J. Stöhr, R. Jaeger, P. J. Durham, and J. B. Pendry, *Phys. Rev. Lett.*, **51**, 2052 (1983).

213. C. W. Bauschlicher, S. P. Walch, P. S. Bagus, and C. R. Brundle, *Phys. Rev. Lett.*, **50**, 864 (1983); C. W. Bauschlicher, Jr., and P. S. Bagus, *Phys. Rev. Lett.*, **52**, 200 (1984).

214. V. Bortolani, A. Franchini, F. Nizzoli, and G. Santoro, *J. Electron Spectrosc.*, **29**, 219 (1983); G. Allan, *J. Electron Spectrosc.*, **29**, 61 (1983).

215. J. E. Demuth, N. J. DiNardo, and G. S. Cargill, III, *Phys. Rev. Lett.*, **50**, 1373 (1983).

216. R. Smeenk, Relaxation, Reconstruction and Oxidation of Nickel Surfaces, Ph.D. Thesis, University of Utrecht, Holland (1982); J. W. M. Frenken, J. F. van der Veen, and G. Allan, *Phys. Rev. Lett.*, **51**, 1876 (1983).

217. H. D. Hagstrum and G. E. Becker, *Phys. Rev. Lett.*, **22**, 1054 (1969).

218. J. E. Demuth, D. W. Jepsen, and P. M. Marcus, *Solid State Commun.*, **13**, 1311 (1973).

219. C. B. Duke, N. O. Lipari, and G. E. Laramore, *J. Vac. Sci. Technol.*, **12**, 222 (1975).

220. Y. Gauthier, D. Aberdam, and R. Baudoing, *Surf. Sci.*, **78**, 339 (1978).

221. G. B. Fisher, *Surf. Sci.*, **62**, 31 (1977).

222. T. T. Anh Nguyen and R. C. Cinti, *Surf. Sci.*, **68**, 566 (1977).

223. E. W. Plummer, B. Tonner, N. Holzwarth, and A. Liebsch, *Phys. Rev. B*, **21**, 4306 (1980).

224. S. Anderson, *Surf. Sci.*, **79**, 385 (1979).

225. D. H. Rosenblatt, J. G. Tobin, M. G. Mason, R. F. Davis, S. D. Kevan, D. A. Shirley, C. H. Li, and S. Y. Tong, *Phys. Rev. B*, **8**, 3828 (1981).

226. H. H. Brongersma, *J. Vac. Sci. Technol.*, **11**, 231 (1974).

227. S. P. Walch and W. A. Goddard, III, *Solid State Commun.*, **23**, 907 (1977); *Surf. Sci.*, **72**, 645 (1978).

228. T. H. Upton and W. A. Goddard, III, *CRC Crit. Rev. Solid State and Mater. Sci.*, **10**, 261 (1981).

229. Pei-Lin Cao, D. E. Ellis, and A. J. Freeman, *Phys. Rev. B*, **25**, 2124 (1982).

230. S. Brennan, Ph.D. Thesis, Stanford University 1982 (unpublished) and SSRL Report 82/03.

231. P. H. Citrin, D. R. Hamann, L. F. Mattheiss, and J. E. Rowe, *Phys. Rev. Lett.*, **49**, 1712 (1982).

232. E. Zanazzi, F. Jona, D. W. Jepsen, and P. M. Marcus, *Phys. Rev. B*, **14**, 432 (1976).

233. H. S. Greenside and D. R. Hamann, *Phys. Rev. B*, **23**, 4879 (1981). D. R. Hamann, L. F. Mattheiss, and H. S. Greenside, *Phys. Rev.*, **24**, 6151 (1981).

234. G. Lamble and D. A. King, *Phil. Trans. R. Soc. Lond.*, **A318**, 203 (1986).

235. F. Comin, P. H. Citrin, P. Eisenberger, and J. E. Rowe, *Phys. Rev. B*, **26**, 7060 (1982).

236. A. Salwen and J. Rundgren, *Surf. Sci.*, **53**, 523 (1975).

237. H. D. Shih, F. Jona, D. W. Jepsen, and P. M. Marcus, *Phys. Rev. Lett.*, **36**, 798 (1976).

238. J. E. Demuth, *J. Colloids Interface Sci.*, **58**, 184 (1977).

239. F. Forstmann, W. Berndt, and P. Büttner, *Phys. Rev. Lett.*, **30**, 17 (1973).

240. For review of semiconductor research, see Proceedings of Conferences on Physics of Compound Semiconductor Interfaces (PCSI), *J. Vac. Sci. Technol.*, **17** (1980); **19** (1981); **21** (1982).

241. H. Ibach, K. Horn, R. Dorn, and M. Lüth, *Surf. Sci.*, **38**, 433 (1973).

242. A. Kahn, D. Kanani, P. Mark, P. W. Chye, C. Y. Su, I. Lindau, and W. E. Spicer, *Surf. Sci.*, **87**, 325 (1979).

243. C. B. Duke, R. J. Meyer, and P. Mark, *J. Vac. Sci. Technol.*, **17**, 971 (1980).

244. C. A. Swarts, W. A. Goddard, III, and T. C. McGill, *J. Vac. Sci. Technol.*, **17**, 982 (1980).

245. W. A. Goddard, III, J. J. Barton, A. Redondo, and T. C. McGill, *J. Vac. Sci. Technol.*, **15**, 1274 (1978).

246. H. Ibach and J. E. Rowe, *Phys. Rev B*, **9**, 1951 (1974); **10**, 710 (1974).

247. J. E. Rowe, G. Margaritondo, H. Ibach, and H. Froitzheim, *Solid State Commun.*,

20, 277 (1976).

248. J. M. Hill, D. G. Royce, C. S. Fadley, L. F. Wagner, and F. J. Grunthaner, *Chem. Phys. Lett.*, **44**, 225 (1976).

249. C. M. Garner, I. Lindau, J. N. Miller, P. Pianetta, and W. E. Spicer, *J. Vac. Sci. Technol.*, **14**, 372 (1977).

250. S. Fujiwara, M. Ogata, and M. Nishijima, *Solid State Commun.*, **21**, 895 (1977).

251. R. S. Bauer, J. C. McMenamin, H. Petersen, and A. Bianconi, in *Proceedings of the International Topical Conference on SiO₂ and Its Interfaces*, S. Pantelides (Ed.), Pergamon, New York, 1978, p. 401.

252. I. T. McGovern, A. W. Parke, and R. H. Williams, *Solid State Commun.*, **26**, 21 (1978).

253. C. M. Garner, I. Lindau, C. Y. Su, P. Pianetta, and W. E. Spicer, *Phys. Rev. B*, **19**, 3944 (1979).

254. M. Chen, I. P. Batra, and C. R. Brundle, *J. Vac. Sci. Technol.*, **16**, 1216 (1979).

255. C. Y. Su, P. R. Skeath, I. Lindau, and W. E. Spicer, *J. Vac. Sci. Technol.*, **18**, 843 (1981).

256. R. S. Bauer, R. Z. Bachrach, H. W. Sang, Jr., G. V. Hansson, and W. Göpel, SSRL Report 81/02, p. VII-5.

257. R. Ludeke and A. Koma, *Phys. Rev. Lett.*, **34**, 1170 (1975); A. Koma and R. Ludeke, *Phys. Rev. Lett.*, **35**, 107 (1975).

258. R. S. Bauer, J. C. McMenamin, R. Z. Bachrach, A. Bianconi, L. Johansson, and H. Petersen, *Inst. Phys. Conf. Ser.*, **43**, 797 (1978).

259. A. Bianconi, *Surf. Sci.*, **89**, 41 (1979); A. Bianconi and R. S. Bauer, *Surf. Sci.*, **99**, 76 (1980).

260. P. Pianetta, I. Lindau, C. M. Garner, and W. E. Spicer, *Phys. Rev. Lett.*, **35**, 1356 (1975); *Phys. Rev. Lett.*, **37**, 1166 (1976).

261. P. Pianetta, I. Lindau, C. M. Garner, and W. E. Spicer, *Phys. Rev. B*, **18**, 2792 (1978); W. E. Spicer, P. Pianetta, I. Lindau, and P. W. Chye, *J. Vac. Sci. Technol.*, **14**, 885 (1977).

262. R. Ludeke and A. Koma, *Crit. Rev. Solid State Sci.*, **5**, 259 (1975); *J. Vac. Sci. Technol.*, **13**, 241 (1976).

263. R. Ludeke, *Solid State Commun.*, **21**, 815 (1977).

264. C. R. Brundle and D. Seybold, *J. Vac. Sci. Technol.*, **16**, 1186 (1979).

265. C. Y. Su, I. Lindau, P. R. Skeath, P. W. Chye, and W. E. Spicer, *J. Vac. Sci. Technol.*, **17**, 936 (1980).

266. E. J. Mele and J. D. Joannopoulos, *Phys. Rev. Lett.*, **40**, 341 (1978); *Phys. Rev. B*, **18**, 699 (1978).

267. W. A. Goddard, III, J. J. Barton, A. Redondo, and T. C. McGill, *J. Vac. Sci. Technol.*, **15**, 1274 (1978).

268. J. J. Barton, C. A. Swarts, W. A. Goddard, III, and T. C. McGill, *J. Vac. Sci. Technol.*, **17**, 164 (1980).

269. G. Lukovsky and R. S. Bauer, *Solid State Commun.*, **31**, 931 (1979).

270. M. Schlüter, J. E. Rowe, G. Margaritondo, K. M. Ho, and M. L. Cohen, *Phys. Rev. Lett.*, **37**, 1632 (1976).

271. J. E. Rowe, G. Margaritondo, and S. B. Christman, *Phys. Rev. B*, **16**, 1581 (1977).

272. P. H. Citrin, J. E. Rowe, and P. Eisenberger, *Phys. Rev. B*, **28**, 2299 (1983).

273. P. H. Citrin and J. E. Rowe, *Surf. Sci.*, **132**, 205 (1983).

274. *Thin-Films—Interdiffusion and Reactions*, J. M. Poate, K. N. Tu, and J. W. Mayer (Eds.), Wiley, New York, 1978.

275. *Thin Films and Interfaces*, Materials Research Society Symposia Proceedings, Vol. 10, P. S. Ho and K. N. Tu (Eds.), North-Holland, New York, 1982.

276. R. Dalwen, *Introduction to Applied Solid State Physics*, Plenum, New York, 1980.

277. For a recent review see, M. Schlüter, in *Thin Films and Interfaces*, Materials Research Society Symposia Proceedings, Vol. 10, P. S. Ho and K. N. Tu (Eds.), North-Holland, New York, 1982, p. 3.

278. W. E. Spicer, I. Lindau, P. Skeath, C. Y. Su, and P. Chye, *Phys. Rev. Lett.*, **44**, 420 (1980).

279. J. C. Phillips, *J. Vac. Sci. Technol.*, **11**, 947 (1974).

280. J. M. Andrews and J. C. Phillips, *Phys. Rev. Lett.*, **35**, 56 (1975).

281. J. L. Freeouf, *Solid State Commun.*, **33**, 1059 (1980).

282. J. L. Freeouf, *J. Vac. Sci. Technol.*, **18**, 910 (1981).

283. J. J. Lander and J. Morrison, *Surf. Sci.*, **2**, 553 (1964).

284. J. J. Lander, *Progr. Solid State Chem.*, **2**, 26 (1975).

285. C. B. Duke, A. Paton, R. J. Meyer, L. J. Brillson, A. Kahn, D. Kanani, J. Carelli, L. J. Yeh, G. Margaritondo, and A. D. Katnani, *Phys. Rev. Lett.*, **46**, 440 (1981).

286. Y. Terada, T. Yoshizuka, K. Oura, and T. Hanawa, *Surf. Sci.*, **114**, 65 (1982).

287. W. S. Yang and F. Jona, *Solid State Commun.*, **42**, 49 (1982).

288. G. LeLay, G. Quentel, J. P. Faurie, and A. Masson, *Thin Solid Films*, **35**, 273 (1976); **35**, 289 (1976).

289. Y. Gotoh and S. Ino, *Jpn. J. Appl. Phys.*, **17**, 2097 (1978).

290. M. Saitoh, F. Shoji, K. Oura, and T. Hanawa, *Jpn. J. Appl. Phys.*, **19**, L421 (1980).

291. M. Saitoh, F. Shoji, K. Oura, and T. Hanawa, *Surf. Sci.*, **112**, 306 (1981).

292. R. Tromp, E. J. Van Loenen, M. Iwami, R. Smeenk, and F. W. Saris, in *Thin Films and Interfaces*, Materials Research Society Symposia Proceedings, Vol. 10, P. S. Ho and K. N. Tu (Eds.), North-Holland, New York, 1982, p. 155.

293. D. Denley, R. S. Williams, P. Perfetti, D. A. Shirley, and J. Stöhr, *Phys. Rev. B*, **19**, 1762 (1979).

294. S. Kiyono, Y. Hayasi, S. Kato, and S. Mochimaru, *Jpn. J. Appl. Phys.*, **17**, Suppl. 17-2, 212 (1978).

295. R. D. Leapman, L. A. Grunes, and P. L. Fejes, *Phys. Rev. B*, **26**, 614 (1982).

296. J. Stöhr, R. Jaeger, G. Rossi, T. Kendelewicz, and I. Lindau, *Surf. Sci.*, **134**, 813 (1983).

297. R. W. Bower, D. Sigurd, and R. E. Scott, *Solid State Electron.*, **16**, 1461 (1973).

298. G. Ottaviani, K. N. Tu, and J. W. Mayer, *Phys. Rev. B*, **24**, 3354 (1981).

299. S. Okada, K. Oura, T. Hanawa, and K. Satoh, *Surf. Sci.*, **97**, 88 (1980).

300. K. Oura, S. Okada, Y. Kishikawa, and T. Hanawa, *Appl. Phys. Lett.*, **40**, 138 (1982).

301. T. Narusawa, W. M. Gibson, and A. Hiraki, *Phys. Rev. B*, **24**, 4835 (1981).

302. P. E. Schmid, P. S. Ho, H. Föll, and G. W. Rubloff, *J. Vac. Sci. Technol.*, **18**, 937 (1981).

303. P. S. Ho, T. Y. Tan, J. E. Lewis, and G. W. Rubloff, *J. Vac. Sci. Technol.*, **16**, 1120 (1979).

304. P. S. Ho, P. E. Schmid, and H. Föll, *Phys. Rev. Lett.*, **46**, 782 (1981).

305. W. Krakow, in *Thin Films and Interfaces*, Materials Research Society Symposia Proceedings, Vol. 10, P. S. Ho and K. N. Tu (Eds.), North-Holland, New York, 1982, p. 111.

306. W. D. Buckley and S. C. Moss, *Solid State Electron.*, **15**, 1331 (1972).

307. For a review see G. W. Rubloff and P. S. Ho, in *Thin Films and Interfaces*, Materials Research Society Symposia Proceedings, Vol. 10, P. S. Ho and K. N. Tu (Eds.), North-Holland, New York, 1982, p. 21.

308. G. Rossi, R. Jaeger, J. Stöhr, T. Kendelewicz, and I. Lindau, *Phys. Rev. B*, **27**, 5154 (1983).

309. O. Bisi and C. Calandra, *J. Phys. Rev. C*, **14**, 5479 (1981) and C. Calandra (unpublished results).

310. F. Wehking, H. Beckermann, and R. Niedermayer, *Surf. Sci.*, **71**, 364 (1978).

311. A. McKinley, R. H. Williams, and A. W. Parke, *J. Phys. C*, **12**, 2447 (1979).

312. G. V. Hansson, R. Z. Bachrach, R. S. Bauer, and P. Chiaradia, *Phys. Rev. Lett.*, **46**, 1033 (1981).

313. G. Dufour, J. M. Mariot, A. Masson, and H. Roulet, *J. Phys. C*, **14**, 2539 (1981).

314. G. Rossi, I. Abbati, L. Braicovich, I. Lindau, and W. E. Spicer, *Surf. Sci.*, **112**, L765 (1981).

315. E. Bauer and H. Poppa, *Thin Solid Films*, **12**, 167 (1972).

316. G. LeLay, M. Manneville, and R. Kern, *Surf. Sci.*, **72**, 405 (1978).

317. K. Spiegel, *Surf. Sci.*, **7**, 125 (1967).

318. Y. Terada, T. Yoshizuka, K. Oura, and T. Hanawa, *Surf. Sci.*, **114**, 65 (1982).

319. G. LeLay, G. Quentel, J. P. Faurie, and A. Masson, *Thin Solid Films*, **35**, 273 (1976); **35**, 289 (1976).

320. Y. Gotoh and S. Ino, *Jpn. J. Appl. Phys.*, **17**, 2097 (1978).

321. J. A. Venables, J. Derrien, and A. P. Janssen, *Surf. Sci.*, **95**, 411 (1980).

322. K. Oura, T. Taminaga, and T. Hanawa, *Solid State Commun.*, **37**, 523 (1981).

323. D. Bolmont, Ping Chen, and C. A. Sebenne, *J. Phys. C*, **14**, 3313 (1981).

324. J. Derrien, G. LeLay, and F. Salvan, *J. Phys. (Paris) Lett.*, **39**, L287 (1978).

325. J. P. Gaspard, J. Derrien, A. Cros, and F. Salvan, *Surf. Sci.*, **99**, 183 (1980).

326. M. Saitoh, F. Shoji, K. Oura, and T. Hanawa, *Jpn. J. Appl. Phys.*, **19**, L421 (1980).

327. M. Saitoh, F. Shoji, K. Oura, and T. Hanawa, *Surf. Sci.*, **112**, 306 (1981).

328. J. D. Van Otterloo, *Surf. Sci.*, **104**, L205 (1981).

329. T. Hoshino, *Surf. Sci.*, **121**, 1 (1982).

330. V. Barone, G. DelRe, G. LeLay, and R. Kern, *Surf. Sci.*, **99**, 223 (1980).

331. G. Apai, J. F. Hamilton, J. Stöhr, and A. Thompson, *Phys. Rev. Lett.*, **43**, 165 (1979).

332. P. Eisenberger and G. S. Brown, *Solid State Commun.*, **29**, 481 (1979).

333. G. Beni and P. M. Platzman, *Phys. Rev. B*, **14**, 1514 (1976).

334. B. K. Teo, in B. K. Teo and D. C. Joy (Eds.), *EXAFS Spectroscopy, Techniques and Applications*, Plenum, New York, 1981, p. 13.

335. P. Roubin, D. Chandesris, G. Rossi, J. Lecante, M. C. Desjonquères and G. Tréglia, *Phys. Rev. Lett.*, **56**, 1272 (1986).

336. M. Bader, A. Puschmann, C. Ocal and J. Haase, *Phys. Rev. Lett.*, **57**, 3273 (1986).

337. P. A. Lee and J. B. Pendry, *Phys. Rev. B*, **11**, 2795 (1975). B. K. Teo, *J. Am. Chem. Soc.*, **103**, 3990 (1981).

338. J. Stöhr, R. Jaeger, and J. J. Rehr, *Phys. Rev. Lett.*, **51**, 821 (1983).

339. D. Arvanitis, K. Baberschke, L. Wenzel and U. Döbler, *Phys. Rev. Lett.*, **57**, 3175 (1986).

340. E. Zanazzi and F. Jona, *Surf. Sci.*, **62**, 61 (1977).

341. D. A. Fischer, U. Döbler, D. Arvanitis, L. Wenzel, K. Baberschke and J. Stöhr, *Surf. Sci.*, **177**, 114 (1986).

342. J. F. Hamilton and R. C. Baetzold, *Science*, **205**, 1213 (1979).

343. P. H. Citrin, Proceedings of the International Conference on EXAFS and Near Edge Structure IV, Fontevraud, France, July 1986; Journal de Physique (to be published).

344. U. Döbler, M. Farle, D. Arvanitis and K. Baberschke (to be published).

345. K. Baberschke, U. Döbler, L. Wenzel, D. Arvanitis, A. Baratoff, and K. H. Rieder, *Phys. Rev. B*, **33**, 5910 (1986).

346. U. Döbler, K. Baberschke, J. Stöhr, and D. A. Outka, *Phys. Rev. B*, **31**, 2532 (1985).

347. J. H. Onuferko and D. P. Woodruff, *Surf. Sci.*, **95**, 555 (1980).

348. U. Döbler, K. Baberschke, J. Haase, and A. Puschmann, *Phys. Rev. Lett.*, **52**, 1437 (1984).

349. M. D. Crapper, C. E. Riley and D. P. Woodruff, *Phys. Rev. Lett.*, **57**, 2598 (1986).

350. A. Puschmann, J. Haase, M. D. Crapper, C. E. Riley, and D. P. Woodruff, *Phys. Rev. Lett.*, **54**, 2250 (1985).

351. A. Puschmann and J. Haase, *Surf. Sci.*, **144**, 559 (1984).

352. A. Puschmann, K. C. Prince, J. Haase, G. Paolucci, and A. M. Bradshaw (to be published).

353. D. Chandesris, P. Roubin, G. Rossi, and J. Lecante, *Surf. Sci.*, **169**, 57 (1986).

354. F. Comin, J. E. Rowe, and P. H. Citrin, *Phys. Rev. Lett.*, **51**, 2402 (1983).

355. R. G. Jones, S. Ainsworth, M. D. Crapper, C. Somerton, and D. P. Woodruff, *Surf. Sci.*, **152**, 443 (1985).

APPENDIX: SURFACES STUDIED BY SEXAFS

Adsorbate/Substrate	LEED Pattern	Absorption Edge (eV)	Detection Technique[a]	Model Compound	Model Nearest-Neighbor Distance (Å)	Model Number of Nearest Neighbors	SEXAFS Site[b]	SEXAFS Nearest-Neighbor Distance (Å)	References	LEED Site[b]	LEED Nearest-Neighbor Distance (Å)	References
O/Al(111)	1×1	K(530)	TY	α-Al$_2$O$_3$	1.915	4		1.79 ± 0.05	71, 72	3FH	2.26	159
	1×1	K(530)	TY	α-Al$_2$O$_3$	1.915	4		1.76 ± 0.03[c]	73	3FH	2.12 ± 0.05	160
	1×1	K(530)	TY	Amorphous Al$_2$O$_3$	1.92	4		1.92 ± 0.05	164	3FH	2.20 ± 0.04	161
O/LaAl$_2$	1×1[e]	K(530)	TY	α-Al$_2$O$_3$	1.915	4	3FH-TU	1.75 ± 0.03[e]	73	3FH	1.80 ± 0.02[d]	162
	Disordered	K(530)	TY	α-Al$_2$O$_3$	1.915	4		1.88 ± 0.03	72, 73	3FH	1.83 ± 0.02	162
O/Si(111) 2×1	Disordered	K(530)	SY	SiO$_2$	1.61	2		1.80 ± 0.05	344			
	Disordered	K(530)	SY					1.65 ± 0.03	68, 69			
O/Ni(100)	p(2×2)	K(530)	PAY	NiO	2.084	6	4FH	1.96 ± 0.03	75	4FH	1.98 ± 0.04	197
	c(2×2)	K(530)	PAY	NiO	2.084	6	4FH	1.95 ± 0.03	75	4FH	1.98 ± 0.04; 1.76 ± 0.02	192, 209
O/Ni(110)	1×1[f]	K(530)	PAY	NiO	2.084	6	4FH	1.98 ± 0.03[f]	211			
	p(2×1)	K(530)	PAY	NiO	2.084	6	2FB[g]	1.85 ± 0.03	345			
O/Cu(100)	c(2×2)	K(530)	PAY	CuO or Cu$_2$O	1.96	4	4FH	1.94 ± 0.04	346	2FB	1.90	347
O/Cu(110)	p(2×1)	K(530)	PAY		1.84	4	2FB[g]	1.84 ± 0.02	336, 348			
CH$_3$O/Cu(100)		K(530)	PAY			4	Mixed[g]	1.97 ± 0.04	144			
HCO$_2$/Cu(100)		K(530)	PAY				4FH[g]	2.38 ± 0.03	144			
		K(530)	PAY				g	1.99 ± 0.10	349			
HCO$_2$/Cu(110)		K(530)	PAY				g	1.98 ± 0.07	350			
O/GaAs(110)	Disordered	K(530)	PAY	GaN	1.94	4		1.60 ± 0.10	70			
O/Ag(110)	p(2×1)	K(530)	PAY	Ag$_2$O	2.05	4	2FB	2.06 ± 0.05	351			
O$_2$/Ag(110)		K(530)	PAY	Ag$_2$O	2.05	4	g	2.30 ± 0.05	352			
O$_2$/Pt(111)		K(530)	PAY	α-PtO$_2$	2.07	3	g	2.15 ± 0.04	211			
S/Ni(100)	c(2×2)	K(2470)	EAY	NiS	2.394	6	4FH	2.23 ± 0.02	74	4FH	2.19 ± 0.06	192
Cl/Si(111) 7×7	7×7	K(2820)	EAY	SiCl$_4$	2.02	1	1FA	2.02 ± 0.02	272			
Cl/Si(111) $\sqrt{19} \times \sqrt{19}$	1×1	K(2820)	EAY	SiCl$_4$	2.02	1	1FA	1.98 ± 0.02	272			

Cl/Ge(111) 2×8	1×1	K(2820)	EAY	GeCl₄	2.08	1	1FA	2.07 ± 0.02	272			
Cl/Cu(100)	c(2 × 2)	K(2820)	EAY	CuCl	2.341	4	4FH	2.37 ± 0.02	231	4FH	2.64 ± 0.07	232
Cl/Ag(110)	c(2 × 2)	K(2820)	TY	AgCl	2.774	6	4FH	2.69 ± 0.03	234			
Cl/Ag(111)	√3 × √3	K(2820)	TY	AgCl	2.774	6	3FH[g]	2.70 ± 0.01	234			
Co/Cu(111)	1×1	K(7710)	TY	Co	2.51	6	3FH	2.47 ± 0.03	335, 353			
Ni/Si(111)	1×1	K(8335)	EAY	NiSi₂	2.336	8	[g]	2.37 ± 0.03	354			
Pd/Si(111) 7×7	1×1	L₂(3330)	EAY	Pd₂Si	2.42	4	Pd₂Si[g]	2.42 ± 0.02	117			
Ag/Si(111) 7×7	√3 × √3	L₂(3525)	EAY	(Pd₂Si [Theory]	2.42	4	3FH	2.48 ± 0.04	117, 296	3FH		318
	Disordered	L₂(3525)	EAY	InP [Theory]	2.541	4	3FH	2.48 ± 0.05	296			
I/Si(111) 7×7	7×7	L₃(4560)	TY	SiI(CH₃)₃	2.46 ± 0.02	1	1FA	2.44 ± 0.03	66			
I/Ge(111) 2×8	1×1	L₃(4560)	TY	GeI₄	2.50 ± 0.03	1	1FA	2.50 ± 0.04	66			
Te/Si(111) 7×7	7 × 7	L₃(4340)	TY	SiI(CH₃)₃	2.46 ± 0.02	1	2FB	2.44 ± 0.03	66			
	√3 × √3	L₃(4340)	TY	SiI(CH₃)₂	2.46 ± 0.02	1	2FB	2.51 ± 0.04	273			
	3 × 1 and 2 × 2	L₃(4340)	TY	SiI(CH₃)₂	2.46 ± 0.02	1	2FB	2.51 ± 0.04	273			
Te/Ge(111) 2×2	2 × 2	L₃(4340)	TY	GeI₄	2.50 ± 0.03	1	3FHA	2.45 ± 0.03	66			
Te/Cu(100)	p(2 × 2)	L₃(4340)	TY	Cu₂Te	2.667	4	4FH	2.62 ± 0.02	235	4FH	2.48 ± 0.10	236
Te/Cu(111)	2√3 × √3	L₃(4340)	TY	Cu₂Te	2.667	4	6FS	2.69 ± 0.02	235			
I/Ni(100)	√3 × √3	L₃(4560)	TY	NiI₂	2.78	3	[g]	2.74 ± 0.02	355			
I/Cu(111)	√3 × √3	L₃(4560)	TY	γ-CuI	2.617	4	3FH	2.66 ± 0.02	65			
I/Cu(100)	p(2 × 2), √3 × √3	L₃(4560)	TY	γ-CuI	2.617	4	4FH	2.69 ± 0.02	65			
I/Ag(111)	√3 × √3	L₃(4560)	EAY	γ-AgI	2.803	4	3FH	2.87 ± 0.03	36	3FH	2.80 ± 0.15	239

[a] EAY = elastic Auger electron yield, TY = total electron yield, PAY = partial Auger electron yield, SY = secondary electron yield.

[b] 3FH = threefold hole, 4FH = fourfold hole, 1FA = onefold atop, 2FB = twofold bridge, 3FHA = threefold hole atop a second layer atom, 3FS = threefold substitutional, 3FH-TU = threefold hollow tetrahedral underlayer, 6FS = sixfold substitutional.

[c] Corresponding to 50-L oxygen exposure.

[d] Corresponding to 90-L oxygen exposure. All other values correspond to 100–150-L exposure.

[e] 50-L oxygen heated to 200°C.

[f] Corresponding to a thin epitaxial NiO layer.

[g] See original reference.

CHAPTER

11

XANES SPECTROSCOPY

A. BIANCONI

Dipartimento di Fisica
Universita di Roma "La Sapienza"
Rome, Italy

11.1. INTRODUCTION

11.1.1. Historical Perspective of X-Ray Absorption Spectroscopy

X-ray spectroscopy provided in its golden era, at the beginning of this century, the basic experimental information on the electronic structure of atoms and thus contributed to the development of the quantum theory of atoms (1). In these last years focusing of scientific interest on understanding the properties of condensed matter in its more complex forms (e.g., proteins, glasses, alloys, and surfaces) through the study of the role of local order—Local atomic arrangement and local electronic properties—has determined the growing of x-ray absorption spectroscopy.

The advances in x-ray spectroscopy are related to the development of x-ray sources. The limitations in the energy range and intensity of standard x-ray tubes, used to obtain the first absorption spectra (2–4), were overcome by the use of electron synchrotron ($E < 1$ GeV) sources for soft x-ray studies in the 1960s and early 1970s. This research was concerned with the field of atomic, molecular, and solid state physics and good review papers were written (5–7). The advent of electron storage rings $E > 1$ GeV in the 1970s, first developed at the Frascati laboratories, as a stable intense synchrotron radiation source for x-ray spectroscopy has determined the expansion of the scientific domain of x-ray absorption spectroscopy for local structure determination at selected atomic sites in very complex systems.

Since 1920 any structure in the x-ray absorption spectra of condensed materials observed near the inner-shell absorption threshold has been known as a

Kossel structure (8) and has been assigned to bound excited states. The observed features beyond about 20 eV have commonly been referred to as a Kronig structure (9), in honor of the first investigator who proposed their explanation as being due to local structure. This division between structures below and above 20 eV was maintained for a long time. It was used to separate the low-energy range determined by the electronic structure (2–4) from the high-energy range determined by the local spatial structure. The revival of interest in the Kronig structure began with the success of one-electron, single scattering, short-range order theory (10–14) to explain the Kronig structure. It was renamed EXAFS (extended x-ray absorption fine structure), and it is now currently used as a tool for the determination of local structures in complex materials. The energy range of x-ray absorption spectra that has been currently used for EXAFS analysis extends above the values of the photoelectron wave vector $k = 3$–$4\ \text{Å}^{-1}$.

The interpretation of the strong absorption features in the continuum part of the spectra of condensed materials above the photoionization energy in a range of 30 to 60 eV remained unclear. It has been shown that this part of the x-ray absorption spectrum called XANES (x-ray absorption near-edge structure) of biological molecules (15), surfaces (16), solids, and solutions (17) is determined by the atomic geometrical arrangements in a local cluster around the absorbing atom via the multiple scattering of the excited photoelectron. Multiple scattering in XANES gives the higher-order atomic correlation functions while the single scattering in EXAFS gives only the first-order pair correlation function.

In the literature there is much confusion concerning the definition of absorption threshold and XANES energy range; in fact three different absorption edges can be defined:

1. The "absorption threshold," The energy of the lowest energy state reached by the core excitations.
2. The "absorption jump edge" or "rising edge," the energy where the absorption coefficient is at half-height of the atomic absorption jump, that is, the difference between the absorption coefficient above and below threshold.
3. The "continuum threshold" or "ionization threshold," that is, the energy where the electron is ejected in the continuum, that is, the vacuum level in atoms and molecules, the Fermi level in metals, and the bottom of the conduction band in insulators.

Examples of the strong and sharp XANES peaks above the continuum threshold and below the beginning of the weak EXAFS oscillations in the absorption spectra of condensed molecular complexes, are shown in Figs. 11.1 and 11.2. Figure 11.1 shows a typical XANES spectrum where the XANES resonances

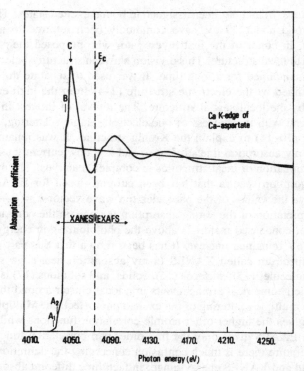

Figure 11.1. Calcium K-edge x-ray absorption spectrum of calcium aspartate showing the strong XANES features and the weak EXAFS oscillations.

have been normalized to the high energy atomic absorption α_A determined by the usual polynomial fitting of the weak EXAFS oscillations. The relative absorption α/α_A is usually plotted, where α is the measured absorption after pre-edge subtraction, in order to compare the intensities of XANES peaks between different samples. In Fig. 11.2 $(\alpha - \alpha_A)/\alpha_A$ is plotted, showing a relative variation of the absorption coefficient of about 30% in the XANES region (in some spectra it can be larger than 100%), compared to the EXAFS modulations being smaller than 4%. The spectral features of XANES have been interpreted as due to multiple-scattering resonances of the low kinetic energy photoelectron (15–19). The full multiple-scattering theory first applied to small molecules (20–22) has been shown to be the most general theoretical framework able to interpret the XANES in crystalline, amorphous, chemical, and biological materials (23–29). For crystalline materials the band structure approach, which properly takes into account photoelectron inelastic scattering and matrix element effects and which gives the local partial density of states at the selected atomic site, has

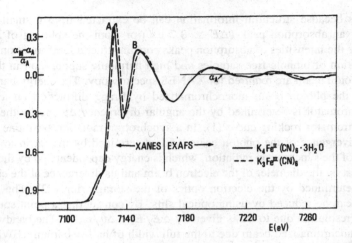

Figure 11.2. Iron K-edge x-ray absorption spectra of $K_3 Fe(CN)_6$ and $K_4Fe(CN)_6$ showing XANES resonances and EXAFS oscillations.

been used to interpret the XANES of simple metals (30, 31). It can be shown that the multiple scattering and band structure approaches are two equivalent ways to calculate the same unoccupied density of states.

The physical origin of the absorption features in the first 10-eV energy range above the absorption threshold is different in different classes of materials: Rydberg states in atoms, bound valence states or bound multiple-scattering resonances in molecules, core excitons in ionic crystals, unoccupied local electronic states in metals and insulators, atomiclike resonances in solids, multiplet splitting, many-body singularities, and multielectron configuration interactions.

We can actually distinguish three parts in a x-ray absorption spectrum:

1. The first part extending over about 8 eV, called "edge region," "threshold region," or the low-energy XANES region.

2. The region of multiple scattering in the continuum, called the "XANES region."

3. The region of single scattering at higher energies, called the "EXAFS region."

11.2. EXPERIMENTAL ASPECTS

While low-resolution x-ray absorption spectra ($\sim$ 6-eV bandwidth) are sufficient for EXAFS data analysis, high-resolution spectra are required for XANES data analysis because structural information can be extracted from a small energy shift of an absorption peak ($\Delta E \geq 0.2$ eV) or from the splitting of a peak. Because the intensities of absorption peaks contain structural information careful preparation of pinhole free samples and high harmonic suppression in the incident photon beam are required for XANES spectroscopy. The energy bandwidth ΔE of the photon beam monochromatized by Bragg diffraction on a crystal monochromator is determined by the angular divergence $\Delta \theta$ and by the crystal monochromator rocking curve (1). In a synchrotron radiation beamline the angular divergence of the photon beam $\Delta \theta$ is determined by the intrinsic vertical spread of the synchrotron radiation, which is energy dependent and by the source size, that is, the diameter of the electron beam and its divergence at the emission point determined by the electron optics of the storage ring (32). The angular spread can be reduced by using optical slits. By considering a point source the energy resolution due to $\Delta \theta$ is given by $\Delta E / E = \Delta \theta / \tan \theta$. The bandwidth of the monochromatized beam due to the full width of half-maximum (FWHM) of the total reflective profile of the monochromator crystal or rocking curve is given by

$$\omega_n(\lambda) = 2.12 \, r_0 \left(\frac{\lambda}{n+1} \right)^2 \left(\frac{N F_h}{\pi \sin 2\theta} \right)$$

for a single Bragg reflection of polarized x-rays, where r_0 is the electron radius, N is the atomic density, F_h is the scattering crystal structure factor: $F_h = F(\sin \lambda / d) / (h^2 + k^2 + l^2)$, where h, k, and l are the Miller indices of the reflection plane, θ is the Bragg angle, λ is the wavelength, and n is the order of harmonic present in the beam. Therefore, the crystal energy resolution is given by

$$\frac{\Delta E}{E} = \frac{2.12}{2\pi} \, r_0 \, \frac{2d^2}{(n+1)^2}$$

which is nearly energy independent.

In Fig. 11.3 the energy bandwidths due to the rocking curve and the angular spread have been plotted for the Si(111) and Si(220) monochromators and for an angular spread $\Delta \theta = 4.10^{-5}$ rad, which is the value at the Frascati x-ray beamline using a 0.5-mm vertical exit slit. The experimental bandwidth is clearly given by the sum of the two contributions that are compared with the intrinsic width of $1s$ shells due to the core hole lifetime.

The resolution can be improved either by changing the crystal or the reflection plane. In the "channel-cut" monochromator, which is often used with synchrotron radiation, the two parallel reflections reduce the tails of the rocking curve, thus increasing the resolution but leaving the harmonics content as that of a

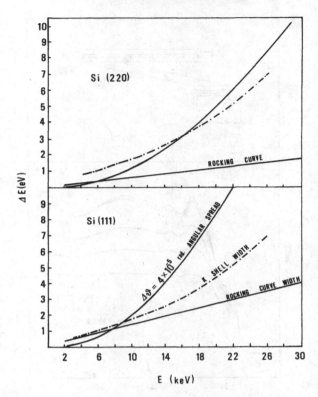

Figure 11.3. Energy bandwidth ΔE of the monochromatic x-ray beam determined by the angular spread of the photon beam $\Delta\theta = 4 \times 10^{-5}$ rad and by the rocking curve of the crystal. The bandwidths for the silicon reflections Si(111) and Si(220) and the intrinsic width of the K shells due to core hole lifetime are plotted as a function of the $1s$ binding energy.

single reflection. High-resolution crystal monochromators have been described using antiparallel reflections (33–35). Recently Calas and Petiau (36) have measured high-resolution XANES spectra at the iron K-edge using higher-order reflections. Order sorting monochromators suitable for synchrotron radiation have been realized. The device uses two crystals and harmonic rejection is achieved by detuning one crystal with respect to the other. Because the bandwidth $\omega_n(\lambda)$ is much narrower for the harmonics than for the fundamental at a given Bragg angle, misaligning the two crystals will cause the intensity of harmonics to drop off much more rapidly than the intensity of the fundamental (cf. Chapter 4, Section 4.5.2.1). The dramatic effect of the harmonic contamination on the XANES of vanadium is shown in Fig. 11.4 (37). The higher harmonics content in the synchrotron radiation beam is due to the intense con-

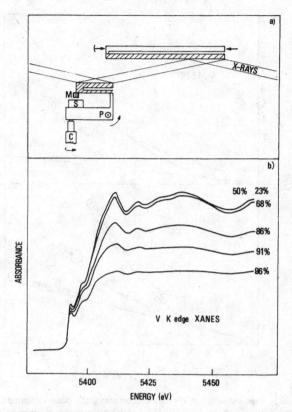

Figure 11.4. (*a*) Order sorting Si(220) monochromator. *M* is a ferrite magnet influenced by the solenoid *S*, activated to misalign the two crystals. (*b*) XANES of vanadium metal foil measured with different amounts of harmonic contamination in the monochromatized beam. The percentages refer to I/I_0, where I_0 is the intensity when the two crystals are aligned (37).

tinuum of the primary beam extending towards high energies. Therefore, the harmonic contamination of the monochromatic beam can be drastically reduced by lowering the kinetic energy of the electrons in the storage ring or in a wiggler beamline by lowering the wiggler magnetic field. The EXAFS monochromators discussed in Chapter 3 are currently used for XANES experiments. The high resolution required for XANES experiments will be strongly improved in the future by using many pole wigglers or modulators (38), new high-brillance synchrotron radiation sources like the future European 6-GeV storage ring, and monochromators with antiparallel reflections. Recently, the dispersive EXAFS system using a focusing curved monochromator developed by Flank et al. (39) reached the same resolution as the standard channel cut monochromators and open up the field of time resolved XANES experiments. In the soft x-ray range vacuum beamlines are used and crystal monochromators like InSb(111) and beryl give good resolution in the range 2000–800 eV. From 800 to 100 eV grazing incidence monochromators with gratings are used.

Special detection methods have been used for XANES. In a vacuum beamline, gas absorption experiments on molecules or atoms of low-Z elements require special gas cells (40). The experimental setup for surface XANES studies (16) is similar to SEXAFS studies (cf. Chapter 10), where the x-ray absorption is measured by recording the intensity of emitted electrons or ions. Surface XANES of clean surfaces have been measured by recording the Auger electrons in the kinetic energy range of low escape depth, approximately 50 eV (41). The XANES of thick single crystals were measured by recording the intensity of an emitted UV fluorescence line with an optical photomultiplier normal to the x-ray beam (42) and scanning the x-ray energy. The XANES of very dilute biological systems are measured by the x-ray fluorescence detection methods. It should be remarked that because XANES spectral features are much stronger than EXAFS features, XANES studies can be performed in more diluted systems.

11.3. ATOMIC ABSORPTION

Multiple-scattering resonances in the XANES of molecules and condensed materials are modulations of the atomic x-ray cross sections of the absorbing atom, therefore, it is important to start with a short overview of atomic x-ray absorption.

11.3.1. Atomic X-Ray Absorption Over a Large Energy Range

The quantitative determination of atomic x-ray absorption coefficients has been the object of extensive research since the work of Barkla and Sadler in 1909 (43) because of its importance in plasma physics, atmospheric physics, space physics, and astrophysics. The atomic absorption coefficient α (cm^{-1}) is related to the atomic cross section for absorption σ (barn/atom) by the relation $n\sigma = \alpha$, in which n is the atomic density in units of 10^{24} atoms cm^{-3}. Extensive work has been carried out on the compilation of tables of atomic x-ray absorption cross sections over the full x-ray range that is now available (44–46). Figure 11.5 shows the total atomic cross section of copper over the range 10 eV to 100 GeV. The calculated values are in good agreement with available data taken with very low-energy resolution.

In the interaction of x-rays with atoms two main effects must be distinguished: The photoionization cross section or photoelectric cross section, where a photon

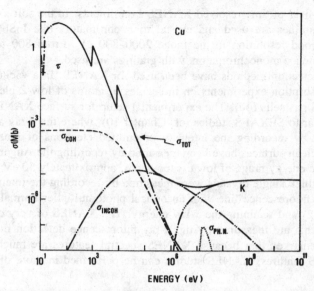

Figure 11.5. Total cross section of copper (solid line) and theoretical photoelectric cross section (— · —·), coherent scattering (— — —) σ_{coh} and incoherent scattering (— · · —)σ_{incoh} and pair production cross section (K).

is absorbed and an electron is excited, and the scattering cross section, where the x-ray photon is deflected by an atom with or without loss of energy. Figure 11.5 shows the contribution of the two types of scattering processes, the incoherent or Compton scattering (σ_{incoh}), and the coherent or Rayleigh scattering (σ_{coh}). At energies below about 10 keV the cross section for photoionization is equal to the total photoabsorption cross section since processes such as Compton scattering and bremsstrahlung are negligible. Fano and Cooper have reviewed (5) the theory of atomic photoionization over the range from 10 to 10 keV. Large interest has been focused on the photoionization of the noble gases. Figure 11.6 shows the photoionization of argon and krypton in the soft x-ray range (47) and in the x-ray range (43–45).

11.3.2. Atomic X-Ray Cross Section of Deep 1s and 2p Shells: Continuum Absorption and Rydberg States

Figure 11.6 shows that the K- and L-edges of argon and krypton exhibit a simple absorption jump at the core absorption thresholds. The atomic absorption can be fitted by the empirical Victoreen relation, over a large energy range: $\alpha = C\lambda^3 - D\lambda^2 + A$, where λ is the photon wavelength and where C and D are functions of atomic number Z, changing abruptly at the absorption edges. High-

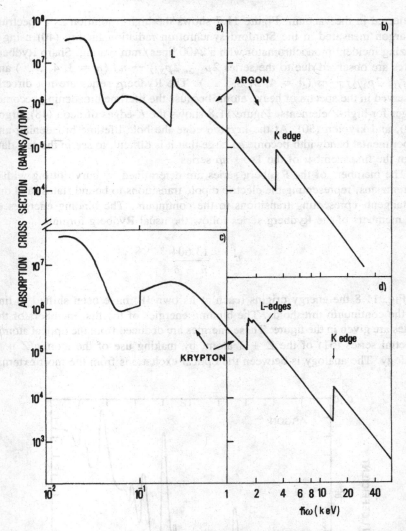

Figure 11.6. X-ray absorption cross section for argon [panels (*a*) and (*b*)] and krypton [panels (*c*) and (*d*)] (43–45, 47).

resolution spectra show a departure from this simple formula, mostly because of bound Rydberg states and multiple ionization, which induce structure in the edge region.

Below the continuum threshold discrete peaks appear due to the Rydberg series, that is, the discrete electron bound states $1s \rightarrow np$ merging in the continuum at the photon energy of the ionization potential (IP) where the electron is ejected in the vacuum. Figure 11.7 shows the high-resolution $L_{2,3}$ spectrum of argon measured at the Stanford synchrotron radiation facility (40) using a grazing incident monochromator with a 2400 lines/mm grating. Sharp Rydberg states are observed due to the series $2p_{1/2}, 2p_{3/2} \rightarrow nd$ ($n = 3, 4, \ldots$) and $2p_{1/2}, 2p_{3/2} \rightarrow ns$ ($n = 4, 5, 6, \ldots$). The Rydberg series are not directly observed in the spectra of heavy atoms because the lifetime broadening becomes larger for high-Z elements. Figure 11.8 shows the K-edges of neon (48), argon (49), and krypton (50). At the krypton edge the hole lifetime broadening and experimental bandwidth become so large that it is difficult to see in the raw data even the first member of the $1s \rightarrow np$ series.

The members of the Rydberg series are determined by curve fitting with n Lorentzians, representing the electric dipole transitions to bound states, and one arctangent representing transitions to the continuum. The binding energies of the members of the Rydberg series follow the usual Rydberg formula

$$E_n = \frac{13.604}{n - \delta}$$

In Fig. 11.8 the energy origins (each at its own IP) have been shifted to line up the continuum thresholds. The binding energies of the first members of the series are given in the figure. These energies are deduced from the optical atomic spectral series (51) of the $Z + 1$ atom, by making use of the atomic $Z + 1$ analogy. The analogy is between the optical excitations from the most external

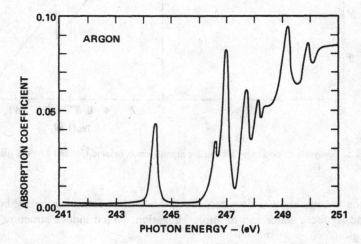

Figure 11.7. High-resolution spectrum of the Rydberg bound states at the $L_{2,3}$-edge of argon. The spin–orbit splitting is 2.03 eV (40).

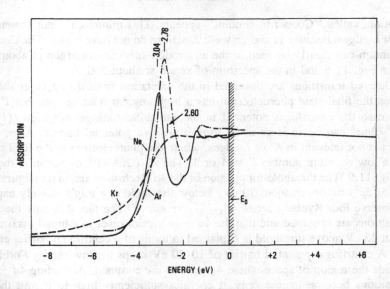

Figure 11.8. High resolution K spectra of neon (48), argon (49), and krypton (50) showing the Rydberg bound states that become broader with increasing $1s$ binding energy due to lifetime broadening; the spectra are aligned at the energies of ionization potentials E_0.

occupied valence orbitals to the unoccupied highly excited final-state orbitals in the $Z + 1$ atom on the one hand, and the core excitations in the Z atom to the same final-state orbitals on the other hand. In fact, in these excitations the excited photoelectron is in the same final state and the core hole in core excitations of the Z atom is effectively replaced by the extra positive electronic charge in the nucleus of the $Z + 1$ atom in optical excitations.

11.3.3. Atomic X-Ray Cross Section of Shallow $3p$, $3d$, and $4d$ Shells: Cooper Minima, Centrifugal Barrier Effects, and Interchannel Interactions

Although in this chapter we are concerned only with K- or L-edges, it is important to give a brief account of large deviations from the flat photoelectric cross section, according to the Victoreen rule, in the first 100 eV above the edges of shallow $3p$, $3d$, and $4d$ inner shells. These deviations can easily be seen in the soft energy range in Fig. 11.6. They are determined by the overlap between the core radial wave function and the radial wave function of the excited photoelectron with wave vector k.

If the initial state radial wave function has a node, a negative interference occurs some tens of electron volts above threshold and a minimum in the absorption, called "Cooper minimum," appears. This minimum is not present in K- or L-edges, because 1s and 2p wave functions do not have nodes. The Cooper minimum can clearly be seen in the absorption spectrum of argon at about 50 eV in Fig. 11.6 and in the spectrum of xenon at about 200 eV.

Delayed transitions are observed in the excitation from d or f inner shells, where the final-state photoelectron has a high angular quantum number l'. In that case the centrifugal potential in the radial Schrödinger equation $l(l' + l)\hbar^2/2mr^2$ can be so large as to form a positive potential barrier. Also, this effect is not relevant in K- or L-edges, where the photoelectron in the final state has a low quantum number $l' = 1$ or 2. The $N_{4,5}$ XANES of xenon is shown in Fig. 11.9. The threshold for ejection of the 4d electron of xenon (configuration $4d^{10}5s^25p^6$) occurs at about 70 eV. Below threshold one might naively expect to observe four Rydberg series $4d_{3/2,5/2} \rightarrow np, nf$. The fact that only the d–p transitions are observed and that the 4d cross section rises to a broad maximum about 35 eV above threshold is explained in terms of a centrifugal barrier effect (5). A centrifugal potential barrier of 10–20 eV keeps the low-energy f orbitals outside the region of space where 4d orbitals are confined. According $4d \rightarrow \epsilon f$ transitions become intense only at energies sufficiently high to permit the f-electron wave function to penetrate inside the barrier, thus overlapping the 4d wave function. Although one electron Herman–Skillman calculations predict the main feature, a better agreement with experiment is found when the exchange interaction between the ϵf photoelectron and the 4d hole is included (52). An even better agreement is found when using many-body approaches (53, 54), which take into account interchannel correlations. It is important to note that

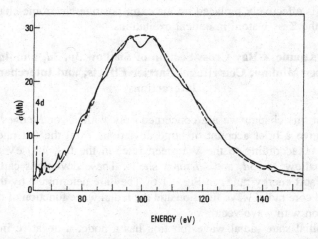

Figure 11.9. Photoabsorption cross section at the 4d threshold for gaseous and solid xenon (55).

the spectra of gaseous and solid xenon measured by Haensel et al. (55) and shown in Fig. 11.9 are almost identical, proving that such strong atomic effects on the photoionization cross section are essentially unaltered in the solid and that the electron scattering by neighboring atoms in the solid causes only the weak modulations of the atomic phtotoionization cross section.

The role of exchange interaction between holes and photoelectrons is strongest between shells with equal principal quantum number and hence affects $4d \rightarrow 4f$ and $4p \rightarrow 4d$ transitions. The exchange interaction plays a dominant role at the $4d$ edge of rare earth metals (56), where it splits the $4d^9 4f^{N+1}$ excited configurations and raises some multiplets by approximately 20 eV. In accordance with the dominant role of atomic exchange interaction at the $4d$ threshold the spectra of cerium in the solid and vapor phases were indeed found similar (57). The cerium $4d$ absorption spectra of a series of cerium intermetallic compounds have recently been measured to obtain information on $4f$ occupation, because they provide a local atomic probe of the electronic configuration (58).

The $3p$ threshold spectrum of transition metals is dominated by a strong broad asymmetric absorption band. The atomic manganese $3p$ spectrum has been measured by Bruhn et al. (59) and the broad asymmetric absorption band above threshold, which remains similar in the solid phase, has been interpreted by Davis and Feldkamp (60) as being the interference between different photoionization channels, the discrete $3p^6 3d^5 4s^2 \rightarrow 3p^5 3d^6 4s^2$, and the continuum transitions $3p^6 3d^5 4s^2 \rightarrow 3p^6 3d^4 4s^2 \epsilon f$.

A detailed review of atomic absorption spectroscopy using synchrotron radiation in the soft x-ray range has been published by Connerade (61).

11.3.4. Multielectron Excitations

The one-electron excitations from an inner-core level are always superimposed on the continuum of electronic transitions from levels with smaller binding energy. Figure 11.10 shows the partition of the total photoionization cross section of neon into its components of single ionization in $2p$, $2s$, and $1s$ subshells (62). Because of the smooth line shape of the partial cross section of levels less deep than the $1s$ core level, their contribution is generally removed in the measured inner-shell spectra at a K-edge by the "pre-edge background subtraction" procedure. The pre-edge absorption is fitted over a large energy range with a polynomial and its extrapolation above the core threshold is subtracted from the measured total absorption cross section (cf. Chapter 6, Section 6.2.4). The breakdown of the one-electron approximation is demonstrated by the importance of "multiple photoionization processes" to the total photoionization cross section, shown in Fig. 11.10. In multiple photoionization processes two or more electrons are excited by a single photon. These processes have been observed

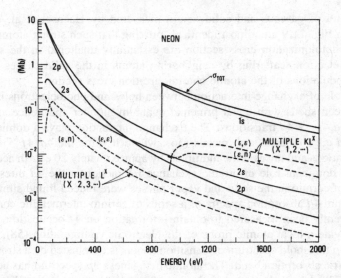

Figure 11.10. Partition of the total photoionization cross section of neon into its components of single ionization in $2p$, $2s$, and $1s$ subshells, 'multiple ionization, and simultaneous excitation and ionization in various subshells (62).

in x-ray emission, photoemission, Auger spectroscopy, and ion charge distribution studies (63).

In the one-electron approximation only one core electron is allowed to interact with the incoming photon and all the other $N - 1$ electrons (passive electrons) stay in their old orbits, either in those of the neutral atom (initial-state frozen configuration) or in the new orbits of the ionized atom (final-state frozen configuration or fully relaxed configuration). Multiple ionization processes are related to the rearrangement of passive electrons in which the electrons adjust rapidly from the initial states to the final states (64, 65). Following interaction with the radiation and inner-shell ionization the electrons of the atom may find themselves in various states of excitation. When such a state is a bound excited state, it is called shake-up and when it is in the continuum it is called shake-off. While shake-off processes produce a broad continuum contribution to the total cross section, shake-up processes produce absorption peaks in the near-edge structure (cf. Chapter 1, Section 1.5.3.2) (66–68). In a multielectron description of the initial and final states the shake-up peaks arise because of mixing of multielectron configurations in the final states.

At the threshold of two-electron excitations a sharp peak has been observed in the argon K-XANES (49). The peak at 19 eV above threshold has been assigned to two holes $1s$, $3p$, and two electrons in the Rydberg $4p$ orbitals. The series of peaks in the range 19–40 eV has been associated with several final-state configurations also involving $2p$, $3s$ holes. The final-state configurations have been identified experimentally by measurements of the x-ray emission spectra above and below the threshold of multielectron excitations. The neon K-spectrum shows two peaks at 53.1 and 38.1 eV, above the first main Rydberg line $1s \rightarrow 3p$, associated with the $1s$, $2p$ holes, and $3p^2$ electrons in the final state (69).

Recently, the interest for the XANES of metal atoms has been growing. With the development of furnaces for x-ray absorption experiments it has been possible to measure the K-XANES of sodium vapor (70, 71). Sodium having a $3s$ electron in an uncomplete external shell provides a prototype of atomic systems, different from rare gases, with incomplete shells. In these systems two electron excitations are very close to one-electron Rydberg states and just above the continuum threshold. The series of excited states with two holes in the $1s$ and $3s$ shells in sodium has been fully resolved by Tuillier et al. (71).

11.4. XANES OF MOLECULES

11.4.1. Multiple-Scattering Resonances in the Continuum

The x-ray absorption spectra of molecules can be clearly separated into the "discrete part" below and the "continuum part" above the continuum threshold E_0. The continuum threshold E_0 can be determined experimentally by x-ray photoemission experiments (XPS or ESCA), in fact E_0 is the ionization potential of the core level and thus it is equal to the photon energy required to eject a core electron into the vacuum. Figure 11.11 shows the K-edge photoabsorption spectra of carbon in CH_4 and CF_4 (72) plotted on the same energy scale after aligning continuum thresholds. On this energy scale the discrete part of the spectrum is at negative energies, while the continuum part is at positive energies. The XANES spectrum of CH_4 is similar to an atomic spectrum with no structure above E_0, while the XANES of CF_4 exhibits strong resonances B and C in the continuum before the beginning of weak broad EXAFS oscillations. Similar spectra have been measured for GeH_4 and $GeCl_4$ (73), showing in the continuum no resonances for GeH_4 and two for $GeCl_4$ as confirmed by a more recent experiment with synchrotron radiation (74). It is important to note that the silicon K-XANES of $SiCl_4$ (75) is very similar to that of $GeCl_4$ and CF_4, showing a strong peak in the discrete part of the spectrum and two resonances in the continuum.

When La Villa and Deslattes (76) measured and compared the K absorption spectra of sulfur in H_2S and SF_6 early in 1965 they pointed out the presence of

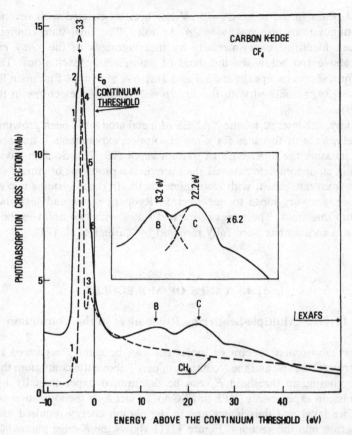

Figure 11.11. K absorption spectra of CH_4 and CF_4. The zero of the energy scale is at the ionization potentials, that is, the continuum thresholds. The continuum part of the spectra is at positive energies (96).

a strong bound localized resonance in SF_6. Zimkina and Vinogradov (77) were the first to point out that the existence of strong absorption maxima on the high-energy part of the $L_{2,3}$ ionization threshold in the continuum of inorganic molecules (SF_6, SiF_4, $SiCl_4$, BF_3, BCl_3, NF_3, and others) was a general phenomenon. Gianturco et al. (78) succeeded in reproducing the number and energy positions of the two main peaks in the SF_6 $L_{2,3}$ spectrum by identifying them as the virtual orbitals of a Hartree–Fock LCAO calculation. The calculation showed that the excited states are localized within a region of space about the size of the molecule. Nefedov (79) and Dehmer (80) introduced the concept of the effective potential barrier, meaning a region of space on the outer rim of the molecule where the potential is repulsive. This was similar to the familiar concept of centrifugal barriers in atomic physics, which occurs in the effective potential for electronic states with high ($l > 2$) angular momentum (cf. Section 11.3.2). Nefedov and Dehmer suggested that such a barrier could be provided by:

1. Direct Coulomb repulsion due to electron excess, located on the periphery of the molecules on electronegative ligands.
2. Local exchange interaction.
3. Centrifugal effects in the $l' > 2$ final states.
4. A pseudopotential, the barrier of which could arise from the requirement for the photoelectron to be orthogonal to occupied orbitals on electronegative ligands (i.e., a repulsive force due to the Pauli exclusion principle).

The presence of localized states in the continuum within the molecular region was demonstrated by molecular orbital calculations for BF_3 (81) and SO_4^{2-} (82). These calculations were able to predict the energy position of quasibound resonances in the continuum, but although they found a barrier in their "molecular effective potential," they were not able to determine the physical origin of this barrier. However, the dominant role of a Coulomb barrier was ruled out by the experiment on the K-XANES of dinitrogen (N_2) (83), where the bonding is clearly covalent. The K absorption spectrum of dinitrogen shown in Fig. 11.12 shows a clear resonance B in the continuum. Dehmer and Dill (20, 84), using the multiple-scattering $X\alpha$ method (MS-$X\alpha$), were able to calculate the absorption cross section in the continuum shown in Fig. 11.12, and they obtained a good quantitative agreement with experiments. In the dinitrogen spectrum the resonance in the measured total photoionization cross section arises because of a strong resonance at approximately 1 Ry above E_0 in the $l' = 3$, σ_u photoionization channel, that is, for the $4f$ photoelectron wave function excited along the molecular axis. The centrifugal effects on the $l' = 3$ final states provide a potential barrier whose height is approximately $l(l + 1)/R^2$ (Ry), if R (measured in atomic units) is defined as the radius of the cluster beyond which the potential is negligible. This explains the resonance in the σ_u channel. The final-state wave function of the photoelectron in the continuum has been calculated at, below, and above the resonance energy, showing a large amplitude enhancement within the molecule, at the resonance energy (85).

Dehmer and Dill pointed out that this resonance has the same origin as "shape resonances" observed in electron scattering by molecules in the gas phase at 10–40-eV electron kinetic energies (86–88). These σ resonances appear at higher energies than the well studied π shape resonances at low energy (~ 1–5 eV) in

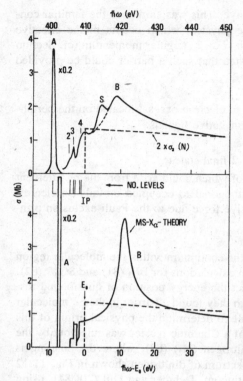

Figure 11.12. K absorption spectrum of dinitrogen (N_2) (83). The dashed line is the expected two atom absorption cross section above the continuum thresholds. In the lower panel the calculated absorption spectrum (20) shows the shape resonance in the continuum and the discrete molecular bound states in agreement with the molecular optical $Z + 1$ analogy (i.e., the valence excited states of NO). Peak S is due to a two-hole, two-electron excitation.

electron–molecule scattering [e.g., the π_g d-wave ($l = 2$) resonance at 2.4 eV in $e - N_2$ scattering]. The shape resonances are very sensitive to the changes of molecular geometry and can even couple with nuclear motion (89). The same effect is observed in XANES where the resonances above threshold are very sensitive to atomic displacements. The difference between the shape resonances in electron–molecule scattering experiments and the resonances in XANES is due both to the Coulomb attractive interaction between the electron and the core hole in the x-ray absorbing atom and to the $N - 1$ passive electrons that go into the fully relaxed orbitals in the final-state potential.

The success of the one-electron MS-$X\alpha$ calculations, where the XANES spectra are interpreted by electron multiple scattering by atoms in the molecules in a static final-state potential, without having to introduce dynamic processes and many-body effects, has been demonstrated by a careful theoretical calculation on $GeCl_4$ which, like other tetrahedral molecules, shows a spectrum similar to that of CF_4 shown in Fig. 11.11. With their self-consistent calculation

Natoli et al. (22) were able to calculate the two resonances in the continuum and the strong peak in the discrete part of the spectrum as shown in Fig. 11.13. The actual calculation for $GeCl_4$ ruled out that the Coulomb (a) and exchange (b) interactions are the source of the potential barrier, because the sum of the Coulomb potential (including nuclear attraction) and the exchange potential is negative everywhere (26). Sachenko et al. (90) and Levinson et al. (91) have calculated the two very strong resonances in the continuum above the sulfur $L_{2,3}$-edge of SF_6, which are stronger than the bound excited states, by means of the multiple-scattering model. Levinson et al. (91) pointed out that these resonances exist even in the absence of charge transfer and that in SF_6 the charge transfer, as described by the self-consistent potential, actually is opposite to what would be necessary if it were to be a major reason for the existence of the resonances. Also, the centrifugal barrier found in dinitrogen is not necessary to sustain a multiple scattering resonance (MSR) in the continuum. In fact, sometimes one finds resonances even for low-l values ($l = 1, 2$) and bigger R, as in the case of $GeCl_4$ (22) where the centrifugal barrier is definitely insufficient to sustain a quasibound state. Therefore, the concept of a potential barrier due to Coulomb (a) or exchange (b) or even due to centrifugal (c) effects is in many cases insufficient to describe the general phenomenon of MSR in XANES.

The general framework capable of describing MSR in all situations from molecules to extended media is the theory of potential scattering. In this theory the concepts that are valid in the case of a single scattering center with spherical symmetry are extended to the case of photoelectron scattering by a collection of centers, such as atoms around the absorbing atom in molecules and condensed systems (23, 24, 26, 28, 29). For a further discussion of the theoretical multiple-scattering approach we refer the reader to Chapter 2.1, Section 2.3.

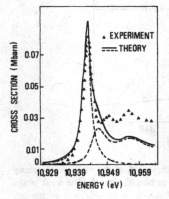

Figure 11.13. Experimental (triangles) and computed (heavy solid line) K-edge XANES of $GeCl_4$ (22). The dashed curves represent the bound state and continuum contribution separately, while the heavy curve gives the sum of the two. The energy placement of the experimental points has been adjusted for a reasonable fit.

11.4.2. Effect of Molecular Geometry and Interatomic Distances on Multiple-Scattering Resonances

The shape of XANES spectra—concerning the number of multiple-scattering resonances (MSR) and their relative intensities—is similar for each type of molecular symmetry, as shown by the similarity in the K spectra of CF_4 and $GeCl_4$. Therefore, the geometry of a molecule can easily be recognized by the shape of its XANES spectrum. The spectra of the K-edges are very different from those of the $L_{2,3}$-edges because the resonances are due to multiple scattering of the $l' = l + 1$ electron, that is, the p-photoelectron in the K-edge and d-photoelectron at the L_2- and L_3-edges, respectively.

The energy positions of the MSR peaks are strongly dependent on interatomic distance. In the case of diatomic molecules it can be shown (18, 92) that the value of the photoelectron wave vector k, at the energy of the resonance is related to the interatomic distance R by the rule $k_r R$ = constant. Therefore, the contraction (expansion) of interatomic distance shifts the energy of the resonance

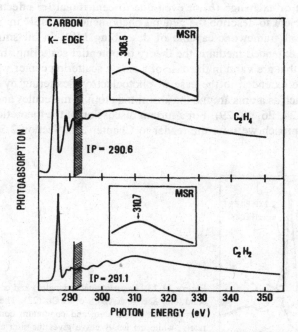

Figure 11.14. Carbon K-XANES of C_2H_4 and C_2H_2. The energy shift of the multiple-scattering resonance MSR gives a measure of the variation of the carbon–carbon distance R. The weak peaks above the ionization potential (IP) and below the MSR are shake-up or two-electron excitations (72).

to higher (lower) energy. This effect can be clearly observed in Fig. 11.14 where the XANES spectra of C_2H_2 and C_2H_4 (92) are reported. The σ_u–MSR shifts toward higher energies by $\Delta E = 4$ eV, going from the C—C distance $R = 1.32$ Å in C_2H_4 to $R = 1.2$ Å in C_2H_2. Such a large shift show the sensitivity of MSR to interatomic distance, in accordance with the $k_r R =$ constant rule, where the photoelectron wave vector k is given by $2m/\hbar \, (\hbar\omega - E_0 - \overline{V})^{1/2}$, E_0 is the binding energy of the core level or the ionization potential and $\overline{V}$ is the average interstitial potential.

Recently nonself-consistent multiple-scattering calculations (93) have shown that the angular resolved XANES of $C_2H_n (n = 2, 4, 6)$ can be predicted and the main MSR resonance in the continuum appears for photon polarization in the direction of the C—C axis (z), as is shown in Fig. 11.15. The MSR moves toward lower energies in the theoretical as well as in the experimental data and it drops partially in the discrete part of the spectrum for C_2H_6.

In a range of 20% variation of the interatomic distance, where the energy dependence of scattering phase shifts is negligible, a linear relation between k and $1/R$ has been found in agreement with experimental data. The dependence of MSR from interatomic distance has been studied in a large set of molecules by Sette et al. (94).

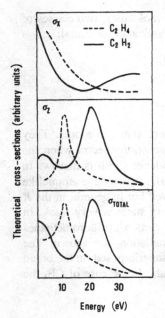

Figure 11.15. Calculated carbon K-XANES spectra for oriented C_2H_4, C_2H_2 with their axis along the z direction for longitudinal ($\mathbf{E}\|\mathbf{z}$), transverse ($\mathbf{E}\|\mathbf{x}$) polarizations and at the bottom the unpolarized absorption spectrum (93). It is clearly shown how the MSR in the z direction shifts at lower energies with increasing C—C distance according to $(E - \overline{V})R^2 =$ constant.

In order to overcome the problem of $\overline{V}$ determination to extract the interatomic distance R of an unknown molecule (by using the fact that a similar relation for the scattering T matrix gives the position of valence bound states, E_b) we have derived the relation between the bond length R and the energy position of the continuum MSR, E_p, referred to a bound state that takes the simple form ($E_p - E_b)R^2 = $ constant where all quantities are directly measurable (93).

In a linear diatomic molecule the bound state is the π_g resonance giving the main sharp peak at K threshold in the discrete part of the absorption spectrum. In C_2H_2 it is at -4.6 eV below the IP because of the attractive potential of the C $1s$ hole in the final state. In the ground state the resonance is located at a few electron volts ($+2.6$ eV in C_2H_2) in the continuum (95). Because the resonance is mostly atomiclike it is weakly affected by interatomic distance.

In conclusion it can be stated that the XANES can be predicted by the multiple-scattering one-electron theory. The number and relative intensities of the resonances are determined by the molecular symmetry, and the energy separations between the resonances are very sensitive to variation of interatomic distances.

11.4.3. Bound Excited States in Molecules: Rydberg States, Valence Orbitals, and the Molecular $Z + 1$ Analogy

In the discrete part of the x-ray absorption spectra of molecules, two classes of bound excited states can be distinguished below the ionization potential:

1. Rydberg states.
2. Valence orbitals and bound resonances.

Rydberg states in molecules are analogous to Rydberg states in atoms. They appear as sharp peaks in the soft x-ray K and L absorption spectra when the instrumental energy bandwidth and the intrinsic core-level width is narrow. In the spectra of molecules like CH_4 (Fig. 11.16) and SiH_4, which have atomiclike character, the Rydberg excitations can be well observed but in general in the K and L spectra of molecules the intensity of the Rydberg states is very weak. In Fig. 11.16 the bound excited states at the K-edge of carbon in fluoromethanes measured by Brown et al. (96), using synchrotron radiation, are shown. The Rydberg peaks labeled with numbers 1–6 in CF_4 are superimposed on the broad and very intense transitions to empty molecular orbitals. In the case of CF_4 the

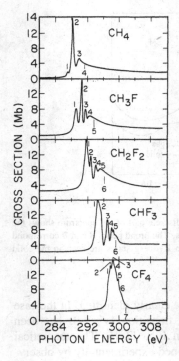

Figure 11.16. Bound excited states in the *K*-XANES for methane and fluoromethanes. Vertical lines 4–7 indicate ionization energies from XPS measurements. Going from CH_4 to CF_4 the intensity of the Rydberg series decreases. In CF_4 and broad intense $1s \rightarrow t_2$ transition dominates the discrete part of the spectrum (96).

broad intense peak is due to a $1s \rightarrow t_2$ transition to a bound state similar to the bound MSR observed and calculated by Natoli et al. in $GeCl_4$ (22). In Fig. 11.17 the Rydberg states at the $L_{2,3}$-edge of SiF_4, are shown (97). These appear as weak structures superimposed on the two bound molecular states *A* and *B*.

Rydberg states extend far beyond the ligands in real space. The extreme weakness of Rydberg states indicates that they are effectively excluded from the region of the central atom in molecules like SF_6, CF_4, and SiF_4. In fact the transitions to localized molecular orbitals are very strong because of the large overlap with the inner shell. Where such transitions occur, the discrete part of the spectrum is dominated by these broad intense peaks that appear to reduce drastically the Rydberg series and to "steal" intensity up to 50 eV in the continuum. It is, therefore, necessary to measure high resolution and high signal-to-noise spectra to find Rydberg series in many molecules. Nakamura et al. (98) and Gluskin et al. (99) succeeded in discovering the four extremely weak Rydberg series at the S–$2p$ threshold of SF_6. This is in contrast to the case of molecules like CH_4, SiH_4 (100), and H_2S (76), where the intense absorption

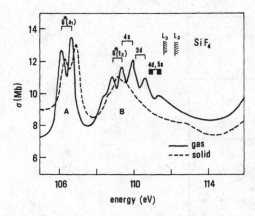

Figure 11.17. $L_{2,3}$-edge spectra of SiF_4 gas and solid. The discrete part of the spectrum shows the Rydberg states $4s$, $5s$, $3d$, $4d$, and their spin-orbit components. The broad peaks A and B correspond to the valence final-state orbitals a_1 and t_2, respectively. No Rydberg states are seen in the solid spectrum (97).

peaks due to the atomiclike Rydberg series can be clearly identified. In the case of the CH_4 K-edge a dipole forbidden Rydberg transition $1s \rightarrow 3s$ has been identified as the first peak 1 of the spectrum in Fig. 11.16. The theoretical assignment by Bagus et al. (101) has been confirmed experimentally by observing a decrease of its intensity by 19% in the CD_4 spectrum due to a different vibronic coupling mechanism (96, 102).

The energy positions of peaks due to bound excited valence states below E_0 can be predicted with a typical error of less than 1 eV and the final state can be identified by using the "molecular $Z + 1$ analogy" (103). This analogy assumes a correspondence between the final states of core excitations and the optical excited states due to molecular valence excitations in the corresponding molecule, in which the absorbing Z atom is replaced by the $Z + 1$ atom and the other atoms remain at the same distance and bond angles. This analogy assumes that the system of $(N - 1)$ passive electrons is fully relaxed following the positive core hole excitation; moreover the potential of the absorbing atom is the same, to a first approximation, whether a core electron is removed or the nuclear charge on the corresponding atom is increased by 1. This "molecular $Z + 1$ analogy" has been found to be correct in a series of molecules. In the case of dinitrogen, shown in Fig. 11.12, the residual molecular core, following the dinitrogen $N_2(1s)$ excitation, is similar to the oxygen core of the NO molecule. The valence electron spectrum of NO and the discrete dinitrogen K-edge have structures separated by the same energies (83, 103).

11.4.4. Chemical Shift of Bound Excited States and the Effective Charge on the Absorbing Atom

It is well established that the effective atomic charge of an atom in a molecule can be measured from the chemical shift of the core-level binding energy obtained by XPS. We have investigated whether it is possible to extract similar information from the x-ray absorption edge. The XANES spectra do not show any characteristic feature at the ionization energy because the bound excited states overlap close to the ionization energy and often nearly all the oscillator strength is stolen by discrete bound states and multiple-scattering resonances in the continuum. Therefore, it is not possible to directly measure the core binding energy, like in XPS. However, we have shown in the case of fluoromethanes that the chemical shift of the first excited state closely follows the core-level chemical shift (cf. Fig. 11.18). This is related to the negligible change of the binding energy of the excited state relative to the continuum threshold if the final state in all fluoromethanes is nearly the same. Another piece of experimental evidence that the chemical shift of the first bound excited state at the edge is related to the effective atomic charge is shown in the XANES spectrum of N_2O (Fig. 11.19). In this linear molecule the effective charges on the central

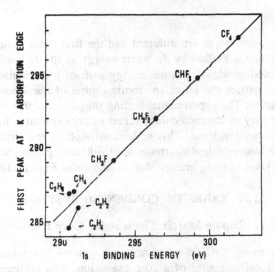

Figure 11.18. Energy of the first bound excited state in the XANES measured at the K-edge of hydrocarbons and fluoromethanes versus the binding energy of the $1s$ level measured by x-ray photoemission spectroscopy (XPS) in the gas phase. For fluoromethanes the chemical shifts measured by XANES and by XPS are in agreement.

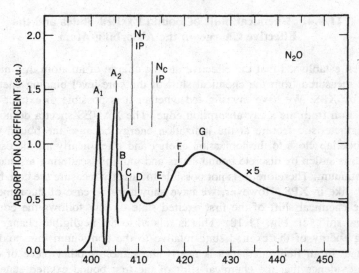

Figure 11.19. The mixed state of the N_T—N_C—O molecule (N_2O) as probed by XANES. The splitting of the first excited state A is the same as the difference in the binding energies of the $1s$ levels of the terminal N_T and central N_C nitrogen atoms, which have a different effective charge (83).

and terminal nitrogen atoms are different and the first strong excited state like peak A in dinitrogen is shifted by the same energy as the $1s$ binding energy.

Finally it should be stated that the energy shift of a particular bound state, which is used to extract the effective atomic charge of the absorbing atom, is not a direct measure. The experimental finding suggests that this method is valid only if the symmetry of the molecular excited valence orbital is the same in the series of molecules considered. This is demonstrated by the chemical shifts of the first excited states of hydrocarbons with different geometries that do not follow the core-level binding energy shift as shown in Fig. 11.18.

11.5. XANES OF CONDENSED SYSTEMS

11.5.1. Dipole Matrix Element for Core Excitations

The interpretations of XANES of condensed systems has taken a long time because of the local character of a core excitation. The difficulty in properly describing the local part of the photoelectron wave function at the x-ray absorbing site in the range of low kinetic energies up to 50 eV has been a barrier for the development of XANES spectroscopy. Recent progress, which has allowed XANES to develop to a quantitative method for local structure determination (25), is due to the identification of the XANES features in the continuum, in the range of 10–50-eV kinetic energy of the excited photoelectron, as "multiple-scattering resonances" within a small atomic cluster of neighbor atoms similar to the MSR in molecules discussed previously (15–17).

The angular part of the photoelectron wave function is selected by the dipole matrix element with $\Delta l = \pm 1$. However, the oscillator strength $f(n, l \rightarrow E, l + 1)$ for excitation of a photoelectron with kinetic energy E and angular quantum number $l + 1$, is generally larger than $f(n, l \rightarrow E, l - 1)$ by one or two orders of magnitude.

For XANES only the shape of the radial wave function of the photoelectron on the absorbing site is relevant since the initial state wave function ψ_c of the atomic core level is zero outside the spatial region of the atomic core. Therefore, the product $\psi_c(r) r \psi_E(r)$ is also zero outside this region and the shape of $\psi_E(r)$ inside this region determines the XANES features. Therefore, XANES probes the radial wave function on the site of the absorbing atom, with an angular symmetry defined by the photoelectron ($l' = 1$ at the K- or L_1-edges and $l' = 2$ at the $L_{2,3}$-edges). The presence of neighboring atoms modifies the wave function of the excited photoelectron by multiple scattering and therefore modulates the atomic absorption cross section.

11.5.2. Comparison Between EXAFS and XANES

In Chapter 2 one-electron theories (22–31) of XANES are described. Direct structural information can be extracted from XANES which, like EXAFS, is due to the scattering of the excited photoelectron by neighboring atoms. Therefore, it is useful to discuss the comparison between EXAFS and XANES and the differences in information that can be obtained. The simplicity of EXAFS is an advantage, but contributes a limitation at the same time. Generally, information on coordination geometry (bond angles) cannot be extracted and serious limitations appear in nonordered structures (104). Bond angles can only be determined by EXAFS through the focusing effect, when first and second neighbors are present in almost collinear fashion (cf. Chapter 1, Section 1.5.2) (105). In the EXAFS region the wave function of the excited photoelectron can be described by a simple theory. That is to say that the high kinetic energy photoelectron, extracted from the absorber (the central atom), is weakly backscattered by one of the neighbor atoms in a single-scattering process. This gives information about local structures only in terms of atomic radial distribution (distances) around the central atom within a range of about 4 Å (short range).

XANES contains information of the stereochemical details (coordination geometry and bond angles), which are particularly important for complex systems such as proteins, characterized by weak order low symmetry. In the photoionization process the low kinetic energy (10–40 eV) excited photoelectron is strongly backscattered by neighboring atoms, generating multiple-scattering processes shown in Fig. 11.20. It is because of this multiple scattering involving several atoms that XANES is informative about relative positions of the neighboring atoms. The multiple-scattering pathways can be classified according to the number of scattering events (n). Moreover (n) indicates the order of correlation function probed by the absorption [the ($n = 2$) pair correlation function is probed by EXAFS]. In the full multiple-scattering regime all the orders (infinite in principle) contribute to XANES. However, because of inelastic scattering the contribution of the longest pathways is reduced. In conclusion XANES is determined by higher-order correlation functions of the distribution of neighboring atoms, while EXAFS gives only the first-order pair-correlation function.

11.5.3. The XANES Multiple-Scattering Energy Range

The definition of the energy E_c separating the multiple strong scattering regime, XANES, from the single weak scattering regime, EXAFS, is somewhat arbitrary. In fact a transition region between the two regimes is expected. However, the multiple-scattering resonances can often be recognized because they are stronger and sharper then the broad weak EXAFS oscillations (see Figs. 11.1 and 11.2), and cannot be explained by EXAFS theory. In simple molecules where neighboring atoms are at a single interatomic distance R from the central atoms, like N_2, CO, or CF_4, a minimum energy range of XANES can be defined.

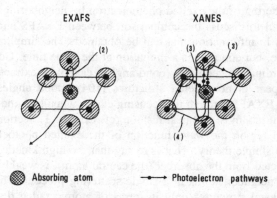

Figure 11.20. Pictorial view of photoelectron scattering processes in the single-scattering regime, EXAFS, and in the multiple-scattering regime, XANES. In EXAFS the photoelectron is scattered only by a single neighbor atom, in XANES all the scattering pathways, classified according to the number of scattering events (3), (4), (5), . . . contribute to the absorption cross section.

In fact the single-scattering EXAFS approximation is expected to break down where the photoelectron wavelength is larger than the interatomic distance, or if its wave vector $k < k_c = 2\pi/R$. A pictorial view of the final states at different energies is shown in Fig. 11.21. At the energy of a maximum in the XANES the photoelectron radial wave function in the continuum should be confined to the molecular region. This criterion shows that the XANES range is expected to expand in the continuum energy range as the interatomic distances decrease. For large molecules and condensed materials the value of k_c is defined by the shortest interatomic distance d of the cluster of neighbor atoms determining the XANES (16–18). $k_c = 2\pi/d$ or $E_c(\text{eV}) = 151/d^2(\text{Å}) - \overline{V}$ (where $\overline{V}$ is the interstitial potential).

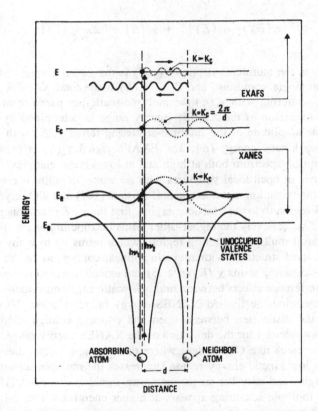

Figure 11.21. Pictorial view of the final-state wave functions in the core excitation in a diatomic molecule at high energies, EXAFS region, and in the low-energy, XANES region. The dotted curves are the wave functions of the emitted photoelectron.

The empirical low-energy cutoff for the photoelectron wave vector k_c in current EXAFS data analysis is usually put at 3–4 Å^{-1} in many inorganic and organic systems, where the minimum interatomic distances are in the range 1.5–2 Å. These values of k_c satisfy the condition $k_c d = 2\pi$ we have given for a minimum XANES energy range.

The XANES energy range can be defined quantitatively as the region where the amplitude of the XANES signal due to multiple-scattering pathways with order n larger than 2 (as shown in Fig. 11.20) can be detected. If the maximum modulus ρ of the multiple-scattering matrix satisfies the condition $\rho < 1$, the absorption coefficient can be expanded (106, 107) in a series over the EXAFS and XANES range as

$$\alpha(E) = \alpha_0(E) \left[1 + \chi_2(E) + \sum_{n=3} \chi_n(E) \right]$$

Here $\alpha_0(E)$ is the atomic absorption, $\chi_2(E)$ is the EXAFS signal in spherical wave due to single scattering, and $\Sigma_{n=3} \chi_n(E)$ is the total XANES signal due to multiple scattering with $\chi_n(E)$ the multiple-scattering pathway of order $n(n \geq 3)$. The extension of the XANES energy range is determined by the attenuation of the amplitude of the multiple-scattering terms $\chi_n(E)$ with increasing photoelectron kinetic energy (E). The EXAFS signal $\chi_2(E)$ is present in the entire absorption spectrum both at high and at low kinetic energies.

If one goes in open local structures with no atoms in colinear configuration (such as in the transition metal tetrahedral clusters (107) and SiO_2 crystals) from the EXAFS region to lower kinetic energies, first the $\chi_3(E)$ component at about 100 eV, and successively the higher-order terms become important. Finally one reaches the full multiple-scattering region, where terms up to many orders are relevant. In local structures with atoms in colinear configurations the amplitude of multiple-scattering terms $\chi_n(E)$ $(n \geq 3)$ can extend to very high energies but negative interference effects between multiple-scattering contributions $\chi_n(E)$ and $\chi_{n+1}(E)$ determine the limited XANES energy range to about 50 eV (107). Therefore, the distinction between open and colinear configurations in local clusters is not relevant for the definition of the XANES energy range. There are some unique cases like Cu K-edge, where the very low backscattering probability in the low kinetic energy region suppresses the multiple-scattering signal at low energies and only the single-scattering signal is dominant (108). However, some multiple scattering appears at higher energies where the scattering amplitude is larger.

11.5.4. The Size of the Cluster of Neighboring Atoms

The effect of neighboring atoms of the XANES of condensed systems can be identified by experimental comparison with the atomic absorption in the continuum of the same absorbing atomic species. Figure 11.22 compares the $L_{2,3}$-XANES of solid argon with the $L_{2,3}$-XANES of the argon atom in the gas phase. This was one of the first experiments done by Haensel et al. (109) at the DESY synchrotron radiation facility (Hamburg, W. Germany). In the spectrum of solid argon the flat atomic absorption is strongly modulated by the effect of the neighboring atoms. The large peaks, A, B with their fine structure are determined by scattering of the low-energy photoelectron $0 < E < 20$ eV by neighbor atoms.

Figure 11.23 shows the manganese K-XANES of a $MnCl_2$ solution measured

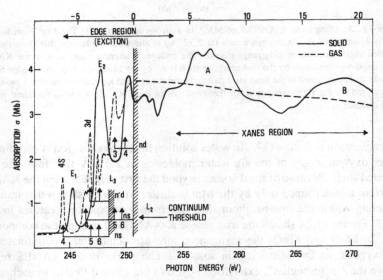

Figure 11.22. Absorption spectra of solid and gaseous argon at the $L_{2,3}$-edge with a 0.3-eV energy resolution (109). Above the continuum threshold for the L_2 and L_3 levels, separated by the spin-orbit splitting of 2.03 eV, multiple-scattering resonances (MSR) A and B appear in the XANES. The weak structures indicate the effect of several neighboring atomic shells. In the discrete energy part of the gas spectrum the Rydberg series $2p_{3/2} \rightarrow ns$, $2p_{3/2} \rightarrow nd$, $2p_{1/2} \rightarrow ns$, $2p_{3/2} \rightarrow nd$, and the continuum threshold energies are indicated. In solid argon core excitons E_1 and E_2 appear with a reduced binding energy in comparison with the first Rydberg lines.

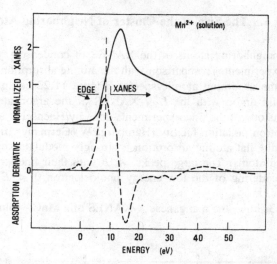

Figure 11.23. Manganese K-XANES of Mn^{2+} in aqueous solution (17). The Mn^{2+} ion is octa-hedrally coordinated by six oxygen atoms (at 2.17 Å) of the water molecules that form the first hydration shell. Because of long-range disorder the multiple-scattering resonances in the XANES region are determined only by the first neighbor shell (i.e., by the MnO_6 cluster). In the edge region the peaks can be assigned to the final-state molecular orbitals t_{2g} and t_{1u} because the $Z + 1$ molecular analogy predicts the right separation for these orbitals in the FeO_6 cluster with the same metal-ligand distance.

by synchrotron radiation (17). In water solution the manganese ion is coordinated by six oxygen atoms of the six water molecules at 2.17 Å that form the first hydrated shell. Because of the disorder beyond the first neighbor shell the XANES spectrum is determined only by the MnO_6 cluster. It can be shown that multiple scattering within the first neighbor shell determines the spectral features in Fig. 11.23. Figure 11.24 shows the manganese K-XANES of manganese compounds (17). In MnO and MnO_2 the manganese ions are octahedrally coordinated by six oxygens, as the Mn^{2+} ion in solution. The peaks in the XANES region (above the "edge region" extending ~ 8 eV) are quite different in each spec-trum, demonstrating the effects of further neighbor shells beyond the first shell. In crystalline systems the ordered atomic distribution induces a splitting of the main peaks and a fine structure appears. It was experimentally shown that if further shells appear in the Fourier transform of EXAFS oscillation also in the XANES the effects of further shells can be observed. The effect of the coordination geometry within the first coordination shell on the XANES determines the gross feature of the spectrum. This is shown in Fig. 11. 24,

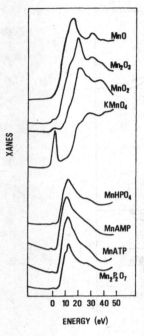

Figure 11.24. Manganese K-XANES of manganese compounds. In MnO and MnO_2 the manganese ions are octahedrally coordinated by six oxygen ions. The multiple-scattering resonances are due to several neighbor shells as shown by the mutual differences and the comparison with the XANES of the MnO_6 cluster (Fig. 11.23). The effect of further neighbor shells is a splitting of the broad XANES features, giving additional fine structure. The effect of the ion coordination geometry is shown by the large difference both in the XANES and in the edge region between the $KMnO_4$ spectrum, where manganese has a tetrahedral coordination, and the manganese oxides. The lower part shows the spectra of molecular compounds where manganese is bound to phosphate groups (PO_4).

where the XANES spectrum of $KMnO_4$ (where the manganese ion has a tetrahedral coordination) can be easily distinguished from the other spectra of octahedral coordinated sites.

In molecular complexes and in biological systems, the metal ions M are coordinated by molecular groups like phosphate (PO_4) (Fig. 11.24), carbonyl (CO), carboxyl (COO^-) hystidine (C_3N_2), imidazole rings (CN_4), cyanide (CN), and molecular groups G. In these systems we have found experimental evidence that the XANES is determined by big clusters MG_n involving a large number of atoms. The relative orientation and distances from the central atoms of the molecular groups induce a change in the XANES features. In Fig. 11.25 the calcium K-XANES of calcium formate and of the Ca–EGTA complex are shown. The XANES features C and D above the shoulder B are due to the cluster $Ca(COO)_6$. The difference between the two spectra (energy shifts of peaks C and D as well as changes in their relative intensity) is due to a different orientation of the COO groups. Figure 11.26 shows the calculated XANES spectra of a $Ca(COO)_8$ cluster, including successively the first, second, and third shells. It is clear that while the first shell gives only the main peak the effect of further shells determines the structures B and C at its high-energy side.

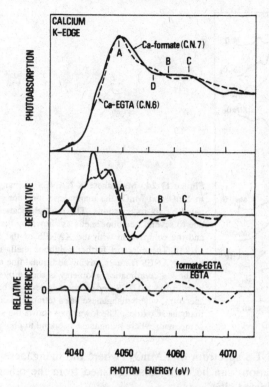

Figure 11.25. Calcium K-XANES of calcium formate and calcium EGTA complexes. In both systems the XANES above peak B are due to the clusters $Ca(CO_2^-)_6$. In Ca—EGTA, Ca^{2+} is coordinated by six oxygens of CO_2^- in the unidentate configuration, but in calcium formate one of the six carboxyl groups has a bidentate configuration, in which both oxygen ions are coordinated to Ca^{2+}. The change in the atomic positions of the COO^- groups gives a shift of peaks C and D, as well as changes in the relative intensities.

The joint effect of the inelastic mean free path for the photoelectron, dropping to less than 10 Å at kinetic energies above 10 eV (16), and the core hole lifetime reduces the number of neighbor atoms contributing to elastic scattering to a small cluster. For kinetic energies lower than ~10 eV the photoelectron undergoes inelastic scattering with valence electrons (at binding energies less than 10 eV) and only core electrons at higher binding energies contribute to the elastic scattering. Therefore, the elastic scattering is essentially determined by the positions of neighbor atoms and weakly affected by valence electron distribution.

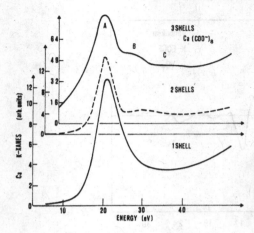

Figure 11.26. Multiple-scattering calculations for a cluster of Ca(COO)$_8$. The calculation for a small cluster including only the first shell shows a single resonance, other features are induced by further shells.

Recently, the variation of the vanadium K-XANES in VO$_2$ due to the metal–insulator phase transition has been measured (19, 110). The relative difference between the XANES spectra of the metal and the semiconductor is shown in Fig. 11.27. The effect of the phase transition on the XANES is a small increase in broadening in the metal phase. On the contrary, large effects are observed in the first 8–10-eV edge region, where the electronic structure dominates. The increase in the broadening of the XANES in the metal phase can be assigned to the increase of photoelectron–valence electron inelastic scattering in the metal phase.

Some general conclusions have been derived from a large set of experiments on XANES multiple-scattering features:

1. The continuum XANES features are due mainly to the atomic distribution of neighboring atoms, with a minor role of the electronic structure of the system (the distribution of valence electrons in the occupied levels). This point makes XANES closer to EXAFS than the optical spectroscopy. Moveover, this point introduces a distinction between the edge region (~8 eV), where atomic effects and the electronic structure play the major role, and the multiple-scattering XANES region.

2. In XANES the geometrical distribution of atoms, that is, bond angles and relative atomic positions (not generally probed by EXAFS) in the environment of the absorbing atom, is of importance.

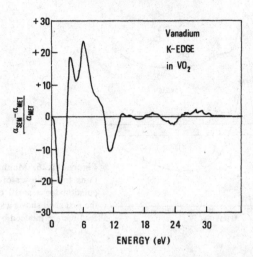

Figure 11.27. Relative difference between the vanadium K-XANES of VO_2 in the metal ($T = 400$ K) and semiconductor ($T = 300$ K). The zero is fixed at the first weak $1s \rightarrow 3d$ peak. The electronic phase transition induces large changes in the 8–10-eV edge region and only a very weak broadening in the XANES region (110).

3. The main physical effect determining the XANES of condensed systems is photoelectron multiple scattering. This gives MSR with an enhancement of the photoelectron wave function at the atomic absorption site. Therefore, the XANES peaks have the same physical origin as the multiple-scattering resonances or shape resonances in molecules. The difference between K, L_1 and L_2, L_3-XANES is due to the photoelectron angular momentum determined by the dipole selection rule ($l' = 1$ and $l' = 2$, respectively).

4. The size of the cluster of neighbor atoms, that determines the local part of the photoelectron wavefunction by multiple scattering, is limited by the lack of order beyond the first coordination shell, as in EXAFS. The core hole lifetime and the inelastic photoelectron–valence electron scattering determine the finite size of the cluster in ordered systems.

11.5.5. The Effect of the First Neighbors and of Interatomic Distance

The connection between the XANES of condensed systems and the shape resonances or multiple-scattering resonances in molecules can be seen in Fig. 11.28 where the silicon $L_{2,3}$-XANES of SiF_4(97) is shown. According to X_α multiple-

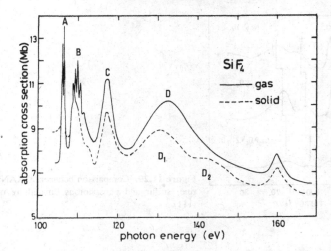

Figure 11.28. Silicon $L_{2,3}$-XANES of SiF_4 in the solid and gas phase. The peaks C and D in the continuum are due to shape resonances. In the solid the multiple-scattering resonances B, C, D, and D are clearly related with the molecular resonances. Atoms in further shells determine the presence of the resonance D (97).

scattering calculations the peaks C and D are assigned to shape resonances e and t_2, respectively, according to their final state symmetry. Going from the gas phase to the solid phase the XANES spectrum in the continuum is weakly modified. The only effect of atoms beyond the first coordination shell is the small shoulder D_2 at 142 eV appearing on the high-energy side of the peak D in the molecular spectrum. Therefore, this is a case where the XANES of the solid is practically determined by the first coordination shell. The presence of a tetrahedral coordination can be easily identified by the presence in the $L_{2,3}$-XANES of low-Z elements of the two strong resonances C and D and by the two bound excited states A, B in the edge region.

The K-XANES of low-Z elements with tetrahedral coordination should be similar to the carbon K-XANES of CF_4 shown in Fig. 11.11. The K-XANES of chlorine, sulfur, and phosphorus tetrahedrally coordinated in ClO_4, SO_4, and PO_4 molecular clusters are shown in Fig. 11.29 (111). The spectra show the typical two weak resonances B and C in the continuum as in CF_4 and $GeCl_4$ molecules. The shift of the resonances B and C is due to the decrease of the interatomic distance between the oxygens and the central atom (P—O, S—O, and Cl—O distances are 1.54 Å, 1.49, and 1.44 Å, respectively).

Kutzler et al. (23) have used a modified X_α self-consistent multiple-scattering theory to calculate both the XANES spectra in the continuum and the bound excited states in the edge region of transition metal complexes. Figure 11.30

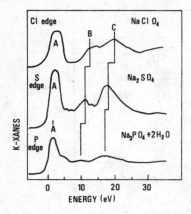

Figure 11.29. Comparison between K-XANES of chlorine, sulfur and phosphorous dinated by oxygen ions (111).

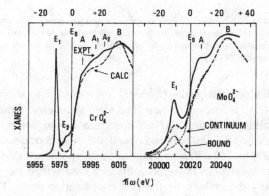

Figure 11.30. Calculated versus experimental molybdate (MoO_4^{2-}) and chromate (CrO_4^{2-}) K-XANES. The uppermost solid lines are experimental data. The continuum threshold E_0 has been calculated from the total energy difference between the excited state and the ionized state (23). In both spectra the energy position of the calculated data has been adjusted to align the theoretical and experimental bound state, which has a Lorenzian line shape. In the molybdate both the calculated continuum and discrete cross section (dotted lines) are plotted. The calculated total cross section (dashed line) is the sum of the two contributions (23).

shows the theoretical and experimental molybdenum and chromium of molybdate and chromate compounds, where the transition metal ions are tetrahedrally coordinated by four oxygens. Two multiple-scattering resonances A and B appear in the continuum. The X_α self-consistent calculation considers only the first condition shell. While the experimental curve of MoO_4^{2-} exhibits only two resonances in the continuum A and B, the spectrum of CrO_4^{2-} exhibits four peaks where the peaks A_1 and A_2 are not predicted by the theory. These features are probably due to further atoms beyond the first shell not taken into account by the theory. The theory is able to predict the position of the continuum threshold E_0 and shows that the first peak E_1 is a bound excited state. This calculation demonstrates that the one-electron approximation using a final-state fully relaxed potential can explain all the main XANES features both in the continuum and in the discrete part of the spectrum. Because of the dipole approximation only transitions to final states of t_2 molecular symmetry of the tetrahedral cluster are taken into account.

When the XANES calculation has been performed for a cluster like MO_4 (M = transition metal) it can be easily extended to similar clusters with different interatomic distances. Using the rule $(E_r - E_b)R^2$ = constant (93) where E_r is the energy of the resonance and E_b is the energy of a bound state at threshold, the energy position of the multiple-scattering resonance can be predicted. This is shown in Fig. 11.31 where the K-XANES of the 3d transition metal ions

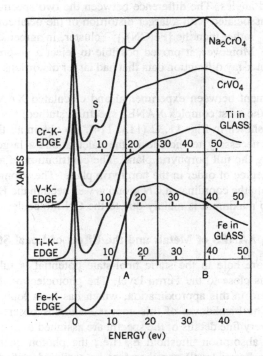

Figure 11.31. K-XANES of transition metals in tetrahedral coordination. The energy scales have been scaled by a factor $1/R^2$ and the zero has been fixed at the $1s \rightarrow 3d$ bound resonance at threshold (112).

Cr^{6+}, V^{5+}, Ti^{4+} ions with four oxygens in tetrahedral coordination are shown (112). The energy scales have been scaled with a factor $1/R^2$, thus aligning peaks A and B.

The application of the nonself-consistent multiple-scattering calculation of XANES, described in Chapter 2, for the interpretation of the molecular complexes $K_4Fe(CN)_6$ and $K_3Fe(CN)_6$ (25) was successful, not only to reproduce the experimental spectra but also to determine the geometrical distortion of the $Fe(CN)_6$ cluster in $K_4Fe(CN)_6$. This work is described in Section 2.5 of Chapter 2 and demonstrates for the first time XANES at a level of a quantitative method for local structure determination. Here we will only briefly mention the main conclusions of this work. Most importantly it was found that the iron K-XANES spectra (cf. Fig. 11.2) could only be explained if all carbon and nitrogen neighbors were included. The second shell of nitrogen atoms proved very important. Furthermore, the position of the strong XANES peaks A and B moves toward higher energy with contraction of the $Fe-C$ distance and their splitting depends on the $C-N$ distance. The relative intensity and line shape of XANES features depend on bond angles. The difference between the two spectra shown in Fig. 11.2 has been associated with a larger distortion of the octahedral symmetry in the $[Fe(CN)_6]^{4-}$ rather than the $[Fe(CN)_6]^{3-}$ cluster, in agreement with neutron diffraction data. Moreover, it proved possible to reject a proposed structure of $K_4Fe(CN)_6$ from x-ray diffraction data that had larger distortions from octahedral symmetry.

Good agreement between experimental and calculated XANES spectra has been found for the most complex XANES spectrum studied: the iron K-XANES of hemoglobin shown in Fig. 11.32 (113, 114). To obtain all the experimental features it was necessary to include in the calculations a larger cluster of 30 atoms, including the full porphyrin plane. The contribution of a large cluster is due to a large degree of order in the porphyrin plane. The strong effect of shape resonances within the coordinated CN group is well seen in the HbCN spectrum, where the sharp maximum is mainly due to the CN molecule.

11.5.6. XANES of Metals and the Effect of Local Structure

In metals the core hole in the static final-state potential is fully screened by valence electrons close to the Fermi level. The photoelectron–hole interaction is reduced to zero in this approximation, which has been found to be valid in most spectra. The breakdown of this approximation is observed in very few cases and only very fine details of the spectra are assigned to many-body effects. The continuum absorption threshold E_0 (i.e., the photon energy of the core excitation to the Fermi level) can therefore be easily identified in the experimental curves as the first derivative maximum.

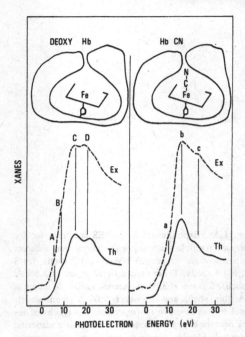

Figure 11.32. Experimental iron K-XANES of deoxyhemoglobin (Hb) and of Hb with a bound CN molecule (HbCN) dashed curves. The theoretical iron K-XANES for a cluster of 30 atoms around the absorbing iron (including the porphyrin plane in the protein and the proximal hystidine group) are shown as solid curves. The zero of the energy scale is fixed at the $1s \rightarrow 3d$ localized excited state that appears as a weak dipole-forbidden peak at threshold.

Greaves et al. (115) have demonstrated that the K-XANES of copper metal can be explained by the multiple-scattering approach. The K-XANES of copper is shown in the upper part of Fig. 11.33. The feature A is close to the Fermi level. The multiple-scattering calculations in real space for face-centered cubic (fcc) clusters starting with one shell and running up to four shells are shown in the lower panel going from calculated spectrum (a)–(d). The calculation of the three-shell cluster gives quite good agreement with the experiment. By adding the fourth shell the XANES spectrum (curve d) is only weakly affected. An energy independent broadening parameter has been used in the calculations to take the effects of inelastic electron–electron scattering and core hole lifetime into account. The experimental feature A can be obtained by removing the broadening in the calculation (dashed line) to take account of the long mean free path of photoelectrons at very low kinetic energies.

The XANES features depend mainly on the atomic arrangements of neighboring atoms. This is demonstrated by the close similarity between the XANES, measured over a large energy range, in metals with the same crystal symmetry. Figure 11.34 shows the experimental results by Grunes (116), where the XANES of fcc nickel and copper are very similar. Because the multiple-scattering res-

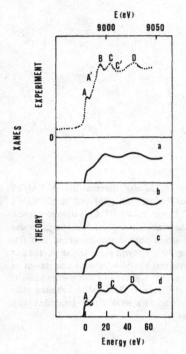

Figure 11.33. Experimental K-XANES spectrum for fcc copper and multiple-scattering calculations for different cluster sizes of fcc copper (a) 1 shell; (b) 2 shells; (c) 3 shells; (d) 4 shells. These consist of 12 atoms at 2.55 Å (first shell), 6 atoms at 3.61 Å (second shell), 24 atoms at 4.42 Å (third shell), and 12 atoms at 5.09 Å (fourth shell). The broken line in curve (d) represents the results of removing near the edge the broadening for inelastic scattering and core hole lifetime, which otherwise was kept constant over the XANES range (115).

onances are expected to shift in energy with changes in the interatomic distances, as we have described for hydrocarbons following the rule $k \cdot d$ = constant, the XANES of different elements have been plotted with different energy scales scaled by a factor proportional to their corresponding values of $1/d^2$.

The band structure approach for the interpretation of XANES has been developed by Müller et al. (30–31) to take into account the "local" and "partial" (of selected angular momentum) character of the final-state wave function and of the photoelectron–electron inelastic scattering. They have calculated a "local partial density of states," including the matrix element and inelastic electron–electron interaction, which can be quite different from the "total density of states" of the crystal. Their approach starts from the electronic band structure of the crystal and arrives at a local density of states for the excited photoelectron, which is the same as the result of the multiple-scattering approach starting from a small cluster of three shells. The agreement between the Müller calculations and the experiments is very good, as shown in Fig. 11.34. In their approach, the calculation of the oscillator strength $f_{nl,\,El'}$, is according to the Golden Rule

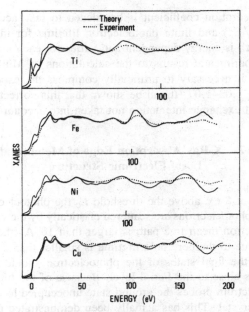

Figure 11.34. Calculated (solid lines) and measured (dotted lines) K-XANES of $3d$ transition metals (116).

$$f_{nl, El'} \propto P(E) \, D_{l'}(E)$$

The partial density of states is defined by

$$D_{l'}(E) \simeq 2 \sum_{kn} \delta(E - E_{\bar{k}n}) \sum_m |\langle l', m | \psi_{\bar{k}n} \rangle|^2$$

where the final states $\psi_{\bar{k}n}$ are band states with reduced wave vector $\bar{k}$ and band index n. The effective dipole matrix element $P(E)$ is given by

$$P(E) \sim \left| \frac{\langle \psi_c | r | \psi_{l'}(E) \rangle}{\langle \psi_{l'}^2(E) \rangle^{1/2}} \right|^2$$

where the partial wave $\psi_{l'}(E)$ is a solution of the radial Schrödinger equation with energy E for a single muffin-tin sphere and ψ_c is the core wave function.

The obtained absorption coefficient is broadened to take account of the core hole lifetime $(1/\Gamma_e)$ and finite photoelectron lifetime for inelastic scattering $[1/\Gamma_x(E)]$, which is energy dependent and varies for each material.

To fit the experimental results to the calculations of Müller et al. Grunes (116) has found it necessary to artificially compress the experimental energy scale by a factor 1.02–1.07. It can be shown that this correction is due to the energy dependent exchange interaction not taken into account into the theory.

11.5.7. X-Ray Absorption Edge of Metals and the Local Electronic Structure

In the region about 8 eV above the threshold E_0 the photoelectron wavelength is very long, the photoelectrons are scattered elastically by the valence electrons, and the photoelectron mean free path is larger than 10 Å (the broadening parameter mainly due to the core hole lifetime is smaller than 2 eV). Therefore, the edge region the final states of the photoelectron should be described as unoccupied bands close to the Fermi level. Because of the fully screened core hole the edge spectrum probes the ground-state unoccupied local partial density of states of the crystal. This has actually been demonstrated by many authors for $3d$ transition metals (30, 31, 115–118), calcium (119), $3d$ transition metal compounds (120, 121), $4d$ metals (122), platinum (123, 124), and rare earth compounds (125).

Because of the selection rules $\Delta l = +1$ and $\Delta j = +1$ the edge spectra select the angular momentum l' and the total angular momentum j of the local density of states. This is shown in Fig. 11.35 where the L_2- and L_3-edges of palladium are plotted. In the high-energy XANES region the difference between the two spectra is due to the difference in the scattering phase shift by neighboring atoms between the $l' = 2$ (L_3-edge) and $l' = 1$ (L_1-edge) excited photoelectron. The large differences in the edge region are due to the metal electronic structure near the Fermi level. The L_1 spectrum of palladium shows the typical threshold line shape expected for core transitions to the Fermi level. It is similar to the K-edge spectra of other transition metal elements (Figs. 11.33 and 11.34). Because of the large p–d mixing of conduction bands at the Fermi level the $l' = 1$ photoelectron finds a large density of dipole allowed delocalized p-like final states. If the density of unoccupied states above the Fermi level is described as a steplike function and the core hole as a state with width Γ due to its intrinsic lifetime, the absorption edge can be fitted with a function

$$F(\hbar\omega) = A \arctan\left(2\,\frac{(\hbar\omega - E_0)}{\Gamma}\right)$$

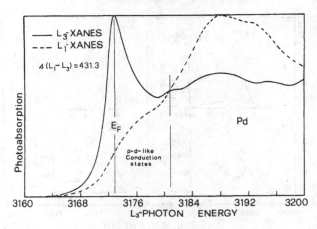

Figure 11.35. The effect of final-state photoelectron angular symmetry. The L_1 and L_3-XANES spectra of palladium show the effect of the different final-state channels $l' = 2$ and $l' = 1$, respectively (143).

Because of the presence of the fine structure above the continuum threshold E_0 the fitting of experimental curves with this function is expected to be good in the low-energy side up to the inflection point, which gives the position of the Fermi level E_F.

The absorption cross section in the K-edge region of transition metals is strongly influenced by p–d hybridization, because in zero order the unoccupied bands are d bands. In systems where the p–d hybridization is weak the p-like density of states is strongly reduced and only weak structures are observed in the first 8 eV. In that case the arctan line shape of the threshold is not easily observed because of the weak absorption structures. In Fig. 11.36 the weak features at the K-edge sprectra of both titanium and iron in the alloy TiFe, measured by Motta et al. (121), are shown. The difference between the two experimental spectra shows that the local electronic structure on the titanium and iron sites is probed, as the peak at about 6 eV above threshold is stronger at the titanium edge than at the iron edge. The comparison with the four theoretical local (on titanium and iron sites) and partial (p-like and d-like) density of electronic states spectra show clearly the reliability of the edge measurement to obtain fine details of the local electronic structure. The two peaks are related to the t_{2g} and e_g bands of the system but the splitting between the two maxima of the partial-local density of states is different for each spectrum.

The L_3-edges of transition metals show a strong peak at threshold (Fig. 11.35). The expected arctan function is strongly modified by the allowed high density of states. Clearly as the hybridization of d states becomes weaker the

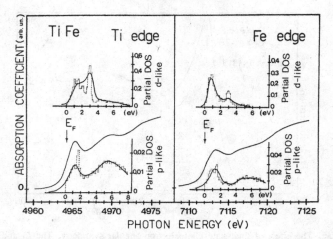

Figure 11.36. The measured titanium K-edge and iron K-edge of metallic TiFe are compared with the local partial (p-like) density of states at the iron and titanium sites. The splitting between the t_{2g} and e_g derived conduction bands can be different in the partial p-like and d-like density of states (120).

peak at the L_3 threshold is enhanced and the L_1 or K threshold is depressed. The strong peak at the L_3 edge of transition metals is a "metallic white line." Its large intensity is due to the atomiclike character of the d resonance in transition metals. The study of these "white lines" has attracted a large interest since the first experiments of x-ray absorption. They were called white lines because at the time when this phenomenon was first observed, x-rays were detected by photographic film and the large absorption peak appeared as an unexposed "white line" on the negative (126). White lines have been observed at the L_3 edges of first transition period elements by x-ray absorption using standard x-ray sources (127, 128), by synchrotron radiation using the grazing incidence monochromator "grasshopper" (129), and by electron energy loss (130) with a ~ 1-eV resolution. There are several reasons for the fact that there is a paucity of x-ray absorption L_3-edge data of the $3d$ transition metals. They lie in an energy range (400–1000 eV) where it is difficult to monochromatize the (synchrotron) radiation by crystals or by grazing incidence monochromators, because of radiation damage on organic crystals and low resolution given by grazing incidence monochromators. Recently, the use of a beryl crystal monochromator (131, 132) in a vacuum synchrotron radiation beamline has allowed high-resolution spectra to be obtained in this energy region.

The L_3-edge of the $4d$ transition metals ruthenium, rhodium, and palladium lie in the 3-keV energy range. Their $2p_{1/2}$-$2p_{3/2}$ spin–orbit splitting is about 150 eV, and the best resolution, obtained with a Si(111) monochromator and a 100-μm exit slit, is 0.4 eV (122). The L_3-edges of the $5d$ transition metals tantalum, tungsten, and platinum have been studied since the early time of x-ray spectroscopy and recently platinum (123) and tantalum (133) have been studied by synchrotron radiation. These edges lie in the 10 keV energy range where vacuum beamlines are not required, the L_2-L_3 spin–orbit splitting is very large (~ 1.5 keV), and the typical energy resolution using a Si(220) monochromator is $\Delta E \approx 1.5$ eV. The result of the one-electron theory of the unoccupied local partial ($l = 2$) density of states by Müller et al. (30) is compared with the L_3-XANES spectrum of palladium in Fig. 11.37 and a very good agreement has been found.

Changes in the shape of the white lines with increasing atomic number are determined by the localization and hybridization of the d unoccupied states and by the progressive filling of the d band. A large part of the $2p$ inner-shell oscillator strength is in the strong $2p \rightarrow nd$ atomiclike resonance, therefore, the XANES at higher energy due to local structure effects is much less conspicuous than in the K- and L_1-edge spectra. The detailed analysis of the L_3 white line shows that in palladium the one-electron calculation gives an account of the experimental data both concerning its intensity and width. In palladium, nickel, and platinum, where the unoccupied final nd states have a narrow bandwidth of 0.2–0.4 eV, the white line maximum corresponds to the Fermi level E_F. For elements with a wide unfilled nd band it is more difficult to locate the Fermi level on the rising edge of the experimental XANES spectrum.

The least-squares analysis of the tantalum L_3-edge (133) in terms of (a) a Lorentzian profile and (b) a Breit–Wigner–Fano type formula, has been carried out and both appeared to provide a good fit to the line shape. The latter line shape was derived by Fano (5) and applied to the case of interconfiguration

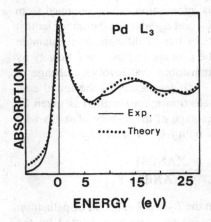

Figure 11.37. Comparison of the high-resolution experiment and the one-electron Müller et al. calculation of the L_3-XANES of palladium (122).

interaction in autoionizing states of helium. Its formula is given by

$$F(E) = \frac{a_2(q + \varepsilon)^2}{(1 + \varepsilon^2)} + b_2$$

with $\varepsilon = 2(E - E_0)/\Gamma$, where ε is the reduced energy (in units of half the width Γ) measured from the resonance position E_0 (not the peak), q is a line shape parameter, and a_2 and b_2 are constants. The Fano formula is applicable whenever two (or more) processes of different energy dependence contribute to the same complex scattering amplitude.

The details of the line shape of the white line are determined by the local density of states of the unoccupied d bands. It is important to remark that the long high-energy tail determining in Fig. 11.37 the asymmetry of the L_3 white line in palladium can be assigned to p–d hybridization of one-electron states up to 10 eV above the Fermi level.

The line shape of the nickel L_3 white line has been the object of large interest (128, 130) and the latest measurements show that the experimental width (3 eV corrected for the instrumental energy resolution) is larger than the predicted theoretical value 2.4 eV (117) taking into account the core hole lifetime, $\Gamma_e = 0.48$ eV, and hot electron broadening for the 0.1-eV bandwidth of the $3d$ unoccupied final states. On the contrary, a good agreement between one-electron predictions has been found in the case of the L_3 white line of palladium (122). Also the width of the L_3 white line in vanadium is in good agreement with APW calculations of Papaconstantopoulos (134, 135), showing the importance of including the finite lifetime of the final excited electron.

Because of the selection rule $\Delta j = \pm 1$ and $\Delta S = 0$ the edge spectroscopy provides detailed information on the total angular momentum j of the unoccupied electronic states above the Fermi level. This information can be obtained from the comparison of L_3 with L_2-edges. In fact L_3- and L_2-edges of the same element probe the same local $l' = 2$ density of states but of different total quantum number j. The $j = \frac{3}{2}$ density is probed by the L_2-edge and the $j = \frac{5}{2}$ density by the L_3-edge. We can neglect the $\Delta j = 0$ transitions that involve a change of spin from $s = \frac{1}{2}$ to $-\frac{1}{2}$. Because of the $2j + 1$ degeneracy of the initial core states the L_3/L_2 intensity ratio in the one-electron approximation is given by the statistical ratio $2:1$. Therefore, the percentage of the density of states with an angular momentum $j = \frac{5}{2}$ relative to that with $j = \frac{3}{2}$ is given by

$$\frac{D_{l'=2;j=5/2}(E)}{D_{l'=2;j=3/2}(E)} = \frac{(L_3 - \text{XANES})}{2(L_2 - \text{XANES})}$$

In Fig. 11.38 the relative difference between the L_2 and L_3 spectra of palladium from ref. (122) has been plotted. The two spectra have been shifted by the

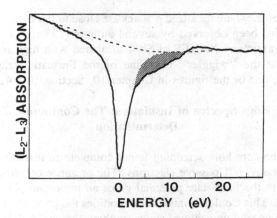

Figure 11.38. The effect of the total angular momentum j of the final-state photoelectron in the conduction band can be seen in the difference spectrum $\alpha_{L_2}(\hbar\omega - 156.7 \text{ eV}) - \alpha_{L_3}(\hbar\omega)/2.1$. The negative peak is due to the higher intensity of the L_3 white line. The dashed line in the background is due to L_3-EXAFS oscillations at the L_2-edge.

measured spin-orbit splitting (156.7 eV) and the L_2-$L_3/2$ difference has been plotted, because of the presence of some weak EXAFS oscillations of the L_2-edge at the L_3-edge. The difference shows the predominance of the $j = \frac{5}{2}$ states, not only in the spike at the Fermi level but also in the p-like conduction band up to about 8 eV, that is to say, over the whole edge region where some d-character should be present due to hybridization. This result demonstrates that the large asymmetric tail on the high energy of the white line in palladium is due to the presence of d-character up to 8 eV above the Fermi level. Above this region the L_3/L_2 intensity ratio is 2.1, very close to the statistical ratio.

In tantalum (133) and in rare earth compounds (125), where the $5d$ band is nearly completely empty, the L_3/L_2 intensity ratio is 2. The width of the L_2 white line is larger than that of the L_3 white line, because of the additional Auger $L_2 L_3 V$ Coster–Kronig decay mechanism for the L_2 hole. A large deviation from the statistical L_3/L_2 ratio has been found for the $3d$ elements (130). Since the $2p$ spin–orbit splitting is small (from 5 to 17 eV from scandium to nickel) it is impossible to measure the intensity ratio of the absorption cross sections at high energies, as for palladium. The $\frac{5}{2}$–$\frac{3}{2}$ spin–orbit splitting of the $3d$ resonance in the conduction band is very small in nickel. Therefore, the anomalous L_3/L_2 intensity ratio cannot be assigned to a different $j = \frac{3}{2}$ and $j = \frac{5}{2}$ density of states but to a breakdown of the one-electron picture due to multielectron configuration interaction. In a recent paper, Wendin has shown that this is the case where the spin-orbit separation is comparable to the exchange energy (136).

Finally, we want to point out that as the K-edges of germanium, antimony and other systems, where localized p states are close to the continuum threshold, white lines have been observed by several authors (123). Figure 11.39 shows the L_1 and L_3 antimony-XANES of SbSI measured with high resolution in the 4-keV range at the "wiggler" beamline of the Frascati storage ring (137). Compare also, one of the figures in Chapter 10, Section 10.5.4.

11.5.8. Edge Spectra of Insulators: The Continuum Threshold Determination

In insulators the core hole screening is not complete in the fully relaxed static final state of the $N - 1$ passive electrons. The presence of partial screening of the core hole in the final-state potential gives an important photoelectron–core hole attraction. This Coulomb interaction modifies the photoelectron states compared with the ground unoccupied states (without the core hole). Bound excited states are formed below the continuum threshold E_0. The continuum absorption threshold E_0 in condensed systems is no longer the well-defined core level ionization potential (IP) of molecules. In insulating systems, E_0 is the photon energy required to excite the core electron to the energy of the bottom of the conduction band.

Since at the continuum threshold a weak density of states is expected, generally no characteristic feature can be identified at E_0 in the XANES spectra.

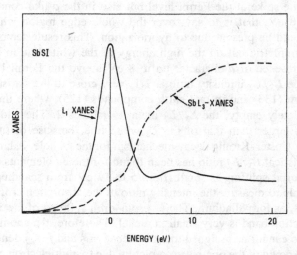

Figure 11.39. The antimony XANES of semiconducting SbSI shows the opposite behavior from transition metal edges (compare the L_3- and L_1-edges of palladium in Fig. 11.35). The L_1-edge exhibits a symmetric white line due to p-like final states and the L_3-edge exhibits the typical line shape of dipole forbidden transitions at threshold.

Several experimental methods have been used to determine E_0. Parratt and Jossem (138) proposed a method using the joint analysis of x-ray emission and optical absorption spectroscopy. The energy E_B of the transitions from the top of the valence band to the core level is measured from x-ray emission spectra. Second, the band gap E_g between the valence and conduction band is measured from optical absorption spectra. The sum $E_B + E_g$ gives the photon energy E_0 of the threshold of transitions from the core level to the conduction band. By this method Parratt and Jossem have identified the first sharp and strong absorption peak at the potassium K-edge in KCl as a bound excited state at -3.2 eV and have assigned it to a "core exciton."

Gudat et al. (139) and Pantelides et al. (140, 141) proposed a second method used in the soft x-ray range. Gudat et al. measured both the K-edge XANES of lithium in LiF by the total electron yield method, and the x-ray photoelectron spectra (XPS) of both the valence band and of the $1s$ core level. Using synchrotron radiation they have been able to measure both spectra on the same evaporated film and with the same soft x-ray monochromator, thereby avoiding energy calibration problems. In Fig. 11.40 the lithium $1s$ level, the valence band (XPS data), the $\varepsilon_2(\omega)$ curve (from UV reflectance spectrum), and the lithium K-edge-XANES are plotted in the same figure. The continuum threshold E_0 at 63.8 eV is obtained by the sum of the band gap E_g (the energy separation between the bottom of the conduction band and the top of the valence band) and the binding energy of the $1s$ level (the energy separation in the XPS spectrum between the top of the valence band and the $1s$ core level). The errors in E_0 obtained by this method are generally in the range 0.2–0.5 eV. Because of calibration problems, the errors in the estimate of E_0 can be much larger if the binding energy of the core level and the XANES spectrum are measured by different research groups with different monochromators.

A third method has been suggested, which is particularly useful in the case of metal oxides (142, 143) and can be used in the x-ray range. The chemical shift of the same atomic core line, in XPS experiments, going from the metal of the element to the compound under study, can be easily measured

$$E_B(\text{INS}) - E_B(\text{MET}) = \Delta_{\text{XPS}}$$

The XANES of the compound and of the metal are measured in the same experiment. The x-ray absorption threshold in metals gives directly the binding energy of the core level below its Fermi level ($E_0(\text{MET})$).

The energy difference E_D between the Fermi level of the metal and the conduction band in the insulator is given by the measurement of the metal–insulator potential discontinuity for the conduction band, which can be measured by electron-tunneling experiments. When this value is not available, E_D can be

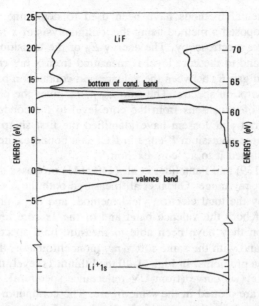

Figure 11.40. Data of optical and electron spectroscopy are composed to reveal the effect of core-photoelectron interaction in the lithium K-XANES of LiF (139). The zero of the energy scale on the left side is chosen in such a way that it gives the photon energy of the fundamental UV absorption spectrum (top left). The energy scale on the right side gives the photon energy of the lithium $1s$ core absorption in LiF (top right). The dashed line is the top of the valence band and the E_0 level is at the solid line.

estimated as half the optical band gap E_g, $E_D = E_g/2$. Therefore, the $E_0(\text{INS})$ can be obtained as

$$E_0(\text{INS}) = \Delta_{\text{XPS}} + E_D + E_0(\text{MET})$$

This method for E_0 determination can also be used when only the Δ_{XPS} values of different core levels are available, since Δ_{XPS} is nearly the same for all deep core levels within approximately 0.5 eV. An example of E_0 determination by this method is shown in Fig. 11.41 for the L_1 and L_3-XANES spectra of PdO (143).

Let us now discuss the basic difference between the value of E_0 in solids and in atoms. The value of E_0 in solids is some electron volts lower than the IP of the corresponding atomic species in the gas phase. See, for example, the recent experiments on the comparison of the L_3 spectra of gas and solid rare earths (144). In Figs. 11.22 and 11.28 the L_3-XANES spectra of SiF$_4$ in the gas and solid phases are shown. In the gas phase the $E_0 = \text{IP}$ is set at the limit of the

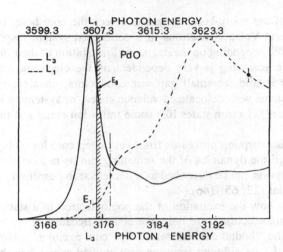

Figure 11.41. L_3-XANES and L_1-XANES of palladium in PdO. PdO is an insulator in which palladium ions are coordinated by four oxygens in a planar configuration. The E_0 value has been obtained by the third method described in the text using the XPS and XANES spectra as a reference: E_0 (INS) $= \Delta_{xps} + E_D + E_0$(MET). The symmetric excitonic white line at the L_3-edge is the $2p \rightarrow 4d$ core exciton and the peak E_1 is the dipole forbidden (quadrupole allowed) $2s \rightarrow 4d$ core exciton in the L_1-edge spectrum.

Rydberg series at about 112 eV. In the solid the Rydberg states are quenched and using XPS and optical data the E_0 is set at about 108 eV, that is, at -4-eV lower energy. The variations between the continuum threshold, E_0 in a crystal and the IP in the gas phase for a core level depend on different chemical environments and valence configurations. Moreover the polarization (or relaxation) energy due to polarization around the core hole and correlation effects within the absorbing atom reduce E_0 in solids (145).

11.5.9. Bound Excited States and Core Excitons in the Edge Region of Insulators

In the edge region of insulators bound excited states appear below the continuum threshold E_0 over a range of 5 to 10 eV, therefore, the interpretation of the edge region of nonmetals should take account of final-state effects. In the one-electron elastic approximation the interaction of the $(N - 1)$ passive electrons with the core hole is separated from the photoelectron–core hole interaction.

Following the creation of the core hole all the $(N - 1)$ electrons (core and valence electrons excluding the core excited electron) are excited in the new orbitals in the final state. This is a many-body effect that can be described as the response of the many-body system to screen the core hole. It is called the relaxation effect. Valence electrons in a condensed system respond to the localized core hole depending on the degree of localization of their orbitals. Therefore, core hole screening is very dependent on the electronic structure of the systems. This is true for small gap semiconductors, covalent systems, ionic insulators, systems with delocalized valence states, or systems with highly correlated (localized) electron states like some transition metal and rare earth compounds.

In the photoabsorption processes from very deep core levels in a completely filled main shell the dynamics of the screening can be neglected, and therefore the photoabsorption can be described as a one-electron transition in a static fully relaxed potential (23, 65, 146).

Considering now the excitation of the photoelectron in a static potential we analyze the photoelectron–core hole interaction. Because of the localization of the core hole the "bound excited states" or "core excitons" below the continuum threshold E_0 are different from the well-studied excitons in the UV spectra of insulators where the hole is in a delocalized valence band. In core excitons the deep core hole is fixed at an atomic site. Given the atomic local character of the core hole, the core hole attraction is mainly acting on the local atomic or molecular orbital components of the unoccupied wave function. The orbitals localized in the spatial region of the absorbing atom are pulled below the continuum threshold E_0 and form bound excited states or core excitons. Therefore, the photoelectron core hole interaction is responsible for the final-state localization on the absorbing atom and the edge region of insulators is probing only the local structure of the systems through the excited electronic states confined in a small cluster around the absorbing atom.

The local character of the final states of core excitons can be seen in Fig. 11.22 where the core excitons E_1, E_2 in solid argon are close to the first $4s$ and $3d$ Rydberg states of the series in the atomic spectra. Only the first Rydberg states, which have the smallest radius, appear as core excitons $2p \rightarrow 4s$, E_1 and $2p \rightarrow 3d$, E_2 in the solid state. The other Rydberg states with a large radius disappear following condensation. The energy shift of the core excitons from the gas phase to the solid phase is due to reduction of the binding energy of the bound state in the solid because of the dielectric response of the condensed system, that is, the core hole screening by valence electrons. Altarelli et al. (147) have interpreted the core excitons in solid argon in the framework of the intermediate coupling exciton theory. They found a 3.4-eV binding energy for the core excitons, in good agreement with the experiment.

Very good theoretical review papers on the problem of core excitons in semiconductors (148) and on the general aspect of core excitons by Sugano (149) and by Kotani et al. in ref. (6) have been published. The experimental aspects of soft x-ray core excitons have been well described by Kunz (150) and

Brown (7). At the moment there is general agreement that, in the case of large gap insulators, because of the core hole attraction, the final state of the core excitons can be described by molecular orbitals of the cluster formed by the absorbing atom and its first neighbor atoms. This idea was first put forward by Satoko and Sugano (151) in the framework of ligand field theory for transition metal ions and it is discussed extensively in Sugano's review paper (149). In the case of ionic alkali halides Aberg and Dehmer (152) have applied the same kind of interpretation. Recently, Kunz et al. (153) also have given an interpretation of core excitons in alkali halides by a local approach. In the case of rare gas solids Baroni et al. (154) have obtained very good agreement with experiments in the calculation of binding energies of core excitons by a local (atomiclike) approach, including the screening due to the electronic structure of the crystal.

A distinction can be made between solids discussed previously, with delocalized conduction bands that exibit core excitons in the final states, and systems where localized molecular orbitals form the lowest unoccupied conduction band. In these systems the localized unoccupied bands become bound states in the final state like the bound orbitals in the discrete part of the XANES of molecules. In the silicon $L_{2,3}$-XANES of solid SiF_4 (Fig. 11.28), the Rydberg states disappear going from the gas phase to the solid phase, but the bound excited molecular valence states remain in the solid at about the same energy. Therefore, in the solid the final-state orbitals of the peaks A and B are the same a_1 and t_2 valence orbitals of the SiF_4 cluster as in the gas phase. These states can easily be identified using the molecular $Z + 1$ analogy as the orbitals of the cluster PO_4^{3-} (97). This can be considered as a prototype of systems where in the ground state the unoccupied states are localized valence orbitals. The ground-state orbitals are pulled down in energy by the core hole but they move nearly rigidly, and the excited states can directly be associated with the ground-state orbitals. The excited states are well-localized molecular states as is demonstrated by the experimental evidence that they are practically not affected by the position of the continuum threshold E_0. In fact the peak B remains practically the same in the gas and in the solid, although in the gas it was a bound state below IP and in the solid it is in the continuum region above E_0.

The edge spectra of transition metal compounds are of great interest because of their relevance in many fields. In transition metal compounds the unoccupied localized orbitals, derived from the nd-metal orbitals, can be described by ligand field theory. On the other hand, for core excitons, the final-state orbital is localized on the absorbing metal ion M and therefore, the excited states can be described as valence states of the molecular cluster formed only by the metal M and its ligands $L_n(ML_n)$. Sugano was the first to propose and quantitatively describe such core excitons or better "bound states" as the core hole–valence orbital excited states (151).

The edge spectra of transition metal compounds can be classified according to the type of symmetry of the $3d$-derived molecular orbitals of the ML_n clusters. In octahedrally coordinated transition metals (ML_n) the molecular orbitals are usually classified as t_{2g} and e_p, and for tetrahedral clusters (ML_4) as a_1 and t_2. The t_{2p} and e_g orbitals are mostly d-like and therefore, they appear as weak prepeaks at the K-edges of transitions metals (see Fig. 11.24) because they are dipole forbidden transitions. Also for square planar coordination the molecular orbitals are mostly d-like and therefore the corresponding peaks are very weak at the $K(L_1)$ edges, while very strong exciton "white lines" appear at the L_3-edges. This is the case for the PdO edge shown in Fig. 11.41. The transitions from $1s$ ($1 = 0$) to d-type orbitals ($l = 2$) are dipole forbidden, but since the x-ray photon wavelength is close to the interatomic spacing it could be partially quadrupole allowed. Hahn et al. (155) have experimentally demonstrated the presence of a quadrupole transition at the copper-edge by measuring angular resolved absorption spectra with polarized synchrotron radiation of a single crystal.

Grunes (116) has performed a detailed comparison between the titanium L_3- and K-edges and oxygen K-edge of TiO_2. The results of his experiment on TiO_2 are summarized in Fig. 11.42. In TiO_2 both molecular orbital (MO) and band structure approaches give a similar density of unoccupied states and the e_g, t_{2p}, a_{1p}, and t_{1u} bands can be identified in all spectra. The different core-level edges give the same energy separation between the experimental peaks. This shows that the core–electron interaction is nearly the same for all states (as in SiF_4) leaving the energy separation between the excited states as in the ground state.

The fine structure of the iron K-edges of several compounds has been resolved by Calas and Petiau by using high-resolution crystal monochromators (36). The resolved features can be fully interpreted on the basis of MO theory of excited states. They show that the details of the weak features at threshold of the K spectra (generally called prepeaks because they occur several electron volts below the absorption jump edge, or rising edge) can be used to determine the local coordination of metal ions in complex systems (see Fig. 11.43). When the $3d$-derived orbitals have $l = 1$ character at the central atom, such as the t_2 orbitals in tetrahedral geometry, the transitions from the K or L levels are dipole allowed and a white line appears. This is shown in Fig. 11.23 at the K-edge of $KMnO_4$ and in Fig. 11.31.

Kutzler et al. (23) have been able to obtain a quantitative theoretical interpretation of the line shape and binding energy of the dipole allowed $1s \rightarrow t_2$ bound state in tetrahedrally coordinated transition metal complexes in the framework of the one-electron approximation. Figure 11.30 shows the calculated absorption cross section for bound final states of the tetrahedral clusters. The presence of the core hole increases the localization of the t_2 orbital and enhances

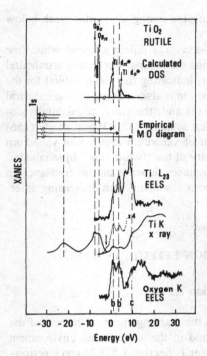

-30 -20 -10 0 10 20 30
Energy (eV)

Figure 11.42. TiO$_2$ XANES spectra in the edge region about 10 eV above the oxygen K-, titanium $L_{2,3}$-, and titanium K-edges. The first absorption peaks are aligned. The energy spacings are in good agreement with MO calculations for the TiO$_6$ cluster, as well as with the conduction band density of states (DOS) of the crystal, shown in the upper part of the figure. The arrow in the K-edge spectrum indicates an excited state with no correspondence with the ground-state electronic structure (116).

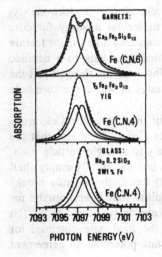

Figure 11.43. High-resolution spectra of the prepeaks or 1s → 3d-like transitions at the iron K-threshold of iron-compounds with tetrahedral (CN4) and octahedral (CN6) coordination by Calas and Petiau (36).

the absorption cross section, thus explaining the strong absorption peak below the continuum threshold.

Figure 11.31 shows that the $1s \rightarrow 3d$-like (t_2) dipole allowed white line appears at threshold of formally $3d^0$ transition metal ions. For iron in tetrahedral coordination the white line is much weaker, indicating that the t_2 orbital for the Fe($3d^5$) ion is occupied. However, the transition is always stronger in tetrahedral iron complexes than in octahedral complexes and, therefore, tetrahedrally co-ordinated iron can be identified in complex systems (37). Wong et al. (156) have carried out an extensive investigation of vanadium K-edges in vanadium compounds and have shown that the intensity of the "white line" in tetrahedral and five coordinated square pyramidal complexes depends on the covalency in the metal–ligand bond and therefore becomes weaker with increasing inter-atomic distance.

11.6. MANY-BODY RELAXATION EFFECTS IN XANES

11.6.1. Relaxation Effects

Many-body final-state effects are due to the rearrangement of the rest of the electrons (passive electrons) in the atom and in the surrounding environment following the core hole excitation (cf. Chapter 1, Section 1.5.3.2) and electron-electron correlation in the final and initial states. Most of the XANES spectra can be interpreted in the frame of the von Barth–Grossman final-state rule (157), which states that the wave function of the excited photoelectron is determined by the final-state potential of the core hole and the relaxed $N - 1$ electrons. As examples we mention the success of the $Z + 1$ molecular analogy for core bound excited states and of one-electron theories to explain all the major feature of XANES. The final-state rule has also been tested in metals in a detailed analysis of $M_{2,3}$-edges of transition metals (158). This analysis showed that the XANES final states at threshold are fully relaxed states in which the core hole is completely screened by conduction electrons.

Recently, Stern and Rehr (159) have pointed out that the excited electrons should be orthogonal to all initially occupied states (160). They find that the final-state rule, which reduces the XANES effectively to a single-particle prob-lem, is valid only when the photoelectron is excited into an initially empty shell (band). When the photoelectron fills a shell (band) the final-state rule breaks down and the one-electron density of states of the initial state is appropriate for the interpretation of XANES. Following Stern and Rehr, in systems with $3d^9$(Ni or Cu^{2+}) or $4d^9$(Pd) ground-state configurations, the final-state potential for core transitions at the edge should be the ground-state potential. In agreement with this theoretical result the high-resolution L_3-XANES spectrum of palladium (122) shows good agreement between the experiment and one-electron density of the state of the initial state, both for the white line $2p^6 4d^9 \rightarrow 2p^5 4d^{10}$ and for the whole 40-eV energy range of XANES (see Fig. 11.37).

Thus, in the case of systems where the first unoccupied band is completely empty (CaF_2) or where the valence band is filled completely by the excited photoelectron (palladium), the XANES is described by one-electron excitation in the final- or initial-state potential, respectively. For partially filled bands, final-state effects may be of importance.

In the EXAFS regime at high electron kinetic energies multielectron "shake-off" excitations, as discussed in atomic absorption cross sections, reduce the EXAFS amplitude (161). The theoretical analysis of the dynamical response of the passive electron gas as a function of the velocity of the photoelectron leaving the absorption atom has been the object of theoretical analysis (65, 146, 162, 163).

11.6.2. The Many-Body Infrared Singularity and Plasmon Satellites

The many-body problem of core-level photoionization in a uniform electron gas has attracted much interest since the work of Nozières and De Dominicis (164) and Mahan (165, 166). At the $L_{2,3}$ threshold of free electron metals like aluminum and sodium a small skewed spike appears, over a 1-eV energy range (see Fig. 11.50), which is not predicted by the one-electron density of states. This is due to a continuous spectrum of low-energy electron–hole excitations at the Fermi level, which follow the core hole excitation. This effect has been called "infrared singularity." Several theoretical approaches have been developed (164–169) including density of states, phonon interaction, spin-orbit coupling, and effective electron–hole exchange interaction. A good agreement with experiments (170–172) has been found at threshold, over a small range of $\Delta E \approx 200$ meV, and band structure effects become quite important in the off-threshold region. At the $L_{2,3}$-edge of metallic lithium the dominant role of the final-state density of states over a 2-eV energy range has been found (171, 173). The IR singularity is relevant at the $2p$ threshold of low-Z free electron metals but it becomes negligible for high-Z elements because the short lifetime of the core hole gives broad threshold features over some electron volts. Also, in transition metal or rare earth atoms, which have a high density of states at the Fermi level due to localized nd bands, the effects of the IR singularity on inner-shell threshold line shape are negligible.

Because of the sudden creation of a core hole in a metal, if the energy is available, a plasmon excitation in the free electron gas of valence conduction electrons could be expected. Although many authors have attempted the identification of intense peaks above threshold as plasmon satellites there is actually no evidence for the presence of strong satellite peaks in the XANES. Senemaud has recently identified a very weak feature in the K-XANES spectra of magnesium and aluminum (174) at the photon energy $E_F + E_p$ (where E_F is the energy threshold for core excitations to the Fermi level and E_p is the plasmon energy at zero-momentum plasmon) in agreement with theoretical calculations (175). The intensity of the plasmon satellites is so weak that the probability of its excitation in XANES is certainly negligible in comparison with XPS.

11.6.3. Shake-Up Satellites

Parratt (2) pointed out many years ago that two holes could be excited in core-level photoionization of solids as in atoms and molecules. Photoemission measurements in gas and solids have revealed satellites besides the main peak, which were identified as "shake-up" satellites (176, 177). In the "shake up" a valence electron is excited to an unoccupied orbital because of the sudden creation of the core hole potential, followed by relaxation of all other passive electrons from ground-state orbitals to orbitals of the doubly ionized atoms. The shake-up peaks occur at higher binding energy than the main core line in the XPS spectrum and the energy separation between the satellite and the main peak gives the energy in the shake-up electron. In the framework of multielectron configuration interaction (149, 177) the shake-up satellite arises from mixing of final-state configurations, called FSCI (final-state configuration interaction).

In the study of atomic absorption spectra we have seen that extensive studies on two-hole and two-electron excitations have been carried out. In the K-XANES of small molecules like N_2, CO, and C_2H_2 shake-up peaks have been found above the continuum threshold. In the K-XANES of these molecules shown in Figs. 11.12, 11.14, and 11.44 the shake-up peaks appear in the spectral region above the continuum threshold E_0 and below the multiple-scattering resonance. Compare, for example, the peak S in the XANES spectrum of dinitrogen in Fig. 11.12.

In the XANES of condensed systems evidence for multielectron satellites is very rare. The only clear evidence of a multielectron satellite or shake up has been reported by Kutzler et al. (23), because it is not predicted by one-electron theory at the K-threshold of Na_2CrO_4 (shown both in Figs. 11.31 and 11.30).

Stöhr and Jaeger (178) have measured the surface carbon and nitrogen K-XANES of carbon monoxide and dinitrogen chemisorbed on single crystal metal surfaces. The shake-up peak in the gas phase XANES spectra of N_2 and CO (shown in Fig. 11.44) is suppressed by the chemisorption phase, even though better resolved surface XANES spectra are necessary for a definitive assignment.

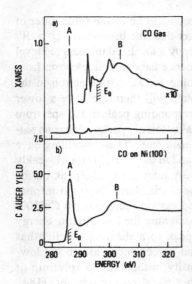

Figure 11.44. Carbon K-XANES of carbon monoxide molecules in the gas phase and chemisorbed on the Ni(100) surface (178).

The results of an extensive investigation on both XPS and XANES of rare earth insulating compounds (179) has shown that in many systems where very strong shake-up peaks appear in XPS no shake-up peaks appear in XANES. The quenching of shake-up peaks in XANES of solids can be understood from the difference in the excited final states in XPS and XANES spectroscopies. In XPS the photoelectron is excited with a very high kinetic energy (300–1000 eV) into an energetic propagating state and does not play a role in shielding the core hole. In the x-ray absorption the excited photoelectron is excited just above the Fermi energy, filling the first empty levels. The photoelectron in this case partially shields the core hole from neighboring atoms and the charge transfer shake up is expected to be significantly weaker than in the photoemission case. Moreover in the XANES final state there is one more electron than in the XPS final state, therefore the electron–electron Coulomb repulsion partially compensates the electron-hole attraction.

11.6.4. Shake-Down Satellites

The core induced relaxation process in systems with localized electronic states is of particular interest. Localized electronic states have a narrow energy bandwidth W and large electron correlations. In these bands the valence electron-electron Coulomb repulsion described by the Hubbard correlation parameter U can be larger than the bandwidth, $U > W$. The valence electron–core hole interaction is measured by the Coulomb attraction Q. Because both Q and U

are one-center Coulomb integrals, they are expected to be of the same order of magnitude and consequently Q also can be larger than the bandwidth, $Q > W$. Following core hole excitation in XPS spectroscopy a localized unoccupied level can be pulled below the top of the occupied valence band and an electron from a nearby atom can fill the relaxed level providing for an extra electron in the final state to screen the core hole. The final state will therefore have a lower binding energy than the main line and the corresponding peak in the spectrum is called a "shake-down" satellite (176). "Shake-down" peaks appear as satellites at the low-binding energy side of the main line. These satellites have been observed in rare earth intermetallics (180) and in the XPS spectra of weakly chemisorbed molecules on metal surfaces. A very intense satellite peak on the low-binding energy side of the main $1s$ core level has been observed in photoemission spectra of dinitrogen on Ni(100) and carbon monoxide on Cu(100) surfaces (181, 182). A different interpretation assigning the low-binding energy peak to the normal line and the high-energy peak to a shake-up satellite has been proposed (183). However, there is general agreement to assign the low- and high-energy peaks of about the same intensity in the $1s$ XPS spectrum of dinitrogen on nickel to the well screened (with a metal charge transfer) (184–186) and unscreened final states.

Stöhr and Jeager (178) have measured the K-shell surface XANES of carbon monoxide and dinitrogen chemisorbed on Ni(100). While the XPS spectrum excited with $\hbar\omega = 500$-eV photons shows the screened and unscreened peaks, the dinitrogen K-XANES of dinitrogen does not show any evidence of satellite structure at the K-edge.

Evidence of quenching of shake-down satellites has been found in a L_3-XANES study of intermetallics of cerium and lanthanum (187). The $3d$-XPS core lines (180) show a strong shake-down peak assigned to a well-screened final state. A weak shake-down peak has been found on the low-energy side of the metallic $2p \rightarrow \overline{5, \varepsilon}\, d$ white line at the threshold in the L_3 spectra of $LaPd_3$ and $CePd_3$, while the high-Z rare earth isomorphous compounds, $SmPd_3$ and $NdPd_3$, exhibit a simple white line shown in Fig. 11.45 and 11.46. The local electronic structure in these compounds is similar and therefore the shoulder at -3 eV below the main line has been assigned to the final state $2p^5 4 f^{n+1} \overline{5, \varepsilon}\, d^1$ and the main line to the $2p^5 4 f^n \overline{5, \varepsilon}\, d^1$ state. In the initial state the f^{n+1} configurations above the Fermi level, but following the core hole creation it is pulled below the Fermi level. By subtracting the one-electron cross section, using the normalized $NdPd_3$ spectrum, the intensity of the shake-down peak in $LaPd_3$ is obtained. It is 12% of the main line, to be compared with the $3d^9 4 f^{n+1}$ shake-down line in x-ray photoemission spectra (XPS), which is 70% of the main line. The different intensity of the shake-down peak observed in XANES and XPS spectra clearly indicates the smaller relevance of this final-state effect in XANES because of the different type of final excited state. These results

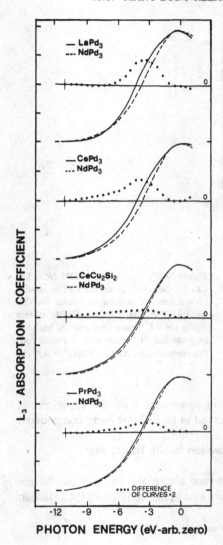

Figure 11.45. L_3-threshold of rare earth in RPd_3 (R = La, Ce, Pr, Nd, Sm). The spectra have been normalized at the maximum of the $2p \rightarrow \overline{5, \epsilon d}$ white line and the energies have been shifted to align the maxima. The $LaPd_3$ and $CePd_3$ spectra exhibit a shoulder on the low-energy side at -3 eV below the main line. This shoulder is assigned to a shake-down satellite ($4f^{n+1}$ final state). The difference spectra of each spectrum with $NdPd_3$ are given, showing the variation of the intensity of the shake-down feature.

show that a comparison between XPS and XANES can aid in identifying effects in core spectroscopies determined by initial-or, and final-state multielectron configuration interaction (176, 177, 188).

The shake-down peak disappears in heavy rare earth compounds because the $4f$ level becomes more localized and less hybridized. It has recently been shown that the white lines of the rare earth metals from samarium to lutetium are fully

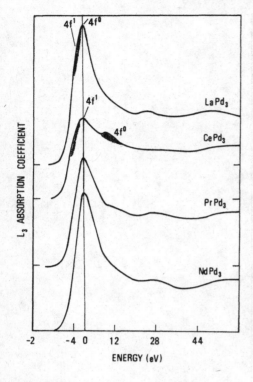

Figure 11.46. L_3-XANES of RPd$_3$ (R = La, Ce, Pr, Sm, Nd) compounds. The similar crystal structure determines similar XANES spectra. The mixed valent CePd$_3$ system shows the $4f^1$ (main line) and the high-energy satellite ($4f^0$) due to initial-state configuration interaction $4f^1(5d\,6s)^3 \leftrightarrow 4f^0(5d\,6s)^4$

described by one-electron theory (189), in agreement with the experimental finding that no shake-down peaks are observed in high-Z rare earth compounds.

11.6.5. Configuration Interaction in the Initial State

In correlated electron systems the ground-state electronic structure can be described by mixing of multielectron configurations in the initial-state ISCI (initial-state configuration interaction).

Examples of such systems are the following:

1. Valence fluctuating rare earth compounds. In these materials, such as CePd$_3$ (190) and TmSe (125), the $4f$ level at the Fermi level is degenerate with the $5d6s$ delocalized band, and the ground state fluctuates between $4f^n(5d6s)^m \leftrightarrow 4f^{n-1}(5d6s)^{m+1}$ configurations of the valence electrons.

2. Rare earth compounds with localized valence states formed by mixing of the $4f$ and ligands orbitals. It has been shown recently that systems

like CeO_2 (191), $LaPd_3$ (192), $Ce(SO_4)_2$ (179), and other formally $4f^0$ compounds exhibit a large mixing between the ligand L orbitals and the $4f$ level of the rare earth metal. These materials are described by a mixing of configurations: $4f^0 \leftrightarrow 4f^1L$, where L indicates a hole on the ligand.

3. Transition metal compounds, such as VO_2 (119) and NiO (193), with localized t_{2g} and e_g bands, respectively. NiO can be described as mixing of $3d^8$ and $3d^9L$ configurations. In VO_2 the valence t_{2g} band is formed by a localized vanadium $3d$ band, called $d_{\parallel}$, and a delocalized π band. The ground state can be described by mixing between virtual valence band configuration: $d_{\parallel}^1\pi^0 \leftrightarrow d_{\parallel}^0\pi^1$.

Systems 2 and 3 can be called interatomic intermediate valence systems. In these systems the Coulomb interaction Q between the photoexcited core hole and the localized level, occupied by the passive valence electrons, is responsible for pulling down the energy of the localized level in the final state (149). In the final state a single hole–photoelectron configuration, such as the $2p^5\overline{5,\varepsilon}\,d^{n+1}$ final state responsible for the white line in the L_3 edges of the rare earth compounds, shows a replicate splitting due to the two possible configurations when the passive electrons are included. In valence fluctuating rare earth compounds the replicate splitting is 8–9 eV, equal to the energy by which the $4f$ level is lowered by the core hole (194). The relative intensities of the two white lines give the relative occupation probabilities of the two configurations in the ground state only in the case where there is no hybridization between the localized and delocalized orbitals (188). This method allows a simple and direct measure of the intermediate valence v in heavy Z rare earth intermediate valence compounds (125).

The relative intensity of the absorption peaks due to replicate splitting is not a good measure of the ground-state occupation numbers of the valence configurations if there is hybridization between the localized and delocalized orbitals, and where the final-state hybridization is largely different from that in the ground state. $CePd_3$ is a good example of interatomic f–d hybridization and the extracted $4f$ occupation number is expected to be smaller than that in the ground state (180, 188, 190). In XPS the variation of the hybridization in the final state is much larger than in XANES as shown by the $4f^{n+1}$ shake-down satellite, due to final state configuration interaction in $CePd_3$, which is much stronger in XPS than in XANES, as discussed previously. Fortunately the difference between the occupation numbers deduced by XANES and the ground-state occupation numbers does not seem to be large. Recent experiments by several groups show that there is good agreement between the occupation numbers deduced by XPS and XANES in valence fluctuating materials despite the different final states in the two spectroscopies. Moreover the numbers deduced by XANES seem to be close to that deduced by other methods not involving core hole excitation, which can be explained by the fact that there will only be a small final-state mixing between two configurations separated by a large energy splitting (8 eV).

Figure 11.46 shows the L_3 spectrum of RPd_3 intermetallics over a large energy range. For the valence fluctuating systems the two white lines indicated by f^0 and f^1 are due to the splitting of the $2p \rightarrow \overline{5, \epsilon}\, d$ transition, because of the final-state configurations $2p^5 4f^1 (5d6s)^3 \overline{5, \epsilon}\, d^1$ and $2p^5 4f^0 (5d6s)^4 \overline{5, \epsilon}\, d^1$.

Also systems 2 and 3, which are considered not to be valence fluctuating systems, show splitting of multielectron configurations as shown in Fig. 11.47.

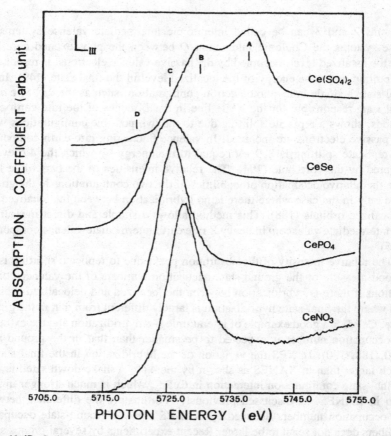

Figure 11.47. Cerium L_3-XANES of insulating $Ce(SO_4)_2$ ($4f^0$) and cerium ($4f^1$) compounds. The cerium $4f^1$ compounds exhibit strong final-state satellite peaks in XPS spectra therefore, evidence of quenching of shake-up peaks in XANES is obtained. On the contrary, the $4f^1$ and $4f^0$ peaks in $Ce(SO_4)_2$ are due to initial-state configuration interaction.

CeO_2 is a prototype of these materials and its XANES shows well-resolved $4f^0$ and $4f^1$ final states (195). Recently $Ce(SO_4)_2$ has been shown to exhibit the same properties (179). No shake-down or shake-up peaks are observed in the other integer valent $4f^1$ systems in Fig. 11.47, which on the contrary show strong satellites in the $3d$ XPS spectra. The oxygen K-edge of NiO and the vanadium K-edge of VO_2 show also the presence of multielectron configuration interaction. Figure 11.48 shows the splitting of the $1s \rightarrow t_{2g}$ transition in the vanadium K-XANES of metallic VO_2. In the spectrum of semiconducting VO_2 a single final-state configuration ($d_\parallel^1$) is observed (110).

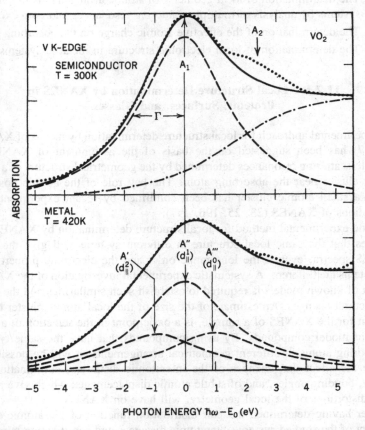

Figure 11.48. The $1s \rightarrow t_{2g}$ transition in VO_2 at the vanadium K-edge. The change of the absorption spectrum from the semiconductor phase (upper panel) to the metal phase (lower panel) is shown. The metallic phase shows a broadening due to splitting of the final-state configurations characterized by different occupation numbers of the relaxed, localized $d_\parallel$ band. The curve fitting results are indicated by dashed lines (110).

11.7. APPLICATIONS OF XANES

The physics of the photoionization processes in XANES has been developed in the past years concurrent with the application of XANES in the study of local order in complex systems. XANES experiments on a large variety of systems have contributed to our understanding of XANES and as a result of this the XANES method has grown to a quantitative tool for local structure determination in the field of material science, surface science, biology, and chemistry.

We distinguish three types of XANES applications:

1. The determination of local geometry of atomic arrangements in complex systems by analysis of multiple-scattering resonances and core excitons.
2. The determination of the effective atomic charge on the absorbing atom.
3. The determination of local electronic structure in metallic systems.

11.7.1. Local Structure Determination by XANES in Proteins, Surfaces, and Glasses

An experimental approach for local structure determination by means of XANES (15, 17) has been suggested on the basis of the assignment of XANES to multiple-scattering resonances determined by the geometrical structure of a small atomic cluster near the absorbing atom. The key role of the atomic positions within a small atomic cluster has been confirmed by recent experiments and calculations of XANES (23, 25, 196, 197).

In the experimental method for local structure determination by XANES one assumes that the same local structure in different systems will give the same XANES spectra, even if the long-range order and the electronic properties of the systems are different. A systematic experimental investigation of the XANES spectra of known models is required to establish such similarities and the effect of structural changes. An estimate of the size of the local atomic cluster that is relevant for the XANES of a sample, is a basic point in the selection of a good series of model compounds. By using compounds that have the same type of neighboring atoms in different geometrical arrangements, it will be possible to deduce the effect that changes of the interatomic distances, coordination geometry, bonding angles, and of subtle atomic displacements, which give rise to small distortions of the local geometry, will have on XANES.

After having determined the pair distribution function of the atomic arrangement of the studied system (interatomic distances and coordination numbers) by the EXAFS analysis, XANES provides further geometrical information not probed by EXAFS. Therefore, XANES can probe differences in local structures of systems that exhibit the same EXAFS spectra. The limitations of the method occur because the similar XANES spectra do not rule out differences in structure. Moreover, the few XANES features are determined by the relative positions of many atoms in the local cluster, ranging from 4 to 30 atoms in different systems. However, fortunately XANES is very sensitive to small atomic displacements and it is very difficult to find two identical spectra, even of very similar compounds. On the other hand high demands are made on the signal-to-noise ratio and energy resolution of XANES spectra, if very small differences between XANES spectra have to be identified. In conclusion the XANES can be used to determine the local geometrical structure by comparison with model compounds, to select between different local structures proposed on the basis of other experiments, and to reveal induced structural changes.

The success of multiple-scattering calculations of XANES has been a key step in the development of XANES as a quantitative tool for structure determination. The parameters in the theory that are related to the atomic scattering amplitudes are determined by calculations of the XANES of model compounds. The effects of atomic displacements, distortions, and so on, on XANES are studied theoretically to determine the origin of the differences between the XANES spectrum of the sample under study and the spectra of model compounds. A great improvement in quality of extracted information may come from measurements of the angular resolved XANES or polarized XANES on single crystals, which can give unique information on details of local geometries (198, 199).

The structures of active sites in metalloproteins and biological molecules have been studied by XANES. Thus the structure of the calcium site in Ca–ATP has been studied (15). The EXAFS analysis of calcium ions in biological molecules faces strong limitations due to the spread of interatomic distances in the first coordination shell. Therefore, XANES has been used for local structure determination of calcium sites in troponin-C (200) and calmodulin (201). Calcium binding in milk (milk calcium phosphate) has been found to have the calcium phosphate brushite type structure (202). The XANES of potassium in frog blood cells (203) has been measured. The effect of the pH on the local structure of Mn—ATP complexes and the differences in manganese binding between Mn—ATP and Mn—AMP complexes (204) have been determined by XANES.

Recently the angular resolved XANES of single crystals of proteins have been measured. Scott et al. (205) have measured the polarized XANES spectra of oriented plastocyanin single crystals and they have found a very large angular dependence of XANES features due to the largely asymmetric site. Figure 11.49 shows the XANES with the electric field polarization along the heme normal, the two strong resonances due to CO, peaks C_1 and C_2 are identified. The polarized spectra have been calculated by multiple-scattering theory (113, 114) in very good agreement with experiments. Quantitative XANES analysis has allowed the determination of subtle geometrical distortions in hemoglobin that

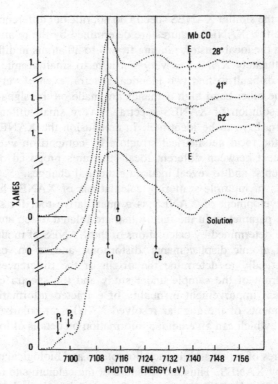

Figure 11.49. Polarized x-ray absorption spectra of a carbonmonoxy–myoglobin single crystal. In the upper part the iron K-XANES with the angle θ between the electric vector E and the heme normal $\hat{n}$ are shown. The solution spectrum is shown at the bottom of the figure. C_1 and C_2 are resonances due to the carbon monoxide molecules (206).

are not detected by EXAFS. Both the rotation of the bonding angle between the ligand molecules O_2, CO, CN, and the porphyrin plane (113, 207) and the iron displacement out of the porphyrin plane in deoxyhemoglobin (114) have been determined by XANES.

In the field of catalysis Vlaic et al. (208) have studied the $Cu/ZnO/Al_2O_3$ system used for the conversion of CO and H_2O to H_2 and CO_2 at low temperature. The cooper and zinc K-XANES of fresh catalyst show the local structure of free oxides. Upon reduction the copper-XANES indicates structural modifications at the copper site while the zinc site is unaffected.

In the field of amorphous systems XANES spectroscopy is expected to be of importance. The presence of technetium metal clusters in borosilicate glasses has been identified (209). The variation of vanadium coordination as a function of the V_2O_5 concentration in $(V_2O_5)_x(P_2O_5)_{1-x}$ amorphous oxides has been determined (210). Iron in glasses has been extensively studied by Petiau and co-workers (211, 212). Amorphous metallic Pd—Ge alloys have been investigated by comparison with a theoretical XANES analysis by Gaskell et al. (213).

In the field of surface science, surface XANES spectroscopy has demonstrated that it is able to give further information on surface structure beyond that obtained by other experimental methods (16). The study of the transition from the chemisorption to the oxidation phase upon oxygen exposure to aluminum single crystals was the first surface XANES experiment (214). Figure 11.50 shows the evolution of surface XANES at aluminum sites on the surface, (selected by recording the intensity of interatomic Auger electrons) as a function of oxygen exposure. The transition from the chemisorption to the oxidation phase can be observed. While other methods, such as LEED (Low Energy Electron Diffraction), probe long-range order and electron spectroscopies probe the electronic structure, surface XANES gives direct structural information on the local geometry in non-ordered surfaces also. The formation of SiO_2-like oxide (formed by SiO_4 units) on the surface of the Si(111) surface exposed to oxygen at room temperature and the formation of a SiO-like surface oxide on

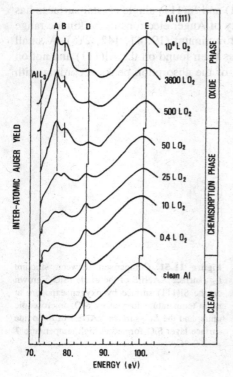

Figure 11.50. Evolution of the aluminium surface $L_{2,3}$-XANES as a function of oxygen exposure on the Al(111) surface (214). The transition from the atomic oxygen chemisorption phase to the oxide phase can be easily identified. The shape of the L_3-XANES of the surface oxide indicate an unexpected octahedral coordination for aluminium in amorphous oxide layer, confirmed by surface EXAFS.

the silicon surface when the surface temperature rises above 700°C, has been
determined by XANES (215). Figure 11.51 shows the L_3 spectra of surface
silicon oxides. While the spectrum of the oxide grown at low temperature is
clearly characteristic of SiO_4 tetrahedral clusters the oxide grown at high temperature shows the clear variation due to the local order in SiO. The angular
resolved surface XANES of bromine on graphite (216) and of carbon monoxide
and dinitrogen on Ni(100) (178) have been measured (see Fig. 11.44). The
angular dependence of the intensities of the multiple-scattering resonances of
the diatomic molecules has allowed a good direct determination of the bonding
angles of the chemisorbed molecules. The multiple-scattering XANES theory
has been applied to the study of oxygen chemisorption on Ni(100) (217). Both
theory and experiment show the strong sensitivity of XANES for structural
details of surface sites. In the case of chemisorbed diatomic molecules (or small
hydrocarbons) surface EXAFS cannot provide a measure of the interatomic
distances in the molecule because the EXAFS oscillations are too weak. But it
has been shown that surface XANES can give the variation of the interatomic
distance in the molecule upon chemisorption when a theoretical analysis is made
of the energy shift of the multiple-scattering resonances (92, 218).

The surface XANES of clean Al(111) and Si(111) single crystal surfaces has
been measured by recording the intensity of Auger electrons in the kinetic range
50–90 eV where the escape depth is at minimum (19, 41, 142, 215). A small
relaxation of the interatomic distance has been found on the Al(111) and not on
the Al(100) surface (41) from the shift of the absorption peaks according with
the rule $K_r d = $ constant (92, 93).

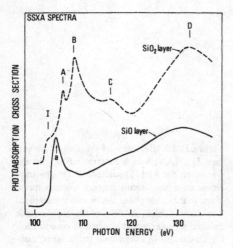

Figure 11.51. Comparison between silicium
$L_{2,3}$-surface XANES of the oxide (SiO_2) grown
on the Si(111) surface by oxygen exposure at
room temperature formed by SiO_4 microscopic
units, and the $L_{2,3}$-surface XANES of the oxide
surface layer SiO formed at high temperature T
= 700 °C.

11.7.2. Structural Information from Bound States at Threshold in Proteins and Glasses

Core excitons, or bound excited states, at the threshold of nonmetallic systems are due to well-localized final states within the first coordination shell of the absorbing atom. Therefore, bound states are very sensitive probes of the coordination geometry of the absorbing atom. The best studied excitations are those at the K-edge of transition metals (see Figs. 11.36, 11.43, and 11.48). These bound state excitations below the continuum threshold E_0 are sometimes said to appear in the pre-edge region. The $1s \rightarrow 3d$ transition is generally weak, being dipole forbidden ($\Delta 1 = 2$), but in tetrahedral complexes the lack of a center of inversion allows a dipole transition to the t_2 unoccupied orbitals to take place. Also the presence of a short double bond in the coordination of the metal ion leads to the loss of a center of inversion and therefore to an intense $3d$-derived peak. This effect has been used to determine the presence of tetrahedral coordination or of double bonds in a large variety of systems. In biochemistry it has for instance been applied to the study of many systems. Here we cite the studies of the iron site in rubredoxin (219), of the molybdenum site in xanthine oxidase (220), and of the iron site in azido–methemeerythrin (221) (cf. Chapter 7).

Wong et al. (156) performed an extensive investigation of the integrated intensity A of the $1s \rightarrow 3d$ derived core exciton in vanadium compounds. He has found that the intensity varies with the coordination geometry according to $A \propto [\Sigma_i C_i R_i^2 \exp(-\alpha_i R_i)]^2$, where C_i are the LCAO coefficients of the final state MO, R_i the bond distances, and α_i the orbital exponents in the Slater-type orbitals of the ith type ligand. This means that the intensity A of the localized core exciton depends on the size of the molecular cage defined by the short double $V=O$ bond. Calas and Petiau (36) have shown that the relative intensity of the two observed bound states (t_{2g} and e_g in octahedral coordination) in iron minerals is strongly dependent on the final-state symmetry determined by the coordination geometry.

The analysis of the intensity of the $3d$ core exciton has allowed a simple and clear determination of the coordination of transition metal ions in oxide glasses. In binary silicate glasses titanium has been found by Sandstrom et al. (222) to occup fourfold coordinated sites. In vanadium phosphate glasses, vanadium has been found to occupy both octahedral and square pyramid sites (210). In both experiments the measurement of the core exciton intensity has allowed an estimate of the percentage of the fourfold and fivefold coordinated sites occupied by the transition metals as a function of the metal concentration in the glasses (see Fig. 11.52).

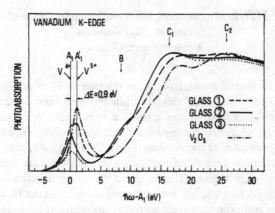

Figure 11.52. Example of the local structure determination from the intensity of the $1s \rightarrow 3d$ core excitons (peak A) in glasses. The figure shows the XANES spectra of iron–vanadium–phosphate glasses with increasing V_2O_5 concentration going from 1 to 3. The intensity of peak A increases with the vanadium concentration, demonstrating the increasing number of V^{5+} sites with V=O double bonds in the glasses 2 and 3. In V^{4+} sites the core exciton is at 1-eV lower energy than in V^{5+} sites.

11.7.3. The Chemical Shift and the Effective Atomic Charge

The determination of the effective charge on the absorbing atom from the chemical shift of the x-ray absorption threshold is an old but interesting topic (2, 4, 7). The main problem is that a direct measure of the ionization threshold from XANES is not possible because in XANES spectra there is no signature at the ionization threshold energy. Therefore, XANES is not a direct method for the determination of the binding energy of core levels. In contrast to XANES x-ray photoemission (XPS or ESCA) is a direct probe of core-level binding energies. Be that as it may, XANES has the advantage that it does not require a vacuum for the experiments and is not surface sensitive, thereby allowing the study of systems like proteins, which cannot be studied by ESCA.

We have already seen in the XANES of molecules (Section 11.4.4 and Figs. 11.18 and 11.19) that the energy shift of the first bound excited state at the absorption threshold follows the binding energy shift of the core level. It has recently been pointed out that in oxides the energy shift of the first core exciton at the absorption threshold of a metal ion is linearly dependent on the core-level binding energy, that is on the atomic effective charge (17, 36, 156) for ions in similar coordination geometry. Thus the relative change form a majority of V^{4+} ions at the low V_2O_5 concentrations to a majority of V^{5+} ions at high V_2O_5 concentration in vanadium–phosphate glasses has been found (210) (see Fig. 11.52).

Because the core exciton appears in the K-XANES of transition metals as a very weak structure (being dipole forbidden), often the chemical shift of the rising absorption edge (absorption jump edge) or the energy of the first strong multiple-scattering resonance (which is generally the maximum absorption peak above threshold) are measured. In this case there is no linear relationship between the measured shift and the effective atomic charge on the ion. In fact, the absorption jump edge is determined by the threshold of dipole allowed transitions, therefore it is dependent on the d-sp mixing between the unoccupied orbitals. The energy of multiple-scattering resonances is strongly dependent on interatomic distance, therefore the chemical shift of the multiple-scattering resonance is much larger than that of the core exciton (17). Often the variation of effective charge and interatomic distance goes together, where the effective charge on an atom is increased (and thus the core-level binding energy is increased), the interatomic distance to neighboring atoms decreases, and all these effects move the multiple-scattering resonances rapidly toward higher energies (see Fig. 11.31).

In spite of its limitations the x-ray absorption chemical shift has attracted a large interest in the field of metalloproteins for the determination of changes of the effective charge on the metal ion is relation to the function of the protein. Several such studies are discussed in Chapter 7, for instance, studies on molybdenum in nitrogenase (Chapter 7, section 7.3.3.1), on copper in cytochrome oxidase (Section 7.3.2.3), and on copper and zinc in superoxide dismutase (Sections 7.3.2.4 and 7.3.4.3).

The large shift of 3.5 eV between the first multiple-scattering resonances of oxygenated and deoxyhemoglobin, observed by several groups, has recently been assigned to an increase of the positive effective charge of the Fe(II) ion in the oxygenated protein (223, 224). Comparison of multiple-scattering calculations with high-resolution data of the chemical shift of the core exciton of hemoglobin indicated that the contribution of the change in the iron effective charge to the total chemical shift amounted to about 1.0 eV. The contribution of the contraction of the $Fe-N_\rho$ distances, when going from deoxy to oxyhemoglobin, was found to be approximately 2 eV (225).

11.7.4. Determination of the Local Electronic Structure in Metallic Systems

The edge region of the XANES of metallic systems probes the local partial electronic structure as was shown in Section 11.5.7. This information is particularly important for complex systems such as catalysts, alloys, valence fluctuating compounds, and for systems that exhibit metal–insulator phase transitions. In these systems the local electronic properties play a key role and there are no other experimental methods that can provide the same type of information. Most of the research in this field has been concentrated on the $2p \rightarrow nd$ white lines at the L_3 thresholds of transition metals and of rare earth compounds. From these white lines one can probe:

1. The number of vacant d states in transition metals from the intensity variation of the white lines in different systems or under different conditions.
2. The total angular character of the unoccupied conduction band states from the intensity ratio of the L_3 and L_2 white lines.
3. The intermediate valence v in mixed valent rare earth compounds from the replicate splitting of the white line as discussed in Section 11.6.5.

The x-ray absorption spectrum of Cu—Ni solid solutions has attracted the interest of Azaroff and Das for many years (226). The magnetization of Cu_xNi_{1-x} alloys decreases with increasing copper concentration until at 60 at. % copper the alloy is no longer magnetic. The decrease of the number of nickel $3d$ holes has been associated with the change in magnetic properties. The nickel L_3 white lines (227) did not show any change in intensity up to 70 at.% copper concentration. Munoz et al. (228) were able to calculate the L_2 and L_3 spectra as a function of copper concentration and they found a good agreement with experiment. Their calculations indicate that the effect of the change of the number of $3d$ holes is too small to be observed by XANES. Recently, palladium L_3-edge spectra of Pd—Ag alloys (229) have been measured and evidence has been found that the number of $4d$ holes/palladium declines upon alloying with silver.

Studies of the intensities of white lines in the XANES spectra of transition metal catalysts are described in Chapter 8, Section 8.4.3. In these studies a correlation has been made between white line intensity and d-orbital vacancy. A change in d-orbital vacancy may be due to a change in the effective atomic charge, as well as due to s–d rehybridization.

We have shown in Section 11.6.5 that the replicate splitting of the $2p \rightarrow \overline{5, \varepsilon} d$ white line in the L_3-XANES of rare earth compounds provides a measure of the intermediate valence v of mixed valence systems. In fact one of the first experiments showing the mixed valence state of rare earth compounds was an x-ray absorption experiment (230). Samarium, europium, thulium, and ytterbium mixed valence systems fluctuate between the $4f^n(5d6s)^m$ (rare earth^{3+}) and the $4f^{n+1}(5d6s)^{m-1}$(rare earth^{2+}) configurations, in which each rare earth ion has an integer number of localized $4f$ electrons. The intermediate valence is defined as $v = 3 - r$, where r is the ratio $r = A/(A+B)$ of the amplitudes of the two electronic configurations $A(4f^n)$ and $B(4f^{n+1})$ present in the mixed valence ground state. The measuring time of the intermediate valence by XANES can be estimated from the lifetime of the excited state corresponding with the white line. The full width of the white line in TmSe is 7.4 eV (125), which gives a lifetime of the order of 10^{-16} s. Therefore, the L_3-XANES is a fast local probe in comparison to the characteristic fluctuation time of the local electronic charge on the single rare earth site in a system like TmSe, which is of the order of 10^{-14} s. Since the studies of Launois et al. (231) and Martin et al. (232), which showed that synchrotron x-ray absorption spectra provided a fast method to obtain interatomic distances and intermediate valence from the same experiment, an extensive investigation of many mixed valence systems has been carried out (125, 190, 233–237). Also, experiments with standard x-ray sources have been performed (238, 239). While the value of the intermediate valence deduced from XANES has been found to be systematically smaller than that deduced from crystal lattice parameters, it has been found to be very close to that deduced from x-ray photoemission and magnetic measurements. Recently, interest has focused on the cerium compounds that can show both $4f$–$5d$ hybridization as well as $4f$–L (ligand orbitals) mixing. As has been discussed in Section 11.6.5, in these systems the measured intermediate valence seems to be very close to the ground-state value.

Because XANES spectroscopy does not require vacuum for experimentation,

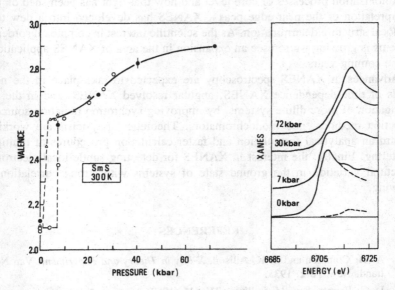

Figure 11.53. Determination of the mixed valence v in SmS by L_3 absorption edge spectra as a function of pressure. The insert shows the L_3 samarium spectra at different pressures. The triangles indicate data taken at decreasing pressure. The first-order phase transition at low pressure is followed by a continuous increase of the valence up to 80 kbar (234).

studies of phase transitions induced by pressure on a crystal can quite easily be performed. The $\gamma \rightarrow \alpha$ phase transition in cerium (235, 236), the first-order transition in SmS (234) (see Fig. 11.53), and the second-order phase transition in $EuPd_2Si_2$ (234) have been studied as a function of pressure.

Metal–insulator phase transition studied by XANES started with the experiment of Slowik and Brown (240) on amorphous Mg_xBi_{1-x} alloys. They found sharp core excitons in the range of magnesium concentrations $0.57 < x < 0.60$ at the $L_{2,3}$-edge of magnesium. Outside this range a broad metallic threshold is observed. The metal–insulator transition in the vanadium oxides V_2O_3 (241) and VO_2 (110) K-XANES has been observed (Fig. 11.48). Although only a few experiments on phase transitions have been performed up to now XANES seems to be a very promising technique for the determination of the subtle variations of local electronic structure in such systems.

11.8. CONCLUSIONS

The introduction of synchrotron radiation has made it possible to measure high-resolution x-ray absorption spectra and to investigate very dilute systems such as proteins and surfaces. We have seen a rapid growth of our understanding of photoionization processes of core level and now that light has been shed on the interpretation of the near-edge peaks, XANES has developed into a new tool for local structure determination. As the scientific interest in complex disordered systems is growing we foresee an expansion in the area of XANES applications in the coming years.

Advances in XANES spectroscopy are expected to take place in the new fields of time-dependence XANES, angular resolved XANES, and in the investigation of more dilute systems, by improving sychrotron radiation sources, detection methods, and monochromators. Theoretical physicists are working toward an analytical formulation and faster calculation procedures of multiple scattering. Finally, the interest in XANES for detecting subtle local electronic structural variation in the ground state of systems with strong correlation is growing.

REFERENCES

1. A. H. Compton and S. K. Allison, *X-Ray in Theory and Experiment*, Van Nostrand, New York, 1935.

2. L. G. Parratt, *Rev. Mod. Phys.*, **31**, 616 (1959).

3. L. V. Azaroff, *Rev. Mod. Phys.*, **35**, 1012 (1963).

4. L. V. Azaroff and D. M. Pease, in *X-Ray Spectroscopy*, L. V. Azaroff (Ed), McGraw-Hill, New York, 1974, p. 296.

5. U. Fano and J. W. Cooper, *Rev. Mod. Phys.*, **40**, 441 (1968).

6. C. Kunz (Ed.), *Synchrotron Radiation*, Topics in Current Physics, Vol. 10, Springer-Verlag, Berlin, 1978.

7. F. C. Brown, in *Synchrotron Radiation Research*, H. Winick and S. Doniach, (Eds.), Plenum, New York, 1980, p. 61.

8. W. Kossel, *Z. Phys.*, **1**, 119 (1920).

9. R. de L. Kronig, *Z. Phys.*, **70**, 317 (1931).

10. D. E. Sayers, E. A. Stern, and F. W. Lytle, *Phys. Rev. Lett.*, **27**, 1204 (1971).

11. B. M. Kincaid and P. Eisenberger, *Phys. Rev. Lett.*, **34**, 1361 (1975).

12. C. A. Ashley and S. Doniach, *Phys. Rev. B*, **11**, 1279 (1975).

13. P. A. Lee and J. B. Pendry, *Phys. Rev. B*, **11**, 2795 (1975).

14. B. K. Teo and P. A. Lee, *J. Am. Chem. Soc.*, **101**, 2815 (1979).

15. A. Bianconi, S. Doniach, and D. Lublin, *Chem. Phys. Lett.*, **59**, 121 (1978).

16. A. Bianconi, *Appl. Surf. Sci.*, **6**, 392 (1980).

17. M. Belli, A. Scafati, A. Bianconi, S. Mobilio, L. Palladino, A. Reale, and E. Burattini, *Solid State Commun.*, **35**, 355 (1980).

18. A. Bianconi, *EXAFS for Inorganic Systems*, S. S. Hasnain and C. D. Garner (Eds.), Daresbury Report DL/SCI/R17 E, 1981, p. 13.

19. A. Bianconi, in *EXAFS and Near Edge Structure*, A. Bianconi, L. Incoccia, and S. Stipcich (Eds.), Springer-Verlag, Berlin, 1983, p. 118.

20. J. L. Dehmer and D. Dill, *J. Chem. Phys.*, **65**, 5327 (1976).

21. M. G. Lynch, D. Dill, J. Siegel, and J. L. Dehmer, *J. Chem. Phys.*, **71**, 4249 (1979).

22. C. R. Natoli, D. K. Misemer, S. Doniach, and F. W. Kutzler, *Phys. Rev. A*, **22**, 1104 (1980).

23. F. W. Kutzler, C. R. Natoli, D. K. Misemer, S. Doniach, and K. O. Hodgson, *J. Chem. Phys.*, **73**, 3274 (1980).

24. P. J. Durham, J. B. Pendry, and C. H. Hodges, *Solid State Commun.*, **38**, 159 (1981); *Comput. Phys. Commun.*, **25**, 193 (1982).

25. A. Bianconi, M. Dell'Ariccia, P. J. Durham, and J. B. Pendry, *Phys. Rev. B*, **26**, 6502 (1982).

26. C. R. Natoli, in *EXAFS and Near Edge Structure*, A. Bianconi, L. Incoccia, and S. Stipcich (Eds.), Springer-Verlag, Berlin, 1983, p. 43.

27. J. B. Pendry, in *EXAFS and Near Edge Structure*, A. Bianconi, L. Incoccia, and S. Stipcich (Eds.), Springer-Verlag, Berlin, 1983, p. 4.

28. P. J. Durham, in *EXAFS and Near Edge Structure*, A. Bianconi, L. Incoccia, and S. Stipcich (Eds.), Springer-Verlag, Berlin, 1983, p. 37.

29. F. Fujikawa, T. Matsuura, and H. Kuroda, *J. Phys. Soc. Jpn.*, **52**, 905 (1983).

30. J. E. Müller, O. Jepsen, and J. W. Wilkins, *Solid State Commun.*, **42**, 365 (1982).

31. J. E. Müller and J. W. Wilkins, *Phys. Rev. B*, **29**, 4331 (1984).

32. P. Pianetta and I. Lindau, *Nucl. Instrum. Methods*, **152**, 155 (1978).

33. M. Hart, *Rep. Prog. Phys.*, **34**, 435 (1971).

34. J. H. Beaumont and M. Hart, *J. Phys. E*, **7**, 823 (1974).

35. K. Kohra, M. Ando, T. Matsushita, and H. Hashizume, *Nucl. Instrum. Methods*, **152**, 161 (1978).

36. G. Calas and J. Petiau, *Solid State Commun.*, **48**, 625 (1983).

37. G. N. Greaves, G. P. Diakun, P. D. Quinn, M. Hart, and D. P. Siddons, *Nucl. Instrum. Methods*, **208**, 335 (1983).

38. A. Bienenstock and H. Winick, *Phys. Today*, p. 48 (June 1983).

39. A. M. Flank, A. Fontaine, A. Jucha, M. Lemonnier, in *EXAFS and Near Edge Structure*, A. Bianconi, L. Incoccia, and S. Stipcich (Eds.), Springer-Verlag, Berlin, 1982, p. 405.

40. R. Z. Bachrach, A. Bianconi, and F. C. Brown, *Nucl. Instrum. Methods*, **152**, 53 (1978).

41. A. Bianconi and R. Z. Bachrach, *Phys. Rev. Lett.*, **42**, 104 (1979).

42. A. Bianconi, D. Jackson, and K. Monahan, *Phys. Rev. B*, **17**, 5543 (1977).

43. C. G. Barkla and C. A. Sadler, *Philos. Mag.*, **17**, 739 (1909).

44. J. H. Hubbel and W. J. Viegele, *NBS Technical Note*, **901** (1976); J. H. Hubbel, *J. Phys. Paris* **32**, Colloq. **C4**, 10 (1971).

45. W. H. Mc Master, N. Kerr-Del Grande, J. H. Mallet, and J. H. Hubbel, *Compilation of X-ray Cross Sections*, National Technical Information Service, Springfield, VA; University of California, Livermore, Lab. Report UCRL 50174.

46. J. H. Hubbel, *Radiat. Res.*, **70**, 58 (1977).

47. D. J. Kennedy and S. T. Manson, *Phys. Rev. A*, **5**, 227 (1972).

48. F. Wuilleumier, *J. Phys. Paris Colloq.*, **32**, 88 (1971).

49. R. D. Deslattes, R. E. La Villa, P. L. Cowan, and A. Henins, *Phys. Rev. A*, **27**, 923 (1983).

50. M. Breining, M. H. Chen, G. E. Ice, F. Parente, B. Crasemann, and G. S. Brown, *Phys. Rev. A*, **22**, 520 (1980).

51. C. E. Moore, *Atomic Energy Levels*, NSRDS-NBS 35, USA Government Printing Office, Vol. I, 1971.

52. F. Combet Farnoux, *J. Phys.*, **31**, C4, 203 (1970).

53. M. Ya Amusia, N. A. Cherenkov, and L. V. Chernysheva, *Sov. Phys. JEPT*, **33**, 90 (1971); *Phys. Lett.*, **59A**, 191 (1976).

54. G. Wendin, *J. Phys. B.*, **6**, 42 (1973).

55. R. Haensel, G. Keitel, P. Schreiber, and C. Kunz, *Phys. Rev.*, **188**, 1375 (1969).

56. J. L. Dehmer, A. F. Starace, U. Fano, J. Sugar, and J. W. Cooper, *Phys. Rev. Lett.*, **26**, 1521 (1971).

57. H. W. Wolff, R. Bruhn, K. Radler, and B. Sonntag, *Phys. Lett.*, **59A**, 67 (1976).

58. D. J. Peterman, J. H. Weaver, and M. Croft, *Phys. Rev. B*, **25**, 5530 (1982).

59. R. Bruhn, B. Sonntag, and H. W. Wolff, *Phys. Lett. A*, **69**, 9 (1978); *J. Phys. B*, **12**, 203 (1979).

60. L. C. Davis and L. A. Feldkamp, *Phys. Rev. A*, **17**, 2012 (1978); F. Combet-Farnoux and M. Ben Amar, *Phys. Rev. A*, **21**, 1975 (1980).

61. J. P. Connerade. *Contemp. Phys.*, **19**, 415 (1978).

62. F. Wuilleumier and M. O. Krause, *Phys. Rev. A*, **10**, 242 (1974).

63. M. O. Krause, *J. Phys. Paris Colloq.*, **32**, 67 (1971); T. A. Carlson and M. O. Krause, *Phys. Rev. A*, **140**, 1057 (1965).

64. M. Ya Amusia, in *Atomic Physics 5, Proceedings of V International Conference on Atomic Physics*, R. Marrus, M. Prior, and H. Shugart (Eds.), Plenum, New York, p. 537.

65. G. Wendin, in *EXAFS and Near Edge Structure*, A. Bianconi, L. Incoccia, and S. Stipcich (Eds.), Springer-Verlag, Berlin, 1983, p. 29.

66. H. W. Schnopper, *Phys. Rev.*, **131**, 2558 (1963).

67. F. Wuilleumier, *J. Phys. Paris*, **26**, 776 (1965).

68. G. Bradley Armen, T. Aberg, Kh. T. Karim, J. C. Levin, B. Crasemann, G. Brown, M. H. Chen, and G. E. Ice, *Phys. Rev. Lett.*, **54**, 182 (1985).

69. J. M. Esteva, B. Gauthe, P. Dhez, and R. C. Karnatak, *J. Phys. B*, **16**, L263 (1983).

70. R. E. La Villa, *Phys. Rev. A*, **19**, 1999 (1979).

71. M. H. Tuillier, D. Laporte, and J. M. Esteva, *Phys. Rev. A*, **26**, 372 (1982).

72. A. Bianconi, R. Z. Bachrach, and F. C. Brown, unpublished.

73. H. Glaser, *Phys. Rev.*, **82**, 616 (1951).

74. B. M. Kincaid and P. Eisenberger, *Phys. Rev. Lett.*, **34**, 1361 (1975).

75. D. L. Mott, *Phys. Rev.*, **144**, 94 (1966).

76. R. E. La Villa and R. D. Deslattes, *J. Chem. Phys.*, **44**, 4399 (1966).

77. T. M. Zimkina and A. S. Vinogradov, *J. Phys. Paris Colloq.*, **32**, C4-3 (1971).

78. F. A. Gianturco, C. Guidotti, and U. Lamanna, *J. Chem. Phys.*, **57**, 840 (1972).

79. V. I. Nefedov, *J. Struct. Chem. USSR*, **11**, 277 (1970).

80. J. L. Dehmer, *J. Chem. Phys.*, **56**, 4496 (1972).

81. B. Cadioli, U. Pincelli, E. Tosatti, U. Fano, and J. L. Dehmer, *Chem. Phys. Lett.*, **17**, 15 (1972).

82. G. L. Bendazzoli and P. Palmieri, *Theor. Chim. Acta*, **36**, 77 (1974).

83. A. Bianconi, H. Petersen, F. C. Brown, and R. Z. Bachrach, *Phys. Rev. A*, **17**, 1907 (1978).

84. J. L. Dehmer and D. Dill, *Phys. Rev. Lett.*, **35**, 213 (1975).

85. D. Loomba, S. Wallace, and D. Dill, *J. Chem. Phys.*, **75**, 4546 (1981).

86. J. D. Dehmer and D. Dill, *J. Chem. Phys.*, **65**, 5327 (1983).

87. D. Dill, J. Welch, J. L. Dehmer, and J. Siegel, *Phys. Rev. Lett.*, **43**, 1236 (1979); M. G. Lynch, D. Dill, J. Siegel, and J. L. Dehmer, *J. Chem. Phys.*, **71**, 4249 (1979).

88. J. A. Tossell and J. W. Davenport, *J. Chem. Phys.*, **80**, 813 (1984).

89. J. L. Dehmer and D. Dill, in *Electronic and Atomic Collisions*, N. Oda and K. Takayanagi (Eds.), North-Holland, Amsterdam, 1980, p. 195.

90. V. P. Sachenko, E. V. Polozhentsev, A. P. Kovtun, Yu. F. Migal, R. V. Vedrinsky, and V. V. Kolesnikov, *Phys. Lett. A*, **48**, 169 (1974).

91. H. J. Levinson, T. Gustafson, and P. Soven, *Phys. Rev. A*, **19**, 1089 (1979).

92. A. Bianconi, M. Dell'Ariccia, A. Gargano, and C. R. Natoli, in *EXAFS and Near Edge Structure*, A. Bianconi, L. Incoccia and S. Stipcich (Eds.), Springer-Verlag, Berlin, 1983, p. 57.

93. M. Dell'Ariccia, A. Gargano, C. R. Natoli, A. Bianconi, Frascati INFN Report LNF-84/51 (P).

94. F. Sette, J. Stohr, and A. P. Hitchcock, *Chem. Phys. Lett.*, **110**, 517 (1984).

95. M. Tronc, R. Azria, and Y. Lecoat, *J. Phys. B*, **17**, 2327 (1980) and references therein; J. A. Tossel, *J. Phys. B*, **18**, 387 (1985).

96. F. C. Brown, R. Z. Bachrach, and A. Bianconi, *Chem. Phys. Lett.*, **54**, 425 (1978).

97. H. Friedrich, B. Pittel, P. Rabe, W. H. E. Schwarz, and B. Sonntag, *J. Phys. B*, **13**, 25 (1980).

98. M. Nakamura, Y. Morioka, T. Hayaishi, E. Ishiguro, and M. Sasanuma, in *III International Conference on VUV Radiation Physics*, Tokyo, Y. Nagai (Ed.), 1971.

99. E. S. Gluskin, A. A. Krasnoperova, and V. A. Mazalov, *J. Struct. Chem. USSR*, **1**, 185 (1976).

100. W. Hayes and F. C. Brown, *Phys. Rev. A*, **6**, 21 (1972).

101. P. S. Bagus, K. Krauss, and R. E. La Villa, *Chem. Phys. Lett.*, **23**, 13 (1973).

102. H. P. Hitchcock, M. Pocock, and C. E. Brion, *Chem. Phys. Lett.*, **49**, 125 (1977).

103. W. H. E. Schwarz, *Angew. Chem. Int.*, **13**, 454 (1974).

104. P. Eisenberger and B. Lengeler, *Phys. Rev. B*, **22**, 3551 (1980).

105. B. K. Teo, *J. Am. Chem. Soc.*, **103**, 3990 (1981); A. Bianconi, in *EXAFS and Near Edge Structure*, A. Bianconi, L. Incoccia, S. Stipcich (Eds.), Springer-Verlag, Berlin, 1983, p. 11.

106. A. Bianconi, J. Garcia, A. Marcelli, M. Benfatto, C. R. Natoli, and I. Davoli, *J. de Phys.* (Paris), **46**, 39-101 (1985).

107. M. Benfatto, C. R. Natoli, A. Bianconi, J. Garcia, A. Marcelli, M. Fanfoni, and I. Davoli, *Phys. Rev. B*, **34**, 5774 (1986).

108. J. E. Muller and W. L. Schaich, *Phys. Rev. B*, **27**, 6489 (1983).

109. R. Haensel, G. Keitel, N. Kosuch, U. Nielsen, and P. Schreiber, *J. Phys. Paris Colloq.*, **32**, C4, 236 (1971).

110. A. Bianconi, *Phys. Rev. B*, **26**, 2741 (1982).

111. C. Sugiura, Y. Fujino, and S. Kiyono, Technol, *Rep. Tohoku Univ.*, **34**, 107 (1969).

112. A. Bianconi, E. Fritsch, G. Calas, J. Petiau, *Phys. Rev. B*, **32**, 4292 (1986).

113. P. J. Durham, A. Bianconi, A. Congiu-Castellano, S. S. Hasnain, L. Incoccia, S. Morante, and J. B. Pendry, *EMBO J.*, **2**, 1441 (1983).

114. A. Bianconi, E. Burattini, A. Congiu-Castellano, M. Dell'Ariccia, P. J. Durham, A. Giovannelli, and M. Barteri, *FEBS Lett.*, **178**, 165 (1984).

115. G. N. Greaves, P. J. Durham, G. Diakun, and P. Quinn, *Nature (London)*, **294**, 139 (1981).

116. L. A. Grunes, *Phys. Rev. B*, **27**, 2111 (1983).

117. F. Szmulowicz and D. M. Pease, *Phys. Rev. B*, **17**, 3341 (1978).

118. S. Wakoh and Y. Kubo, *Jpn. J. Appl. Phys.*, **17**, Suppl. 17–2, 193 (1978).

119. J. W. Mc Caffrey and D. A. Papaconstantopoulus, *Solid State Commun.*, **14**, 1055 (1974).

120. A. Balzarotti, M. De Crescenzi, and L. Incoccia, *Phys. Rev. B*, **25**, 6349 (1982).

121. N. Motta, M. De Crescenzi, and A. Balzarotti, *Phys. Rev. B*, **27**, 4712 (1983).

122. M. Benfatto, A. Bianconi, I. Davoli, L. Incoccia, S. Mobilio, and S. Stizza, *Solid State Commun.*, **46**, 367 (1983).

123. M. Brown, R. E. Peierls, and E. A. Stern, *Phys. Rev. B*, **15**, 738 (1977).

124. L. F. Mattheis and R. E. Dietz, *Phys. Rev. B*, **22**, 1663 (1980).

125. A. Bianconi, S. Modesti, M. Campagna, K. Fisher, and S. Stizza, *J. Phys. C*, **14**, 4737 (1981).

126. J. Veldkamp, *Physica*, **2**, 25 (1935); Y. Cauchois and I. Manescu, *C. R. Acad. Sci.*, **210**, 172 (1940).

127. Y. Cauchois and C. Bonnelle, *C. R. Acad. Sci.*, **245**, 1230 (1957).

128. C. Bonnelle, *Ann. Phys. Paris*, **1**, 439 (1966).

129. D. Denley, R. S. Williams, P. Perfetti, D. A. Shirley, and J. Stöhr, *Phys. Rev. B*, **19**, 1762 (1979).

130. R. D. Leapman, L. A. Grunes, and P. L. Fejes, *Phys. Rev. B*, **26**, 614 (1982).

131. M. Lemonnier, O. Collet, C. Depautex, J. M. Esteva, and D. Raoux, *Nucl. Instrum. Methods*, **152**, 109 (1978).

132. J. M. Esteva, R. C. Karnatak, J. C. Fuggle, and G. A. Sawatzky, *Phys. Rev. Lett.*, **50**, 910 (1983).

133. P. S. P. Wei and F. W. Lytle, *Phys. Rev. B*, **19**, 679 (1979).

134. P. A. Papaconstantopoulos, D. J. Nagel, and C. Jones-Bjorklund, *Int. J. Quantum Chem. Symp.*, **12**, 497 (1978).

135. J. R. Andersen, D. A. Papaconstantopoulos, L. L. Boyer, and J. E. Schriber, *Phys. Rev. B*, **20**, 3172 (1979).

136. G. Wendin, *Phys. Rev. Lett.*, **53**, 724 (1984).

137. G. D'Alba, P. Fornasini, and E. Burattini, *J. Phys. C*, **16**, L1091 (1983).

138. L. G. Parratt and E. L. Jossem, *Phys. Rev.*, **97**, 916 (1959); *J. Phys. Chem. Sol.*, **2**, 67 (1959).

139. W. Gudat, C. Kunz, and H. Petersen, *Phys. Rev. Lett.*, **32**, 1370 (1974).

140. S. T. Pantelides and F. C. Brown, *Phys. Rev. Lett.*, **33**, 298 (1974).

141. S. T. Pantelides, *Phys. Rev. B*, **11**, 2391 (1975).

142. A. Bianconi, *Surf. Sci.*, **89**, 41 (1979).

143. I. Davoli, S. Stizza, M. Benfatto, A. Bianconi, V. Sessa, and C. Furlani, *Solid State Commun.*, **48**, 479 (1983).

144. G. Materlick, B. Sonntag, and M. Tausch, *Phys. Rev. Lett.*, **51**, 1300 (1983).

145. B. Johansson and M. Matersson, *Phys. Rev. B*, **21**, 4427 (1980).

146. G. S. Brown and S. Doniach, in *Synchrotron Radiation Research*, H. Winick and S. Doniach (Eds.), Plenum, New York, 1980, p. 353.

147. M. Altarelli, W. Andreoni, and F. Bassani, *Solid State Commun.*, **16**, 143 (1975).

148. F. Bassani, *Appl. Opt.*, **19**, 4093 (1980).

149. S. Sugano, in *Spectroscopy of the Excited State*, B. Di Bartolo (Ed.), NATO Advanced Study Institute, Plenum, New York, 1975, p. 279.

150. C. Kunz, *J. Phys. Paris*, **39**, *C4*, 103 (1978).

151. C. Satoko and S. Sugano, *J. Phys. Soc. Jpn.*, **34**, 701 (1973).

152. T. Aberg and J. L. Dehmer, *J. Phys. C*, **6**, 1450 (1973).

153. A. B. Kunz, J. C. Boisvert, and J. O. Woodruff, *J. Phys. C*, **15**, 5037 (1982).

154. S. Baroni, G. Grosso, and G. Pastori Parravicini, *Phys. Rev. B*, **22**, 6440 (1980), and references cited therein.

155. J. E. Hahn, R. A. Scott, K. D. Hodgson, S. Doniach, S. R. Dejardins, and E. J. Salomon, *Chem. Phys. Lett.*, **88**, 595 (1982).

156. J. Wong, R. P. Messmer, D. H. Maylotte, and F. W. Lytle, *Phys. Rev. B*, **30**, 5596 (1984).

157. U. von Barth and G. Grossmann, *Solid State Commun.*, **32**, 645 (1979); *Phys. Rev. B*, **25**, 5150 (1982).

158. R. E. Dietz, E. G. McRae, and J. H. Weaver, *Phys. Rev. B*, **21**, 2229 (1980).

159. E. A. Stern and J. J. Rehr, *Phys. Rev. B*, **27**, 3351 (1983).

160. L. C. Davis and L. A. Feldkamp, *Phys. Rev. B*, **23**, 4269 (1981).

161. E. A. Stern, S. M. Heald, and B. Bunker, *Phys. Rev. Lett.*, **42**, 1372 (1979); *Phys. Rev. B*, **21**, 5521 (1980).

162. C. Noguera and D. Spanjaard, *J. Phys. F*, **11**, 1133 (1981).

163. J. J. Rehr and S. H. Chou, in *EXAFS and Near Edge Structure*, A. Bianconi, L. Incoccia, and S. Stipcich (Eds.), Springer-Verlag, Berlin, 1983, p. 22.

164. P. Nozières and C. T. De Dominicis, *Phys. Rev.*, **178**, 1097 (1969).

165. G. D. Mahan, *Phys. Rev.*, **163**, 612 (1967).

166. G. D. Mahan, *Phys. Rev. B*, **21**, 1421 (1980).

167. J. D. Dow and C. P. Flynn, *J. Phys. C*, **13**, 1341 (1980).

168. L. Olivera and J. W. Wilkins, *Phys. Rev. B*, **24**, 4863 (1981).

169. W. Hansch and W. Ekardt, *Phys. Rev. B*, **24**, 5497 (1981).

170. P. H. Citrin, G. K. Wertheim, and M. Schluter, *Phys. Rev. B*, **20**, 3067 (1979).

171. T. A. Callcot, E. T. Arakawa, and D. L. Ederer, *Phys. Rev. B*, **18**, 6622 (1978); **17**, 5185 (1977).

172. W. Hansch and W. Ekardt, *Phys. Rev. B*, **25**, 7815 (1982).

173. H. Petersen, *Phys. Rev. Lett.*, **35**, 1363 (1975).

174. C. Senemaud, *Phys. Rev. B*, **18**, 3929 (1978).

175. S. M. Bose and P. Longe, *Phys. Rev. B*, **18**, 3821 (1978).

176. G. Wendin, "Breakdown of One-electron Pictures in Photoionization Spectra," *Structure and Bonding*, Vol. 45, Springer-Verlag, Berlin, 1981, p. 1.

177. D. A. Shirley, in "Photoemission in Solids I," L. Ley and M. Cardona (Eds.), *Topics in Applied Physics*, Springer-Verlag, Berlin, 1978, p. 165.

178. J. Stöhr and R. Jaeger, *Phys. Rev. B*, **26**, 4111 (1982).

179. A. Bianconi, A. Marcelli, M. Tomellini, and I. Davoli, *J. Magn. Magn. Mat.*, **47/48**, 209 (1985).

180. J. C. Fuggle, E. U. Hillebrecht, Z. Zolmerek, R. Lassert, C. Freiburg, O. Gunnarsson, and K. Schonhammer, *Phys. Rev. B*, **22**, 7330 (1983).

181. J. C. Fuggle, E. Umbach, D. Menzel, K. Wendelt, and C. R. Brundle, *Solid State Commun.*, **27**, 65 (1978); J. C. Fuggle and D. Menzel, *Surf. Sci.*, **72**, 33 (1978).

182. E. Umbach, *Surf. Sci.*, **117**, 482 (1982).

183. R. P. Messmer, S. H. Lanson, and D. R. Salahub, *Solid State Commun.*, **36**, 265 (1980); H. J. Freund and E. W. Plummer, *Phys. Rev. B*, **23**, 4859 (1981).

184. K. Schonhammer and O. Gunnarsson, *Solid State Commun.*, **26**, 399 (1978); *Phys. Rev. B*, **18**, 6608 (1978); O. Gunnarsson and K. Schonhammer, *Phys. Rev. Lett.*, **41**, 1608 (1978); **42**, 195 (1979).

185. N. D. Lang and A. R. Williams, *Phys. Rev. B*, **16**, 2408 (1977).

186. A. Kotani and Y. Toyozawa, *J. Phys. Soc. Jpn.*, **37**, 912 (1974).

187. A. Bianconi, A. Marcelli, I. Davoli, S. Stizza, and M. Campagna, *Solid State Commun.*, **49**, 409 (1984).

188. S. J. Oh and S. Doniach, *Phys. Rev. B*, **26**, 2085 (1982).

189. G. Materlik, J. E. Muller, and J. W. Wilkins, *Phys. Rev. Lett.*, **50**, 267 (1983).

190. A. Bianconi, M. Campagna, and S. Stizza, *Phys. Rev. B*, **25**, 2477 (1982).

191. A. Fujimori, *Phys. Rev. B*, **27**, 3992 (1983); **28**, 4489 (1983).

192. A. Marcelli, A. Bianconi, I. Davoli, and S. Stizza, *EXAFS and Near Edge Structure III*, Springer Proceedings in Phys., Vol. 2, Springer-Verlag, Berlin, 1984, p. 52.

193. A. Fujimori and F. Minami, *Phys. Rev. B*, **30**, 957 (1984).

194. J. F. Herbst and J. W. Wilkins, *Phys. Rev. B*, **26**, 1689 (1982).

195. K. R. Bauchspiess, W. Boksch, E. Holland-Moritz, H. Launois, R. Pott, and D. Wohlleben, in *Valence Fluctuations in Solids*, L. M. Falicov, W. Hanke and M. B. Maple (Eds.), North-Holland, Amsterdam, 1982 p. 417.

196. F. W. Kutzler, D. E. Ellis, T. I. Morrison, G. K. Shenoy, P. J. Viccaro, P. A.

Montano, E. H. Appelman, L. Stein, M. J. Pellin, and D. M. Gruen, *Solid State Commun.*, **46**, 803 (1983).

197. G. S. Knapp, B. W. Veal, H. K. Pan, and T. Klippert, *Solid State Commun.*, **44**, 1343 (1982).

198. F. W. Kutzler, R. A. Scott, J. M. Berg, K. O. Hodgson, S. Doniach, S. P. Cramer, and C. H. Chang, *J. Am. Chem. Soc.*, **103**, 6083 (1981).

199. A. Bianconi in *EXAFS and Near Edge Structure III*, Springer Proceedings in Physics, Vol. 2, Springer-Verlag, Berlin, 1984, p. 167.

200. A. Bianconi, A. Giovannelli, L. Castellani, S. Alemá, P. Fasella, B. Oesh, and S. Mobilio, *J. Mol. Biol.*, **165**, 125 (1983).

201. A. Bianconi, A. Giovannelli, I. Ascone, S. Alemá, P. Durham, and P. Fasella, in *EXAFS and Near Edge Structure*, A. Bianconi, L. Incoccia, and S. Stipcich, (Eds.), Springer-Verlag, Berlin, 1983, p. 355.

202. S. S. Hasnain, in *EXAFS and Near Edge Structure*, A. Bianconi, L. Incoccia, and S. Stipcich (Eds.), Springer-Verlag, Berlin, 1983, p. 330.

203. H. W. Huang, S. H. Hunter, W. K. Warburton, and S. C. Moss, *Science*, **204**, 191 (1979).

204. M. Belli, A. Scafati, A. Bianconi, E. Burattini, S. Mobilio, C. R. Natoli, L. Palladino, and A. Reale, *Nuovo Cimento D*, 2, 1281 (1983).

205. R. A. Scott, J. E. Hahn, S. Doniach, H. C. Freeman, and K. O. Hodgson, *J. Am. Chem. Soc.*, **104**, 5364 (1982).

206. A. Bianconi, A. Congiu-Castellano, P. J. Durham, S. S. Hasnain, and S. E. Phillips, *Nature*, **318**, 685 (1985).

207. A. Congiu-Castellano, A. Bianconi, M. Dell'Ariccia, A. Giovannelli, E. Burattini, and P. J. Durham in *EXAFS and Near Edge Structure III*, Springer Proceedings in Physics, Vol. 2, Springer-Verlag, Berlin, 1984, p. 164.

208. G. Vlaic, J. C. Bart, W. Cavigiolo, and S. Mobilio, *Chem. Phys. Lett.*, **76**, 453 (1980).

209. M. Antonini, C. Caprile, A. Merlini, J. Petiau, and F. R. Thornley, *EXAFS and Near Edge Structure*, A. Bianconi, L. Incoccia, and S. Stipcich (Eds.), Springer-Verlag, Berlin, 1983, p. 261.

210. A. Bianconi, A. Giovannelli, I. Davoli, S. Stizza, L. Palladino, O. Gzowski, and L. Murawski, *Solid State Commun.*, **42**, 547 (1982).

211. J. Petiau, G. Calas, P. Bondot, C. Lapeyere, P. Levitz, and G. Loupias, *EXAFS for Inorganic Systems*, S. S. Hasnain and C. D. Garner (Eds.), Daresbury, Report DL/SC1/R17E, 1981, p. 127; J. Petiau and G. Calas, in Proceedings of the V International Conference on the Physics of Non Crystalline Sol., Montpelier, 1982.

212. G. Calas and J. Petiau, *Bull. Mineral.*, **106**, 33 (1983).

213. P. H. Gaskell, D. M. Glover, A. K. Livesey, P. J. Durham, and G. N. Greaves, *J. Phys. C*, **15**, L597 (1982).

214. A. Bianconi, R. Z. Bachrach, and S. A. Flodstrom, *Phys. Rev. B*, **19**, 3879 (1979).

215. A. Bianconi and R. S. Bauer, *Surf. Sci.*, **99**, 76 (1980).

216. S. M. Heald and E. A. Stern, *Phys. Rev. B*, **17**, 4069 (1978).

217. D. Norman, P. J. Durham, J. B. Pendry, J. Stohr, and J. Jaeger, *EXAFS and Near Edge Structure*, A. Bianconi, L. Incoccia, and S. Stipcich (Eds.), Springer-Verlag, Berlin, 1983, p. 146; *Phys. Rev. Lett.*, **51**, 2052 (1983).

218. J. Stöhr, J. L. Gland, W. Eberhardt, O. Outka, R. J. Madix, F. Sette, R. J. Koestner, and U. Doebler, *Phys. Rev. Lett.*, **51**, 2414 (1983).

219. R. G. Shulman, P. Eisenberger, B. K. Teo, B. M. Kincaid, and G. S. Brown, *J. Mol. Biol.*, **124**, 305 (1978).

220. T. D. Tullius, D. M. Kurtz, Jr., S. D. Conradson, and K. O. Hodgson, *J. Am. Chem. Soc.*, **101**, 2776 (1979).

221. W. A. Hendrickson, M. S. Co, J. L. Smith, K. O. Hodgson, and G. L. Klippenstein, *Proc. Natl. Acad. Sci. USA*, **79**, 6255 (1982).

222. D. R. Sandstrom, F. W. Lytle, P. S. P. Wei, R. B. Gregor, J. Wong, and P. Schultz, *J. Non-Cryst. Sol.*, **41**, 201 (1980).

223. S. Pin, B. Alpert, and A. Michalowicz, *FEBS Lett.*, **147**, 106 (1982).

224. S. Morante, M. Cerdonio, S. Vitale, A. Congiu-Castellano, A. Vaciago, G. M. Giacometti, and L. Incoccia, *EXAFS and Near-Edge Structure*, A. Bianconi, L. Incoccia, and S. Stipcich (Eds.), Springer-Verlag, Berlin, 1983, p. 352.

225. A. Bianconi, A. Congiu Castellano, M. Dell'Ariccia, A. Govannelli, E. Burattini, and P. J. Durham, *Biochem. Biophys. Research Commun.*, **131**, 98 (1985).

226. L. V. Azaroff and B. N. Das, *Phys. Rev. A*, **134**, 747 (1964).

227. B. Cordts, D. M. Pease, and L. V. Azaroff, *Phys. Rev. B*, **22**, 4692 (1980).

228. M. C. Munoz, P. J. Durham, and B. L. Gyorffy, *J. Phys. F*, **12**, 1497 (1982).

229. B. Cordts, D. M. Pease, and L. V. Azaroff, *Phys. Rev. B*, **24**, 538 (1981).

230. E. E. Vainshtein, S. M. Blokhin, and Yu. B. Paderno, *Sov. Phys. Solid State*, **6**, 2318 (1965).

231. H. Launois, M. Rawiso, E. Holland-Moritz, R. Pott, and D. Wohlleben, *Phys. Rev. Lett.*, **44**, 1271 (1980).

232. R. M. Martin, J. B. Boyce, J. W. Allen, and F. Holtzberg, *Phys. Rev. Lett.*, **44**, 1271 (1980).

233. G. Krill, J. P. Kappler, A. Meyer, L. Abadli, and M. F. Ravet, *J. Phys. F*, **11**, 1713 (1981).

234. J. Rohler, G. Krill, J. P. Kappler, and M. F. Ravet, *EXAFS and Near Edge Structure*, A. Bianconi, L. Incoccia, S. Stipcich (Eds.), Springer-Verlag, Berlin, 1983 p. 213.

235. B. Lengeler, J. E. Muller, and G. Materlik, *Phys. Rev. B*, **28**, 2276 (1983).

236. J. Rohler, D. Wohlleben, J. P. Kappler, and G. Krill, *Phys. Lett. A*, **103**, 220 (1984).

237. R. Ingalls, J. M. Tranquada, J. E. Whitmore, and E. D. Crozier, in *Physics of Solids Under High Pressure*, J. S. Schilling and R. N. Shelton (Eds.), North-Holland, Amsterdam, 1981, p. 67.

238. R. Nagarajan, E. V. Sampathkumaran, L. C. Gupta, A. Vijayaraghavan, Bhak-darshan, and B. D. Padalia, *Phys. Lett. A*, **81**, 397 (1981); **84**, 275 (1981).

239. S. Nakai, C. Sugiura, S. Kunii, and T. Suzuki, *Jpn. J. Appl. Phys.*, **17**, Suppl. 17-2, 197 (1978).

240. J. H. Slowik and F. C. Brown, *Phys. Rev. Lett.*, **29**, 934 (1972).

241. A. Bianconi and C. R. Natoli, *Solid State Commun.*, **27**, 1177 (1978).

INDEX